As the most fundamental life process on earth, photosynthesis is the focus of a vast body of research, spanning studies of femtosecond reactions at the molecular level through to field studies requiring a whole season of observation. This is the first advanced-level treatment which spans the broad range of the topic within a single volume, so providing a uniquely comprehensive, authoritative and self-contained source book. Written by an international team of experts, the volume opens by considering the cell and molecular biology of chloroplasts, followed by a section which presents the latest information on the biochemistry and physiology of photosynthesis. These chapters are then complemented by coverage of more ecological and applied aspects, such as photosynthesis and global climate change, and crop productivity.

# Photosynthesis: A Comprehensive Treatise

# Photosynthesis
## A Comprehensive Treatise

Edited by A. S. RAGHAVENDRA

*Department of Plant Sciences, School of Life Sciences,*
*University of Hyderabad, Hyderabad 500046, India*

PUBLISHED BY THE PRESS SYNDICATE OF THE UNIVERSITY OF CAMBRIDGE
The Pitt Building, Trumpington Street, Cambridge CB2 1RP, United Kingdom
CAMBRIDGE UNIVERSITY PRESS
The Edinburgh Building, Cambridge CB2 2RU, United Kingdom
40 West 20th Street, New York, NY 10011–4211, USA
10 Stamford Road, Oakleigh, Melbourne 3166, Australia



First published 1998

Printed in the United Kingdom at the University Press, Cambridge

Typeset in Monotype Ehrhardt 9/12 pt.

*A catalogue record for this book is available from the British Library*

*Library of Congress Cataloguing in Publication data*

Photosynthesis: a comprehensive treatise/edited by A. S. Raghavendra.
p. cm.
Includes bibliographical references (p. ) and index.
ISBN 0 521 57000 X (hardbound)
1. Photosynthesis. I. Raghavendra, A. S.
QK882.P538 1997
572′.46–dc21 97–4035 CIP

ISBN 0 521 57000 X hardback

To

my family

Rama, Radha & Raul

*for their understanding*

# Contents

# Contributors

H. BAUWE
Biochemistry and Physiology Department,
IACR-Rothamsted, Harpenden AL5 2JQ, UK

G. A. BERKOWITZ
Department of Plant Science, College of Agriculture and Natural Resources, University of Connecticut, 1376 Storrs Road, Storrs, CT 06269–4067, USA

W. R. BEYSCHLAG
Lehrstuhl fur Experimentelle Okologie und Okosystembiologie, Universität Bielefeld, Universitätstr.25, W4-107, 33615 Bielefeld, Germany

W. BILGER
Julius-von-Sachs-Institut für Biowissenschaften, Universität Würzburg, Mittelerer Dallenbergweg 64, D-97082 Würzburg, Germany

P. BÖGER
Department of Plant Physiology and Biochemistry, University of Konstanz, D-78434 Konstanz, Germany

P. R. CHITNIS
Department of Biochemistry and Biophysics, Iowa State University, Ames, IA 50011–1020, USA

C. CRITCHLEY
Department of Botany, The University of Queensland, QLD 4072, Australia

B. G. DRAKE
Smithsonian Environmental Research Centre, PO Box 28, Edgewater, MD 21037, USA

B. W. DREYFUSS
Department of Chemistry and Biochemistry, University of California-Los Angeles, Los Angeles, CA 90024–7009, USA

R. T. FURBANK
CSIRO Division of Plant Industry, GPO Box 1600, Canberra ACT 2601, Australia

P. GARDESTRÖM
Department of Plant Physiology, University of Umea. S-901 87 Umea, Sweden

M. A. GONZÀLEZ-MELER
Smithsonian Environmental Research Center, PO Box 28, Edgewater, MD 21037, USA

W. Z. HE
Department of Plant Biology, University of California, Berkeley, CA 94720-3120, USA

D. HEINEKE
Institut für Biochemie der Pflanze der Georg-August Universität Göttingen, Untere Karspüle 2, 37073 Göttingen, Germany

J. K. HOOBER
Department of Botany and Center for the Study of Early Events in Photosynthesis, Arizona State University, Tempe, AZ 85287–1601, USA

H. HORMANN
Julius-von-Sachs Institut für Biowissenschaften, Universität Würzburg, Mittlerer Dallenbergweg 64, D-97082 Würzburg, Germany

S. C. HUBER
USDA/ARS/Crop Science and Botany Department, North Carolina State University, 3127 Ligon Road, Raleigh, NC 27695–7631, USA

N. HUNER
Department of Plant Science, University of Western Ontario, London, Ontario, N6A 5B7, Canada

V. HURRY
Botanisches Institut, Universität Heidelberg, Im Neuenheimer Feld 360, D-69120 Heidelberg, Germany

R. ISHII
Laboratory of Crop Sciences, Faculty of Agriculture, The University of Tokyo, Bunkyo-ku, Tokyo 113, Japan

J. JACOB
Smithsonian Environmental Research Center, PO Box 28, Edgewater, MD 21037, USA; *Present address*: The Rubber Research Institute of India, Kottayam 686 009, Kerala, India

S. KAPOOR
Centre for Gene Research, Nagoya University, Chikosa-ku, Nagoya 464–01, Japan

U. LIEBL
BIP-CNRS, UPR 9036, 31 Chemin Joseph-Aiguier, 13402 Marseille Cedex 20, France

U. LÜTTGE
Institut für Botanik, FB 10, Technische Hochschule Darmstadt, Schnittspahnstr. 3, D-64287 Darmstadt, Germany

R. MALKIN
Department of Plant Biology, University of California, 111 Koshland Hall, Berkeley, CA 94720–3120, USA

D. T. MORISHIGE
Department of Biochemistry and Biophysics, Texas A&M University, College Station, TX 77843–2128, USA

U. MÜHLENHOFF
Biologie II, Universität Freiburg, Schanzlestr.1, 79104 Freiburg, Germany

C. NEUBAUER
Julius-von-Sachs Institut für Biowissenschaften, Universität Würzburg, Mittlerer Dallenbergweg 64, D-97082 Würzburg, Germany

W. NITSCHKE
BIP-CNRS, UPR 9036, 31 Chemin Joseph-Aiguier, 13402 Marseille Cedex 20, France

D. J. OLIVER
Department of Botany, Iowa State University, Ames, IA 50011–1020, USA

G. ÖQUIST
Department of Plant Physiology, University of Umea, S-901 87 Umea, Sweden

K. PADMASREE
Department of Plant Sciences, School of Life Sciences, University of Hyderabad, Hyderabad 500 046, India

R. W. PEARCY
Section of Plant Biology, University of California, Davis, CA 95616-8537, USA

A. S. RAGHAVENDRA
Department of Plant Sciences, School of Life Sciences, University of Hyderabad, Hyderabad 500 046, India

S. RAWSTHORNE
Brassica and Oilseeds Research Department, John Innes Centre, Colney, Norwich NR4 7UJ, UK

R. J. RYEL
Department of Rangeland Resources and the Ecology Center, Utah State University, Logan, UT 84322–5230, USA

G. SANDMANN
Biosynthesis Group, Institute of Botany, Johann Wolfgang Goethe-Universität, PO Box 11932, D-60054 Frankfurt, Germany

H. SCHEER
Botanisches Institut der Universitat Munchen, Menzinger Str. 67, D-80638 Munchen, Germany

U. SCHREIBER
Julius-von-Sachs-Institut für Biowissenschaften, Universität Würzburg, Lehrstuhl für Botanik 1, Mittelerer Dallenbergweg 64, D-97082 Würzburg, Germany

G. SCHULTZ
Botanical Institut, Veterinary School of Hannover, D-30559 Hannover, Germany

E. SELSTAM
Department of Plant Physiology, University of Umea, S-901 87 Umea, Sweden

T. D. SHARKEY
Department of Botany, University of Wisconsin-Madison, 430 Lincoln Drive, Madison WI 53706–1381, USA

M. SUGIURA
Center for Gene Research, Nagoya University, Chikosa-ku, Nagoya 464–01, Japan

J. WHITMARSH
Photosynthesis Research Unit, ARS/USDA, Department of Plant Biology and Center for Biophysics and Computational Biology, University of Illinois at Urbana-Champaign, IL 61801, USA

# Preface

Photosynthesis is an important area of research and is a core component of graduate and postgraduate courses in plant biology and the related disciplines of agriculture, horticulture or biotechnology. As a teacher and researcher, I had long felt that a single advanced-level textbook on photosynthesis was not available although there are a number of well-known series, each of several volumes, on the subject.

This volume has been planned to be a comprehensive textbook, presenting a state-of-the-art picture of the different aspects of photosynthesis. Each topic is written by a specialist, acknowledged internationally for his or her expertise in the field. The chapters have been chosen to cover cell biology, molecular biology, physiology, biochemistry, agronomy, environmental biology and biotechnology as they relate to photosynthesis, as well as a few other special topics. The book is aimed at students, and at faculty and research scientists in universities, research institutions and industry.

As photosynthesis is of an interdisciplinary nature and of contemporary interest, the literature on photosynthesis is extensive. It is therefore almost impossible to limit the size of a book expected to give a comprehensive coverage of the subject. I chose to make a hard editorial decision and requested the authors to limit their citations of original references to a minimum and include only reviews which were immediately relevant. I wish to express my personal apology for not being able to cite several excellent articles, due to the limitation on space.

I hope that this book will fill the gap between the longstanding demand and the availability of a comprehensive/concise textbook on photosynthesis, as well as being an excellent source of information and a stimulation for further research in this fascinating field of plant biology.

The idea and the preparation were encouraged by several of the photosynthesis experts. I express my gratitude and sincere appreciation to Professors V. S. Rama Das (who also introduced me to the subject), and to David Walker, Ulrich Heber, Hans Heldt, Govindjee, A. Gnanam and Dr P. V. Sane, for their significant role in shaping my career in photosynthesis and for being instrumental in the success of the present book.

I thank Professor G. Mehta, Vice-Chancellor of our University, and Professor A. R. Reddy, Dean of the School of Life Sciences, for their encouragement of my academic pursuits. I also thank Ms. K. Padmasree and Ms K. Parvathi, my graduate students, for their help at several stages in the preparation of this book.

Finally, I thank Dr Maria Murphy and her team at Cambridge University Press for their excellent assistance and execution of this project.

*A. S. Raghavendra*
*May 1997*

# Abbreviations

A = antheraxanthin *or* assimilation
AAN = aminoacetonitrile
ABA = abscisic acid
ACCase = acetyl-CoA carboxylase
AGPase = ADP-glucose pyrophosphorylase
ALA = α-amino-levulinic acid
ALS = acetolactate synthase
AMPA = aminomethylphosphonic acid
ANT-2p = 2-(3-chloro-4-trifluoromethyl)anilino-3,5-dinitrothiophene
BCAA = branched chain amino acids
Bchl = bacteriochlorophyll
Bphe = bacteriopheophytin
BY = biological yield
CA = carbonic anhydrase
CABP = carboxyarabinitol 1,5-bisphosphate
Calvin cycle = *see* PCRC
CAM = crassulacean acid metabolism
CaMV = cauliflower mosaic virus
CC = core complex
CCCP = carbonylcyanide-*m*-chlorophenylhydrazone
CE = carboxylation efficiency
CGR = crop growth rate
ch-cpn = chloroplast chaperonin
Chl = chlorophyll
chlp = chloroplast
$C_i$ = internal leaf [$CO_2$]
cp-FBPase = chloroplastic fructose bisphosphatase
cpn = chaperonin
CPS = canopy photosynthesis
CPTA = [2-(4-chlorophenylthio)-triethylamine HCl]
CRS = canopy respiration
cyano = cyanobacteria
cyt = cytochrome
D1 = protein member of the heterodimeric centre of PS
DAHS = 3-deoxy-D-arabinoheptulosonate-7-phosphate
DCCD = N,N′-dicyclohexylcarbodiimide
DCMU = 3-(3,4-dichlorophenyl)-1,1-dimethyl urea
DGDG = digalactosyl-diacylglycerol
DHAP = dihydroxyacetone phosphate
DHP = dihydrodipicolinate
DIC = dissolved inorganic carbon
DMAPP = dimethylallyl pyrophosphate
ELIP = early light-inducible protein
$E_M$ = midpoint oxidation–reduction potential
EPR = electron paramagnetic resonance
EPS = epoxidation state of the xanthophyll cycle pigments
EPSPS = 5-enolpyruvylshikimate-3-phosphate synthase
ER = endoplasmic reticulum
ESR = electron spin resonance
ET = electron transport
EXAFS = extended X-ray absorption fine structure
EY = economic yield
F26BP = fructose 2,6-bisphosphate
F6P = fructose 6-phosphate
FBP = fructose 1,6-bisphosphate
FBPase = fructose 1,6-bisphosphatase

Fd = ferredoxin
FNR = ferredoxin-$NADP^+$ oxidoreductase
FPP = farnesyl pyrophosphate
Fru-2,6-$P_2$ = fructose-2,6-bisphosphate
GAP = glyceraldehyde 3-phosphate
GAPDH = glyceraldehyde-3-phosphate dehydrogenase
GDC = glycine decarboxylase
GOGAT = glutamine 2-oxoglutarate aminotransferase
GGPP = geranylgeranyl pyrophosphate
GPP = geranyl pyrophosphate
Gln = glutamine
Glu = glutamate
GOGAT = glutamine:2-oxoglutarate aminotransferase
GOX = glyoxylate
GPP = gross primary production
GS = glutamine synth(et)ase
HI = harvest index
HMG-CoA = 3-hydroxy-3-methylglutaryl-CoA
IDH = isocitrate dehydrogenase
IML = intermittent light
IPP = isopentyl pyrophosphate
α-KGA = α-ketoglutarate
LAI = leaf area index
LAR = leaf area ratio
LEDR = light-enhanced dark respiration
LHC = light-harvesting complex
Lhcb = apoprotein of LHC
LHCP = light-harvesting chlorophyll-binding protein
LIE = light intercepting efficiency
LPS = leaf photosynthesis
LS = large subunit
LSC = large single copy
MAP = Mehler ascorbate peroxidase
MDH = malate dehydrogenase
ME = malic enzyme
MGDG = monogalactosyl-diacylglycerol
MVA = mevalonate
MYA = million years ago
NAD-ME = NAD-malic enzyme
NADP-MDH = NADP-linked malic dehydrogenase
NADP-ME = NADP-malic enzyme
NEP = net ecosystem production
NiR = nitrite reductase
NR = nitrate reductase
OAA = oxaloacetate
2-OG = 2-oxoglutarate
P680 = primary donor/reaction centre of photosystem II
P700 = primary donor/reaction centre of photosystem I
PAM = pulse amplitude modulation
PAR = photosynthetically active radiation
PBG = porphobilinogen
PC = plastocyanin
Pchlid = protochlorophyllide
PCK = phosphoenolpyruvate carboxykinase
PCO = photosynthetic carbon oxidation
PCR = photosynthetic carbon reduction
PCRC = photosynthetic carbon reduction cycle *or* Calvin cycle
PDC = pyruvate dehydrogenase complex
PEP = phosphoenolpyruvate
PEPC = phosphoenolpyruvate carboxylase
PFD = photon flux density
PG = phosphatidyl-diacylglycerol
3-PGA = 3-phosphoglyceric acid *or* 3-phosphoglycerate
Phe a = pheophytin a
Pheo = pheophytin
phyta = 2-en-1-ol-phytol
pLHCP = precursor of LHCP
PLP = pyridoxal-5′-phosphate
POR = Pchlid-oxidoreductase
PPdK = pyruvate-orthophosphate dikinase
PPi = inorganic pyrophosphate
PPP = phytyl diphosphate
PQ = plastoquinone
$PQH_2$ = plastoquinol (fully reduced form of plastoquinone)
PRK = phosphoribulokinase
PS = photosystem

pSSU = precursor of SSU
Rc = carboxylation resistance
RC = reaction centre
Rr = transportation resistance
Rs = stomatal resistance
Rt = total resistance
rubisco = ribulose-1,5-bisphosphate carboxylase-oxygenase
RuBP = ribulose 1,5-bisphosphate
SAM = S-adenosyl-methionine
SBPase = sedoheptulose-1,7-bisphosphatase
SHMT = serine hydroxymethyl transferase
SLW = specific leaf weight
SPS = sucrose-6-phosphate synthase
SQDG = sulphoquinovosyl-diacylglycerol
SSC = small single copy
SSU = small subunit of rubisco
SuSy = sucrose synthase
TCA cycle = tricarboxylic acid cycle
TPP = thiamine pyrophosphate
TPT = triose-phosphate translocator
triose-P = triose-phosphate
UDPase = UDP glucose pyrophosphorylase
UQ = ubiquinone
V = violaxanthin
VPD = vapour pressure difference
WOC = water oxidation centre
WUE = water use efficiency
XANES = X-ray absorption near-edge structure
$Y_z$ = tyrosine residue on D1 acting as the donor to P680
Z = zeaxanthin

PART I

# Cell and molecular biology of chloroplasts

# 1 Chloroplast structure and development

J. K. HOOBER

## INTRODUCTION

Plastid structure is highly variable and flexible. Its plasticity is revealed by the variety of its forms within various photosynthetic organisms and under different environmental conditions. Chloroplasts respond to changes in light, nutrients, water and temperature. Although there is a generalized – even idealized – ultrastructure in mature, fully nurtured leaves, and indeed the basic features of the chloroplast in many plants are similar to the idealized model (for reviews, see Coombs & Greenwood, 1976; Hoober, 1984; von Wettstein, Gough & Kannangara, 1995), the specific arrangement of the thylakoid membrane system varies from tissue to tissue and plant to plant. The overall structure of the organelle, which also includes envelope membranes, the stromal matrix, lipid globules, starch grains and pyrenoid bodies, results from combinations of these constituents. Consequently, rather than simply a generalized structure, the chloroplast must be considered within its dynamic context.

## EVOLUTIONARY CONTEXT

Evidence suggests that photosynthesis is as old, or nearly so, as life itself (Awramik, 1992; Blankenship, 1992). Nisbet, Cann & Van Dover (1995) proposed that chemotrophic thermophiles near hydrothermal vents were predecessors to photosynthetic organisms containing long-wavelength, bacteriochlorophyll-type pigments 3500 to 3800 million years ago (MYA). Fossils of several taxa of filamentous, cyanobacterium-like microorganisms that most likely were photosynthetic, were dated to 3465 MYA (Schopf, 1993). Such diversity suggests that more primitive photosynthetic organisms predated these forms of life. Energy transducing mechanisms that involved membrane-bound electron transport and generation of proton gradients driven by absorption of light energy were probably present in these earliest cells (Castresana & Saraste, 1995).

It seems generally accepted (Lake, 1994; Doolittle *et al.*, 1996) that eukaryotes developed from the archaebacterial lineage 1800 to 2000 MYA, perhaps by fusion of eubacterial and archaebacterial cells (Margulies, 1996). The last common ancestor of plants and animals existed about 1000 MYA (Doolittle *et al.*, 1996), which may set the time of entrapment of a photosynthetic prokaryotic cell by a eukaryotic host. Seed-bearing plants appeared about 360 MYA and reached sufficient complexity to diverge into monocots and dicots, which on the basis of nucleotide sequences of mitochondrial DNA occurred 200 MYA (Laroche, Li & Bousquet, 1995). While tremendous genetic changes were occurring in the nucleus of the host during this time, most of the genes of the original endosymbiont had already been transferred to the nucleus and the plastid maintained its basic characteristics. To what extent the transfer of DNA from plastid to nucleus influenced evolution of the nuclear genome is an interesting question. Particularly intriguing is the mechanism by which much of the genome of the original 'chloroplast' was transferred through multiple membranes to the nucleus.

Photosynthetic complexes, in particular the reaction centres, trace a monophylogenetic origin of thylakoid membranes back through cyanobacteria to the photosynthetic bacteria (for reviews see Blankenship, 1992; Meyer, 1994; Vermaas, 1994; Wolfe & Hoober, 1996). Evidence continues to mount for the proposal

(Schimper, 1883; Cavalier-Smith, 1982) that plastids arose from a primary endosymbiotic event in which a photosynthetic organism, similar to a cyanobacterium, was enclosed by an eukaryotic host. Phylogenetic relationships of extant plastids reveal strong support for a monophyletic lineage rooted within the cyanobacterial family (Bhattacharya & Medlin, 1995), although to accommodate this conclusion, lateral transfer of genes between organisms may be required (Reith, 1995). The monophyletic lineage of the chloroplast implies that common elements should prevail in all plastid types. From the prokaryotic cyanobacteria upward, photosynthetic reaction centres contain chlorophyll *a* as the photochemically active pigment. Modern cyanobacteria contain protein–chlorophyll *a* complexes and phycobilisomes as light-harvesting antenna (Grossman *et al.*, 1995). Although eukaryotic rhodophytes, which also contain chlorophyll *a* and phycobilins in antenna complexes, and eukaryotic chlorophytes, which contain chlorophyll *a* and *b* in light-harvesting complexes and no phycobilisomes, are only distantly related according to molecular phylogenies, polypeptides of the light-harvesting complexes from these organisms are immunologically related (Wolfe *et al.*, 1994). Moreover, this relatedness extends to the chromophytes that contain chlorophyll *a* and *c* and the unique carotenoid fucoxanthin. Bhattacharya & Medlin (1995) concluded that the original ancestor of plastids contained chlorophylls *a*, *b* and *c* and phycobilins. The diversity of contemporary plastids may have resulted in part by the loss of characters as well as divergence influenced by the eukaryotic hosts in which they existed. Because of, first, the markedly smaller size of the chloroplast genome as compared with the bacterial genome and, secondly, the generally consistent complement of genes within the organelle, genomic reduction following the initial endosymbionic event may have occurred before extensive diversification of plant cells.

The structure of the chloroplast, in its most basic form, is thus expected to be a product of its origin as well as its particular host. Photosynthetic membranes in cyanobacteria exist as multiple lamellae that traverse the length of the cell or form concentric layers encircling the protoplasm (Nierzwicki-Bauer, Balkwill & Stevens, 1983). In the steps leading to the chloroplast, the lamellar system became differentiated into appressed and unappressed regions. Also important in eukaryotic cells are the membranes that enclose the organelle, the envelope. Because cyanobacteria are gram-negative organisms, with an outer membrane in addition to the plasmalemma, an endocytic event would yield three membranes surrounding the engulfed cytoplasm, with the outermost membrane derived from the host's plasma membrane as the endocytic vacuole formed (Cavalier-Smith, 1982).

The scenario described above would provide the structure found in current euglenophytic and dinophytic algal families, which contain plastids surrounded by three membranes. Plastids in green algae and higher plants are surrounded by an envelope of two membranes, which implies that one of the original three disappeared. The inner membrane is apparently derived from the plasma membrane of the endosymbiont. Characteristics of the outer membrane are more similar to the outer membrane of gram-negative bacteria than to a plasmalemma (Douce & Joyard, 1990), and thus the membrane of the endocytic vacuole may have been lost (Cavalier-Smith, 1982). Plastids in several groups of algae, including those in the chrysophyte, phaeophyte, cryptomonad and chlorarachniophyte families, are surrounded by four membranes that originated from a secondary endosymbiotic event in which a eukaryotic host engulfed another eukaryotic, plastid-bearing cell. The third and fourth membranes would then have derived from the plasma membrane of the algal endosymbiont and the membrane of the endocytic vacuole of the second host, respectively. These additional membranes are referred to as the chloroplast endoplasmic reticulum, and the outermost membrane of the reticulum in some algal families is continuous with the outer membrane of the nuclear envelope (Gibbs, 1981). Cryptomonad and chlorarachniophyte algae contain the residual nucleus, designated the nucleomorph, of the primary host (Gillott & Gibbs, 1980; Erata *et al.* 1995), which has remained transcriptionally active (Douglas *et al.*, 1991; McFadden *et al.*, 1994). The lack of a close evolutionary relationship between nuclear genomes of the host organisms suggests that the secondary endocytic event may have occurred numerous times, and thus the host organisms are less related than the plastids they contain (Bhattacharya & Medlin, 1995).

The additional membranes surrounding the chloroplast consequently have evolutionary, but probably not functional, significance. However, over the course of evolutionary time, the additional membranes apparently influenced communication between the nuclear–cytosolic and chloroplast compartments and required development of different mechanisms for cytoplasmically synthesized products to enter the plastid. The genome of chloroplasts bounded by four membranes generally include additional genes such as *RBCS* that encodes the small subunit of ribulose bisphosphate carboxylase/oxygenase (Reith, 1995), which in chlorophytes is within the nuclear genome (Dean, Pichersky & Dunsmuir, 1989). Recent work has clearly shown that in *Euglena*, which has three membranes around the chloroplast, the *RBCS* gene is in the nucleus but the gene product, as well as the apoproteins (Lhcb) of the thylakoid light-harvesting chlorophyll *a/b*–protein complex (LHCII) associated with photosystem II (PSII), enters the chloroplast by a path different from the mechanism elucidated for protein import into plastids surrounded by only two membranes (Schnell, Kessler & Blobel, 1994; Hirsch *et al.*, 1994; Keegstra *et al.*, 1995; Soll, 1995). In the latter case, an import apparatus has been characterized that comprises proteins of the outer and inner membranes of the envelope. This complex threads individual proteins containing an N-terminal basic, *hydrophilic* transit peptide through a channel into the plastid stroma. In *Euglena* the proteins are synthesized in the cytoplasm as polyproteins containing as many as eight mature Lhcb sequences within a single molecule. Instead of a transit sequence, an N-terminal, *hydrophobic* extension serves as a signal sequence to target the polyprotein to the endoplasmic reticulum (Sulli & Schwartzbach, 1996). The proteins then are transferred to the Golgi (Osafune, Schiff & Hase, 1991) and subsequently to the plastid, where they are processed to the mature form. Although the details of the latter steps are not clear, vesicular traffic from the Golgi to the plastid may be required to get proteins through the additional, outer membranes.

## DEVELOPMENT OF CHLOROPLAST STRUCTURE

The fundamental structure of the chloroplast is an envelope enclosing a system of thylakoid membranes, which develops from a simple double-membrane enclosed compartment. This precursor organelle, usually about 1 μm in diameter, often exhibits a few invaginations of the inner membrane (Fig. 1.1). The 'proplastid,' in the appropriate host cell, has the capacity to develop into a chloroplast. The developmental processes, as well as the structure and function of the mature organelle, are strongly influenced by the environment. Although many algae and gymnosperms are capable of synthesizing chlorophyll in the dark, because they contain subunits of a light-independent NADPH: protochlorophyllide oxidoreductase encoded by plastid genes *chlB*, *chlL* and *chlN*, and also develop chloroplasts in the dark (Bauer, Bollivar & Suzuki, 1993), a few algal strains – some of them mutants selected for this property – and the angiosperms make chlorophyll only in the light (Suzuki & Bauer, 1995). Chlorophyll is not only required for func-

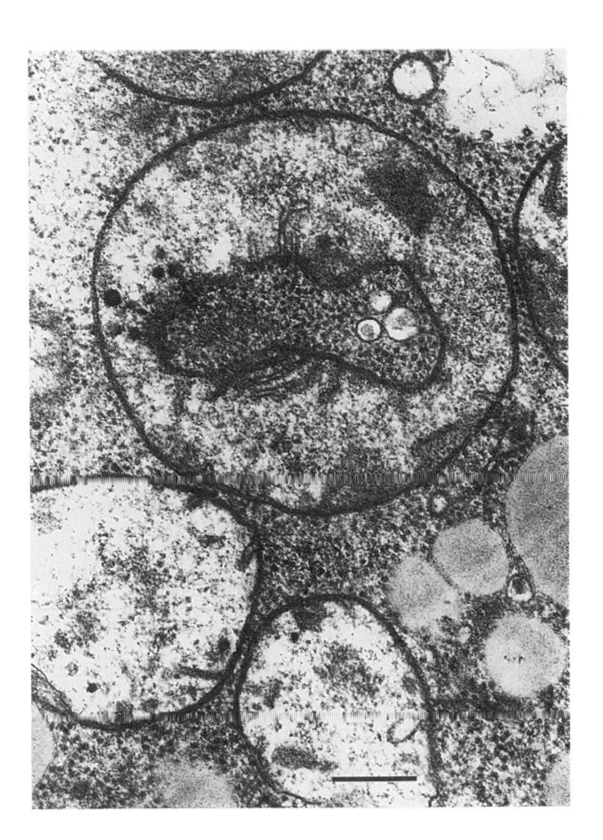

**Figure 1.1.** Micrograph of proplastids in the shoot apex of a *Zea mays* seed. The largest proplastid in the centre of the Figure has a finger of cytoplasm protruding through the section. Extending into the stroma from the proplastid envelope are numerous invaginations of the inner membrane. Bar, 250 nm. (Micrograph courtesy of Richard N. Trelease.)

tion of the chloroplast but also for its complete development. Most angiosperms develop an etioplast when seedlings are grown for the first several days in the dark. The etioplast contains an elaborate reticular structure, the prolamellar body, composed of NADPH : protochlorophyllide oxidoreductase as the predominant protein in a ternary complex with its two substrates, NADPH and protochlorophyllide (Lindsten, Ryberg & Sundqvist, 1988; Böddi *et al.*, 1989). More than 90% of the lipid of the prolamellar body is mono- and digalactosyl diglyceride in a ratio of 1.6 : 1 (Selstam & Sandelius, 1984). The preponderance of monogalactosyl diglyceride, a lipid that promotes non-lamellar structures (Brentel, Selstam & Lindblom, 1985), seems to be a factor in generating the branched structure of the prolamellar body. Exposure of etioplasts to light results in conversion of protochlorophyllide to chlorophyllide *a* and dispersal of the prolamellar body into prothylakoid membranes (for review see Reinbothe *et al.* 1996).

Development into a chloroplast requires biogenesis of the extensive thylakoid membrane system, the most prominent membrane system in the plant cell. A question of importance is how this system is assembled. The uncertainty of the answer to this question is remarkable, given the extensive amount of effort devoted to it. Electron microscopic studies performed over 30 years ago of developing chloroplasts in seedlings exposed to light after fixation with $KMnO_4$ suggested that thylakoid membranes arise by expansion and invagination of the envelope inner membrane (Mühlethaler & Frey-Wyssling, 1959; von Wettstein, 1961). The structure of proplastids (Fig. 1.1), often containing invaginations of the inner membrane, supported this concept. From an evolutionary perspective, expansion of the inner membrane is analogous to formation of photosynthetic chromatophores in photosynthetic bacteria by insertion of the apparatus for photosynthesis into the respiratory plasma membrane (Scherer, 1990). Thylakoid membranes in cyanobacterial cells also seem to connect intermittently with the plasma membrane (Nierzwicki-Bauer *et al.*, 1983). However, use of $OsO_4$ and glutaraldehyde as fixatives for electron microscopy revealed a high level of detail in excellently preserved material, but invaginations of the envelope membrane in developing chloroplasts, or connections of thylakoid membranes with the envelope, were not observed. Whether this outcome resulted from differences in the relative rates of fixation with various chemicals or the manner in which tissues were handled is not clear. Rapid fixation by cryopreservation (Gilkey & Staehelin, 1986) holds promise for reinvestigating this question.

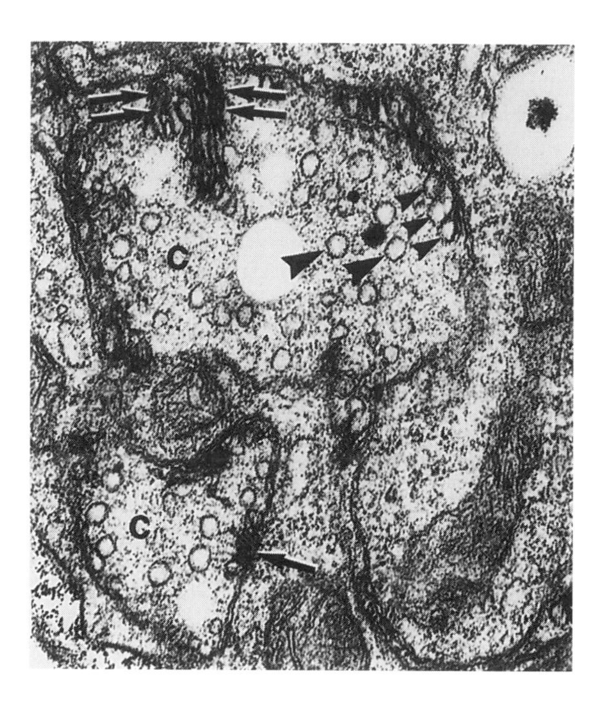


**Figure 1.2.** Micrograph of a portion of the chloroplast during the initial stage of development in a greening cell of *Chlamydomonas reinhardtii* yl. Localized regions of the envelope are engaged in expansion of the envelope membrane, which folds into the stroma (*double arrows*) or generates vesicles (*arrowheads*). The small region of the envelope (*single arrow*) surrounded by typical double-membrane profiles suggests that membrane expansion occurs at sites where protein import occurs (Schnell & Blobel, 1993). *c*, Chloroplast stroma; *m*, mitochondria in the adjacent cytoplasm. (From Hoober *et al.*, 1991.)

Rapid fixation of greening cells of the alga *Chlamydomonas reinhardtii* by injection into cold $OsO_4$ solutions trapped structures in which extensive membrane material emanated from the envelope (Fig. 1.2). Extensions of the inner membrane seemed to be the source of vesicles in the stroma (Hoober, Boyd & Paavola, 1991). The extended regions of the envelope over which membranes formed was consistent with the long, narrow zones of appression of the envelope membranes (Cline, Keegstra & Staehelin, 1985; Cremers *et al.*, 1988), where import of proteins from the cytoplasm occurs (Schnell & Blobel, 1993). Because the inner membrane of the envelope is a major site of lipid synthesis (Joyard, Block & Douce, 1991; Ohlrogge & Browse, 1995), stimulation of these pathways in the light is expected to cause expansion of the membrane. The envelope also appears to be the site of initial chlorophyll synthesis (Douce & Joyard, 1990; Joyard *et al.*, 1990). Membranes associated with the inner membrane contained light-harvesting complexes as detected by binding of antibodies raised against Lhcb (White *et al.*, 1996). Analysis of fluorescence of greening algal cells showed that these initial membranes were photochemically functional (White & Hoober, 1994). Further support for vesiculation of the inner membrane was obtained by electron microscopy of plastids undergoing conversion of chloroplasts to chromoplasts in the pepper plant (Hugueney *et al.*, 1995). Energy-dependent vesiculation of purified envelope membranes, with subsequent fusion of the envelope-derived vesicles with thylakoid membranes, was established by Morré *et al.* (1991). These studies support the envelope as a staging area for thylakoid biogenesis.

During early development, the thylakoid segments line up with a remarkably even distribution along the polar axis of the organelle. As the amount of membrane increases, membranes adhere to form the multilayered structures referred to as grana (Robertson & Laetsch, 1974). In *Chlamydomonas* and other green algae the membranes are appressed to their neighbours over most of their surface (Ohad, Siekevitz & Palade, 1967). Portions of thylakoids in most higher plant chloroplasts differentiate into discrete structures that resemble a stack of coins. Surface views reveal a collection of remarkably regular, circular profiles 0.4 to 0.5 $\mu$m in diameter (Coombs & Greenwood, 1976; Schnell & Blobel, 1993; Albertsson, 1995). In cross-section, the typical pattern is one of stacked thylakoids interconnected by 'stromal lamellae' (Fig. 1.3, see also von Wettstein *et al.*, 1995). Evidence that these local differentiations are reversible, and thus most likely the result of functional aggregation of complexes within the membrane, was provided by experiments in which grana-rich thylakoid membranes, 'destacked' by treatment with chelating agents, formed long lamellar sheets (Steinback, Burke & Arntzen, 1979; Arntzen & Burke, 1980). Replenishment with divalent cations restored the granal structures. Bundle sheath cells in C4 plants generally, although not universally, contain thylakoids that are not differentiated into granal and stromal regions but rather exist primarily as single lamellae (Coombs & Greenwood, 1976). PSII (PSII$\alpha$) is located in the core of the grana, whereas PSI (PSI$\alpha$) is segregated to the periphery. The grana are designed for linear electron transport and the requirement for short-range communication between PSII and PSI centres may be a factor that determines the highly regular diameter of grana. Cyclic electron transport is restricted to stromal lamellae, which are enriched in PSI (PSI$\beta$). The light-absorbing antennae of PSII$\beta$ in stromal lamellae is about half the size of that associated with the granal PSII$\alpha$. Within a number of higher plants, stromal thylakoid membranes consistently account for 20% of the total while grana, including margins and peripheral membranes, account for the remaining 80% of the membrane (Albertsson, 1995). The composition of the membrane in chloroplasts in bundle sheath cells of $C_4$ plants is similar to stromal thylakoids in mesophyll cells and contain less PSII relative to PSI (Bassi, Marquardt & Lavergne, 1995).

The chemical nature of the adherence of thylakoids to form grana is uncertain, but recent hypotheses concerning this property combine the surface-charge theory, in which divalent cations suppress repulsive charges on the membrane surfaces, and the molecular-recognition theory (Stys, 1995). An extensive body of evidence has indicated, although not convincingly, that formation of granal stacks is related to the presence of LHCII. The apoproteins (Lhcb) of LHCII contain a loop enriched in acidic amino acids on the stromal surface of the membrane (Kühlbrandt, Wang & Fujiyoshi, 1994), the same side on which the more basic N-terminal segment resides. Removal of the N-terminal segment by proteolysis abolished stacking of membranes (Mullet, 1983), which suggested that repulsive forces are dominant. The *chlorina f2* mutant strain of barley lacks the ability to synthesize chlorophyll *b* and is unable to assemble the major peripheral LHCII (Król *et al.*, 1995; Preiss & Thornber, 1995). Thylakoid membranes of the *chlorina f2* mutant retain low levels of minor LHCII apoproteins but nearly wild-type levels of LHCI apoproteins (Król *et al.*, 1995). Chloroplasts in this mutant strain still main-

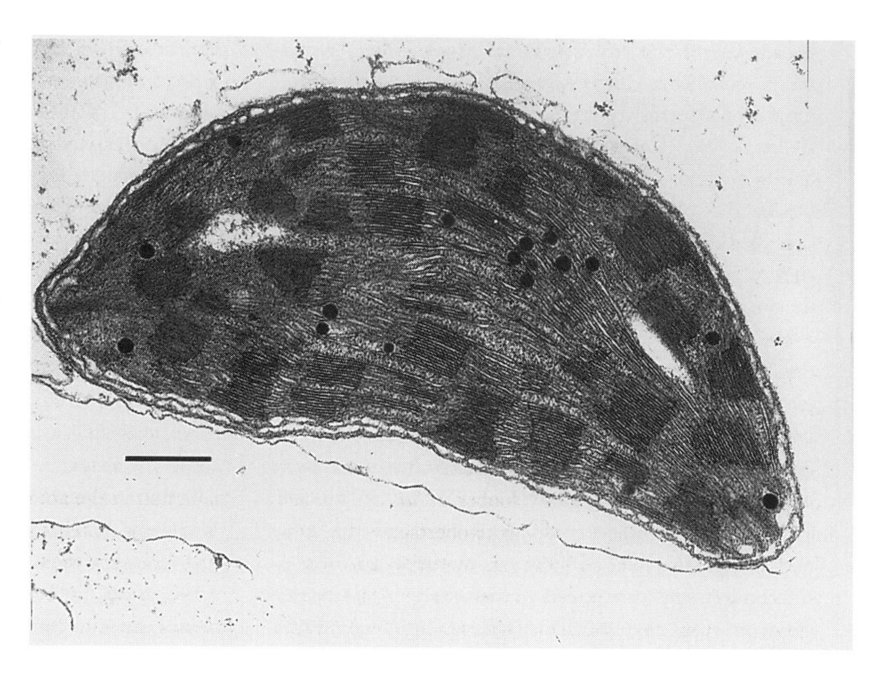

**Figure 1.3.** Ultrastructure of a chloroplast in a mesophyll cell of *Zea mays*. The chloroplast is filled with thylakoids that differentiate to form grana. A vesiculated membrane structure, called the peripheral vesicles, occurs adjacent to the inside surface of the envelope in chloroplasts of these cells (Bachmann *et al.*, 1973). Plastoglobules appear as small darkly stained particles. Bar, 0.5 μm. (From Ristic & Cass, 1992. Used with permission from University of Chicago Press.)

tain variable but significant amounts of grana. A mutant strain of barley, *viridis*-[115], which lacks the PSII core complex yet retains LHCII, forms giant grana approximately five times greater in diameter than those of wild-type with very few intergranal lamellae (Simpson, Vallon & von Wettstein, 1989). In the *ch*-1 mutant of *Arabidopsis*, a strain analogous to the *chlorina f2* mutant of barley, the chloroplasts contain somewhat less thylakoid membranes but the membranes are organized into extensive grana (Murray & Kohorn, 1991). A double mutant of barley, *viridis*-[115]/*chlorina f2*[2800], still retains grana four thylakoids thick and a diameter about four times that of wild-type (Simpson & von Wettstein, 1989). Although appearance of chlorophyll *b* during chloroplast development often coincides with formation of grana, these findings suggest that fully assembled LHCII is not required for stacking and that the amount of the apoproteins, although a contributor to the process, is not the primary factor in appression of adjacent thylakoids. Stys (1995) suggested that stacking is a complex process that involves segregation of photosystems at low concentrations of divalent cations. At higher divalent cation concentrations, localized regions rich in PSII adhere, together with ion-induced complexes between membrane lipids. PSI may tend to collect in unstacked membranes because of insufficient space between appressed membranes for the additional PSI proteins on the stromal membrane surface. Although Stys (1995) suggested a role for digalactosyl diglycerides, a major thylakoid lipid, in grana formation, a mutant strain nearly devoid of this lipid has a slightly higher number of thylakoids per granum than wild-type plants (Dörmann *et al.*, 1995).

When PSI is near PSII, i.e. when the centres are energetically coupled, the kinetically more rapid PSI centres drain excitation energy from PSII and thereby quench its fluorescence (Trissl & Wilhelm, 1993). In thylakoid membranes formed early in development, LHCII apparently transfers energy to both PSII and PSI, a state-2 condition (White & Hoober, 1994). Grana formation therefore, may be required to generate a state-1 condition by separating the photosystems and thereby enhance efficiency of PSII by transferring more energy from LHCII to PSII. Analysis of the fluorescence properties of PSII suggest that excitons travel between multiple PSII units before becoming trapped (Lavergne & Trissl, 1995), which supports the concept of segregated clusters of PSII centers in the core of grana (Albertsson, 1995). Studies of chloroplast development have shown that stacking, and consequently PSII segregation, do not occur until an extensive amount of thylakoid membrane has been made (Ohad *et al.*, 1967; Robertson & Laetsch, 1974; Hoober *et al.*, 1991; von Wettstein *et al.*, 1995). Only the higher plants and the advanced green algae have evolved a lateral segregation of PSII and PSI induced by formation of grana. In the most primitive, green algae, PSI is located within grana with the same frequency as PSII (Song & Gibbs, 1995).

The essential functional structure of the chloroplast is the lamellar thylakoid membrane that encloses a lumen. The hypothesis proposed by Douce & Joyard (1990) that thylakoid biogenesis begins by formation of vesicles and that larger lamellar structures form by the fusion of many smaller compartments has been supported by work with algae (Hoober *et al.*, 1994) and developing higher plant seedlings (Robertson & Laetsch, 1974). The thylakoid system at maturity encloses a single, multifaceted luminal compartment (Albertsson, 1995) as suggested by serial sections of chloroplasts (Paolillo, 1970; Coombs & Greenwood, 1976). In other cases where grana do not form, such as chloroplasts in bundle sheath cells of $C_4$ plants, chromophytic and chrysophytic algae – with loosely associated groups of thylakoids – and rhodophytes whose thylakoids are covered with phycobilisomes (Gantt & Conti, 1966), existence of a continuous intrathylakoid luminal compartment is unlikely. There is no reason from a functional standpoint for the thylakoid lumen to form a continuum. Protons are pumped into the lumen by the flow of electrons through the photosynthetic electron transport pathway in the membrane. Because development of a proton gradient across the membrane is the central element in energy transduction, the process is favoured by minimal luminal volume relative to surface area. In contrast to the spherical chromatophore vesicles in photosynthetic bacteria, the flattened thylakoid lamellar system in cyanobacteria and thus in chloroplasts provides this feature. To this 'sack-like' arrangement was given the Greek word 'thylakoid' (Menke, 1962).

## EFFECTS OF INHIBITORS ON CHLOROPLAST DEVELOPMENT

Gabaculine is a specific, irreversible inhibitor of glutamate-1-semialdehyde aminotransferase (Hoober *et al.*, 1988; Kannangara *et al.*, 1994), the enzyme that catalyses synthesis of 5-aminolevulinate, the precursor of chlorophyll. Chloroplasts in secondary leaves of wheat that formed during treatment with gabaculine had pigment levels (chlorophylls and carotenoids) of about 15% of the control values. Reduction in chlorophyll drastically lowered the amount of thylakoid membranes (Duysen *et al.*, 1993). Volumes in control cells occupied by grana (28%), stromal thylakoids (15%) and stroma (*ca.* 50%) were 10%, 16% and 65%, respectively, in leaves of seedlings treated during greening. When etiolated seedlings were treated in the dark before irradiation, these values were 5%, 13% and 80%, respectively. This morphometric analysis showed that formation of grana was most severely affected, which may be related simply to the reduction in the amount of membranes. Accumulation of LHCII apoproteins in wheat was less sensitive to gabaculine treatment than LHCI apoproteins. Reaction centre proteins (D1, PsaA and PsaB) were not detected in gabaculine-treated chloroplasts (Duysen *et al.*, 1993). These results suggest that chlorophyll is required for stability of the chlorophyll-binding proteins in reaction centres. Absence of any one of the reaction centre apoproteins in *Chlamydomonas* drastically reduced accumulation of other subunits (de Vitry *et al.*, 1989). Inhibition of protein synthesis on plastid ribosomes with chloramphenicol did not change the structure of the prolamellar body in etiolated rye seedlings and had no effect on the dispersion of the prolamellar body during illumination (Prudnikova & Serdyuchenko, 1989). However, development of the thylakoid system was impaired and occasionally abnormal short, thick macrograna formed. Thylakoid membranes that formed in the presence of chloramphenicol during chloroplast development in *Chlamydomonas reinhardtii* contained LHCII, but did not form grana (Hoober, Siekevitz & Palade, 1969).

## USE OF MUTANTS TO STUDY CHLOROPLAST DEVELOPMENT AND STRUCTURE

Further understanding of the role of chlorophyll synthesis in chloroplast development and ultrastructure was obtained by Falbel *et al.* (1996) with mutants of wheat that expressed various levels of Mg-chelatase. The more severe the enzymatic defect, the greater the reduction in the amount of chlorophyll, which concomitantly raised the chlorophyll *a/b* ratio. With less chlorophyll, the chloroplasts contained reduced amounts of thylakoid membranes, and in the most severe deficiencies very little stacking. Interestingly, the abundance of PSII reaction centres was affected much less than that of light-harvesting complexes, which correlated with a higher affinity for chlorophyll *a* that is generally thought to be a property of the reaction centre proteins. Because chlorophyll *b* is only found in light-harvesting complexes, the higher chlorophyll *a/b* ratio correlates with the dramatically reduced level of these complexes.

Activation of phytochrome with red light strongly promotes development of chloroplasts (for review; see Quail *et al.*, 1995). Steps in the subsequent signal transduction mechanism are becoming apparent (Chory, 1993; Deng, 1994; Bowler & Chua, 1994). Regulatory elements upstream of the coding region of *LHCB* genes were characterized as binding sites for proteins that normally repress expression of these genes in the dark (Degenhardt & Tobin, 1996). Enhancement by phytochrome of expression of genes required for chloroplast development seems to involve release from this state of repression. Expression of a number of light-regulated genes is also derepressed in mutant strains in which partial chloroplast development occurs in the dark. Mutations in *COP* (*co*nstitutive *p*hotomorphogenic) and *DET* (*de*-*et*iolated) genes allow *LHCB* gene expression in the dark, even in roots. The *COT/DET* gene products are essential for normal repression of these developmental processes in the dark (Kwok *et al.*, 1996). Plastids in roots of mutant plants develop into immature chloroplasts, with thylakoid membranes and even grana, rather than starch-containing amyloplasts (Chory & Peto, 1990). Several groups of *cot/det*-type mutants have been identified in *Arabidopsis* that differ in the extent of morphological development of the seedling and the plastid (Wei & Deng, 1996). In one group of mutants, prolamellar bodies do not form in the plastids of dark-grown seedlings, and instead extensive formation of membrane material accumulates (Li *et al.*, 1995; Kwok *et al.*, 1996). In another group, prolamellar bodies are present in etioplasts, and developmental events typical of photomorphogenesis are less extensive (see Wei & Deng, 1996, for review). Transformation of *Arabidopsis* with DNA encoding only the N-terminal domain of the COP1 protein interfered with gene repression by normal COP1 in the dark and generated a phenotype including absence of a prolamellar body, similar to strains defective in the *COP1* gene (McNellis *et al.*, 1996). Interestingly, null mutations in the *COP/DET* genes are lethal, which suggests the gene products are more involved in gene regulation than simply acting as repressors.

The *DET2* gene encodes a protein similar to a mammalian steroid 5α-reductase (Li *et al.*, 1996), while the *CPD* gene encodes a cytochrome P450 that is also in the biosynthetic pathway of steroids (Szekeres *et al.*, 1996). Application of the plant steroid brassinolide reversed the photomorphogenic effects of these mutations (Li *et al.*, 1996). These results suggest that the steroid is involved in repression of light-regulated genes and support the concept that light is required normally for derepression rather than engaging an activation mechanism. A mutant of pea was characterized by Frances *et al.* (1992) that was similar to the *cop/det* mutants of *Arabidopsis*. Chloroplasts in the pea mutant grown in the dark also lacked a prolamellar body and contained a significant amount of membrane material. Light, however, is also required to drive synthesis of chlorophyll for the formation and function of stable photosynthetic complexes, and in light-grown plants the chloroplasts in these mutants are essentially normal.

Mutations in the *FUSCA* genes cause accumulation of anthocyanins in *Arabidopsis* cotyledons (Castle & Meinke, 1994) but also arrest development at an early stage. The strains defective in these genes were selected because of their purple colour, but the mutant loci were found to be allelic with the more severe group of *cop/det* mutants (Wei & Deng, 1996). Most of the mutants with a *fusca* phenotype also show pleiotropic photomorphogenic development in the dark. In light grown plants; chloroplasts are nearly spherical, contain a low amount of thylakoid membranes with few grana, and when germinated on media containing sucrose, produce large starch particles (Castle & Meinke, 1994).

A variety of mutations that affect lipid biosynthesis have minimal consequences for the ultrastructure of the chloroplast in *Arabidopsis*. The amount of digalactosyl diacylglycerol was reduced from 16 to 1.2 mole % by mutation at the *dgd1* locus (Dörmann *et al.*, 1995). The relative amounts of other thylakoid lipids increased to compensate for the loss of digalactosyl diacylglycerol, and no gross alteration in phenotype was observed. At the ultrastructural level, the chloroplast had 60% more thylakoid membrane and slightly higher number of thylakoids per granum than wild-type, and the membranes were highly curved and displaced towards the periphery of the chloroplast. The *fadD* mutation in *Arabidopsis*, which reduces the activity of a fatty acid desaturase, causes a large reduction in 18:3 and 16:3 fatty acids in thylakoid membrane lipids, while a compensatory increase in 18:2 and 16:2 fatty acids occurs. This mutant, which also has no obvious phenotypic anomaly, has slightly fewer grana per plastid with fewer thylakoids per granum (McCourt *et al.*, 1987). A single mutation at the *act1* locus leads to loss of the first enzyme of the chloroplast pathway of glycerolipid synthesis, acyl-ACP glycerol-3-phosphate acyltransferase, which transfers an 18:1 fatty acid to make lysophosphatidic acid. The chloroplast is then dependent upon the biosynthetic pathway in the endoplasmic reticulum for the 'eukaryotic' glycerolipids containing 18:3 fatty acids at the *sn*-2 position of glycerolipids rather than the 'prokaryotic' 16:3 fatty acids (Ohlrogge & Browse, 1995). Electron microscopic analyses showed that, although the amount of thylakoid membrane was essentially the same in wild-type and mutant, in the latter the number of thylakoid per granum was reduced from an average of 6.2 in the wild-type to 3.8 in the mutant, with little effect on photosynthetic activity (Kunst, Browse & Somerville, 1989). Single-locus mutations that altered lipid synthesis in *Arabidopsis* did not result in major ultrastructural changes.

Extensive ultrastructural studies were done with a variety of pigment-deficient mutants of barley (Henningsen, Boynton & von Wettstein, 1993) and *Zea mays* (Bachmann *et al.*, 1967, 1973). In strains blocked at various steps in carotenoid or chlorophyll synthesis, a wide variety of gross structural aberrations were observed. Oxidized carotenoids, the xanthophylls, are central components of LHCII (Kühlbrandt *et al.*, 1994) and β-carotene is a component of reaction centres (Bartley & Scolnik, 1995; Bialek-Bylka *et al.*, 1995). Because carotenoids protect the membrane from photooxidative damage (Frank & Cogdell, 1996), thylakoids in carotenoid-deficient mutants can be preserved only in dim light. Some albino mutant strains were able to generate nearly wild-type appearing chloroplasts in dim light but were white in strong light as the result of photodestruction of thylakoid membranes. Yet carotenoid-deficient mutants formed normal prolamellar bodies in etioplasts of seedlings grown in the dark. Barley seedlings treated with norflurazon, an inhibitor of carotenoid synthesis, which reduced carotenoid levels to less than 5% of control levels, also formed normal prolamellar bodies in the dark. Partial development of plastids in control plants was achieved under flashes of white light, but development of treated plastids was arrested at dispersal of the prolamellar body of the etioplast (Bolychevtseva *et al.*, 1995). Although a large number of *xantha* and *albina* mutants of barley show a wide spectrum of structural aberrations of the chloroplast, including few membranes, whorls, disorganized membranes or giant grana, the envelope membranes in most mutant strains appear undisturbed (Henningsen *et al.*, 1993).

Several mutants of the nuclear genome of *Arabidopsis* have been characterized that are developmentally impaired. A mutation in a regulatory locus designated CUE1 (*C*AB *u*nder*e*xpressed) interrupts light-induced expression of structural genes such as the *LHCB* (formerly called *CAB*) genes. Mutant cells contain markedly lowered levels of light-harvesting complex apoproteins in mesophyll cells of leaves. Chloroplasts in mutant cells had less thylakoid membranes, fewer grana and fewer membranes within the granal stack (Li *et al.*, 1995). A mutation at the *pale cress (pac)* locus also leads to impairment of chloroplast and leaf development (Reiter *et al.*, 1994). This pale-green strain accumulates a normal array of pigments but at less than 3% of the wild-type level. In leaf primordia, normal appearing proplastids are present that develop only a small number of thylakoid profiles and generally show no stacking. Mutant plastids formed prolamellar bodies in the dark, which dispersed normally when seedlings containing etioplasts were exposed to light. However, subsequent development was arrested and few thylakoid membranes formed. Transformation of mutant plants with the cloned *PAC* gene restored the wild-type chloroplast ultrastructure.

A fascinating series of papers (Pyke & Leech, 1994; Pyke *et al.*, 1994; Robertson, Pyke & Leech, 1995) described strains of *Arabidopsis* with mutations at several *arc* (*a*ccumulation and *r*eplication of *c*hloroplasts) loci that cause inability of chloroplasts to divide. The number of chloroplasts in wild-type mesophyll cells correlated with size of the cell, with a range of 20 to about 160 chloroplasts per cell. These chloroplasts normally develop from 14 proplastids per cell. In the most extreme mutant, *arc6*, fully expanded leaf cells contain only two large, elongated chloroplasts. Ultrastructure of the thylakoid and differentiation into grana are normal in the mutant strains. The plants show no remarkable phenotype, which indicates the plasticity of chloroplast number and size allowable in plants. These studies provide a basis for elucidating the mechanism of chloroplast division in the expanding cell.

Also interesting is the finding that inactivation of mitochondrial cytochrome oxidase causes impairment of chloroplast development. The non-chromosomal stripe NCS6 mutation in *Zea mays* is a partial deletion of the mitochondrial *cox2* gene that encodes subunit 2 of cytochrome oxidase. Yellow strips contain cells that are homoplasmic, or nearly so, for the mitochondrial deletion. Loss of cytochrome oxidase function leads to tissue containing about 30% of the wild-type chlorophyll level, and chloroplasts containing reduced amounts of thylakoid membranes with little stacking (Gu, Miles & Newton, 1993). In contrast with normal chloroplasts that contained numerous starch grains, those from mutant yellow strips had no starch. A mitochondrial genome rearrangement that inactivates subunit 4 of NADH dehydrogenase in *Zea mays* also causes a reduction in the amount of unappressed thylakoid membranes in bundle sheath chloroplasts in pale-green sectors of maize leaves. The reduced amount of thylakoids in mesophyll cell chloroplasts retain the ability to form grana, but starch was not present in either type of chloroplast (Roussell *et al.*, 1991). The NCS mutant strains are less efficient in energy trapping and $CO_2$ fixation and also deficient in PSI core polypeptides. The mechanism by which mutations in the mitochondrial genome affect chloroplast development is unknown.

## EFFECTS OF ENVIRONMENTAL STRESS ON CHLOROPLAST ULTRASTRUCTURE

Chloroplast structure is affected dramatically by environmental stresses. Intensity of light during plant growth strongly determines the structural and functional properties of chloroplasts. Chloroplasts in pea seedlings that developed under far-red-depleted illumination contained thin grana interconnected by long stromal thylakoids. When illumination was enriched in far-red irradiation (690 to 710 nm range), grana contained more thylakoids and were connected by very short stroma-exposed thylakoids (Melis, 1984). The same ultrastructural differences were observed between chloroplasts from sun-adapted and shade-adapted plants, respectively.

Different plants, and genetic variants of the same plant, show marked differences in their sensitivity to stresses. *Haberlea rhodopensis* and *Ramonda servica*, two homoiochlorophyllous poikilohydric plants, recovered full photosynthetic activity after rewatering for 2 or 8 days, respectively, of plants desiccated for one year (Markovska, Tutekova & Kimenov, 1995). Chloroplasts in desiccated plants were round and lacked an envelope and starch grains, but retained most of the chlorophyll and a lamellar system. After rewatering, the organelle became elliptical, thylakoids and grana became more defined, and starch grains formed. The envelope in *H. rhodopensis* was not restored within the first 48 hours of the recovery period whereas the envelope was clearly visible in *R. serbivca*.

Chloroplasts in mesophyll cells of a drought-resistant line of *Zea mays* were not markedly altered 7 days after the soil containing 13-day old seedlings was dried. Some chloroplasts showed internal disorganization but grana were retained. When, in addition to water stress, the plants were exposed to a 6- or 24-hour treatment at 45 °C, the envelope disappeared. Grana were dramatically reduced at the higher temperature and long undifferentiated thylakoids resulted. After a 3-day recovery period, the chloroplast envelope was restored and grana reformed (Ristic & Cass, 1991). In drought-sensitive *Zea mays* plants, the heat treatment exaggerated the changes in the resistant line. The envelope again disappeared when water-stressed sensitive plants were exposed to the higher temperature and grana were converted to long single lamellae that filled the area of the plastid. When plants were rewatered, the envelope did not reform and the chloroplasts continued to disintegrate. The basis for the difference between resistant and sensitive strains is not clear. However, the resistant strain produced tenfold higher levels of abscisic acid at the higher temperature than the sensitive strain (Ristic & Cass, 1992). Resistant and sensitive strains synthesized 100 kDa, 70 kDa and small (*ca.* 20 kDa) heat shock proteins at 45 °C. The resistant strain synthesized in addition a 45 kDa protein at the higher temperature that was not produced in the sensitive strain (Ristic, Gifford & Cass, 1991).

Fe deficiency severely affected synthesis of proteins in leaves. Reduction in chlorophyll and thylakoid content was accompanied by a highly correlated drop in mRNA for the large and small subunits of ribulose bisphosphate carboxylase/oxygenase and thus in the content of the enzyme (Winder & Nishio, 1995). Supplemental iron allowed nearly normal chloroplast development. In Fe-limited cells of the alga *Dunaliella tertiolecta*, the amounts of several reaction centre proteins, in particular D1, CP43 and CP47, which are synthesized on chloroplast ribosomes, were markedly reduced (Vassiliev *et al.*, 1995). Possibly, deficiency reduced the levels of mRNA for other proteins required for development of the thylakoid system. In contrast, the α subunit of cytochrome $b_{559}$, also a PSII reaction centre component, was ten-fold higher in the Fe-limited cells, although the haem-containing protein was essentially undetectable. A similar accumulation of the α subunit in the absence of PSII reaction centres was also observed in the *viridis*$^{-115}$ mutant of barley (Simpson *et al.*, 1989).

## REFERENCES


Albertsson, P-Å. (1995). The structure and function of the chloroplast photosynthetic membrane – a model for the domain organization. *Photosynthesis Research*, **46**, 141–9.

Arntzen, C. J. & Burke, J. J. (1980). Analysis of dynamic changes in membrane architecture: electron microscopic approach. *Methods in Enzymology*, **69**, 520–38.

Awramik, S. M. (1992). The oldest records of photosynthesis. *Photosynthesis Research*, **33**, 75–89.

Bachmann, M. D., Robertson, D. S., Bowen, C. C. & Anderson, I. C. (1967). Chloroplast development in pigment deficient mutants of maize. I. Structural anomalies in plastids of allelic mutants at the $w_3$ locus. *Journal of Ultrastructure Research*, **21**, 41–60.

Bachmann, M. D., Robertson, D. S., Bowen, C. C. & Anderson, I. C. (1973). Chloroplast ultrastructure in pigment-deficient mutants of *Zea mays* under reduced light. *Journal of Ultrastructure Research*, **45**, 384–406.

Bartley, G. E. & Scolnik, P. A. (1995). Plant carotenoids: pigments for photoprotection, visual attraction, and human health. *The Plant Cell*, **7**, 1027–38.

Bassi, R., Marquardt, J. & Lavergne, J. (1995). Biochemical and functional properties of photosystem II in agranal membranes from maize mesophyll and bundle sheath chloroplasts. *European Journal of Biochemistry*, **233**, 709–19.

Bauer, C. E., Bollivar, D. W. & Suzuki, J. Y. (1993). Genetic analyses of photopigment biosynthesis in eubacteria: a guiding light for algae and plants. *Journal of Bacteriology*, **175**, 3919–25.

Bhattacharya, D. & Medlin, L. (1995). The phylogeny of plastids: a review based on comparisons of small-subunit ribosomal RNA coding regions. *Journal of Phycology*, **31**, 489–98.

Bialek-Bylka, G. E., Tomo, T., Satoh, K. & Koyama, Y. (1995). 15-*cis*-β-Carotene found in the reaction center of spinach photosystem II. *FEBS Letters*, **363**, 137–40.

Blankenship, R. E. (1992). Origin and early evolution of photosynthesis. *Photosynthesis Research*, **33**, 91–111.

Böddi, B., Lindsten, A., Ryberg, M. & Sundqvist, C. (1989). On the aggregational states of protochlorophyllide and its protein complexes in wheat etioplasts. *Physiologia Plantarum*, **76**, 135–43.

Bolychevtseva, Y. V., Rakhimberdieva, M. G., Darapetyan, N. V., Popov, V. I., Moskalenko, A. A. & Kuznetsova, N. Y. (1995). The development of carotenoid-deficient membranes in plastids of barley seedlings treated with norflurazon. *Journal of Photochemistry and Photobiology B: Biology*, **27**, 153–60.

Bowler, C. & Chua, N.-H. (1994). Emerging themes of plant signal transduction. *The Plant Cell*, **6**, 1529–41.

Brentel, I., Selstam, E. & Lindblom, G. (1985). Phase equilibria of mixtures of plant galactolipids. The formation of a bicontinuous cubic phase. *Biochimica et Biophysica Acta*, **812**, 816–26.

Castle, L. A. & Meinke, D. W. (1994). A *FUSCA* gene of *Arabidopsis* encodes a novel protein essential for plant development. *The Plant Cell*, **6**, 25–41.

Castresana, J. & Saraste, M. (1995). Evolution of energetic metabolism: the respiration-early hypothesis. *Trends in Biochemical Sciences*, **20**, 443–8.

Cavalier-Smith, T. (1982). The origin of plastids. *Biological Journal of the Linnean Society*, **17**, 289–306.

Chory, J. (1993). Out of darkness: mutants reveal pathways controlling light-regulated development in plants. *Trends in Genetics*, **9**, 167–72.

Chory, J. & Peto, C. (1990). Mutations in the *DET1* gene affect cell-type-specific expression of light-regulated genes and chloroplast development in *Arabidopsis*.

*Proceedings of the National Academy of Sciences, USA*, **87**, 8776–80.

Cline, K., Keegstra, K. & Staehelin, L. A. (1985). Freeze–fractured electron microscopic analysis of ultrarapidly frozen envelope membranes on intact chloroplasts and after purification. *Protoplasma*, **125**, 111–23.

Coombs, J. & Greenwood, A. D. (1976). Compartmentation of the photosynthetic apparatus. In *The Intact Chloroplast, Topics in Photosynthesis*, vol. 1, ed. J. Barber, pp. 1–51. Amsterdam: Elsevier.

Cremers, F. F. M., Voorhout, W. F., van der Krift, T. P., Leunissen-Bijvelt, J. J. M. & Verkleij, A. J. (1988). Visualization of contact sites between outer and inner envelope membranes in isolated chloroplasts. *Biochimica et Biophysica Acta*, **933**, 334–40.

Dean, C., Pichersky, E. & Dunsmuir, P. (1989). Structure, evolution, and regulation of *RbcS* genes in higher plants. *Annual Review of Plant Physiology and Plant Molecular Biology*, **40**, 415–39.

Degenhardt, J. & Tobin, E. M. (1996). A DNA binding activity for one of two closely defined phytochrome regulatory elements in an *Lhcb* promoter is more abundant in etiolated than in green plants. *The Plant Cell*, **8**, 31–41.

Deng, X.-W. (1994). Fresh view of light signal transduction in plants. *Cell*, **76**, 423–6.

de Vitry, C., Olive, J., Drapier, D., Recouvreur, M. & Wollman, F.-A. (1989). Posttranslational events leading to the assembly of photosystem II protein complex: a study using photosynthesis mutants from *Chlamydomonas reinhardtii*. *Journal of Cell Biology*, **109**, 991–1006.

Doolittle, R. F., Feng, D.-F., Tsang, S., Cho, G. & Little, E. (1996). Determining divergence times of the major kingdoms of living organisms with a protein clock. *Science*, **271**, 470–7.

Dörmann, P., Hoffmann-Benning, S., Balbo, I. & Benning, C. (1995). Isolation and characterization of an *Arabidopsis* mutant deficient in the thylakoid lipid digalactosyl diacylglycerol. *The Plant Cell*, **7**, 1801–10.

Douce, R. & Joyard, J. (1990). Biochemistry and function of the plastid envelope. *Annual Review of Cell Biology*, **6**, 173–216.

Douglas, S. E., Murphy, C. A., Spencer, D. F. & Gray, M. W. (1991). Cryptomonad algae are evolutionary chimaeras of two phylogenetically distinct unicellular eukaryotes. *Nature (London)*, **350**, 148–51.

Duysen, M., Freeman, T., Eskins, K. & Guikema, J. A. (1993). Gabaculine impaired accumulation of pigments and apoproteins in light-harvesting complexes and reaction centers of wheat thylakoids. *Photosynthetica*, **29**, 329–339.

Erata, M., Kubota, M., Takahashi, I. & Watanabe, M. (1995). Ultrastructure and phototactic action spectra of two genera of cryptophyte flagellate algae, *Cryptomonas* and *Chroomonas*. *Protoplasma*, **188**, 258–66.

Falbel, T. G., Meehl, J. B. & Staehelin, L. A. (1996). Severity of mutant phenotype in a series of chlorophyll-deficient wheat mutants depends on light intensity and the severity of the block in chlorophyll synthesis. *Plant Physiology*, **112**, 821–32.

Frances, S., White, M. J., Edgerton, M. D., Jones, A. M., Elliott, R. C. & Thompson, W. F. (1992). Initial characterization of a pea mutant with light-independent photomorphogenesis. *The Plant Cell*, **4**, 1519–30.

Frank, H. A. & Cogdell, R. J. (1996). Carotenoids in photosynthesis. *Photochemistry and Photobiology*, **63**, 257–64.

Gantt, E. & Conti, S. F. (1966). Granules associated with the chloroplast lamellae of *Porphyridium cruentum*. *The Journal of Cell Biology*, **29**, 423–34.

Gibbs, S. P. (1981). The chloroplast endoplasmic reticulum: structure, function and evolutionary significance. *International Review of Cytology*, **72**, 49–99.

Gilkey, J. C. & Staehelin, A. L. (1986). Advances in ultrarapid freezing for the preservation of cellular ultrastructure. *Journal of Electron Microscopic Techniques*, **3**, 177–210.

Gillott, M. S. & Gibbs, S. P. (1980). The cryptomonad nucleomorph: its ultrastructure and evolutionary significance. *Journal of Phycology*, **16**, 558–68.

Grossman, A. R., Bhaya, D., Apt, K. E. & Kehoe, D. M. (1995). Light-harvesting complexes in oxygenic photosynthesis: diversity, control and evolution. *Annual Review of Genetics*, **29**, 231–188.

Gu, J., Miles, D. & Newton, K. J. (1993). Analysis of leaf sectors in the NCS6 mitochondrial mutant of maize. *The Plant Cell*, **5**, 963–71.

Henningsen, K. W., Boynton, J. E. & von Wettstein, D. (1993). *Mutants at* xantha *and* albina *Loci in Relation to Chloroplast Biogenesis in Barley* (Hordeum vulgare L.), Biologiske Skrifter, 42. Copenhagen: The Royal Danish Academy of Sciences and Letters.

Hirsch, S., Muckel, E., Heemeyer, F., von Heijne, G. & Soll, J. (1994). A receptor component of the chloroplast protein translocation machinery. *Science*, **266**, 1989–92.

Hoober, J. K. (1984). *Chloroplasts*. New York: Plenum.

Hoober, J. K., Siekevitz, P. & Palade, G. E. (1969). Formation of chloroplast membranes in *Chlamydomonas reinhardtii y*-1. Effects of inhibitors of protein synthesis. *Journal of Biological Chemistry*, **244**, 2621–31.

Hoober, J. K., Kahn, A., Ash, D. E., Gough, S. & Kannangara, C. G. (1988). Biosynthesis of δ-aminolevulinate in greening barley leaves. IX, Structure of the substrate, mode of gabaculine inhibition, and the catalytic mechanism of glutamate 1-semialdehyde aminotransferase. *Carlsberg Research Communications*, **53**, 11–25.

Hoober, J. K., Boyd, C. O. & Paavola, L. G. (1991). Origin of thylakoid membranes in *Chlamydomonas reinhardtii y*-1 at 38 °C. *Plant Physiology*, **96**, 419–26.

Hoober, J. K., White, R. A., Marks, D. B. & Gabriel, J. L. (1994). Biogenesis of thylakoid membranes with emphasis on the process in *Chlamydomonas*. *Photosynthesis Research*, **39**, 15–31.

Hugueney, P., Bouvier, F., Badillo, A., d'Harlingue, A., Kuntz, M. & Camara, B. (1995). Identification of a plastid protein involved in vesicle fusion and/or membrane protein translocation. *Proceedings of the National Academy of Sciences, USA*, **92**, 5630–4.

Joyard, J., Block, M. A., Pineau, B., Albrieux, C. & Douce, R. (1990). Envelope membranes from mature spinach chloroplasts contain a NADPH : Pchlide reductase on the cytosolic side of the outer membrane. *Journal of Biological Chemistry*, **265**, 21820–7.

Joyard, J., Block, M. A. & Douce, R. (1991). Molecular aspects of plastid envelope biochemistry. *European Journal of Biochemistry*, **199**, 489–509.

Kannangara, C. G., Andersen, R. V., Pontoppidan, B., Willows, R. & von Wettstein, D. (1994). Enzymic and mechanistic studies on the conversion of glutamate to 5-aminolaevulinate. In *The Biosynthesis of the Tetrapyrrole Pigments*, Ciba Foundation Symposium 180, ed. D. J. Chadwick & K. Ackrill, pp. 3–20. Chichester: John Wiley.

Keegstra, K., Bruce, B., Hurley, M., Li, H. & Perry, S. (1995). Targeting of proteins into chloroplasts. *Physiologia Plantarum*, **93**, 157–62.

Król, M., Spangfort, M. D., Hunter, N. P. A., Öquist, G., Gustafsson, P. & Jansson, S. (1995). Chlorophyll *a*/*b*-binding proteins, pigment conversions, and early light-induced proteins in a chlorophyll *b*-less barley mutant. *Plant Physiology*, **107**, 873–83.

Kühlbrandt, W., Wang, D. N. & Fujiyoshi, Y. (1994). Atomic model of plant light-harvesting complex by electron crystallography. *Nature (London)*, **367**, 614–21.

Kunst, L., Browse, J. & Somerville, C. (1989). Altered chloroplast structure and function in a mutant of *Arabidopsis* deficient in plastid glycerol-3-phosphate acyltransferase activity. *Plant Physiology*, **90**, 846–53.

Kwok, S. F., Piekos, B., Miséra, S. & Deng, X.-W. (1996). A complement of ten essential and pleiotopic *Arabidopsis COP/DET/FUS* genes is necessary for repression of photomorphogenesis in darkness. *Plant Physiology*, **110**, 731–42.

Lake, J. A. (1994). Reconstructing evolutionary trees from DNA and protein sequences: paralinear distances. *Proceeding of the National Academy of Sciences USA*, **91**, 1455–9.

Laroche, J., Li, P. & Bousquet, J. (1995). Mitochondrial DNA and monocot–dicot divergence time. *Molecular Biology and Evolution*, **12**, 1151–6.

Lavergne, J. & Trissl, H-W. (1995). Theory of fluorescence induction in photosystem II: derivation of analytical expressions in a model including exciton-radical-pair equilibrium and restricted energy transfer between photosynthetic units. *Biophysical Journal*, **68**, 2474–92.

Li, J.-M., Culligan, K., Dixon, R. A. & Chory, J. (1995). *CUE1*: A mesophyll cell-specific positive regulator of light-controlled gene expression in *Arabidopsis*. *The Plant Cell*, **7**, 1559–610.

Li, J., Nagpai, P., Vitart, V., McMorris, T. C. & Chory, J. (1996). A role for brassinosteroids in light-dependent development of *Arabidopsis*. *Science*, **272**, 398–401.

Lindsten, A., Ryberg, M. & Sundqvist, C. (1988). The polypeptide composition of highly purified prolamellar bodies and prothylakoids from wheat (*Triticum aestivum*) as revealed by silver staining. *Physiologia Plantarum*, **72**, 167–76.

McCourt, P., Kunst, L., Browse, J. & Somerville, C. R. (1987). The effect of reduced amounts of lipid unsaturation on chloroplast ultrastructure and photosynthesis in a mutant of *Arabidopsis*. *Plant Physiology*, **84**, 353–60.

McFadden, G. I., Gilson, P. R., Hofmann, C. J. B., Adcock, G. J. & Maier, U-G. (1994). Evidence that an amoeba acquired a chloroplast by retaining part of an engulfed eukaryotic alga. *Proceedings of the National Academy of Sciences USA*, **91**, 3690–4.

McNellis, T. W., Torri, K. U. & Deng, X. -W. (1996) Expression of an N-terminal fragment of COP1 confers a dominant-negative effect on light-regulated seedling development in *Arabidopsis*. *The Plant Cell*, **8**, 1491–503.

Margulies, L. (1996). Archaeal-eubacterial mergers in the origin of Eukarya: phylogenetic classification of life. *Proceedings of the National Academy of Sciences, USA*, **93**, 1071–6.

Markovska, Y. K., Tutekova, A. A. & Kimenov, G. P. (1995). Ultrastructure of chloroplasts of poikilohydric plants *Haberlea rhodopensis* Friv. and *Ramonda serbica* Panč during recovery from desiccation. *Photosynthetica*, **31**, 613–20.

Melis, A. (1984). Light regulation of photosynthetic membrane structure, organization and function. *Journal of Cellular Biochemistry*, **24**, 271–85.

Menke, W. (1962). Structure and chemistry of plastids. *Annual Review of Plant Physiology*, **13**, 27–44.

Meyer, T. E. (1994). Evolution of photosynthetic reaction centers and light harvesting chlorophyll proteins. *Biosystems*, **33**, 167–75.

Morré, D. J., Morré, J. T., Morré, S. R., Sundqvist, C. & Sandelius, A. S. (1991). Chloroplast biogenesis. Cell-free transfer of envelope monogalactosylglycerides to thylakoids. *Biochimica et Biophysica Acta*, **1070**, 437–45.

Mühlethaler, V. K. & Frey-Wyssling, A. (1959). Entwicklung und Structur der Proplastiden. *Journal of Biophysical and Biochemical Cytology*, **6**, 507–12.

Mullet, J. E. (1983). The amino acid sequence of the polypeptide segment which regulates membrane adhesion (grana stacking) in chloroplasts. *Journal of Biological Chemistry*, **258**, 9941–8.

Murray, D. L. & Kohorn, B. D. (1991). Chloroplasts of *Arabidopsis thaliana* homozygous for the ch-1 locus lack chlorophyll *b*, lack stable LHCPII and have stacked thylakoids. *Plant Molecular Biology*, **16**, 71–9.

Nierzwicki-Bauer, S. A., Balkwill, D. L. & Stevens, S. E., Jr (1983). Three-dimensional ultrastructure of a unicellular cyanobacterium. *Journal of Cell Biology*, **97**, 713–22.

Nisbet, E. G., Cann, J. R. & Van Dover, C. L. (1995). Origins of photosynthesis. *Nature (London)*, **373**, 479–80.

Ohad, I., Siekevitz, P. & Palade, G. E. (1967). Biogenesis of chloroplast membranes. II, Plastid differentiation during greening of a dark-grown algal mutant (*Chlamydomonas reinhardi*). *Journal of Cell Biology*, **35**, 553–84.

Ohlrogge, J. & Browse, J. (1995). Lipid biosynthesis. *The Plant Cell*, **7**, 957–70.

Osafune, T., Schiff, J. A. & Hase, E. (1991). Stage-dependent localization of LHCPII apoprotein in the golgi of synchronized cells of *Euglena gracilis* by immunogold electron microscopy. *Experimental Cell Research*, **193**, 320–30.

Paolillo, D. J., Jr. (1970). The three-dimensional arrangement of intergranal lamellae in chloroplasts. *Journal of Cell Science*, **6**, 243–55.

Preiss, S. & Thornber, J. P. (1995). Stability of the apoproteins of light-harvesting complex I and II during biogenesis of thylakoids in the chlorophyll *b*-less barley mutant *chlorina f2*. *Plant Physiology*, **107**, 709–17.

Prudnikova, I. V. & Serdyuchenko, E. V. (1989). Morphometric analysis of chloroplast ultrastructure in chloramphenicol-treated post-etiolated rye leaves. *Photosynthetica*, **23**, 351–5.

Pyke, K. A. & Leech, R. M. (1994). A genetic analysis of chloroplast division and expansion in *Arabidopsis thaliana*. *Plant Physiology*, **104**, 201–7.

Pyke, K. A., Rutherford, S. M., Robertson, E. J. & Leech, R. M. (1994). *arc6*, a fertile *Arabidopsis* mutant with only two mesophyll cell chloroplasts. *Plant Physiology*, **106**, 1169–77.

Quail, P. H., Boylan, M. T., Parks, B. M., Short, T. W., Xu, Y. & Wagner, D. (1995). Phytochromes: photosensory perception and signal transduction. *Science*, **268**, 675–80.

Reinbothe, S., Reinbothe, C, Lebedev, N. & Apel, K. (1996). PORA and PORB, two light-dependent protochlorophyllide-reducing enzymes of angiosperm chlorophyll biosynthesis. *The Plant Cell*, **8**, 763–9.

Reiter, R. S., Coomber, S. A., Bourett, T. M., Bartley, G. E. & Scolnik, P. A. (1994). Control of leaf and chloroplast development by the *Arabidopsis* gene *pale cress*. *The Plant Cell*, **6**, 1253–64.

Reith, M. (1995). Molecular biology of rhodophyte and chromophyte plastids. *Annual Review of Plant Physiology and Plant Molecular Biology*, **46**, 549–75.

Ristic, Z. & Cass, D. D. (1991). Chloroplast structure after water shortage and high temperature in two lines of *Zea mays* L. that differ in drought resistance. *Botanical Gazette*, **152**, 186–4.

Ristic, Z. & Cass, D. D. (1992). Chloroplast structure after water and high-temperature stress in two lines of maize that differ in endogenous levels of abscisic acid. *International Journal of Plant Science*, **153**, 186–96.

Ristic, Z., Gifford, D. J. & Cass, D. D. (1991). Heat shock proteins in two lines of *Zea mays* L. that differ in drought and heat resistance. *Plant Physiology*, **97**, 1430–4.

Robertson, D. & Laetsch, W. M. (1974). Structure and function of developing barley plastids. *Plant Physiology*, **54**, 148–59.

Robertson, E. J., Pyke, K. A. & Leech, R. M. (1995). *arc6*, an extreme chloroplast division mutant of *Arabidopsis* also alters proplastid proliferation and morphology in shoot and root apices. *Journal of Cell Science*, **108**, 2937–44.

Roussell, D. L., Thompson, D. L, Pallardy, S. G., Miles, D., & Newton, K. J. (1991). Chloroplast structure and function is altered in the NCS2 maize mitochondrial mutant. *Plant Physiology*, **96**, 232–8.

Scherer, L. (1990). Do photosynthetic and respiratory electron transport chains share redox proteins? *Trends in Biochemical Sciences*, **15**, 458–62.

Schimper, A. F. W. (1883). Uber die Entwicklung der

Chlorophyllkörner und Farbkörper. *Botanische Zeitung*, **41**, 105–14.

Schnell, D. J. & Blobel, G. (1993). Identification of intermediates in the pathway of protein import into chloroplasts and their localization to envelope contact sites. *Journal of Cell Biology*, **120**, 103–15.

Schnell, D. J., Kessler, F. & Blobel, G. (1994). Isolation of components of the chloroplast protein import machinery. *Science*, **266**, 1007–12.

Schopf, J. W. (1993). Microfossils of the early Archean Apex chert: new evidence for the antiquity of life. *Science*, **260**, 640–6.

Selstam, E. & Sandelius, A. S. (1984). A comparison between prolamellar bodies and prothylakoid membranes in etioplasts of dark-grown wheat concerning lipid and polypeptide composition. *Plant Physiology*, **76**, 1036–40.

Simpson, D. J. & von Wettstein, D. (1989). The structure and function of the thylakoid membrane. *Carlsberg Research Communication*, **54**, 55–65.

Simpson, D. J., Vallon, O. & von Wettstein, D. (1989). Freeze–fracture studies of barley plastid membranes. VIII. In *viridis*-$^{115}$, a mutant completely lacking photosystem II, OEE1 and the α-subunit of cytochrome *b-559* accumulate in appressed thylakoids. *Biochimica et Biophysica Acta*, **975**, 164–74.

Soll, J. (1995). New insights into the protein import machinery of the chloroplast's outer envelope. *Botanica Acta*, **108**, 277–82.

Song, X.-Z. & Gibbs, S. P. (1995). Photosystem I is not segregated from photosystem II in the green alga *Tetraselmis subcordiformis*. An immunogold and cytochemical study. *Protoplasma*, **189**, 267–80.

Steinback, K. E., Burke, J. J. & Arntzen, C. J. (1979). Evidence for the role of surface-exposed segments of the light-harvesting complex in cation-mediated control of chloroplast structure and function. *Archives of Biochemistry and Biophysics*, **195**, 546–57.

Stys, D. (1995). Stacking and separation of photosystem I and photosystem II in plant thylakoid membranes: a physico-chemical view. *Physiologia Plantarum*, **95**, 651–7.

Sulli, C. & Schwartzbach, S. D. (1996). A soluble protein is imported into *Euglena* chloroplasts as a membrane-bound precursor. *The Plant Cell*, **8**, 43–53.

Suzuki, J. Y. & Bauer, C. E. (1995). A prokaryotic origin for light-dependent chlorophyll biosynthesis of plants. *Proceedings of the National Academy of Sciences, USA*, **92**, 3749–53.

Szekeres, M., Németh, K., Koncz-Kálmán, Z., Mathur, J., Kauschmann, A., Altman, T., Rédei, G. P., Nagy, F., Schell, J. & Koncz, C. (1996). Brassinosteroids rescue the deficiency of CYP90, a cytochrome P450 controlling cell elongation and de-etiolation in *Arabidopsis*. *Cell*, **85**, 171–82.

Trissl, H.-W. & Wilhelm, C. (1993). Why do thylakoid membranes from higher plants form grana stacks? *Trends in Biochemical Sciences*, **18**, 415–19.

Vassiliev, I. R., Kolber, Z., Wyman, K. D., Mauzerall, D., Shukla, V. K. & Falkowski, P. G. (1995). Effects of iron limitation on photosystem II composition and light utilization in *Dunaliella tertiolecta*. *Plant Physiology*, **109**, 963–72.

Vermaas, W. F. J. (1994). Evolution of Heliobacteria: implications for photosynthetic reaction center complexes. *Photosynthesis Research*, **41**, 285–94.

von Wettstein, D. (1961). Nuclear and cytoplasmic factors in development of chloroplast structure and function. *Canadian Journal of Botany*, **39**, 1537–45.

von Wettstein, D., Gough, S. & Kannangara, C. G. (1995). Chlorophyll biosynthesis. *The Plant Cell*, **7**, 1039–57.

Wei, N. & Deng, S.-W. (1996). The role of the *COP/DET/FUS* genes in light control of *Arabidopsis* seedling development. *Plant Physiology*, **112**, 871–8.

White, R. A. & Hoober, J. K. (1994). Biogenesis of thylakoid membranes in *Chlamydomonas reinhardtii yl*. A kinetic study of initial greening. *Plant Physiology*, **106**, 583–90.

White, R. A., Wolfe, G. R., Komine, Y. & Hoober, J. K. (1996). Localization of light-harvesting apoproteins in the chloroplast and cytoplasm during greening of *Chlamydomonas reinhardtii* at 38 °C. *Photosynthesis Research*, **47**, 267–80.

Winder, T. L. & Nishio, J. N. (1995). Early iron deficiency stress response in leaves of sugar beet. *Plant Physiology*, **108**, 1487–94.

Wolfe, G. R. & Hoober, J. K. (1996). Evolution of thylakoid structure. In *Oxygenic Photosynthesis: The Light Reactions*, ed. D. R. Ort & C. F. Yocum, pp. 31–40. Dordrecht: Kluwer Academic Publishers.

Wolfe, G. R., Cunningham, F. X., Durnford, D., Green, B. R. & Gantt, E. (1994). Evidence for a common origin of chloroplasts with light-harvesting complexes of different pigmentation. *Nature (London)*, **367**, 566–8.

# 2 Light-harvesting complexes of higher plants

D. T. MORISHIGE AND B. W. DREYFUSS

## INTRODUCTION

Ubiquitous among photosynthetic organisms are pigment-binding proteins which capture the energy from sunlight used for photochemical charge separation. Oxygenic organisms have two photosystems, the basic energy transducing units, that operate in series to produce ATP and NADPH. Each photosystem consists of (i) a reaction centre core complex (CC) that is necessary and sufficient to carry out primary charge separation and (ii) light-harvesting complexes (LHCs). While they vary greatly in their protein subunit and pigment composition and in their localization with respect to their core complex, all LHCs primarily serve to increase the cross-sectional area utilized for light absorption by a photosystem in two ways: (i) they increase the total number of pigment molecules around a reaction centre and (ii) they broaden the absorption bands by utilizing pigments different from those located in the reaction centre to increase the probability that a photon of an energy not absorbed efficiently by a reaction centre pigment will be captured.

In higher plants, LHCI and LHCII, associated with Photosystem I (PSI) and Photosystem II (PSII), respectively, are composed of biochemically distinct chlorophyll-carotenoid-binding proteins. Chlorophylls (Chls) are the primary light-capturing pigment used to drive the photochemical reactions. Carotenoids, while they contribute in light-harvesting, function primarily to dissipate excess energy from chlorophylls to protect the photosynthetic apparatus from photo-oxidative damage. Proteins non-covalently bind the antenna pigments to both precisely align the pigment molecules to optimize absorption of light at specific wavelengths and to position the pigment molecules so that efficient energy transfer between neighbouring molecules may take place. Mildly denaturing polyacrylamide gel electrophoresis (PAGE) systems have been invaluable in the elucidation of the biochemical composition of the LHCs. Refinement of detergent solubilization and gel electrophoresis techniques used to fractionate and isolate the LHCs has provided an increasingly better means to observe the LHCs in their native state with only minor losses of bound pigments (Fig. 2.1). This brief review will concentrate on the biochemical characterization and biogenesis of the light-harvesting components of vascular plants. This chapter is meant to serve as an introduction to the study of the higher plant LHCs and is certainly not exhaustive. Several other recent reviews have been written covering different areas of LHC research (Bassi, Rigoni & Giacometti, 1990; Jansson, 1994; Paulsen, 1995). In general, the sizes of the LHC proteins stated below are approximations. These values can vary, depending upon the plant species being studied and the separation method employed for size determinations.

## PHOTOSYSTEM I LIGHT-HARVESTING COMPLEXES

The first intact Photosystem I particle was isolated by Mullet and co-workers (Mullet, Burke & Arntzen, 1980), using sucrose gradient centrifugation of Triton X-100 solubilized pea thylakoid membranes. This PSI particle (PSI-110) was composed of approximately 11 protein subunits and contained about 110 Chl molecules per P700. It exhibited the 730 nm longer-wavelength fluorescence maximum at 77 K, characteristic of LHCI (see below). A second PSI preparation, PSI-65, was obtained,

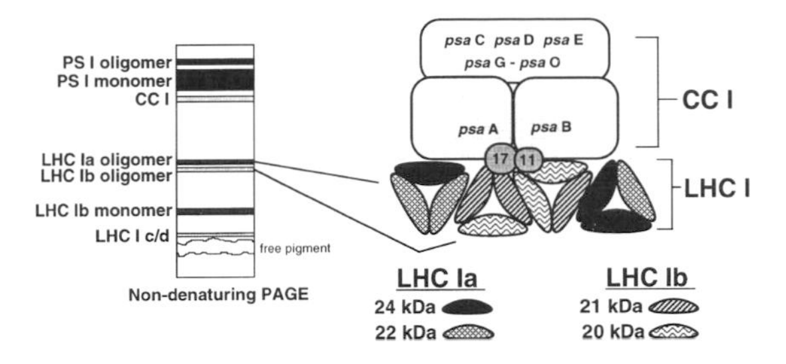


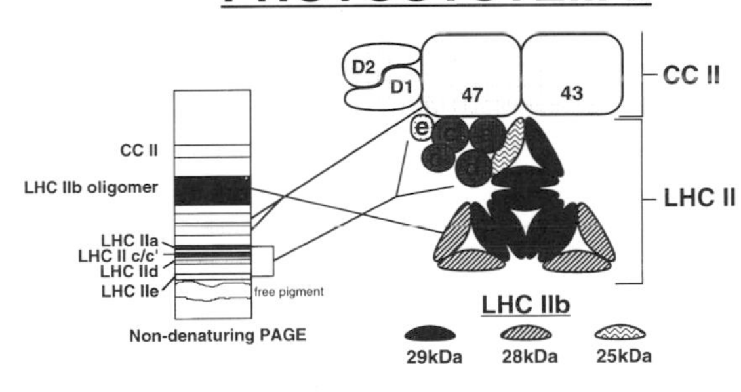


**Figure 2.1.** Models for the organization of the individual light-harvesting components around the core complexes of Photosystems I and II. On the left are idealized depictions of the pigmented complexes fractionated by non-denaturing PAGE. The lines connect each complex in the gel to its corresponding location within the photosystem.

which contained 65 Chls per P700, had six to seven polypeptide subunits, and showed a fluorescence emission maximum at 77 K of about 690 nm, characteristic of CCI. By comparison of the two PSI preparations, it was deduced that polypeptides in the 20–24 kDa range, present in PSI-110 but not in PSI-65, were components of the PSI antenna complex, LHCI. LHCI can be separated into at least two distinct pigment-proteins, LHCIa and LHCIb (for summary see Table 2.1). In total, LHCI binds about 20% of the total chlorophyll. Data suggest that photo-protection via the xanthophyll cycle takes place within LHCI, as well as in LHCII (see below).

### LHCIa

Two proteins of 22 and 24 kDa are found in LHCIa (LHCI-680). The 22 kDa LHCIa protein is encoded by the *Lhca2* gene, while the 24 kDa protein is encoded by *Lhca3*. LHCIa binds Chl *a*, Chl *b* and xanthophylls, but neither neoxanthin nor carotenes. It has the lowest Chl *a*/*b* ratio of all the LHCI components (*ca.* 1.4). This complex has a fluorescence emission maximum of 690 nm at 77 K. Data suggest that it, like LHCIb and LHCIIb (see below), occurs *in situ* in a trimeric form (Knoetzel, Svendsen & Simpson, 1992; Preiss *et al.* 1993; Dreyfuss & Thornber, 1994*b*). The LHCIa oligomer has been further resolved into two distinct pigment complexes LHCI-680A and LHCI-680B, containing the *Lhca3* and *Lhca2* gene products, respectively (Knoetzel *et al.*, 1992). PSI-E is also associated with LHCI-680B, possibly reflecting a functional arrangement *in vivo* (Knoetzel *et al.*, 1992).

### LHCIb

There are two apoproteins of 21 and 20 kDa (*Lhca1* and *Lhca4* gene products, respectively) in the major LHCI pigment–protein complex LHCIb. It contains Chls *a* and *b* with a Chl *a*/*b* ratio of 2.3. Lutein and violaxanthin are the only two carotenoids detected in the complex. LHCIb is responsible for the fluorescence emission maximum at 730–735 nm observed for photosystem I at low temperature in whole leaves. Similar to LHCIa, this complex can be isolated in an oligomeric, probably trimeric, form by sucrose gradient centrifugation and non-denaturing PAGE (Knoetzel *et al.*, 1992; Preiss *et al.*, 1993; Dreyfuss & Thornber, 1994*b*). The individual pigmented subunits have not been separated, probably due to the similar sizes of the apoproteins.

### Additional LHCI complexes

Comparison of the subunit compositions of photosystem I and CCI later revealed that in addition to the apoproteins of the two major LHCIs, two other polypeptides of 11 and 17 kDa were contained in the photosystem I particle, but not in CCI. These proteins were also observed in a purified LHCI holocomplex, suggesting that they are indeed LHCI constituents. There is some evidence that each binds pigment cofactors.

#### *LHCIc*

LHCIc has a single apoprotein of 17 kDa, the N-terminal sequence of which has been obtained and identified

Table 2.1. *Summary of the biochemical characteristics of the major light-harvesting complexes from barley*

| | Apparent size (kDa) | | | | |
|---|---|---|---|---|---|
| | Holocomplex | Apoprotein | Chl *a*/*b* | Gene designation | Alternative name |
| LHCIa | 65/25 | 22,24 | 1.4 | *Lhca*2, *Lhca*3 | LHCI-680 |
| LHCIb | 65/25 | 21, 20 | 2.3 | *Lhca*1, *Lhca*4 | LHCI-730 |
| LHCIc | – | 17 | a > b | *psa*F | psaF |
| LHCId | – | 11 | a > b | – | – |
| LHCIIa | 35 | 31 | 2.25 | *Lhcb*4 | CP29 |
| LHCIIb | 72 | 29, 28, 25 | 1.33 | *Lhcb*1, *Lhcb*2, *Lhcb*3 | CPII |
| LHCIIc | 30 | 29, 26.5 | 1.8 | *Lhcb*5, – | CP27 |
| LHCIId | 24 | 21 | 0.8 | *Lhcb*6 | CP24 |
| LHCIIe | 12 | 11 | 1.4 | – | – |

as *psa*F (Anandan, Vainstein & Thornber, 1989). Controversy exists whether the 17 kDa protein is that of an LHCI apoprotein or that of a non-pigmented component of CCI that functions to bind plastocyanin to the core complex. Like LHCIa and LHCIb, LHCIc apparently binds Chl *a*, maybe small amounts of Chl *b*, and carotenoids (Preiss *et al.*, 1993).

### *LHCId*

The 11 kDa component of the LHCI holocomplex has been termed LHCId, and preliminary evidence indicates that it binds pigment cofactors (see Preiss *et al.*, 1993). Jansson (1994) has speculated that LHCId might be the PSI-E protein associated with LHCI-680B (see above).

## PHOTOSYSTEM II LIGHT-HARVESTING COMPLEXES

Four very similar xanthophyll–chlorophyll *a*/*b*-proteins associated with PSII have been extensively characterized (LHCIIa, b, c and d; Table 2.1). LHCIIb, the most abundant pigmented complex, binds about 45% of the total Chl, while the remaining LHCII constituents bind about 4% each. All of the known LHCII pigment–proteins contain lutein, neoxanthin and violaxanthin in different proportions, but no or very minimal levels of beta-carotene (Peter & Thornber, 1991; Bassi *et al.*, 1993). The minor LHCII components are relatively enriched in carotenoids, especially violaxanthin, and, together with the LHCIs, may contribute significantly in the regulation of non-radiative energy dissipation via the xanthophyll cycle (for review see Demmig-Adams & Adams, 1996)

### LHCIIa

The Chl *a*/*b* ratio of LHCIIa ranges from 2.20–2.25. Its room temperature absorbance spectrum has a maximum in the red at 677 nm with an additional unusual minor peak at 645 nm, characteristic of this pigment–protein (Bassi *et al.*, 1987; Peter & Thornber, 1991). LHCIIa contains the xanthophylls, lutein and violaxanthin, in approximately equal proportions, and is relatively enriched in violaxanthin in comparison with the other LHCII pigment–proteins. Smaller amounts of neoxanthin are also associated with LHCIIa (Peter & Thornber, 1991; Bassi *et al.*, 1993). The single apoprotein of LHCIIa migrates on fully denaturing SDS-PAGE with an apparent size of 30–31 kDa in barley and maize and 29 kDa in spinach and pea. The reason for the difference in molecular weight between species is unknown. Although the protein sequence displays a high degree of similarity with other *Lhc* sequences, a unique 42 amino acid insertion preceding the first predicted membrane spanning region of the protein is observed (Morishige & Thornber, 1992). The function of this motif is not known, but could play a role in the assembly of this particular LHC. It is also interesting to note that the *Lhcb*4 protein sequence exhibits a greater degree of overall similarity with *Lhca* sequences, than with other *Lhcb* sequences (Jansson, 1994).

## LHCIIb

The major chlorophyll-binding protein associated with green plant photosystem II is LHCIIb. Besides serving as the major antenna for PSII, other functions of LHCIIb include the cation-mediated formation of grana stacks and the distribution of excitation energy between photosystems, accomplished through the State 1–State 2 transition. LHCIIb was first observed and isolated as a single, green pigment–protein, CP2. Subsequently, SDS-PAGE systems resolved the LHCIIb complex into two bands. One band migrated with an apparent size of approximately 68 kDa and was believed to be a dimer of the second band, which migrated at 26 kDa. Under the mildest non-denaturing PAGE conditions, LHCIIb migrates solely as a trimer and often as an even higher order oligomer in addition to the trimer. All normal green plant species examined, to date, appear to contain a trimeric form of LHCIIb. The Chl *a/b* ratio of this chlorophyll-protein complex as obtained by Deriphat-PAGE (Peter & Thornber, 1991) is 1.33. Analysis of the LHCIIb crystal structure indicates that each protein binds seven Chl *a* and six Chl *b* molecules, through interaction of the magnesium atoms to polar side chains or to main chain carbonyls of the protein (Kühlbrandt, Wang & Fujiyoshi, 1994). Room temperature absorbance maxima are observed at 653 and 675 nm, corresponding to Chls *b* and *a*, respectively, with a distinct valley between the two peaks (Peter & Thornber, 1991). LHC IIb also contains the xanthophylls lutein, neoxanthin and violaxanthin in a probable molar ratio of five Chls per xanthophyll molecule (Peter & Thornber, 1991; Bassi *et al.*, 1993).

On SDS-PAGE at least three apoproteins of LHCIIb are observed, migrating with apparent sizes of 29, 28–27, and 25 kDa. Early SDS-PAGE systems only resolved two bands of 29 and 28 kDa. Addition of urea to the polyacrylamide gel matrix changes the mobility of various polypeptides, among which is the 25 kDa LHCIIb apoprotein, which co-migrated with the other slightly larger LHCIIb apoproteins in the absence of urea. The multiple 25–29 kDa subunits of LHCIIb do not have a simple stoichiometry, nor does it seem that there are the same number of apoproteins in all plants. Usually the largest apoprotein occurs in an order-of-magnitude greater concentration than the two smaller ones. Although three subunits are clearly present in most plants, as many as eight can be resolved in some instances (Sigrist & Staehelin, 1994).

The earliest LHC genes to be identified and sequenced were those coding for LHCIIb. The different LHCIIb genes can be separated into three different classes (Jansson, 1994). The majority of the identified LHCIIb genes do not contain an intron and are regarded as *Lhcb1* genes. *Lhcb2* genes contain one small intron in the region of the gene coding for the amino terminus of the mature protein. The *Lhcb1* genes generally code for a slightly larger mature protein than the *Lhcb2* genes. *Lhcb3* genes code for a truncated protein, 10–11 amino acid residues shorter than the *Lhcb1* or *Lhcb2* proteins. It is also possible to differentiate each type of *Lhcb* gene from the deduced amino acid sequence near the N-terminus of the protein, where several type-specific motifs are localized. Beyond the first 35 residues of the mature proteins, the sequences are highly conserved. Presently, little is known regarding the functional significance of the different LHCIIb proteins.

## LHCIIc

Originally it was assumed that, since re-electrophoresis of LHCIIb oligomer fractions on a second non-denaturing gel had yielded a monomeric form of LHCIIb, the green chlorophyll–protein band found in the LHCIIb monomer region was solely that of LHCIIb. However, in the mildest non-denaturing gels, which keep all of the LHCIIb in its trimeric or higher oligomeric state, a different pigment–protein (LHCIIc or CP27) has been observed, migrating in the region of the LHCIIb monomer (Bassi *et al.*, 1987; Dunahay, Schuster & Staehelin, 1987; Peter & Thornber, 1991). LHCIIc has different spectral and subunit characteristics than either LHCIIa or LHCIIb, thus substantiating that it is a distinct LHCII pigment–protein. LHCIIc has a Chl *a/b* ratio of 1.8–3.3 (Bassi *et al.*, 1987; Peter & Thornber, 1991). It is relatively enriched in lutein with smaller amounts of neoxanthin and violaxanthin also present

(Peter & Thornber, 1991; Bassi *et al.*, 1993). A room temperature red absorption maxima is observed at 676nm (Peter & Thornber, 1991). Its apoprotein(s) has been reported to be a single non-phosphorylatable polypeptide of 26 kDa in spinach (Dunahay *et al.*, 1987), of 27 kDa in pea, or two subunits of about 29 and 26.5 kDa in barley and maize (Bassi *et al.*, 1987; Peter & Thornber, 1991). Protein microsequencing has shown that the 29 kDa LHCIIc protein from barley is the product of the *Lhcb*5 gene (Morishige & Thornber, 1992). The 26.5 kDa subunit does not react with LHCIIb specific antibodies and is not located in the LHCIIb complex, indicating that it is not a minor co-migrating LHCIIb contaminant. Protein microsequencing of the 26.5 kDa protein shows that it, although related to LHCIIc, is unique (Morishige & Thornber, 1994). These data suggest the presence of an uncharacterized minor pigment–protein, LHCIIc', the product of an unidentified *Lhcb* gene.

## LHCIId

LHCIId (CP24) contains one 21 kDa subunit (Dunahay & Staehelin, 1986), coded by the *Lhcb*6 gene. A red absorption maximum of 668nm has been recorded (Dunahay & Staehelin, 1986; Bassi *et al.*, 1987); however, Peter and Thornber (1991) found their preparation had a considerably longer wavelength maximum at 674nm. Relative to the other LHCs, LHCIId is enriched in Chl *a* with a Chl *a*/*b* ratio of 0.8 (Dunahay & Staehelin, 1986; Peter & Thornber, 1991), When compared to both LHCIIa and LHCIIb, LHCIId apparently has a much higher chlorophyll/protein molar ratio 4.5, 7.8, 10–11, respectively; (Dunahay & Staehelin, 1986), possibly accounting for its absence in pigment deficient mutants (see below). Lutein comprises the largest portion of the xanthophyll content of LHCIId, while its neoxanthin content is much lower than in the other LHCII components (Peter & Thornber, 1991; Bassi *et al.*, 1993).

## Additional LHCII complexes

### *CP22*

CP22, the *psb*S gene product, was originally observed as a non-pigmented protein of 22 kDa in PSII core preparations, relatively devoid of all other LHCII proteins. A single pigment-protein with a single apoprotein of 22 kDa was later isolated by successive rounds of non-denaturing PAGE of detergent extracted PSII-enriched membrane fractions (Funk *et al.*, 1994). The Chl *a*/*b* ratio of the complex after electrophoresis is 1.8. CP22 appears to bind a low amount of carotenoids, although additional carotenoid molecules, as well as Chls, might have been lost during electrophoresis. The room temperature red absorption maxima is 674nm. Fluorescence emission spectrum at 77K reveals a peak at 675nm. CP22 is apparently present in all photosynthetic eukaryotes, but its presence in the CC of cyanobacteria has not been confirmed and is of some controversy. CP22 lies on the periphery of the core complex and might play a role in attaching the LHCII components to the core complex.

### *14 kDa LHCII*

A novel pigmented complex has been isolated from spinach that contains a single apoprotein of 14 kDa (Irrgang *et al.*, 1993). Three Chl *a* and one Chl *b* molecules are bound per protein subunit. The room temperature red absorption maximum for this complex is 670nm. Antibodies produced against the 14 kDa apoprotein do not react with other chlorophyll-binding antenna proteins.

A LHCII component of similar size has been reported in barley and termed LHCIIe (Peter & Thornber, 1991). Whether LHCIIe is equivalent to the pigmented complex described by Irrgang and coworkers (1993) is not clear. LHCIIe has a Chl *a*/*b* ratio of 1.4 and an apoprotein of 12–13 kDa. Relative to the other LHCIIs, LHCIIe is highly enriched in xanthophylls. Because of its high xanthophyll content, LHCIIe has been hypothesized to serve in a photoprotective role for photosystem II or as a carotenoid carrier protein. Alternatively, this protein may be related to an ELIP.

## STRUCTURE OF THE LHC PIGMENT–PROTEINS

The LHC proteins are encoded by a set of nuclear genes that are part of an extended gene family. Each LHCII protein exhibits considerable amino acid sequence similarity to each other and to the LHCI proteins of PSI. At present, the characteristics of an individual LHC protein that functionally distinguishes it as an LHCI or LHCII are unclear. Regions of the proteins that exhibit a higher degree of sequence divergence are no doubt involved in defining the unique characteristics of a specific LHC. Due to sequence similarities, ELIPs and light-harvesting complexes from numerous algal species are all part of a large extended LHC family. Each member contains three putative membrane spanning domains, except CP22 which contains four. The CP22 protein contains a high degree of bilateral symmetry between the two halves of the protein, suggesting that it was derived from a duplication of a two-helix protein. It has been speculated that the progenitor of CP22 lost the fourth C-terminal membrane spanning domain to form the basis of the three helix LHC family (Green & Pichersky, 1994). Recently, a gene encoding a small protein, HLIP (*H*igh *L*ight *I*nducible *P*rotein; *hli*A gene), with a single putative trans-membrane helix has been identified in cyanobacteria (Dolganov, Bhaya & Grossman, 1995). An open reading frame coding for a related protein is also present in the chloroplast genomes of some non-green algae species. The HLIP has a surprising level of sequence similarity to the members of the LHC gene family in those areas of high primary sequence conservation. Thus, it is possible that the HLIP represents the evolutionary progenitor of the LHC gene family.

Refinement of the LHCIIb crystal structure to 3.4 Å resolution is providing exciting insights into the LHCIIb structure and an understanding of the mechanisms involved in energy capture and transfer (Kühlbrandt *et al.*, 1994). Three trans-membrane helices are observed for LHCIIb, agreeing with earlier models derived from hydropathy analysis of the primary structure. Sequence similarity between the areas encompassing the first and third alpha helices is reflected in the two-fold axis of symmetry displayed in the crystal structure. The first and third membrane spans are tilted 32° relative to the membrane normal, while the second membrane span is perpendicular to it. Amino acid residues involved in stabilizing the protein in the membrane and pigment binding have been identified. Two luteins positioned in the centre of the complex are believed to form an internal cross-brace between the first and third membrane spans to stabilize the complex. Although structural data for the other LHC proteins are not available, owing to the high sequence similarity within the extended LHC family, the other members no doubt adopt a similar three-dimensional structure in the thylakoid membrane (Green & Kühlbrandt, 1995).

## BIOGENESIS OF THE LIGHT-HARVESTING APPARATUS

A thorough biochemical characterization of the LHCs has led to the study of their assembly and biogenesis into functional multi-protein complexes. Traditionally, the biogenesis of the light-harvesting apparatus has been examined as it occurs during plastid development. These studies have utilized homogenous populations of developmentally equivalent plastids, obtained either by isolation of a subpopulation of plastids from defined sections of a developing leaf or by arrest of all plastids at an identical developmental stage (e.g. etiolation or intermittent light (IML; 2 min light, 118 min dark) treatments). Mutants that fail to accumulate LHCs, primarily due to deficiencies in pigment biosynthesis, such as the *chlorina* f2 mutant of barley, have also proven useful. At times, the different experimental systems used to study LHC biogenesis have yielded conflicting results. These apparent contradictions are most likely due to the complexity and large number of biochemical processes involved in the biosynthesis of the thylakoid membrane proteins during plastid morphogenesis. The scope of this section will be restricted to discussions on the accumulation of higher plant LHCs as pigmented-proteins and their assembly into multi-subunit complexes, subsequent to chloroplast import. LHC biogenesis will be considered and compared within the context of the different experimental approaches used.

### LHC gene expression

Genes for the best characterized LHC proteins have been identified in different plant species and are members of a large extended gene family. Each gene is located in the nucleus and codes for a precursor protein that is post-translationally imported into chloroplasts. Within the chloroplast they are processed to their mature size, integrated into the thylakoid membrane and associated with pigments (for details see Chapter 5 by Chitnis, this volume and references therein). The expression of the *Lhc* genes is influenced by several intrinsic (e.g. developmental and tissue specific cues) and extrinsic (e.g. light) factors. The levels of mRNA, encoding apoproteins of both LHCI and LHCII, are regulated largely by the red/far-red light photoreceptor phytochrome, but also in part by blue and UV-blue receptors. *Lhc* mRNA levels also undergo diurnal and circadian fluctuations. Steady-state message levels of various members of the *Lhc* gene family vary significantly within, and among, the different subfamilies in response to light and developmental cues. In general, while the steady-state mRNA levels of *Lhc* genes encoding LHCI or the minor LHCII proteins are considerably lower than those of the highly expressed LHCIIb genes, their accumulation parallels that of LHCIIb message during greening of etiolated barley seedlings and slightly precedes the accumulation of the proteins.

## Protein accumulation

Exposure of seedlings to continuous illumination results in a steady accumulation of the LHC proteins. Detectable levels of protein are present four to six hours after exposure to light. Although it would appear that LHC protein accumulation is directly related to mRNA levels owing to their parallel appearance during chloroplast biogenesis, several studies clearly indicate that the two events are not coupled. Transgenic plants expressing *Lhcb*1 antisense constructs exhibit only barely detectable levels of *Lhcb*1 message, while accumulating wild-type levels of protein (Flachmann & Kühlbrandt, 1995). Conversely, in plants grown under IML, in which the plastids are deficient in Chl *b*, but are photosynthetically competent, or in Chl *b* deficient mutants, substantial amounts of LHC mRNA are detectable, while an absence and/or reduced level of LHC proteins, particularly those of LHCIIb, is exhibited. These data indicate that LHC protein levels are not entirely dependent on message levels and are modulated primarily by post-transcriptional factors. The failure to accumulate LHCs in the absence of Chl *b* is due to rapid post-translational degradation of the apoproteins, presumably because of a lack of stabilization by pigments (see below).

During plastid biogenesis, the timing and rate of LHC protein accumulation is directly related to the synthesis of Chl *b*. Chl *b* levels can be modified in IML grown plants by either altering the intensity, duration or period of light exposure. Adjustment of Chl *b* levels, by changing the light conditions, results in a parallel change in levels of one or more of the LHCs. In both the *chlorina* f2 Chl *b*-less barley mutant and IML-treated seedlings, the LHCIIa and LHCIIc/c' apoproteins are present at reduced levels relative to mature plants, but are considerably more abundant than the major LHCIIb polypeptides (29 and 28 kDa) encoded by the *Lhcb*1 and *Lhcb*2 genes, whose levels are severely depleted or absent. In contrast, the 25 kDa LHCIIb subunit (*Lhcb*3 gene product) is present in a wild-type stoichiometry to the reaction centre under limited Chl *b* conditions. LHCIId apparently does not accumulate in the *chlorina* f2 mutant. The elevated levels of the minor LHCII proteins under conditions of Chl *b* deficiency are thought to reflect their closer position to the core complex than the major LHCIIb proteins (see below). Examination of the various LHCIIs during the light-driven greening of *chlorina* f2 has shown that LHCIIa and the 28 and 29 kDa subunits of LHCIIb accumulate during the initial phases of greening, but decrease in abundance during the later phases (Preiss & Thornber, 1995). The apoproteins of all light-harvesting complexes accumulate rapidly upon continuous illumination of IML-treated seedlings. When IML-grown barley or maize is initially exposed to continuous light, the 25 kDa LHCIIb protein, normally found in approximately an order of magnitude less abundance than the dominant 29 kDa LHCIIb protein in wild-type barley, accumulates in the thylakoid membrane to a greater extent than both the 28 and 29 kD apoproteins (Dreyfuss & Thornber, 1994*a*).

Fewer details regarding the accumulation of LHCI apoproteins are known. Synthesis of the LHCI components is delayed in comparison with the LHCII proteins. Following the developmental gradient of plastids along the monocot leaf, the LHCI apoproteins of 20–24 kDa are not detected until midway up the leaf blade and considerably later than the PSI core complex and LHCIIb apoproteins appear. The appearance of the leaf's 77 K fluorescence emission at 735 nm, characteristic of LHCIb within PSI, is observed only in the mature tip segments, indicating that LHCI becomes a major component of photosystem I only after a photochemically functional CCI is established. LHCI present, although at reduced levels, in the thylakoids of the Chl *b*-less *chlorina* f2 barley mutant (Peter & Thornber, 1991; Preiss & Thornber, 1995). All four LHCI apoproteins are initially absent or just barely at the level of detection in IML-grown barley, but begin to accumulate at a steady rate within 2 h of exposure to continuous light (Dreyfuss & Thornber, 1994b).

## Binding of pigments to LHC subunits

Analysis of LHC subunit composition in pigment mutants and from developmental studies of seedlings grown under IML conditions clearly indicate that Chl is required for the stabilization of pigment–proteins in the thylakoid membrane. Pigment binding is probably the primary post-transcriptional event determining LHC protein levels. When limited amounts of pigments are available, the relative abundance of the various LHCII apoproteins may reflect each apoprotein's priority for pigment association or their intrinsic differences in pigment composition. The factors involved in determining a specific level and ratio of pigments found associated with a particular LHC remain unknown.

Greene and co-workers (1988), studying a light-sensitive Chl *b* deficient mutant of maize, suggested that selective partitioning of a limited pool of Chl *b* among the LHC proteins occurs due to intrinsic differences within the LHC apoproteins. The most obvious intrinsic difference is the pigment requirements of the various LHCs, dictated by the primary structure of the specific LHC protein. Other studies have shown that alteration of Chl *a*/*b* ratios by either exposure of seedlings to different IML regimes or in mutants with varying Chl *b* deficiencies, causes an increase in turnover rate of the LHCI and II proteins, coinciding with the extent of Chl *b* deficiency. In pigment mutants with a varying degree of Chl *b* deficiency, the relative abundance of the LHCIIb subunits is directly related to the levels of Chl *b*. LHCIIa and LHCIIc/c′, the dominant species in both IML-grown plants and the *chlorina* f2 mutant, have the highest Chl *a*/*b* ratios, i.e. bind less Chl *b* relative to Chl *a*. LHCIId, absent from the *chlorina* f2 mutant, has the lowest Chl *a*/*b* ratio. Whether the accumulation of LHCIIa and IIc/c′ in conditions of Chl *b* deficiency is due to the stabilization of the apoproteins without bound Chl *b* or to the ability of LHCIIa and IIc/c′ apoproteins to act as more favourable substrates for Chl *b* binding under limited Chl *b* conditions than the 'less-assembled' LHCIIb and IId proteins remains to be determined.

Reconstitution experiments using purified pigment molecules and LHCIIb apoprotein or LHCIIb overexpressed in *E. coli* have demonstrated that a full complement of pigments, including xanthophylls, is required for the stable formation of pigmented LHCIIb (Plumley & Schmidt, 1987; Paulsen, Rumler & Rudiger, 1990). These reconstituted complexes retain the spectral characteristics similar to the native form of the complex. The final ratio of bound pigments in a reconstituted complex is independent of the ratio of pigments in the initial reconstitution mixture. These data suggest that structural motifs in the LHC protein are sufficient to prescribe which specific pigments are to be bound. Pigments apparently induce folding of the LHCIIb protein *in vitro* into a tertiary structure closely resembling native LHCIIb. The requirement of xanthophylls, including lutein, in the reconstitution mixture may reflect the structural role lutein plays in the three-dimensional conformation of the protein. Deletion analysis of the LHCIIb protein reveals that while some segments at both the N- and C-terminal portions of the protein are dispensable, specific domains immediately after the carboxy-proximal membrane spanning region are essential for the formation of stable LHCIIb.

## Assembly of the pigmented multi-protein complexes

Studies on the accumulation of the LHC apoproteins demonstrate a strong relationship between pigment synthesis and apoprotein accumulation, and thus it was assumed that the accumulating apoproteins actually represent pigment complexes. Greening of etiolated plants, although showing variations in timing and rate of accumulation of pigmented LHCII fractions between different species examined, in all cases suggest a parallel between pigment–protein accumulation and the onset and rapid increase of Chl *b* synthesis, indicating that little to no lag period exists between the synthesis of pigments, particularly Chl *b*, and their accumulation within the light-harvesting complexes during chloroplast biogenesis.

Recently, studies have focused on the formation of oligomeric LHC complexes *in vivo*. Pulse-labelling during the first half hour of greening of IML-grown barley indicates that newly synthesized LHCIIb is present in monomeric complexes. During the subsequent chase period, the labeled LHCIIb complexes shift into the trimeric fraction and then into higher order oligomeric forms of LHCIIb (Dreyfuss & Thornber, 1994*a*). Assembly of this higher order complex begins only after a 12 h to 18 h exposure to continuous light, when trimeric LHCIIb complex formation is nearly complete. The higher order oligomer of LHCIIb appears to be composed of three or more trimers containing only the 28 and 29 kDa subunits, as has been previously observed in barley (Peter & Thornber, 1991). The assembly of LHCIIb therefore, appears to occur via the initial pigmentation of the apoproteins to form monomeric complexes. These monomeric complexes serve as intermediates in the formation of trimeric and higher order oligomeric LHCIIb pigmented complexes. Similarly, a fraction of the newly synthesized LHCI apoproteins appear in the thylakoid membranes primarily as monomeric pigment–protein complexes during the early hours of greening of IML-grown barley. Subsequently, these monomers associate into trimeric complexes prior to attachment to the core complex to form the PSI holocomplex (Dreyfuss & Thornber, 1994*b*).

*In vitro* reconstitution of pigments with the LHCIIb apoprotein yields a monomeric pigment–protein complex which can assemble into trimers with properties similar to native LHCIIb trimers. The N-proximal hydrophobic domain of LHCIIb plays an essential role in stabilizing trimeric LHCIIb. Mutagenesis analysis of the N-proximal domain indicates that specific amino acid residues of the *Lhcb*1 gene product are involved in the stabilization of *in vitro* reconstituted LHCIIb trimers and thus forms a potential 'trimerization motif' (Hobe *et al.*, 1995). Lipids, although not initially required for the formation of monomeric complexes, are apparently required for trimer formation; however, the actual formation of the LHCIIb oligomer may not be directly dependent on a specific lipid, rather, lipids are affecting the stability of the final trimeric LHCIIb complex.

Currently, a three-dimensional structure of either photosystem complex from vascular plants is not available, but a general picture is beginning to emerge (Fig. 2.1). Present understanding of the arrangement of the pigment-proteins that constitute LHCII comes from correlating ultrastructural, photochemical and biochemical information with changes that occur either during assembly and disassembly of LHCII or from mutants deficient in one or more LHCII pigment-protein. Presently, there is no unequivocal model for the relative positioning of the LHCII subunits within the higher plant photosystem. Isolation of photosystem II subcomplexes that contain discrete LHCII subunits has led to the conclusion that LHCIIa and LHCIIc are the most tightly associated with CCII (e.g. Bassi *et al.*, 1987; Peter & Thornber, 1991). A subcomplex containing LHCIIa, LHCIId, and the 25 and 29 kDa LHCIIb subunits has been isolated and is believed to serve as a 'connector' between the CCII and the *Lhcb*1/*Lhcb*2 trimers (Fig. 2.1; Dainese & Bassi, 1991; Peter & Thornber, 1991). Those LHCIIb trimers which contain the 25 kDa and 29 kDa subunits are possibly located closer to CCII than the trimers containing the 28 kDa and 29 kDa apoproteins (*Lhcb*1 and *Lhcb*2 gene products), which form the peripheral antenna of PSII (Peter & Thornber, 1991; Sigrist & Staehelin, 1994).

Examination of PSII antenna size and LHCII subunit composition in wild-type and mutant plants with varying antenna components, together with studies of the accumulation of the minor LHCII apoproteins prior to the accumulation of the major LHCIIb complex during chloroplast biogenesis have been used to infer a stepwise formation of the PSII holocomplexes. Current models favour the initial assembly of the minor LHCII pigment–protein complexes with CCII to form a rudimentary PSII complex, followed by the addition of an inner LHCIIb complex and finally, the binding of a peripheral LHCIIb complex to form a PSII holocomplex (Dreyfuss & Thornber, 1994*a*). Assembly of the individual LHCs is probably controlled by spatial constraints in which the inner-most subunits must bind to the CC prior to the assembly of the outer antenna proteins. Thus, the presence of the minor LHCII proteins may be required before the most peripheral LHCIIb can be added to the photosystem.

## CONCLUSIONS

The first pigment–proteins were isolated and characterized in the mid-1960s. Through the years the number of LHCs has continuously increased and new LHC components are still being reported today, some 30 years after the initial reports. The methods used in the analysis of pigment–proteins by mildly denaturing PAGE systems have evolved similarly during this time. The current solubilization and gel electrophoresis protocols used for separation produce only minor amounts of pigment loss, thereby preserving the pigment–protein complexes in a relatively native state. At present, the general biochemical features of the individual LHCs and the three-dimensional structure of LHCIIb are relatively well understood. The antenna systems are complex units in which pigments play a major role in assembly and stability of the LHC proteins. Further dissection of the LHC proteins is required to determine how a particular LHC assembles with a specific photosystem and how binding of specific pigments to a particular LHC protein is accomplished. Attention can now be focused on how the individual LHCs assemble and operate as a single unit to efficiently capture light energy and transfer it to the reaction centre. While many models have been proposed, based on different experimental approaches, a unifying model depicting the arrangement of all the LHCs around a photosynthetic core complex has not been agreed upon. Further biochemical approaches should provide an increasingly better view of the arrangement *in vivo*, but a high resolution crystal structure of the photosystems as a whole will be necessary to complete the picture. Only then will we begin to fully understand and appreciate the intricate and complex interactions among the LHCs and the core subunits which are necessary to bring about optimal photosynthetic charge separation.

## ACKNOWLEDGEMENTS

We thank Dr J. Philip Thornber for his unfailing support and encouragement through the years and dedicate this review to his memory. We also acknowledge the contribution of numerous researchers whose work we were not able to discuss here due to space constraints. DTM is supported by the Cooperative State Research, Education, and Extension Service, US Department of Agriculture (95–37306–2197). BWD is supported by a National Institute of Health Postdoctoral Fellowship (1 F32 GM17483–01).

## REFERENCES

Anandan, S., Vainstein, A. & Thornber, J. P. (1989). Correlation of some published amino acid sequences for photosystem I polypeptides to a 17 kDa LHCI pigment-protein and to subunits III and IV of the core complex. *FEBS Letters*, **256**, 150–4.

Bassi, R., Hoyer-Hansen, G., Barbato, R., Giacometti, G. M. & Simpson, D. J. (1987). Chlorophyll-proteins of the photosystem II antenna system. *Journal of Biological Chemistry*, **262**, 13333–41.

Bassi, R., Rigoni, F. & Giacometti, G. M. (1990). Chlorophyll-binding proteins with antenna function in higher plants and green algae. *Photochemistry and Photobiology*, **52**, 1187–206.

Bassi, R., Pineau, B., Dainese, P. & Marquardt, J. (1993). Carotenoid-binding proteins of photosystem II. *European Journal of Biochemistry*, **212**, 297–303.

Dainese, P. & Bassi, R. (1991). Subunit stoichiometry of the chloroplast photosystem II antenna system and aggregation state of the component chlorophyll *a*/*b*-binding proteins. *Journal of Biological Chemistry*, **266**, 8136–42.

Demmig-Adams, B. & Adams, W. W., III (1996). The role of xanthophyll cycle carotenoids in the protection of photosynthesis. *Trends in Plant Science*, **1**, 21–6.

Dolganov, N. A. M., Bhaya, D. & Grossman, A. R. (1995). Cyanobacterial protein with similarity to the chlorophyll *a*/*b* binding proteins of higher plants: evolution and regulation. *Proceedings of the National Academy of Sciences, USA*, **92**, 636–40.

Dreyfuss, B. W. & Thornber, J. P. (1994*a*). Assembly of the light-harvesting complexes (LHCs) of photosystem II: monomeric LHC IIb complexes are intermediates in the formation of oligomeric LHC IIb complexes. *Plant Physiology*, **106**, 829–39.

Dreyfuss, B. W. & Thornber, J. P. (1994*b*). Organization of the light-harvesting complex of photosystem I and its assembly during plastid development. *Plant Physiology*, **106**, 841–8.

Dunahay, T. G. & Staehelin, L. A. (1986). Isolation and characterization of a new minor chlorophyll *a*/*b*-protein complex (CP24) from spinach. *Plant Physiology*, **80**, 429–34.

Dunahay, T. G., Schuster, G. & Staehelin, L. A. (1987). Phosphorylation of spinach chlorophyll-protein complexes. CPII, but not CP29, CP27, or CP24, is phosphorylated *in vitro*. *FEBS Letters*, **215**, 25–30.

Flachmann, R. & Kühlbrandt, W. (1995). Accumulation of plant antenna complexes is regulated by post-transcriptional mechanisms in tobacco. *The Plant Cell*, **7**, 149–60.

Funk, C., Schroder, W. P., Green, B. R., Renger, G. & Andersson, B. (1994). The intrinsic 22 kDa protein is a chlorophyll-binding subunit of photosystem II. *FEBS Letters*, **342**, 261–6.

Green, B. R. & Pichersky, E. (1994). Hypothesis for the evolution of three-helix Chl *a/b* and Chl *a/c* light-harvesting antenna proteins from two-helix and four-helix ancestors. *Photosynthesis Research*, **39**, 149–62.

Green, B. R. & Kühlbrandt, W. (1995). Sequence conservation of light-harvesting and stress-response proteins in relation to the three-dimensional molecular structure of LHCII. *Photosynthesis Research*, **44**, 139–48.

Greene, B. A., Allred, D. R., Morishige, D. T. & Staehelin, L. A. (1988). Hierarchical response of light-harvesting chlorophyll-proteins in a light-sensitive chlorophyll *b* deficient mutant of maize. *Plant Physiology*, **87**, 357–64.

Hobe, S., Forster, R., Klingler, J. & Paulsen, H. (1995). N-proximal sequence motif in light-harvesting chlorophyll *a/b*-binding protein is essential for the trimerization of light-harvesting chlorophyll *a/b* complex. *Biochemistry*, **34**, 10224–8.

Irrgang, K.-D., Kablitz, B., Vater, J. & Renger, G. (1993). Identification, isolation and partial characterization of a 14–15 kDa pigment binding protein complex of PS II from spinach. *Biochimica et Biophysica Acta*, **1143**, 173–82.

Jansson, S. (1994). The light-harvesting chlorophyll *a/b*-binding proteins. *Biochimica et Biophysica Acta*, **1184**, 1–19.

Knoetzel, J., Svendsen, I. & Simpson, D. J. (1992). Identification of the photosystem I antenna polypeptides in barley. Isolation of three pigment-binding antenna complexes. *European Journal of Biochemistry*, **206**, 209–15.

Kühlbrandt, W., Wang, D. N. & Fujiyoshi, Y. (1994). Atomic model of plant light-harvesting complex. *Nature(London)*, **367**, 614–21.

Morishige, D. T. & Thornber, J. P. (1992). Identification and analysis of a barley cDNA clone encoding the 31 kilodalton LHC IIa (CP29) apoprotein of the light-harvesting antenna complex of photosystem II. *Plant Physiology*, **98**, 238–45.

Morishige, D. T. & Thornber, J. P. (1994). Identification of a novel light-harvesting complex II protein (LHC IIc′). *Photosynthesis Research*, **39**, 33–8.

Mullet, J. E., Burke, J. J. & Arntzen, C. J. (1980). Chlorophyll proteins of photosystem I. *Plant Physiology*, **65**, 814–22.

Paulsen, H. (1995). Chlorophyll *a/b*-binding proteins. *Photochemistry and Photobiology*, **62**, 367–82.

Paulsen, H., Rumler, U. & Rudiger, W. (1990). Reconstitution of pigment-containing complexes from light-harvesting chlorophyll *a/b*-binding protein overexpressed in *Escherichia coli*. *Planta*, **181**, 204–11.

Peter, G. F. & Thornber, J. P. (1991). Biochemical composition and organization of higher plant photosystem II light-harvesting pigment-proteins. *Journal of Biological Chemistry*, **266**, 16745–54.

Plumley, F. G. & Schmidt, G. W. (1987). Reconstitution of chlorophyll *a/b* light-harvesting complexes: xanthophyll-dependent assembly and energy transfer. *Proceedings of the National Academy of Sciences, USA*, **84**, 146–50.

Preiss, S. & Thornber, J. P. (1995). Stability of the apoproteins of light-harvesting complex I and II during biogenesis of thylakoids in the chlorophyll *b*-less barley mutant *chlorina* f2. *Plant Physiology*, **107**, 709–17.

Preiss, S., Peter, G. F., Anandan, S. & Thornber, J. P. (1993). The multiple pigment-proteins of the photosystem I antenna. *Photochemistry and Photobiology*, **57**, 152–7.

Sigrist, M. & Staehelin, L. A. (1994). Appearance of type 1, 2, and 3 light-harvesting complex II and light-harvesting complex I proteins during light-induced greening of barley (*Hordeum vulgare*) etioplasts. *Plant Physiology*, **104**, 135–45.

# 3 Photosystems I and II

W. Z. HE AND R. MALKIN

## INTRODUCTION

Photosynthesis in all oxygen-evolving organisms involves the co-operation of two photosystems, known as PSI and PSII. They function in series in the so-called non-cyclic electron transport chain to oxidize water, reduce $NADP^+$ and generate ATP. PSI can also function independently in the cyclic electron transport pathway to generate ATP. It is widely accepted that PSI and PSII function according to the 'Z-scheme' by which electrons released from water pass through PSII and on to PSI, which generates the strong reductant necessary for $NADP^+$ reduction. The two photosystems are spatially segregated in higher plant chloroplasts. As illustrated in Fig. 3.1, PSII is mainly found in appressed granal regions, while PSI is distributed in non-appressed stromal lamellae and peripheral regions of the grana. The electron transport between the two photosystems is bridged by the cytochrome $b_6$–$f$ complex, which is evenly distributed throughout the thylakoid membranes, and two freely diffusing electron carriers, PQ and PC. An additional protein complex in the thylakoid membranes is the ATP synthase, found only in non-appressed regions. Driven by the $H^+$ gradient built up across the membrane by electron transfer reactions, the ATP synthase converts ADP into ATP.

Each photosystem can be resolved into a photoactive core complex, where light energy is converted into stable chemical products, and a light-harvesting complex (LHC), which are discussed in the preceding chapter. The core complex contains a number of protein subunits that are involved in either binding electron carriers, other pigments, or are involved in interactions with other electron transfer components. In this chapter we will give a state-of-the-art report of our understanding of the molecular organization and functions of those protein subunits and electron transfer cofactors in PSI and PSII. A number of recent reviews focusing on specific aspects of each photosystem are listed at the conclusion of this chapter and should be consulted for more detailed discussion on experiments that led to conclusions described here.

## PSII

The study of PSII has been stimulated by its close homology to the purple bacterial reaction centre for which the atomic structure has been solved (Deisenhofer & Michel, 1991) as well as by the isolation of a highly purified PSII reaction centre preparation (Nanba & Satoh, 1987). Although the reaction centre preparation does not evolve oxygen, it can perform primary charge separation, which can be stabilized with exogenous electron acceptors. This not only confirms the hypothesis that the two PSII subunits, known as D1 and D2, form the PSII reaction centre, but in addition, characterization of this complex has given insights to the physiological condition of photoinhibition through a further understanding of the role of D1 in this process (Aro, Virgin & Andersson, 1993). The unique requirement of water oxidation demands that the PSII primary electron donor, P680, have an extremely high redox potential. With a potential postulated to be in excess of +1.10 V, P680 is the most oxidizing species in living organisms. This unusual property poses two challenging questions: what is the molecular nature of P680 that results in this high redox potential and how can undesired oxidative

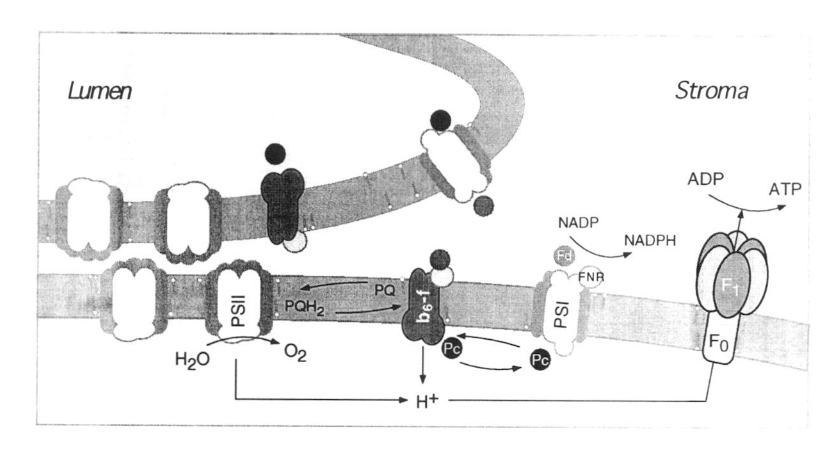


**Figure 3.1.** Diagrammatic representation of the organization of protein complexes in the thylakoid membrane of chloroplasts. PSII complexes are shown in the appressed membrane region of the thylakoid while PSI complexes are localized in the non-appressed regions. The membrane complex linking the two photosystems is the cytochrome $b_6$–$f$ complex.

destruction of pigments or protein residues by P680$^+$ be minimized?

## Protein subunits

PSII is the most complex multimeric protein assembly in thylakoid membranes, with the higher plant complex consisting of at least 18 subunits in addition to the Chl *a*/*b* binding light-harvesting proteins discussed in the preceding chapter. Cyanobacterial PSII has more than 15 subunits. Table 3.1 lists the PSII subunits identified, together with their genes and functions when known. While plant and cyanobacterial PSII show some differences, most subunits are shared by both complexes. Invariably, PSII contains the reaction centre components D1, D2, subunits of cyt b559, as well as the CP47, CP43 and 33 kDa extrinsic protein, as all of these subunits are required for the water oxidation reaction.

### *D1, D2*

The D1 protein, referred to as the $Q_B$-binding, or herbicide-binding protein, has been a subject of interest long before being recognized as a reaction centre protein. D1 and D2 are both very hydrophobic proteins. They tend to appear as diffuse bands in SDS-PAGE with apparent molecular masses of 32 and 34 kDa, respectively. Both are encoded by chloroplast genomes in higher plants and algae.

The protein sequences of D1 and D2 are highly conserved and are homologous to each other. They also show some sequence homology to the L and M subunits of purple bacterial reaction centres. On the basis of this comparison, it was proposed that they were part of the PSII reaction centre (Trebst, 1986) and, in an analogous manner to L and M, that they bind primary reactants. This hypothesis was confirmed by the isolation and characterization of a PSII reaction centre complex that contained only D1, D2, cyt b559 and some small polypeptides (Nanba & Satoh, 1987). A D1–D2 complex free of the cytochrome and PsbI is also available, indicating that the D1–D2 complex is the minimal one required for PSII charge separation.

### *Cyt b559*

Cyt b559 contains two small subunits, α and β, which are encoded by the chloroplast *psbE* and *psbF* genes, respectively. Both have a single transmembrane helix and each contains a conserved histidine residue near the N-terminus. The α-subunit is found to be orientated with its N-terminus on the stromal side of the membrane. The β-subunit has a similar orientation. Although the structure and stoichiometry of cyt b559 in PSII are unresolved, the concept of a hetero-dimeric configuration has gained increasing acceptance. In this scheme, the cyt b559 haem is co-ordinated by two conserved histidine residues from parallel-orientated α- and β-subunits. The haem group would be located near the stromal surface.

The redox potential of cyt b559 is heterogeneous (see Whitmarsh & Pakrasi, 1996). In untreated membranes, the cytochrome is predominantly, if not all, in a high potential form (~+380 mV) and various treatments convert this form to low-potential form (~+50 mV). Cyt

Table 3.1. *PSII protein subunits*

| Proteins ($M_{wt}$) | Genes[a] | Comments |
|---|---|---|
| D1 (32) | *C, psbA* | RC, binds P680, Pheo, $Q_B$, Mn(?) |
| CP47 (47) | *C, psbB* | Chl *a*-binding, Mn(?), MSP |
| CP43 (43) | *C, psbC* | Chl *a*-binding |
| D2 (34) | *C, psbD* | RC, binds P680, Pheo, $Q_A$ |
| cyt b559-α (9) | *C, psbE* | RC, binds haem |
| cyt b559-β (4) | *C, psbF* | RC, binds haem |
| PsbH (10) | *C, psbH* | Phospho-protein |
| PsbI (5) | *C, psbI* | RC |
| n.d.[b] | *C, psbJ* | *orf* in chlp. genome, inactivation affects PSII |
| PsbK (4.3) | *C, psbK* | Dispensable |
| PsbL (4.3) | *C, psbL* | Support $Q_A$ |
| PsbM (4.7) | *C, psbM* | Found in cyano. and *Chlamydomonas* |
| PsbN (4.7) | *C, psbN* | Found in cyano. and *Chlamydomonas* |
| PsbO (33) | *N, psbO* | OEE1, MSP, extrinsic in lumen |
| PsbP (23) | *N, psbP* | OEE2, extrinsic in lumen, in plants/algae |
| PsbQ (17) | *N, psbQ* | OEE3, extrinsic in lumen, in plants/algae |
| PsbR (10) | *N, psbR* | OEE4, intrinsic N terminus in lumen |
| PsbS (22) | *N, psbS* | Chl *a/b*-binding, pigment storage |
| PsbT (5) | *N, psbT* | OEE5, extrinsic in lumen |
| PsbU (9–12) | *c, psbU* | Extrinsic in lumen, in cyano. and red algae |
| Cyt c-550 (15) | *c, psbV* | Extrinsic in lumen, in cyano. and red algae |
| PsbW (6.1) | *N, psbW* | RC, intrinsic, N-terminus in lumen |
| n.d. | *C, psbX* | *orf* in chlp. genome, inactivation affects PSII |
| 4.1 kDa | | Hydrophobic, in cyano. and plants |
| 5 kDa | | Hydrophobic |

[a] N-nuclear encoded; C-chloroplast encoded; c-cyanobacterial/ cyanelle genome only.
[b] Protein is not detected.

b559 of an intermediate potential form has also been reported (~+180 mV). The variation in potential upon biochemical treatment indicates that the protein environment responsible for the high redox potential of cyt b559 is susceptible to external disturbance. It is generally accepted that the high potential form of the cytochrome must be present for active oxygen evolution.

### *CP47 and CP43*

CP47 and CP43 are two Chl *a*-binding proteins in PSII, with 10–15 and 10–12 Chl *a* molecules per protein, respectively. They serve as inner antennae, funnelling excitation energy from peripheral LHC complexes to reaction centre pigments. The *psbB* and *psbC* genes, encoding CP47 and CP43, respectively, have been cloned from a variety of species, and their sequences are highly conserved. According to their sequences, both CP47 and CP43 may have six hydrophobic regions that are long enough to form transmembrane helices (Bricker, 1990). Each has a very large hydrophilic loop between helix V and VI near the C terminus that is located in the thylakoid lumen (Loop E). This loop contains approximately 190 residues in CP47 and 130 residues in CP43. The loops may be involved in binding the 33 kDa PsbO protein. It has been shown that deletion of charged and conserved residues of CP47 in a region located between amino acid residues 370–390 decreases the binding affinity of the 33 kDa subunit. Both CP47 and CP43 are required for the oxygen-evolution activity, presumably providing a proper structural orientation for redoxactive components.

### *Extrinsic proteins*

There are several extrinsic subunits associated with PSII on the lumen side of the thylakoid membrane where the water oxidation reaction takes place. Higher plant PSII has four extrinsic polypeptides: the 33 kDa PsbO, 23 kDa PsbP, 17 kDa PsbQ, and 5 kDa PsbT. They are all nuclear encoded (see Ikeuchi, 1992).

The 33 kDa PsbO protein is ubiquitous to all oxygenic organisms. It can be resolved by high concentrations of $CaCl_2$ or Tris buffer. Biochemical resolution and reconstitution experiments show that the PsbO protein plays an important role in stabilizing the Mn-complex as well as enhancing the oxygen-evolving activity of PSII.

The 23 and 17 kDa subunits are loosely associated with the intrinsic core. Both can be resolved with 1 M NaCl. Their rebinding requires the presence of the PsbO protein. Their removal reduces the oxygen-evolution activity of PSII, but activity can be restored with elevated levels of $Ca^{2+}$ and $Cl^-$. Therefore, these two subunits regulate the access of $Cl^-$ and $Ca^{2+}$ ions that are important cofactors of oxygen evolution.

Cyanobacterial PSII has different extrinsic components. It contains a 15 kDa cyt c550 (PsbV) and a 12 kDa PsbU protein in addition to the PsbO protein, but lacks the PsbP and PsbQ subunits. The 33 kDa subunit in cyanobacteria may also play a slightly different role in maintaining functional configuration of the Mn clusters since cyanobacterial mutants with inactivated *psbO* gene can sustain photoautotrophic growth. This discrepancy can be explained by the presence of the cyt c550 component in cyanobacterial PSII. It seems to complement the function the 33 kDa subunit. It has been shown that deletion of either gene yields autotrophs, but deletion of both genes causes the loss of oxygen-evolving activity.

### *Low molecular mass components*

There are a number of low molecular mass ($\leq$ 10 kDa) hydrophobic polypeptides associated with PSII (see Barry, Boerner & de Paula, 1994). Some are found to be tightly bound to the D1/D2 complex, such as the 5 kDa PsbI protein and the 6.1 kDa PsbW protein. The PsbI protein is found both in higher plants and cyanobacteria, while the PsbW protein is only detected in higher plants. Inactivation of *psbI* gene in *Chlamydomonas* results in increased photosensitivity and reduced amounts of active PSII and oxygen-evolving activity. The PsbW protein is the only reaction center component encoded by the nuclear genome. Other small proteins, such as the PsbH, PsbK, PsbL, PsbM, PsbN and PsbR subunits, are found only in larger oxygen-evolving particles (see Ikeuchi, 1992 and Barry *et al.*, 1994). These subunits do not bind cofactors, and their functions in PSII are not known. Deletion of the respective gene for individual subunits generally gives a less stable PSII complex and slightly reduced oxygen evolving activity. Therefore, they are considered to be dispensable but required for optimal performance of PSII. The two exceptions are the PsaL and PsbH subunits. Mutants with an inactivated *psbL* gene cannot grow photoautotrophically, due to inactivated PSII. The PsbL protein has been shown to be required for an active $Q_A$ configuration. Inactivating the *psbH* gene also impairs electron transfer at the acceptor side of PSII.

## Organization

### *Over-all structure*

The structure of PSII at atomic resolution is not yet determined, although several groups are actively attempting to obtain high-quality crystals for structural analysis. However, biochemical fractionation and chemical cross-linking have provided some information regarding the supramolecular organization of the complex and have led to models for protein subunit organization. A schematic representation of chloroplast PSII is given in Fig. 3.2. At the core of PSII are the D1 and D2 subunits. The $\alpha$- and $\beta$-subunits of cyt b559, the PsbW and PsbI subunits are also tightly associated with D1-D2. CP47, CP43 and the PsbO protein are closely associated with D1-D2. A non-oxygen-evolving core complex with CP47-D1-D2-cyt b559 can be isolated from higher plants, but a larger oxygen-evolving core contains CP47, CP43, the 33 kDa protein and several small subunits.

This figure depicts not only those subunits described above but also shows Chl *a/b* binding proteins including the light-harvesting protein complex. Generally, one copy of each component is found per PSII reaction centre. However, there is evidence for two copies of three extrinsic components per PSII. The stoichiometry of cyt b559 is controversial, as discussed by Whitmarsh and Pakrasi (1996). For simplicity and clarity, we show only one copy of cyt b559 in a heterodimeric configuration and one copy of each extrinsic component.

The overall structure of PSII from both cyanobacteria and higher plants has been studied by electron microscopy in conjunction of single-particle image-averaging analyses (Boekema *et al.*, 1995). The oxygen-evolving core complex, having molecular mass of about 450 kDa and dimensions of 17.2 × 9.7 nm, showed two fold symmetry indicative of a dimeric organization. Confirmation of this came from image analysis of oxygen-evolving cores of PSII isolated from spinach and *Synechococcus* having a mass of 240 kDa. An intact PSII complex associated with LHCII has a mass of about 700 kDa, and electron microscopy revealed it also to be a dimer having dimensions of about 26.8 × 12.3 nm. From comparison with the dimeric core complex, it was deduced that the latter is located in the center of the larger particle, with additional peripheral regions accom-

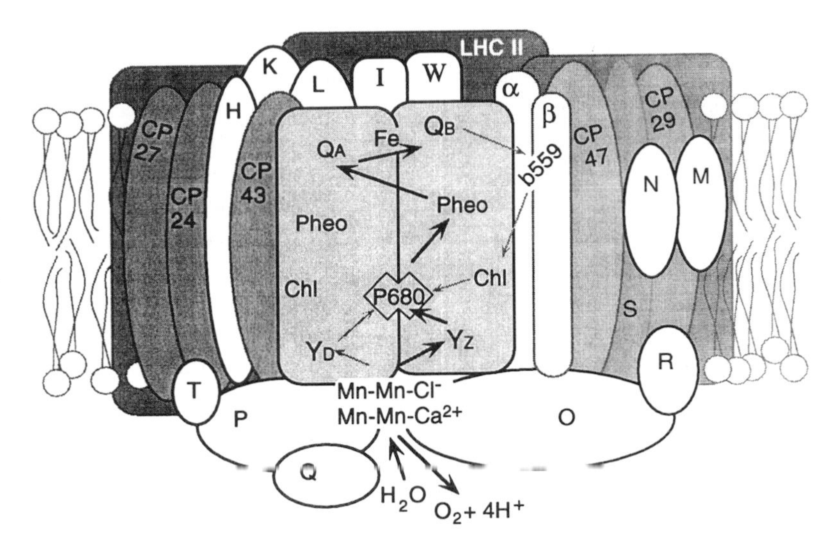


**Figure 3.2.** Molecular organization of chloroplast PSII. Protein components are labelled with the names of the genes encoding these subunits. The two shaded protein subunits are D1 and D2, which are known to bind the indicated pigments or cofactors. Heavy arrows represent the physiological electron transport pathway while thin arrows show alternative electron transport pathways.

modating the Chl *a*/*b* binding proteins. It was suggested that two LHCII trimers are present in each dimeric PSII complex, and each trimer is linked to the PSII core complex by the light-harvesting proteins, CP24, CP26 and CP29. The results also suggest that PSII may exist as a dimer *in vivo*.

### *Reaction centre*

The striking sequence homologies and functional similarities in the acceptor complex of PSII and purple bacterial reaction centres have led to the conclusion that redox active cofactors of PSII, such as P680, Pheo, $Q_A$ and $Q_B$ are bound to a heterodimer of D1 and D2 proteins, analogous to the corresponding reactants in the L and M subunits of purple bacterial reaction centres (Deisenhofer & Michel, 1991). This idea has been confirmed by the isolation of a D1–D2–cyt b559 complex. However, the pigment composition of the isolated PSII reaction center preparation is still an issue of controversy. Generally speaking, the PSII complex contains 4–6 Chl *a*, 1–2 β-carotene, 1 cyt b559 per two Pheo *a*. For comparison, the reaction centres from purple bacteria contain 4 Bchl, 2 Bphe, 1 β-carotene and 2 UQ. PSII seems to contain one or two additional molecules of Chl *a*, one extra β-carotene and one cyt b559, with the quinones being lost during the isolation procedure.

It has been proposed that D1 and D2 each form five transmembrane helices (Trebst, 1986; Deisenhofer & Michel, 1991). As in the bacterial L and M subunits, the His ligands to the special pair chlorophyll *a*, D1-H198 and D2-H197, are conserved in PSII, as are the ligands to the non-haem iron at the acceptor side. The ligand environment for quinone binding is assumed to also be similar to that in the bacterial complex (Trebst, 1986).

Extensive effort has been put into site-directed mutagenesis to test the above described structural model (see Pakrasi, 1995). Mutagenesis of D2-H197 to either Tyr or Leu inactivates the PSII complex. Similarly, mutagenesis of D1-H198 to Leu results in a decreased amount of PSII. In contrast, similar mutations of Mg-ligands of the special pair in purple bacteria yield a Bchl–Bphe heterodimer that is still capable of primary charge separation. Interestingly, mutation of D2-H197 to Gln or Asn, both of which can act as a ligand to Chl, allows a relatively stable assembly of PSII complex, but PSII electron transfer reaction is rapidly inhibited and the midpoint potential of P680 seems to be somewhat lowered. Moreover, mutation of putative non-haem iron ligands in PSII also gives different results from mutations of its bacterial counterpart. The above lines of evidence, plus the difficulty in reconstituting quinones in the isolated PSII reaction centre, indicates that the structural requirements for a functional PSII are much more stringent. This seems reasonable considering the unique water oxidation function of PSII. On one hand, special arrangements will provide the framework for P680 to have an unusually high redox potential. On the other, $P680^+$ needs to be reduced quickly to avoid undesired side-reactions.

The reaction centre of purple bacteria contains two Bchls in addition to the special pair and two Bphes. An analogous composition of four chlorophylls per two Pheos has been reported for the D1–D2–Cyt b559 preparation. However, the histidine ligands to the accessory Chl are not conserved in D1 and D2. Instead, there are two histidine residues, D1-H118 and D2-H118, which are conserved in PSII. Meanwhile, a monomeric chlorophyll, known as $Chl_Z$, is found to be an electron donor to $P680^+$ under certain circumstances. Estimation of the distance between the $Chl_Z$ radical and non-haem iron implies D1-H118 is probably the ligand to $Chl_Z$.

### *Water oxidation centre*

The oxidizing side of PSII involves two tyrosine radicals and the tetrameric Mn-cluster, the latter serving as the water oxidation centre (WOC). The two tyrosine radicals, $Y_Z$ and $Y_D$, are identified as D1-Tyr161 and D2-Tyr160, respectively.

The structure of the Mn-cluster has not been resolved unequivocally. One of the promising models is based on X-ray spectroscopic evidence (Yachandra *et al.*, 1993). In this model, the four Mn atoms are grouped into two di-μ-oxo-bridged dimers. The centre–centre distance between Mn atoms within the dimer is 2.7Å, and the nearest two Mn atoms between the two dimers are 3.3Å apart. The ligands to Mn seem to be mainly nitrogen and oxygen.

It has become clear that the Mn-cluster may be coordinated by D1 and D2, the same subunits that bind components of the photochemical reaction centre, with additional ligands possibly from CP47, CP43 and/or the 33 kDa protein. With the folding models of D1–D2, molecular biologists have attempted to identify specific residues that can provide N- and O-ligands for Mn. Possible ligands are emerging: Asp-170 of D1 and Glu-69 of D2 seem to be important for the functional assembly of PSII. A thorough search of potential ligands in the lumen inter-helical domain and the C-terminus of D1 has been carried out (Chu, Nguyen & Debus, 1995*a,b*), and it has been concluded that D1-His-332, Glu-333, His-337 Asp-342 influence the assembly of Mn-cluster. Therefore, these residues may also ligate the Mn-cluster. In addition, Asp-59, Asp-61 and Asp-342 are potential ligands to $Ca^{2+}$.

In the case of D2, no specific residues near the C-terminus have yet been found to be critical for oxygen evolution. Mutations of all Asn, Asp, Gln, Glu, and His residues in this area result in a photoautotrophic phenotype.

Additional Mn-ligands are probably from CP47 and CP43, since they are required for oxygen-evolving activity of PSII. Deletion of the *psbB* gene prevents the functional assembly of PSII. In contrast, inactivation of the *psbC* gene gives partially functional PSII lacking oxygen-evolution activity.

## Electron transfer

### *Primary charge separation*

PSII reactions are also indicated by arrows in Fig. 3.2. Each quantum of excitation energy trapped by the reaction centre extracts one electron from the primary donor, P680. The electron is passed through a Pheo and $Q_A$ to $Q_B$. $Q_A$ normally accepts only one electron, while $Q_B$ is a two-electron acceptor. Thus the first photochemical event forms $Q_B^-$. Upon a second photochemical event, the $Q_B^-$ becomes fully reduced, protonated and released from the $Q_B$ site as plastoquinol ($PQH_2$) into the quinone pool. The two-electron cycle is completed when an oxidized PQ molecule from the quinone pool binds to the vacant $Q_B$ site.

The chemical nature of P680 remains to be resolved. Based on difference-spectra and redox potential, it involves Chl *a* molecule, and based on the homology with P870, it is likely a dimer of Chl *a*. However, spectroscopic data give conflicting interpretations whether it is a dimer or a monomer. Recently, an oligomeric model has also been proposed to accommodate the special spectral behaviour (Durrant *et al.*, 1995). The high redox potential of P680 remains a paramount challenge for understanding.

On the donor side, the oxidized P680 is re-reduced by accepting an electron from $Y_Z$, which in turn extracts one electron from the Mn-cluster. The Mn-cluster couples one-electron photochemistry with the four-electron water oxidation reaction by accumulating oxidizing equivalents through advances in valence (see below).

There are other redox-active components in PSII that are not directly involved in the linear electron flow from water to PQ. Such components may include the cyt b559, $Y_D$, which is a tyrosine residue in position 161 on the D2 subunit, a non-haem iron (Fe) associated with the quinone acceptors, β-carotene and possibly a monomeric form of Chl *a*. Alternative electron transfer pathways are indicated in Fig. 3.2 as thin arrows.

### *Water oxidation*

The quaternary gating function on the oxidizing side of PSII can be demonstrated in the oscillation of flash-induced $O_2$ yield and can be described in terms of an S-state cycle of PSII. PSII can exist in five redox states, designated $S_0$–$S_4$ depending on the number of stored oxidizing equivalents, with $S_4$ as the most oxidizing state capable of oxidizing water. The S-states are believed to be related to valence states of the tetrameric Mn cluster. Other cofactors that may be involved in the water oxidation reaction include a histidine residue, $Ca^{2+}$ and $Cl^-$ ions. The mechanism of this reaction is unclear in terms of where and when the two $H_2O$ molecules bind to the complex. No intermediates have been detected, and, in addition, the proton releasing pattern is not agreed upon by all groups.

It is widely accepted that the Mn cluster undergoes valence increase upon advance of the S-states, except for the step from $S_2$ to $S_3$. Support for this idea comes from K-edge XANES and EXAFS studies. In $Ca^{2+}$-depleted PSII complexes, oxidation of a histidine or tyrosine residue can be observed during the $S_2$ to $S_3$ transition, but whether this happens *in vivo* when $Ca^{2+}$ is present is still a matter of speculation.

Involvement of $Ca^{2+}$ and $Cl^-$ in the oxygen evolution reaction is well documented (Debus, 1992), but a plausible mechanism for their role is absent. The sequential removal of each of the extrinsic polypeptides, PsbQ, PsbP, PsbO results in an increased requirement for $Cl^-$ and the absence of the 23 kDa results in a specific requirement for $Ca^{2+}$. Removal of $Cl^-$ blocks the electron transfer between the WOC and the primary donor. Spectroscopic evidence shows that chloride is within the bonding circle of the Mn cluster (Yachandra *et al.*, 1993). Interestingly, alteration of the basic residue R448G in CP47 prevents the assembly of functional PSII under $Cl^-$-limiting conditions.

### *Photoinactivation*

PSII has been known to be susceptible to photoinactivation. In high light, the PQ pool becomes over-reduced, and the lifetime of reduced $Q_A$ is increased to 30 s as compared to a tenth of a second normally. In this case, the charge recombination rate increases, as does the yield of Chl triplets. The Chl triplet reacts with molecular oxygen and produces singlet oxygen. Singlet oxygen is a strong oxidant and is known to react with pigments, amino acid residues or lipid. These reactions are believed to decrease PSII activity.

Photoinactivation occurs even under anaerobic or low light conditions. The fact that a highly oxidizing potential is required for extracting protons from water presents an intrinsic problem of PSII. The very oxidizing primary electron donor has the potential to oxidize its neighbour besides its designated partner. Electron donation is usually slow compared to the fast primary charge separation reaction. Under high light or cold climate, limitations in electron donation result in photoaccumulation of $P680^+$, which can then initiate harmful oxidative reactions. This donor side mechanism gains strong support from recent mutagenesis work. A D1-Y161F mutant lacks the native donor to P680, $Y_Z$. Therefore, $P680^+$ has a much longer lifetime under illumination. As expected, the D1 protein in this mutant turns over much faster than in the wild-type control. There are a few lines of defence available in the organism to protect PSII against photoinactivation. They include the quenching of Chl triplets by carotenoid, alternative electron transfer pathways as well as rapid breakdown and *de novo* synthesis of D1.

There are two β-carotene molecules found in the D1–D2–cyt b559 preparation. The carotenoids in the reaction centre can quench the Chl triplet and, therefore, attenuate the yield of singlet oxygen. Upon the loss of one β-carotene, the PSII reaction centre becomes more sensitive to photoinactivation.

The intimate relation between cyt b559 and D1–D2 goes beyond their geometric closeness. Recent investigations have led to the conclusion that there is a cyclic electron transfer pathway, which may play an important role in protecting PSII in the event of over-reduction of the quinone pool under high light conditions (Whitmarsh & Pakrasi, 1995). The oxidation of cyt b559 is suggested to proceed via the photooxidation of a chlorophyll in a sequential pathway of electron transfer:

cyt b559 $\rightarrow$ $Chl_Z$ $\rightarrow$ $P680^+$

Such a scheme would overcome the distance barrier between the cyt and the reaction centre P680.

Rapid turnover of D1 is well documented. Recent studies with *in vitro* systems lead to the conclusion that rapid turnover of D1 is one way for the organism to cope with the continuous loss resulting from photoinactivation. As long as this repair process can keep up with the rate of photoinactivation, the overall rate of photosynthesis is maintained. The sequence of events involved in this repair process remains to be elucidated and is a subject of intense investigation.

## PSI

### Functional aspects and electron carriers

PSI is an integral membrane protein complex that normally functions to transfer electrons from the soluble electron carrier, PC, to the soluble electron carrier, Fd. Under certain environmental conditions in some cyanobacteria and algae, alternative electron donors and acceptors, such as cytochrome *c6* or flavodoxin, can function in place of PC and Fd. In terms of functional activity, PSI is unique in generating highly reducing species that are capable of reducing $NADP^+$ in an energetically favourable reaction. The PSI reductants are the strongest produced in any biological system and pose significant problems in terms of protection against adverse reactions with oxygen. Recent reviews have appeared that consider details of the structure and function of this complex (Golbeck, 1992; Setif, 1992; Chitnis *et al.*, 1995).

The PSI complex contains a number of bound electron carriers, and the identity of these carriers and their $E_m$ values are summarized in Table 3.2. The properties of the respective carriers will be described in more detail.

The most extensively characterized PSI electron carrier is the reaction centre chlorophyll, P700. This pigment undergoes light-induced oxidation, producing $P700^+$. $P700^+$ formation is characterized by absorbance changes in the red and blue spectral regions and by the generation of an EPR free-radical signal indicative of an oxidized chlorophyll species. While there was some disagreement in the early literature concerning the $E_m$ value for the $P700/P700^+$ couple, it is now generally accepted that this value is approximately +500 mV, and that alterations in the $E_m$ may occur that are dependent on detergents used for the preparation of PSI complexes.

Table 3.2. *PSI electron carriers*

| Electron carrier | Function | $E_m$(mV) |
|---|---|---|
| P700 | Reaction centre chlorophyll-dimeric Chl *a* | +500 |
| $A_o$ | Intermed. electron acceptor-monomeric Chl *a* | −1000 (est) |
| $A_1$ | Primary electron acceptor-phylloquinone | −800 (est) |
| $F_X$ | Terminal electron acceptor–4Fe-4S cluster | −730 |
| $F_A$ | Terminal electron acceptor–4Fe-4S cluster | −590 |
| $F_B$ | Terminal electron acceptor–4Fe-4S cluster | −530 |

The molecular nature of P700 is not fully defined. A number of different biophysical techniques have been applied to this analysis, but these early measurements led to conflicting conclusions, such as models based on both monomeric and dimeric structures. Based on the analogy with P870 of the bacterial reaction centre complex, most workers believe that P700 is a Chl *a* dimer and a recent structural analysis of a PSI complex from cyanobacteria has been provided the first evidence for a dimeric structure (Krauss *et al.*, 1993) since a pair of chlorophylls near the lumenal surface of the membrane lying on an axis with one of the PSI electron acceptors has been identified.

The electron lost from P700 is transferred to a series of bound low-potential electron acceptors. The first of these, $A_o$, was initially proposed to be a monomeric Chl *a* molecule that was reduced to the anion form in the light. However, much of the early characterization of $A_o$ was done under strongly reducing conditions where other reduced electron acceptors were also present, and this complicated the interpretations of the results. More recently, time-resolved kinetic measurements on $A_o$ have been done under conditions where the PSI electron acceptors are in the oxidized state prior to light activation (Hastings *et al.*, 1994*a*,*b*). These experiments have not only defined the kinetics of electron transfer through $A_o$ but have also given the best spectral resolution of this carrier and have supported its identity as a monomeric form of Chl *a*. No direct measurements of the $E_m$ of this acceptor have been made, but an indirect estimation has given a value of −1.0 volt. On the basis of these results, it appears that the initial light-induced charge separation in PSI involves only Chl *a* molecules, with a dimer being oxidized and a monomer being reduced in the light.

Electron transfer in PSI proceeds from $A_o$ to $A_1$, the next bound PSI electron acceptor. Evidence based on biochemical and biophysical studies supports the conclusion that phylloquinone (vitamin K1) functions as $A_1$. Biochemical studies have shown that a number of different PSI complexes from a variety of organisms contain two moles of phylloquinone per P700, one of which is linked to the primary photochemistry of the reaction centre complex. The function of the second phylloquinone is not known. Resolution and reconstitution studies are consistent with the proposed role of phylloquinone as $A_1$. Measurements of the spin-polarized EPR signal of the primary reactants in PSI have provided strong evidence for the proposed role of phylloquinone as $A_1$. Direct measurements of the $E_m$ of this component are lacking, but, again, indirect estimations have given a value of approximately −800 mV for $A_1$. From a comparative point of view, it is noteworthy that in common with all other photosynthetic reaction centre complexes, PSI contains a quinone molecule as one of its early electron acceptors, even though in the case of PSI, this acceptor is functioning at an extremely electronegative redox potential.

Electron transfer from $A_1$ proceeds to a series of membrane-bound Fe-S centres. The first acceptor in this series, $F_X$, is unusual in that it is bound by ligands from two different protein subunits in the PSI complex. $F_X$ has been studied primarily by low temperature EPR, although the unusual properties of the centre make detection difficult as extremely low temperatures and high microwave powers are required to detect the characteristic $F_X$ signal. Overlap with the signals from other bound Fe-S centres complicates analysis. Optical absorbance changes in the 400–450 nm region are associated with the reduction of $F_X$, but these are heavily contaminated with absorbance changes from other electron carriers, including P700, $F_A$ and $F_B$. Biochemical analysis of resolved PSI core complexes and other analyses have shown that the $F_X$ cluster is a 4Fe–4S centre. An $E_m$ of −730 mV has been measured for $F_X$ in a purified PSI complex.

The terminal bound electron acceptors in PSI are two iron–sulphur centres known as $F_A$ and $F_B$. The EPR g-values of the two centres are slightly different, and this allows for some differentiation of the two. $F_A$ was found to accumulate after illumination at cryogenic temperatures and has been assumed to be the final electron acceptor in the complex. Although there was some early disagreement on the chemical nature of these iron–sulphur centres, it is now established that both $F_A$ and $F_B$ contain 4Fe–4S centres (see below). The $E_m$ values of $F_A$ and $F_B$ are approximately −530 and −590 mV, respectively. Absorbance changes associated with the reductions of $F_A$ and $F_B$ were used to identify a pigment, known as P430, which is now accepted to be the optical manifestation of the $F_A/F_B$ reduction. It is not possible, however, to distinguish these two clusters on the basis of their optical properties.

## Kinetics of PSI electron transfer reactions

Within the framework of a wide series of studies on the kinetics of primary charge separation in photochemical reaction centres, carried out primarily with bacterial reaction centres, one anticipates that the primary charge separation in PSI will take place in a few picoseconds. Several groups have recently carried out extensive studies of the charge separation resulting in the formation of $P700^+$ and $A_o^-$, and these have indicated that radical pair formation does occur in 1–2 ps (Hastings *et al.*, 1994*a*,*b*; Hecks *et al.*, 1994). It has been shown, in some of these studies, that PSI radical pair formation is limited by trapping of photons in the PSI antenna. Fluorescence measurements have given a time constant of approximately 28 ps for the PSI trapping time (Hastings *et al.*, 1994*a*,*b*), while more recent measurements using photovoltage rise kinetics have given the comparable value of 22 ps for this reaction (Hecks *et al.*, 1994). Thus, although direct measurements of radical-pair formation show charge generation in approximately 20 ps, this reaction is limited by antenna light absorption and excitation energy transfer to the reaction centre. All groups believe this intrinsic rate of charge separation in PSI is less than 5 ps.

The next step in the PSI electron transfer pathway, from $A_o^-$ to $A_1$, has been difficult to evaluate. As noted above, the absorbance bands associated with $A_o$ are poorly defined. In addition, $A_1$, being a quinone, also has poorly resolved absorption spectrum. Finally, as discussed above, it has been found that $A_o^-$ does not become fully reduced in the light, presumably due to its rapid rate of reoxidation, making a complete analysis of its kinetic properties difficult to evaluate. Several recent studies have, however, presented data on this question. In one case, the use of high intensity laser flashes allowed for a more complete reduction of $A_o$ and its subsequent reoxidation in the 20 ps time domain. According to these results, $A_o^-$ was being formed in 28 ps, as noted above, but was reoxidized in 21 ps (Hastings *et al.*, 1994*a*,*b*). Using photovoltaic measurements, which do not have the intrinsic problem of overlapping absorbance changes from other electron carriers, it has been found that $A_o^-$ is formed in 22 ps but the reoxidation rate is approximately 50 ps. The photovoltaic measurements provide additional information in that both steps are electrogenic with the electron transfer from $A_o$ to $A_1$ spanning less of the transmembrane distance available than the electron transfer step from P700 to $A_o$.

Electron transfer beyond $A_1$ in PSI has been difficult to evaluate because of the general problems noted above. These are mainly that the electron acceptors beyond $A_1$ are all bound Fe–S centres that have rather poorly defined absorbance properties, making clear identification of redox changes in these carriers difficult in time-resolved experiments. Several groups have sought experimental ways around these difficulties. Some have used resolved PSI complexes in which some of the Fe–S clusters have been removed. This has allowed for measurements of the rate of $F_X$ reduction in the absence of $F_A$ and $F_B$. Other groups have used pulsed EPR techniques that can detect an electron spin echo signal arising from the $P700^+/A_1^-$ couple. Direct measurements of the kinetics of $A_1^-$ reoxidation, based on absorbance changes in the near ultraviolet region associated with the semiquinone species, have also been reported.

Electron spin echo measurements have given a consistent $A_1^-$ reoxidation rate of approximately 200 ns, which is unchanged when $F_A$ and $F_B$ are removed. The signal diminishes if iron-sulphur centre $F_X$ is removed from the complex. This provides strong evidence that the electron spin echo signal arises from the $P700^+/A_1^-$ pair and that the disappearance of the signal reflects electron transfer from $A_1^-$ to $F_X$.

Optical measurements of the rate of $A_1^-$ oxidation have given a more complicated picture. Originally, a kinetic time constant of approximately 250 ns was reported, but more recent measurements have given biphasic kinetics for this reaction, with half-times of 25 and 150 ns in a spinach PSI complex and 200 ns in a cyanobacterial PSI complex. The faster time measured with the spinach preparation was believed to arise from an altered electron transfer pathway due to structural changes in the PSI complex. If we accept this interpretation, we can define the electron transfer rate between $A_1^-$ and $F_X$ as approximately 200 ns at physiological temperatures.

The kinetics of the reduction of the bound Fe–S centres in PSI have been difficult to define by direct measurements associated with these carriers. As noted above, a component, P430, has been correlated with some of these clusters, and the photoreduction of P430 in PSI was originally shown to occur in less than 100 ns. However, since $F_X$, $F_A$ and $F_B$ all contribute to the absorbance in the 430 nm region, it is not possible to distinguish these three centres in unresolved PSI complexes based on these absorbance changes. While the centres can be resolved using EPR spectroscopy, this technique is not amenable to measurement of rapid kinetics, and, in addition, there is an increasing body of evidence that indicates the pathways of electron transfer in PSI at cryogenic temperatures are not always consistent with pathways at physiological temperatures. As mentioned above, the kinetics of $A_1^-$ oxidation, in the 200 ns time domain, has been interpreted as indicating that $F_X$ reduction occurs with this half-time, but a direct measurement of $F_X$ was not made in these studies. To date, the only direct measurements of $F_X$ are those using a PSI core complex lacking $F_A$ and $F_B$ and it was concluded that $F_X$ was reduced within 5 ns, much faster than the measured rate of $A_1^-$ oxidation estimated by other groups.

Other techniques have also been applied to measuring Fe–S reduction rates. Sigfridsson, Hansson & Brezinski (1995) have used flash-induced voltage changes associated with electrogenic electron transfer events to evaluate electron transfer through the bound Fe–S centres. Electron transfer to $F_B$ was estimated to occur within 1 μs while a 30 μs component was interpreted as arising from the $F_B^-$ to $F_A$ electron transfer. If we assume that the measured rate of electron transfer from $A_1^-$ to $F_X$ occurs in 200 ns, it would appear that electron transfer from $F_X^-$ to $F_B$ is occurring in a comparable time frame, while the subsequent electron transfer from $F_B^-$ to $F_A$ would occur on a much slower timescale, of the order of 30 μs. These kinetic values are consistent with a model for the Fe–S centres derived from the 6 Å resolution structure of a cyanobacterial PSI complex (Krauss *et al.*, 1993), although the identity of which Fe–S cluster is $F_A$ and which is $F_B$ cannot be made from these experiments.

Electron transfer from bound acceptors to the soluble electron carrier Fd is the next step in the PSI chain. The first kinetic measurements for Fd reduction by Hervas, Navarro & Tollin (1992) found a reduction time of approximately 5–10 ms, which was sensitive to ionic strength, in support of the idea that an electrostatic interaction between Fd and PSI is important in this interaction. A recent study of Fd reduction by Setif & Bottin (1994) reported more complex kinetics, with three first-order kinetic components being identified (half-times of 500 ns, 20 μs and 100 μs). The fastest of these times was assigned to electron transfer from $F_A$ or $F_B$ to Fd, and it was suggested that the bound donor was reduced in a time faster than 500 ns. The fastest of these kinetic constants does not agree with the estimates of $F_A$–$F_B$ reduction from photovoltaic measurements described above. The complexity of the kinetics determined for Fd reduction are not fully understood, and it has been proposed that three different PSI–Fd complexes may exist in order to explain the three kinetic components. The molecular nature of these complexes and the nature of the different Fd reduction sites in PSI remain to be resolved.

## PSI protein subunits and their organization

PSI is an integral membrane complex that contains a large number of individual subunits. In the case of the prokaryotic PSI complex, at least 11 polypeptides have been identified by SDS PAGE. In the complex isolated from higher plants, at least three additional non-chlorophyll binding subunits are present, and there are a substantial number of chlorophyll-binding subunits that bind antenna pigments. In cyanobacteria, PSI contains approximately 125 chlorophyll *a* molecules per P700 while the higher plant PSI complex contains approximately 225 chlorophyll *a* + *b* molecules per P700. Advances in our understanding of the function and organization of individual subunits in these respective complexes have been notable in recent years as refined biochemical analyses in conjunction with molecular genetic techniques have proved to be particularly fruitful. Chemical cross-linking and modification has been used to study nearest-neighbour relations among the subunits, the topography of individual subunits and the interaction of PSI with diffusible electron donors and acceptors. Resolution and reconstitution studies have provided a unique approach to our understanding of the role of individual subunits in these complexes. It has been possible to extend these biochemical approaches with the techniques of molecular genetics, particularly with the cyanobacterial PSI complex, by the preparation of specific mutants that are either deficient in a single subunit or contain altered subunits generated by site-directed mutagenesis. The sum total of this work has provided a model for the functional organization of the PSI complex, and a model that incorporates these recent results is shown in Fig. 3.3.

All PSI complexes contain two relatively high molecular mass subunits of approximately 83 kDa (these subunits run anomalously on SDS-PAGE and are generally observed at 66 kDa subunits). These two subunits, PsaA and PsaB are present as single copies to form a PsaA–PsaB heterodimer which binds P700, $A_o$, $A_1$ and $F_X$ as well as approximately 120 chlorophyll *a* molecules. Amino acid sequences for the PsaA and PsaB subunits have shown that each subunit is extremely hydrophobic with approximately 11 transmembrane helices. There are several hydrophilic loops in each protein that extend out from the membrane surface and one of these contains two highly conserved cysteine residues. Evidence has been provided that these four cysteines are the ligands to the Fe–S centre, $F_X$, which is bound in a co-ordinated manner by the PsaA and PsaB subunit. Based on the position of $F_X$, predictions have also been made on which

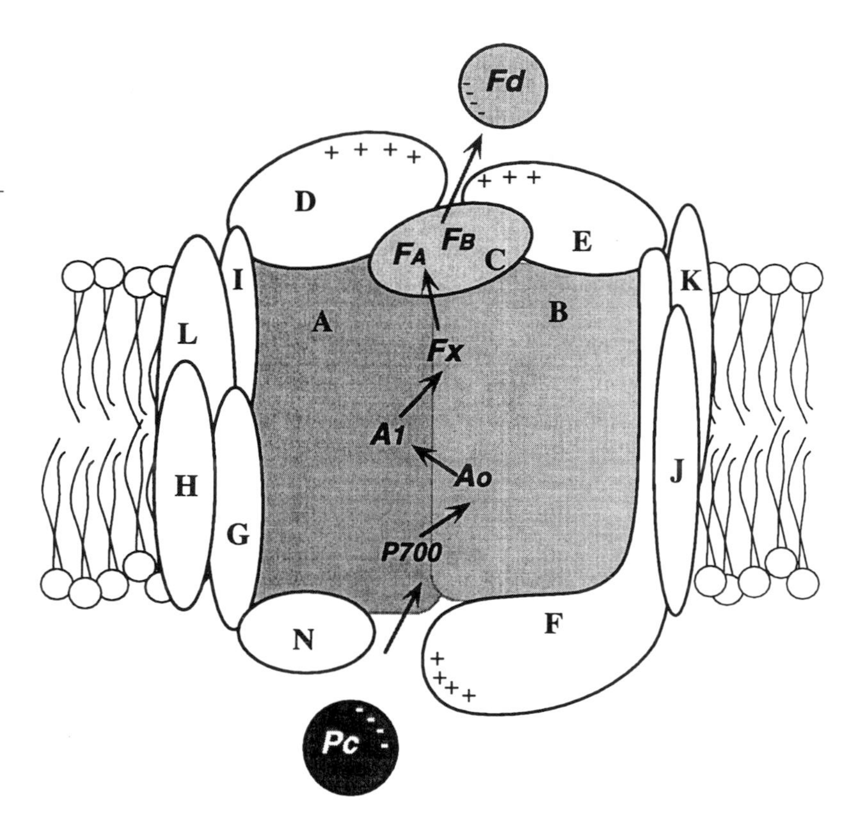


**Figure 3.3.** Molecular organization of chloroplast PSI. Protein components are labelled by the names of the respective genes encoding these subunits. The pathway of electron transfer through the complex is indicated by the arrows.

amino acid residues are involved in the binding of the other prosthetic groups found in PSI.

The terminal electron acceptors in PSI, $F_A$ and $F_B$, are bound by a low molecular mass subunit of approximately 9 kDa, known as PsaC. The PsaC protein is relatively hydrophilic, yet is tightly associated with the PsaA–PsaB heterodimer. The protein has been isolated with its iron–sulphur centre intact, allowing for analysis of these centres in the absence of association with PSI. A series of reconstitution experiments using a genetically modified PsaC protein has identified the respective cysteine ligands for $F_A$ and $F_B$. With the characterization of the three subunits, PsaA, B and C, it has become clear that all the electron carriers in the PSI complex are present in these three subunits and are sufficient for stable charge separation.

The function of the additional non-prosthetic group binding subunits of PSI is less well defined. Two subunits, PsaD and PsaF, have been implicated in the interaction of the PSI complex with its soluble electron transfer partners, Fd and Pc, respectively, but additional functions for these subunits have also been proposed (Chitnis *et al.*, 1995). The results for the PsaF subunit are particularly unclear at this time since cyanobacterial mutants that are deficient in this subunit show an unimpaired photosynthetic function, making a role for this subunit difficult to understand. In the case of PsaD, which has been found to interact with Fd, this subunit is also involved in establishing a proper orientation for the PsaC subunit in the PSI complex. The PsaE subunit has been implicated both in Fd reduction and in cyclic electron transfer around PSI. Using a cyanobacterial PSI mutant, a requirement for PsaL in PSI trimer formation has been found (Chitnis *et al.*, 1995). Other subunits, such as PsaG, H, I, J, K, M and N, have no well-defined functions. Cyanobacterial PSI complexes do not contain the PsaG and H subunits, while these are present in eukaryotic

Table 3.3. *PSI protein subunits*

| Subunit | Mass (kDa) | Function |
|---|---|---|
| PsaA and PsaB | 83.0 and 82.4 | Binds P700, $A_0$, $A_1$, $F_X$ and 100 antenna Chl |
| PsaC | 8,9 | Binds $F_A$ and $F_B$ |
| PsaD | 15.6 | Fd-docking |
| PsaE | 9.7 | Cyclic electron transport and interaction with Fd |
| PsaF | 15.7 | Pc-docking and other functions |
| PsaG | 10 | Linker polypeptide (?) |
| PsaH | 10 | Linker polypeptide (?) |
| PsaI | 4.3 | Interaction with PsaL |
| PsaJ | 4.4 | Interaction with PsaF and E |
| PsaK | 8.5 | Interaction with PsaA and B |
| PsaL | 16.6 | Trimer formation |
| PsaM | 3.4 | Unknown |
| PsaN | 9 | Unknown |

PSI complexes, leading to the suggestion that these subunits may be involved as linkers to antenna-binding subunits only found in eukaryotic PSI complexes.

The properties of the PSI subunits are summarized in Table 3.3.

## Structure–function relationships in PSI

A major advance in our understanding of PSI came when Krauss *et al.* (1993) reported their results on the three-dimensional structure of the cyanobacterial complex. While the resolution is not yet sufficient to detail the molecular architecture of the entire complex, the 6 Å resolution presented a first glimpse of the structure of this complex. The most important information in this analysis was related to the organization of the three Fe–S clusters. The three clusters form an irregular triangle located on the stromal side of the membrane. One cluster that lies closest to the middle of the membrane is assigned to $F_X$ and two additional clusters are found to be 14 and 21 Å from $F_X$. These two clusters are assigned to $F_A$ and $F_B$, although it was not possible to positively identify each cluster. The distance between the presumed $F_A$ and $F_B$ clusters was found to be 12 Å. Other conclusions from the structure are more tentative, but there appears to be a two-fold rotation axis that passes through $F_X$ in an analogous manner to the symmetry that has been demonstrated for the bacterial reaction centre complex. On either side of this axis lie the early electron acceptors, and two chlorophylls that lie close to the two-fold axis at the lumen side of the membrane have been tentatively assigned to P700. A number of membrane protein helices have been tentatively identified as well as some 45 chlorophyll *a* molecules.

One problem that remains to be solved in relation to PSI is the function of the two terminal Fe–S clusters in the PsaC protein. Evidence for both sequential and non-sequential pathways has been presented based on a variety of chemical procedures. It has also been possible to study the Fe–S clusters in the PsaC protein using molecular genetic techniques and, in a recent report, it has been found that *in vitro* reconstituted PSI complexes with an altered $F_B$ can photoreduce Fd/$NADP^+$ at rates that are 70% of those of the wild-type PSI complex. From these results, it would appear that the PSI electron acceptor complex can function efficiently even though one of the two Fe–S centres has been altered. In contrast to this *in vitro* approach to the study of the $F_A$ and $F_B$ clusters, an *in vivo* molecular approach has provided a surprising conclusion concerning these centres (Mannan *et al.*, 1996). A cyanobacterial mutant has been isolated, which has no $F_B$ and its phenotype resembles the wild-type strain in that there is photoautotrophic growth at a rate somewhat lower than that of the wild-type. This result argues that $F_B$ may not be required for electron transfer from PSI to ferredoxin *in vivo* and that the presence of the single cluster, $F_A$, is sufficient for photosynthesis in this organism.

Finally, there are major questions concerning many of the non-prosthetic group-binding subunits in PSI. In studies with cyanobacterial mutants, deletion analysis has shown that many of these subunits are not required for the functioning or assembly of the PSI complex (Chitnis *et al.*, 1995). However, these analyses have been done under optimal growth conditions, and it may be that the requirement for these subunits can only be observed under less optimal conditions. The possible role of these subunits in the stabilization of the PSI complex in a eukaryotic system, where the chloroplast and nuclear genomes co-operate in the ultimate assembly of the complex, is another area where further experimentation is required.

ACKNOWLEDGEMENTS

We would like to thank Drs J. Whitmarsh and J.-R. Shen for sending preprints prior to publication. Work from the authors' laboratory that has been cited was supported by grants from the National Science Foundation.

REFERENCE

Aro, E. M., Virgin, I. & Andersson, B. (1993). Photoinhibition of photosystem II: inactivation, protein damage and turnover. *Biochimica et Biophysica Acta*, **1143**, 113–34.

Barry, B. A., Boerner, R. J. & de Paula, J. C. (1994). The use of cyanobacteria in the study of the structure and function of photosystem II. In *The Molecular Biology of Cyanobacteria*, ed. D. A. Bryant, pp. 217–57. Boston: Kluwer Academic Publishers.

Boekema, E. J., Hankamer, B., Bald, D., Kruip, J., Nield, J., Boonstra, A. F., Barber, J. & Rogner, M. (1995). Supramolecular structure of the photosystem II complex from green plants and cyanobacteria. *Proceedings of the National Academy of Sciences, USA*, **92**, 175–9.

Bricker, T. M. (1990). The structure and function of CPa-1 and CPa-2 in photosystem II. *Photosynthesis Research*, **24**, 1–14.

Chitnis, P. R., Xu, Q., Chitnis, V. P. & Nechushtai, R. (1995). Function and organization of Photosystem I polypeptides. *Photosynthesis Research*, **44**, 23–40.

Chu, H. A., Nguyen, A. P. & Debus, R. J. (1995*a*). Amino acid residues that influence the binding of manganese or calcium to photosystem II. 1. The lumenal interhelical domains of the D1 polypeptide. *Biochemistry*, **34**, 5839–58.

Chu, H. A., Nguyen, A. P. & Debus, R. J. (1995*b*). Amino acid residues that influence the binding of manganese or calcium to photosystem II. 2. The carboxy-terminal domain of the D1 polypeptide. *Biochemistry*, **34**, 5859–82.

Debus, R. J. (1992). The manganese and calcium ions of photosynthetic oxygen evolution. *Biochimica et Biophysica Acta*, **1102**, 269–352.

Deisenhofer, J. & Michel, H. (1991). Structure of bacterial photosynthetic reaction centers. *Annual Review of Cell Biology*, **7**, 1–23.

Durrant, J. R., Klug, D. R., Kwa, S. L. S., Van Grondelle, R., Porter, G. & Dekker, J. P. (1995). A multimer model for P680, the primary electron donor of photosystem II. *Proceedings of the National Academy of Sciences, USA*, **92**, 4798–802.

Golbeck, J. (1992). Structure and function of photosystem I. *Annual Review of Plant Physiology and Plant Molecular Biology*, **43**, 293–324.

Hastings, G., Kleinherenbrink, F. A. M., Lin, S. & Blankenship, R. E. (1994*a*). Time resolved fluorescence and absorption spectroscopy of photosystem I. *Biochemistry*, **33**, 3185–92.

Hastings, G., Kleinherenbrink, F. A. M., Lin, S., McHugh, T. J. & Blankenship, R. E. (1994*b*). Observation of the reduction and reoxidation of the primary electron acceptor in photosystem I. *Biochemistry*, **33**, 3193–200.

Hecks, B., Wulf, K., Breton, J., Leibl, W. & Trissl, H.-W. (1994). Primary charge separation in Photosystem I: a two-step electrogenic charge separation connected with $P700^+A_o^-$ abd $P700^+A_1^-$ formation. *Biochemistry*, **33**, 8619–24.

Hervas, M., Navarro, J. A. & Tollin, G. (1992). A laser flash spectroscopic study of the kinetics of electron transfer from spinach photosystem I to spinach and algal ferredoxins. *Photochemistry and Photobiology*, **56**, 319–24.

Ikeuchi, M. (1992). Subunit proteins of photosystem II. *Botanical Magazine Tokyo*, **105**, 327–73.

Krauss, N., Hinrichs, W., Witt, I., Fromme, P., Pritzkow, W., Dauter, Z., Betzel, C., Wilson, K. S., Witt, H. T. & Saenger, W. (1993). Three-dimensional structure of system I of photosynthesis at 6 Å resolution. *Nature*, **361**, 326–31.

Mannan, R. M., He, W.-Z., Metzger, S., Whitmarsh, J., Malkin, R. & Pakrasi, H. B. (1996). Active photosynthesis in cyanobacterial mutants with directed modification in the ligands for two iron–sulphur clusters in the PsaC protein of Photosystem I. *European Molecular Biology Organization Journal*, **15**, 1826–33.

Nanba, O. & Satoh, K. (1987). Isolation of a photosystem II reaction center consisting of D-1 and D-2 polypeptides and cytochrome b559. *Proceedings of the National Academy of Sciences, USA*, **99**, 153–62.

Pakrasi, H. B. (1995). Genetic analysis of the form and function of photosystem I and photosystem II. *Annual Review of Genetics*, **29**, 755–76.

Setif, P. (1992). Energy transfer and trapping in photosystem I. In *The Photosystems: Structure, Function and Molecular Biology*, ed. J. Barber, pp. 471–99. Amsterdam: Elsevier Scientific.

Setif, P. Q. Y. & Bottin, H. (1994). Laser flash absorption spectroscopy study of ferredoxin reduction by Photosystem I in *Synechocystis* sp. PCC 6803: evidence for submicrosecond and microsecond kinetics. *Biochemistry*, **33**, 8495–504.

Sigfridsson, K., Hansson, O. & Brezinski, P. (1995). Electrogenic light reactions in photosystem I: resolution of electron-transfer rates between the iron–sulphur

centers. *Proceedings of the National Academy of Sciences, USA*, **92**, 3458–62.
Trebst, A. (1986). The topology of the plastoquinone and herbicide binding peptide of photosystem II in the thylakoid membrane. *Zeitschrift für Naturforschung Section C Biosciences*, **41**, 240–5.
Whitmarsh, J. & Pakrasi, H. B. (1996). Form and function of cytochrome b559. In *Oxygenic Photosynthesis: The Light Reactions.*, ed. D. R. Ort & C. F. Yocum, pp. 249–64. Amsterdam: Kluwer Academic Publishers.
Yachandra, V. K., De Rose, V. J., Latimer, M. J., Mukerji, I., Sauer, K. & Klein, M. P. (1993). Where plants make oxygen: a structural model for the photosynthetic oxygen-evolving manganese cluster. *Science*, **260**, 675–9.

# 4 Chloroplast pigments: chlorophylls and carotenoids

G. SANDMANN AND H. SCHEER

## CHLOROPHYLLS AND CAROTENOIDS

The pigments found in the thylakoids of plants are chlorophylls with substituted tetrapyrrol rings and carotenoids, which include $C_{40}$ hydrocarbons and derivatives with hydroxy and epoxy substituents. Both groups of pigments are lipophilic and are tightly associated with each other in the photosynthetic pigment protein complexes. Functionally, they are directly involved in harvesting and transducing of light energy. The atomic model of the light-harvesting complexes from plants and bacteria and from reaction centres reflect the close interaction of chlorophylls and carotenoids.

## NATURE OF CHLOROPHYLLS

Chlorophylls (Chls) are cyclic tetrapyrroles, which contain a fifth isocyclic ring that is biosynthetically derived from the C-13 propionic acid side chain of protoporphyrin, and generally a central magnesium atom and a long-chain terpenoid alcohol at C-$17^3$. Several pheophytins lacking the central magnesium are active in photosynthetic electron transport, and most Chls of the *c*-type, which are widespread in algae, have a free propionic acid side chain. Various aspects of chlorophylls have been treated in the books edited by Scheer (1991), Bryant (1994), Jordan (1991, Deisenhofer and Norris (1993), Chadwick and Ackrill (1995) and Blankenship *et al.* (1995). Generally, only selected work published after these will be cited here explicitly.

Functionally important properties of the Chls are (i) the intense absorption of near-UV, visible and near-infra-red light (300–1050 nm *in situ*), which is essential for light-harvesting, (ii) the one-electron redox chemistry of the macrocycle (the central Mg is inert), which relates to electron transport, (iii) their ready aggregation and (iv) the co-ordination chemistry of the central Mg. The latter two are important factors in the organization of chlorophyll proteins, (v) a functionally important, but potentially deleterious property of chlorophylls is the long lifetime (~ 5 ns) of the first excited singlet states ($S_1$). All energy transducing reactions in photosynthesis have to compete with de-excitation processes like radiationless decay producing only heat. The maximum efficiency of photosynthetic energy storage is therefore, directly related to the $S_1$-lifetime of the Chl. However, there is a concomitant high probability that triplets are formed if energy transduction becomes limiting, e.g. under high or suddenly increasing light. The latter are already highly toxic by themselves, and can produce even more toxic diffusible products like singlet oxygen (see p. 50). Since most of the late chlorophyll precursors starting with protoporphyrin IX and some early degradation products are also phototoxic, photoprotection has to be extended to these metabolites, or else their concentrations have to be maintained very low.

Biosynthesis of Chls in spring and their degradation in autumn at moderate latitudes are the most obvious processes of life visible clearly from outer space. Biosynthetically, the Chls are derived from δ-aminolevulinic acid (ALA) and share intermediate steps with the biosynthesis of haems, the Fe-porphyrins. However, the early steps up to ALA are different in most photosynthetic organisms, and the late steps branch off with the insertion of Mg into protoporphyrin IX. Light is an important factor in controlling the biosynthesis, which in eukaryotes is located in the plastids. Details of the degradation of chlorophylls have only recently begun to

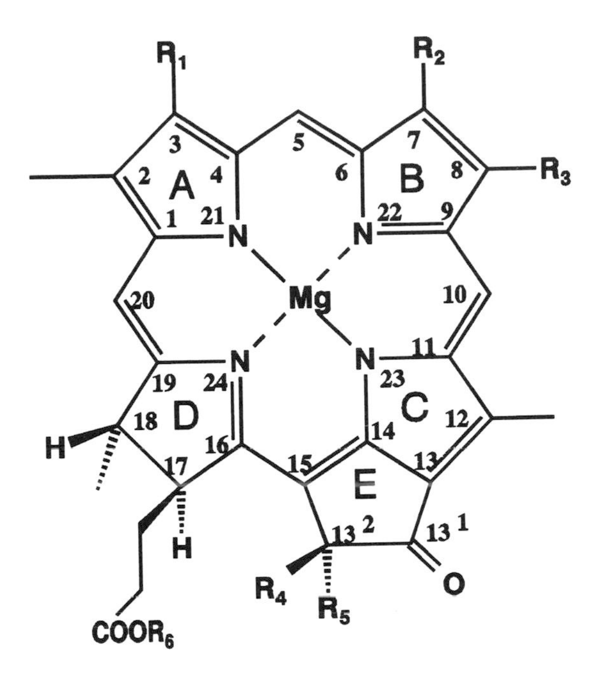


**Figure 4.1.** The general structure of plant chlorophyll. The details of variation among different chlorophyll molecules are presented in Table 4.1. The chlorophylls of chlorin type have a single bond between c-17 and c-18, while chlorophylls c and protochlorophyllides have a double bond between c-17 and c-18.

emerge. They also indicate some similarities with the catabolism of haems.

## STRUCTURE AND FUNCTION OF CHLOROPHYLLS

### Chlorophylls *a,b,c,d*

Chlorophyll *a* (Chl *a*, Fig. 4.1; Table 4.1) contains one saturated ring (D) and is therefore chemically a chlorin (=17,18-dihydroporphyrin). The isocyclic ring carries an asymmetric centre which, as part of a reactive β-keto-ester, is stereochemically quite labile and can convert to the $13^2$-epimer called Chl *a′* (*a*-prime). Chl *b* and *d* (Fig. 4.1) are distinguished from Chl *a* by carrying 7- and 3-formyl-substitutents, respectively.

Chl *c* (Fig. 4.1) is the common name for now more than three structurally closely related pigments. In contrast to Chls *a* and *b*, they have the fully unsaturated porphyrin macrocycle, and they have an acrylic side chain at C-17 which is generally not esterified. Some of the late biosynthetic precursors of chlorophylls are structurally closely related to the *c*-type Chls, so that they may be evolutionarily first.

Chl *a* is present in all organisms capable of oxygenic photosynthesis. A special pair of Chl *a* functions as the primary donor in the reaction centre (RC) of photosystem II (PSII), and a dimer of Chl *a* or possibly Chl *a′* is the primary donor of photosystem I (PSI). Both RCs contain additional Chl *a*-molecules which, as in purple bacterial RCs, are probably involved in electron and triplet energy transfer, and those of PSI also have more than 40 Chl *a* molecules integrated as core antennae.

Chl *a* is also the major pigment in all chlorophyllous light-harvesting complexes (LHC) of oxygenic organisms. However, the intense absorptions at ≈ 680 and 430 nm are quite narrow, and there is only moderate absorptivity in the green spectral region. It is therefore, almost always supplemented by additional light-harvesting pigments. There are three main lines of oxygenic photosynthetic organisms which can be distinguished by these accessory pigments. In the ‘green’ series (green algae, green plants and most prochlorophytes), Chl *a* is accompanied by Chl *b*, which extends the absorption of light from either side into the ‘green hole’. The ‘brown’ series contains the *c*-type chlorophylls. Only a single prokaryotic representative of this class is known (Larkum *et al.*, 1994), but the Chls *c* are widespread among algae. The main visible absorption bands fill the ‘green hole’ of Chl *a* absorption, but the molar extinction coefficients are much smaller than that of Chl *a*, so relatively large amounts are needed. The ‘blue and red’ series contain biliproteins as accessory light-harvesting pigments absorbing in the region of 500–650 nm (classical cyanobacteria, red and cryptophyte algae). *c*-Type chlorophylls have been identified, *together* with biliproteins in the cryptophytes, and more recently with Chl *b* in a few chromophytes like, e.g. *Mantionella squamata* (Schmitt *et al.*, 1994). A fourth line may be constituted by Chl *d*-containing organisms. Large amounts of it have been found in extracts from some *rhodophytes* and, very recently, in a prokaryote (Miyashita *et al.*, 1995). No Chl *d*-containing pigment–protein complex has been characterized.

The majority of the accessory pigments are found in

Table 4.1. *The variation in the structure of chlorophylls, at $R_1$ to $R_6$. The general structure is shown in Fig. 4.1*

| Pigment | $R_1$ | $R_2$ | $R_3$ | $R_4$ | $R_5$ | $R_6$ | Remarks |
|---|---|---|---|---|---|---|---|
| *Chlorin type* | | | | | | | |
| Chl *a* | $C_2H_3$ | $CH_3$ | $C_2H_5$ | H | $COOCH_3$ | $C_{20}H_{39}$ | |
| Chl *a′* | $C_2H_3$ | $CH_3$ | $C_2H_5$ | $COOCH_3$ | H | $C_{20}H_{39}$ | |
| Chlid *a* | $C_2H_3$ | $CH_3$ | $C_2H_5$ | H | $COOCH_3$ | H | |
| Chl *b* | $C_2H_3$ | CHO | $C_2H_5$ | H | $COOCH_3$ | $C_{20}H_{39}$ | |
| Chl *d* | CHO | $CH_3$ | $C_2H_5$ | H | $COOCH_3$ | $C_{20}H_{39}$ | |
| [8-vinyl]-Chl *a* | $C_2H_3$ | $CH_3$ | $C_2H_3$ | H | $COOCH_3$ | $C_{20}H_{39}$ | |
| [8[1]-hydroxy]-Chl *a* | $C_2H_3$ | $CH_3$ | CHOH-$CH_3$ | H | $COOCH_3$ | $C_{20}H_{39}$ | |
| Phe *a* | $C_2H_3$ | $CH_3$ | $C_2H_5$ | H | $COOCH_3$ | $C_{20}H_{39}$ | 1 |
| *Chl. and protochlorophyllides*: | | | | | | | |
| Chl $c_1$ | $C_2H_3$ | $CH_3$ | $C_2H_5$ | (H, $COOCH_3$) | H | 2, 3 | |
| Chl $c_2$ | $C_2H_3$ | $CH_3$ | $C_2H_3$ | (H, $COOCH_3$) | H | 2, 3 | |
| Chl $c_3$ | $C_2H_3$ | $COOCH_3$ | $C_2H_5$ | (H, $COOCH_3$) | H | 2, 3 | |
| Pchlid *a* | $C_2H_3$ | $CH_3$ | $C_2H_5$ | H $COOCH_3$ | H | | |
| [8-vinyl]-Pchlid *a* | $C_2H_3$ | $CH_3$ | $C_2H_3$ | (H, $COOCH_3$) | H | 3 | |

*Remarks:* 1: central Mg replaced by 2H.2: 17-acrylic acid side chain.3: Stereochemistry at C-$13^2$ undetermined.

the LHC of PSII. As an example, Chl *b* amounts to nearly 50% of the chlorophylls in the major LHCII complex(es) of green plants and algae, but is much less abundant in the antenna of PSI (Paulsen, 1995). Both reaction centres lack the accessory pigments, and they are probably also absent in the core antenna complex(es) of both photosystems.

## Structures related to Chlorophyll a

Pheophytin a (Phe a), the demetalated Chl *a*, is ubiquitous among oxygenic photosynthetic organisms. PSII-RC contain two molecules of Phe a, which function as an intermediate acceptor in electron transport, the pigment can therefore, be used as marker for PSII. Chl *a′* ('*a*-prime') likewise appears to be an integral component of PSI of still unknown function. [8-vinyl]-chlorophyll *a* and other [8-vinyl]-pigments ('divinyl-chlorophylls'), have first been characterized besides the respective 'monovinyl' analogues in greening tissue. A number of mature organisms contain small amounts of these pigments, but they are dominant in a mutant of *Zea mays* and, in particular, in some abundant free-living prochlorophytes. A functional significance in photosynthesis is presently unknown.

## Esterifying alcohols

Phytol (phyta-2-en-1-ol) is the most common esterifying alcohol of chlorophylls. Small amounts of chlorophylls esterified with higher unsaturated phytanols are found as precursors in green plants, and frequently as photosynthetically active pigments in bacteria.

The alcohol portion constitutes approximately 30% by weight of the chlorophyll molecule. In moderately polar environments, it hardly affects its chemical and optical properties. However, in hydrophobic or aqueous environments or at polarity boundaries, the alcohols affect profoundly the properties of the pigment. This is evidenced, e.g. by the large chromatographic differences on reverse phases or differences in their aggregation and by alcohol-specific binding in bacterial reaction centres (Scheer & Hartwich in Blankenship *et al.*, 1995). The natural environment of chlorophylls is rather hydrophobic, but the role of the different hydrophobic alcohols is still poorly understood on the molecular level.

# CHLOROPHYLL BINDING SITES

Chlorophylls are (almost) always bound to proteins. Progress in elucidating their structure has been most spectacular in bacteriochlorophyll proteins, but crystal structures of a chlorophyll *a*–peridinin complex (C. Welte, R. G. Hiller *et al.*, unpublished observations), and with somewhat lesser resolution of the major LHCII (Kühlbrandt, Wang & Fujiyoshi, 1994) and PSI-RC (Krauss *et al.*, 1993) have revealed now corresponding details on chlorophyll proteins. Only two topics shall be mentioned here: the ligation to the magnesium and the related question of selection between the chlorophylls and their respective pheophytins in type II RC, and distortions of the tetrapyrrole macrocycle from planarity, which may be important in spectral tuning.

### *Mg-ligation*

The central Mg in chlorophylls is formally four-coordinate to the central N atoms, but co-ordinatively unsaturated and in chlorophyll proteins always five-coordinate. Ligands to the Mg include a number of polar amino acid side chains, but also backbone CO-groups, the CHO-group of *N*-formyl-**met** and even water. The importance of such ligands became first obvious in RC from *Cf. aurantiacus*, where replacement of a highly conserved **his** by **ile** coincides with the replacement of one of the four Bchl *a*s by Bphe *a*. Similar replacements were achieved by a series of site-directed mutations with purple bacterial RC. Whenever a Bchl-binding **his** was substituted with a large hydrophobic amino acid such as **ile**, the respective Bchl was replaced by Bphe, and vice versa. Interestingly, Bchl stays bound when the ligating **his** is replaced by a small amino acid such as **Ala**, which allows the Bchl to carry in its own ligand, e.g. water. Similar results have been obtained with Chl *a* proteins. This selection, rule is supported by exchange experiments with bacterial RC and antennae (Scheer/Hartwich and Loach/Parkes-Loach in Blankenship *et al.*, 1995).

### *Distortions of the macrocycle*

Haem-porphyrins are almost never planar in their binding sites, and this is even more true for the chlorophylls. In particular, the crystal of Bchl-proteins are of sufficiently high resolution to see these distortions in detail. They show considerable variations in the different binding sites, which have not yet been analysed systematically, but may well contribute to fine-tuning of the optical absorptions (Senge, 1993) and redox potentials, and possibly also to selection among the different chlorophylls.

## CHLOROPHYLL METABOLISM

Several recent reviews in the books edited by Blankenship *et al.* (1995), Bryant (1994), Chadwick and Ackrill (1995), Jordan (1991) and Scheer (1991) deal with Chl biosynthesis and should be consulted for further reading. Some aspects on regulation should be emphasized. (i) Chlorophyll proteins are formed by convergent pathways involving the tetrapyrrole moiety, the long-chain alcohol moiety, and the apoproteins. Therefore, they have to be controlled in a co-ordinate fashion. (ii) Fe- and Mg-tetrapyrroles can share common precursors, but they have to be regulated individually, which may involve different pools in eukaryotes. (iii) Any accumulation of porphyrins, either free or complexed with Mg, is potentially lethal to the cell due to photodynamic effects, which are obviously of particular importance in organisms which rely on light for their energy supply. The regulation is consequently complex, but seems to involve mainly the early and very late steps of biosynthesis.

### Early steps to δ-aminolevulinate

δ-Aminolevulinic acid (ALA), the first dedicated intermediate, can be formed by two different pathways. The first involves condensation of succinate and glycine by ALA synthetase. This route is used for Chls by purple photosynthetic bacteria and for haems in many organisms. The second route used for Chls by many photosynthetic organisms and probably by all eukaryotic ones, starts from glutamate. It is activated by condensation to plastidic $tRNA^{Glu}$, which belongs to the pool dedicated normally to protein synthesis, and then reduced in the bound state by $Glu\text{-}tRNA^{Glu}$ -reductase to glutamin-semialdehyde or its cyclic isomer, a hydroxyaminopyranone. The latter is subsequently released and isomerized to ALA in a pyridoxylphosphate-dependent reaction. All enzymes have been cloned and expressed heterologously so they are expected to be characterized in the near future in detail. Biosynthesis of ALA is under light control via phytochrome and possibly other sensors, which seems to by exerted mainly on the reductase (Hori *et al.*, 1996). The reason is most likely to avoid any over-production of phototoxic porphyrins, because externally added ALA is converted into porphyrins without much restrictions.

## Intermediate steps to protoporphyrin IX

Two molecules of ALA are condensed to the mono-pyrrole, porphobilinogen (PBG), by $Mg^{2+}$- or $Zn^{2+}$-dependent ALA-dehydratase. Four of these are subsequently condensed to a linear tetrapyrrole by PGB-deaminase. The X-ray structure of this enzyme (Louie *et al.*, 1992) has substantiated an interesting mechanism. It is first activated by binding two molecules of PBG as a primer for the condensation of four more molecules of PBG to a linear hexapyrrole, from which the terminal tetrapyrrole is then cleaved off. Next, comes a series of decarboxylations by which first the acetic acid side chains at C-2, C-7, C-12 and C-18 are converted to $CH_3$-groups, and then the propionic acid side chains at C-3 and C-8 to vinyl groups. All these reactions take place on the reduced tetrapyrrole in the readily autoxidable porphyrinogen oxidation state. The last step in this phase is an oxidation of the macrocycle to protoporphyrin IX involving six electrons. At least the first part is controlled enzymatically by protoporphyrinogen oxidase. The enzyme is inhibited in plants by a group of herbicides, which results in photodynamic killing by a product of the enzyme, e.g. protoporphyrin IX. This curious behaviour is due to the oxidation of protoporphyrinogen in an uncontrolled fashion and/or by its transport out of the plastid where the product of spontaneous oxidation can no longer be processed further (see Chapter 25 on Herbicides).

The entire process from ALA-to protoporphyrin IX appears to be tightly coupled to avoid leakage of intermediates, which at almost any stage can produce phototoxic porphyrins. It is, however, not tightly controlled (with the possible exception of ALA-dehydratase, Rüdiger, personal communication), because over-production or feeding of ALA does result in the production of excess phototoxic protoporphyrin IX.

## The late steps

Only a few of the enzymes involved in the last steps have been characterized in detail, and many have still not been isolated. However, molecular genetic analysis in plants and in particular in purple bacteria (Alberti *et al.*, in Blankenship *et al.*, 1995) has yielded, during the past years, a wealth of information which makes it likely that this gap will be filled soon.

Insertion of Mg into the tetrapyrrole is the first dedicated step of chlorophyll synthesis, for which three gene products are needed. The ATP-dependent enzyme has a preference for Mg, but also inserts Zn if offered. Zn complexes are also accepted by most of the subsequent enzymes (Helfrich, 1995), so one might expect that some [Zn]-chlorophylls can be found in photosynthetic tissue. Since they are very rare, an unknown discrimination mechanism must exist. Next, the C-13 propionic acid side chain is first regioselectively esterified by methyl transfer from S-adenosyl-methionin (SAM), and subsequently cyclized leading to 3,8-divinyl-pheoporphyrin $a_5$-Mg, which is also called [8-vinyl]-protochlorophyllide or less correctly 'divinyl-protochlorophyllide'. The required enzymes are not yet well characterized. The cyclization step is most likely preceded by the formation of a $13^1$-CO group by an oxygenase.

Somewhere in this region the branching point is expected for the formation of the *c*-type chlorophylls. For all other chlorophylls, ring D of protochlorophyllide *a* (Pchlid *a*) is reduced stereo- and regioselectively. There exist two different Pchlid-oxidoreductases (POR). One is the ubiquitous light-independent POR. The other type of POR is light dependent and there exist two iso-enzymes of it. The light-dependent POR-A, which is present in large amounts in etiolated angiosperms (Holtorf *et al.*, 1995), is currently probably the best studied enzyme of the late steps of Chl biosynthesis. It forms a ternary enzyme–substrate complex with the pigment and NADPH, which is accumulated and stable over long times in the dark. The enzymatic reduction of the tetrapyrrole by NADPH is triggered by light. The ternary complex is an inhibitor of ALA synthesis, so angiosperms accumulate in the dark only small amounts of Pchlid (together with carotenoids) which is all bound to POR. After irradiation of such yellowish etiolated plants, POR A is rapidly decreased to leave a basal level of the light-dependent POR B which appears to be dominant in green angiosperms and sufficient to maintain Chl synthesis. There occur several spectral changes after Pchlid reduction whose nature is uncertain. They have been related to dissociation from the enzyme complex, which may be driven by dephosphorylation, binding to carrier proteins, esterification, and others. The reduction product, chlorophyllide, is a positive effector for formation of the apoproteins for Chl-proteins, and has been implicated in addition to phytochrome and possibly other sensory photoreceptors in controlling protein synthesis and the assembly of the photosynthetic apparatus (Eichacker *et al.*, 1990). During the final steps, the tetrapyrrole and isoprenoid pathways converge. Geranylgeranyl-pyrophosphate (GGPP) is attached to chlid, and three of its four double bonds are reduced in the sequence $\Delta6$, $\Delta10$, $\Delta14$. Only crude preparations of the enzyme(s) involved are hitherto available from plants, and it is not clear whether these reductions take place before or after the condensation, or if both pathways exist in parallel (Rüdiger, 1993).

Branched pathways have been found in green plants for most of the late steps of chlorophyll biosynthesis. Protoporphyrin IX carries two vinyl groups at C-3 and C-8, whereas the final Chl has generally only a single one at C-3. Two (or even more) series of intermediates have been found up to chlid *a* which carry either one or both vinyl groups, indicating that the enzymes catalysing the intervening steps are capable of using substrates from both series. It is then presently unclear at which point the 8-vinyl group becomes reduced. The final product carrying two vinyl-groups, e.g. [8-vinyl]-Chl-*a* has only been rarely observed in nature, but at least two examples show that they are even functional in the native pigment–proteins. Another branching may exist during the conversion of chlid to Chl *a* (see above).

Biosynthesis of Chl *b* involves the transformation of the 7-$CH_3$- to a CHO-group by an oxygenase. The point of this oxygenation is unclear, but mutant analyses in *Chlamydomonas reinhardtii* point to a single enzyme being involved (for references, see Porra *et al.*, 1994). There is also evidence for the reverse transformation (Ito & Tanaka, 1996). No studies are currently available on the enzymes converting the 3-vinyl- to the formyl-group present in Chl *d*.

Almost nothing is known on the biosynthesis of the less abundant chlorophylls including the pheophytins and the 'prime-chlorophylls', e.g. the $13^2$-epimers of the respective Chls. It should be noted, however, that all late enzymes are stereoselective for the natural $13^2$ (S)-epimers, so that epimerase(s) are expected to exist in order to avoid uncontrolled and potentially harmful epimerization.

## Degradation of chlorophylls

Until quite recently, only degradation products of chlorophylls were known in which the tetrapyrrole macrocycle is still intact. These include pigments lacking the $13^2$-$COOCH_3$ group, or the long-chain alcohol. The latter is catalysed by chlorophyllase, one of the earliest enzymes isolated. *In vitro*, it is most active in aqueous acetone; *in situ*, it appears to be associated with Chl proteins and it can attack even pigments bound to their apoproteins. There are also pigments in which the central Mg is removed by enzymatic action or by small, heat-stable molecule(s) (Shioi *et al.*, 1996). It is not clear at present if the demetallated products can be degraded further by the oxygenase(s) which open(s) the macrocycle. All the products hitherto mentioned are phototoxic. This is probably no longer true for an increasing number of open-chain tetrapyrroles (see Fig. 4.2 for examples) which have been isolated over the past years from a green alga and from several higher plants, and which for the first time outline the fate of the chlorophylls before the eventual breakdown to small fragments (Engel, Curty & Gossauer, 1996; Mühlecker & Kräutler, 1996; Matile & Schellenberg, 1996). They all arise from cleavage of the C-4, 5 bond, probably by a mono-oxygenase, if judged from isotope studies, and carry a remarkably stable β-ketoacid system of ring E. This ring-opening of the macrocycle is followed by hydroxylation and further modifications discussed in the references cited. The degrading enzymes appear to be situated in plastid membranes, but not necessarily in the thylakoid.

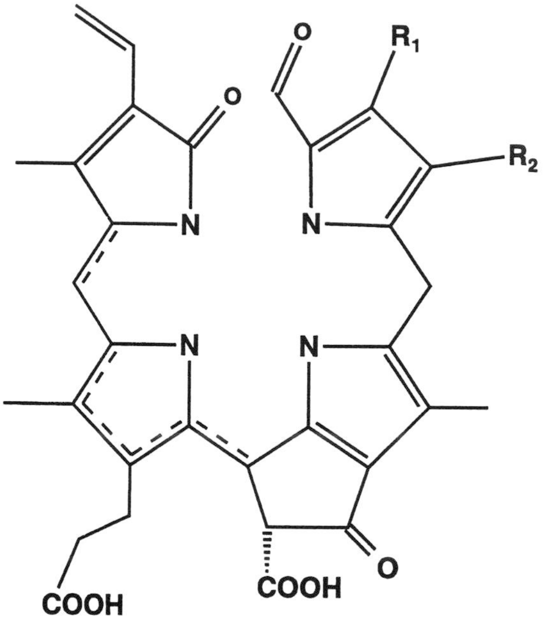

**Figure 4.2.** Type formula for degradation products of chlorophylls *a* ($R_1 = CH_3$) and *b* ($R_1 = CHO$). $R_2$ can be $CH_3$, $CH_2$-O-malonyl or $CH_2$-O-glucosyl. The dashed double bonds indicate bonds which may be single or double.

## STRUCTURE AND FUNCTION OF CAROTENOIDS

There are several new review articles available on carotenoids and their synthesis in bacteria (Armstrong, 1994), cyanobacteria (Hirschberg & Chamovitz, 1994) and also on carotenoid genes and enzymes (Sandmann, 1994). Therefore, this and the following paragraph will mainly focus on carotenoids from plants covering their function and synthesis. Various aspects of plant carotenoids were presented very recently (Bartley & Scolnik, 1995). Due to limited space, older references will not be cited but can be found in the review articles mentioned above.

The carotenoids found in plants are all bicyclic with either two β-ionone rings or an ε- and a β-ring as end groups (Fig. 4.3). The chromophores of α-carotene derivatives consist of nine conjugated double bonds. In β-carotene derivatives, this number is nominally 11. However, the effective conjugated system is less. The contribution of each double bond in both ionone rings is rather weak, because they are sterically hindered to adopt co-planarity with the main polyene chain. Therefore, the spectral properties of α- and β-carotene are quite similar. All plant carotenoids are derived from these two hydrocarbons. The major oxygen-containing xanthophylls are lutein, which is α-carotene-related, and violaxanthin, antheraxanthin, zeaxanthin and neoxanthin, which are all derived from β-carotene by combination of hydroxylation and epoxydation steps. In the case of neoxanthin, an allenic bond additionally is introduced.

Carotenoids exhibit typical spectral properties. Due to their polyenic structure, they absorb light in the visible (blue) range. All the compounds mentioned above show three absorbance maxima (or at least shoulders) with the prominent central one around 445 nm. Light absorbance involves electronic transitions from the ground state $S_0$ to the second excited state $S_2$. There are also higher energy transitions possible, which are reflected by weak absorption bands in the UV range. UV absorbance increases particularly in carotenoids with a double bond in the *cis*-configuration.

The function of carotenoids in photosynthesis is related to their photochemical properties. They can act as accessory light-harvesting pigments and as photoprotective agents, preventing photo-oxidative damage. In each case, excitation energy is transferred to, and from, chlorophyll. In the light-harvesting reaction ground state carotenoid is excited by blue light to its singlet state $S_2$ and the energy then transferred to chlorophyll. It is currently discussed that singlet–singlet energy transfer from carotenoid to chlorophyll should occur via an electron exchange mechanism involving the spectrally overlapping excited $S_1$ states.

Upon sensitization by light, a portion of chlorophyll can undergo intersystem crossing to the triplet state. The lifetime of triplet chlorophyll is relatively high, e.g. in the μs range. This makes it possible for it to react with oxygen which, in its ground state, is a triplet. The resulting singlet oxygen is extremely reactive and can oxidize almost any compound around. Carotenoids protect against these photo-oxidative reactions by quenching the triplet state of chlorophyll. Energy is transferred to ground-state carotenoids and dissipated from the resulting triplet carotenoid as heat. It is suggested that, in a similar way, carotenoids may quench singlet oxygen. The reactions described here make carotenoids essential for any photosynthetic organism. Consequently, carot-

α-Carotene

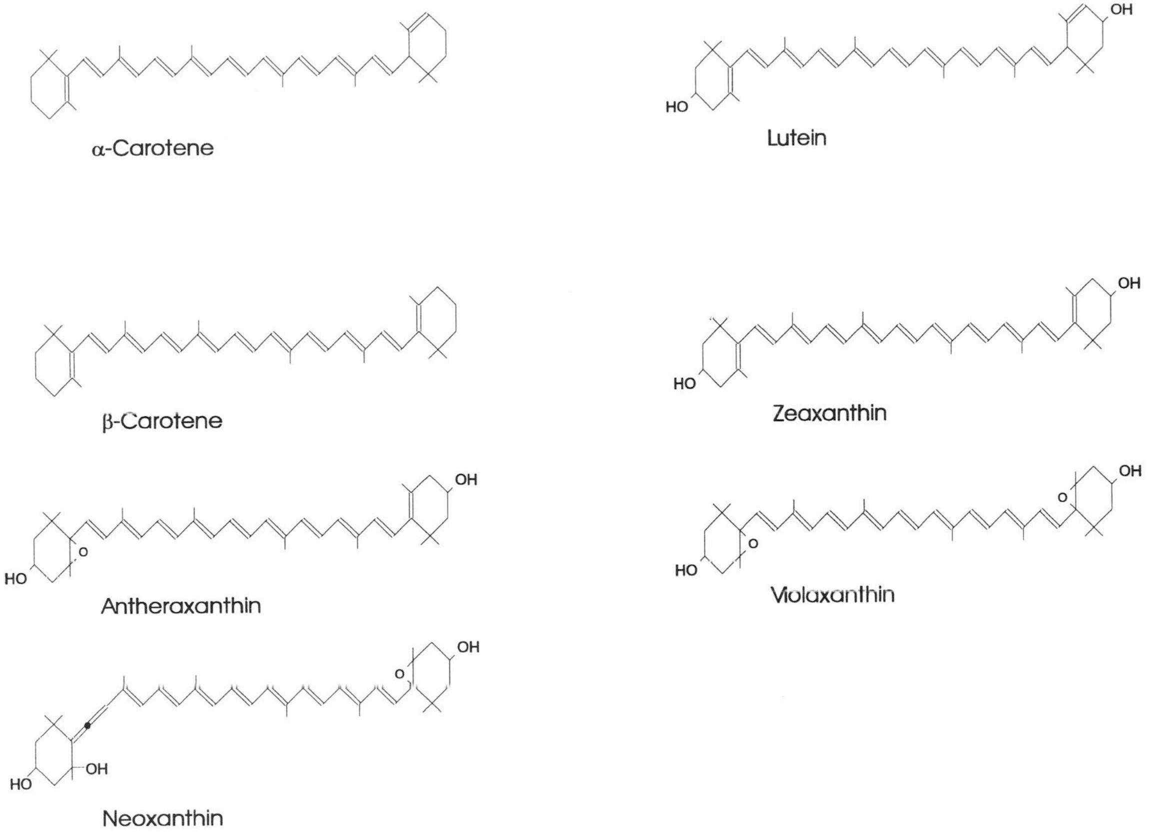

Lutein

β-Carotene

Zeaxanthin

Antheraxanthin

Violaxanthin

Neoxanthin

**Figure 4.3.** Structures of carotenes and xanthophylls present in plant chloroplasts.

enoid depletion in mutants or in inhibitor-treated plants is lethal in the light unless anaerobic conditions are established.

## BIOSYNTHESIS OF CAROTENOIDS

### Pathway

The entire carotenoid biosynthetic pathway is part of the terpenoid metabolism with formation of prenyl pyrophosphates as a common sequence for chain elongations. From the different prenyl pyrophosphates formed (Fig. 4.4), specific routes branch off into various terpenoid end products. Carotenoid biosynthesis relies on the independent synthesis of geranylgeranyl pyrophosphate (GGPP). The reactions involved are 1′-4 condensations between isopentenyl pyrophosphate (IPP) molecules and an allylic pyrophosphate. As in all other prenyl transferase reactions, the mechanism of this head-to-tail joining is basically the same. The carbon bond is formed between the C-4 of IPP and the C-1 of the allylic co-substrate. Simultaneously, inorganic pyrophosphate is liberated from the allylic partner. This condensation reaction involves an allylic carbonium anion as an intermediate which then attacks the 3,4-double bond of IPP.

The colourless carotene phytoene is a $C_{40}$ hydrocarbon with only three conjugated double bonds (Fig. 4.4). Although it is found only in trace amounts, this carotene accumulated in mutants or in the presence of inhibitors which affect its conversion. Enzymatic studies with the purified phytoene synthase from *Capsicum* as well as *E. coli* transformed with a phytoene synthase gene (Misawa *et al.*, 1994*a*) demonstrated that a single gene

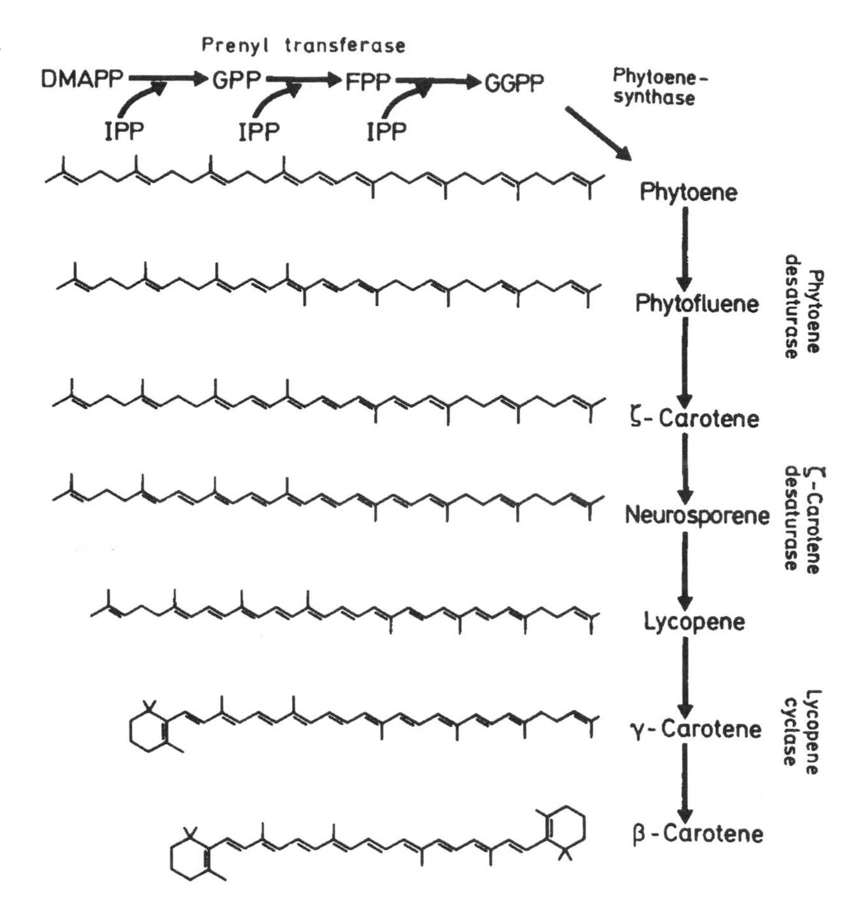


Figure 4.4. Plant carotenoid biosynthetic pathway from dimethylallyl pyrophosphate to β-carotene.

product catalyses the condensation of GGPP to phytoene.

Four dehydrogenation steps are carried out in the conversion of phytoene to the maximally desaturated lycopene (Fig. 4.4). For many years there was a debate on how many enzymes are involved. One opinion was that two distinct enzymes are necessary for the four-step desaturation process to lycopene. The first enzyme was regarded to utilize each half of the symmetrical phytoene molecule as a substrate and to yield ζ-carotene with phytofluene as an intermediate. The second desaturase was supposed to carry out a similar reaction with ζ-carotene to form lycopene via neurosporene. The occurrence of mutants of maize, tomato and *Scenedesmus obliquus* in which ζ-carotene is accumulated as well as inhibitor studies with herbicides that specifically inactivate phytoene desaturation, and others that inhibit ζ-carotene desaturation (see Chapter 25 on Herbicides) supported this view. Direct evidence was obtained by co-transformation of *E. coli* with two different plasmids. One carried the genes *crtB* and *crtE*, which are necessary for the synthesis of the substrate phytoene, and a second plasmid carrying the phytoene desaturase cDNA from tomato. The resulting *E. coli* transformants were pigmented due to the production of ζ-carotene. Similar experiments were carried out with a ζ-carotene desaturase cDNA from *Capsicum* in *E. coli* producing a ζ-carotene background (Albrecht *et al.*, 1995). The reaction product in this case was lycopene. In contrast to algae and higher plants, a single desaturase catalyses the entire desaturation sequence of phytoene to lycopene in bacteria and fungi.

Lycopene is the regular substrate for cyclization reactions. Two types of ionone rings are formed, the β-ring as an end group of both sides of β-carotene and the ε-ring at one side of α-carotene in addition to a β-ring (Fig. 4.3). The mechanisms of cyclization which involves proton attack at C-2 and C-2′ of lycopene is substantially supported by $D_2O$-labelling experiments followed by determination of the position of the deuterated carbon at the β- or ε-ionone ring. The resulting carbonium ion intermediate is stabilized by loss of a proton either from C-1 or C-4 to yield a β- or ε-ring, respectively. As indicated by gene analysis of *Erwinia herbicola*, only one gene product is sufficient for introduction of the two β-ionone rings at both sides of lycopene to form β-carotene via γ-carotene.

Formation of the β- or ε-ring is under different gene control. Introduction of the *B* gene into red (= lycopene-accumulating) tomato resulted in the production of β-carotene at the expense of lycopene, whereas the *Del* gene is responsible for formation of ε-rings. Consequently, we can assume that formation of β-carotene is catalysed by a single enzyme, lycopene cyclase β. In the case of α-carotene, two different enzymes, lycopene cyclase β and lycopene cyclase ε are involved. Mutants of *Arabidopsis* (Pogson *et al.*, 1995) and the green alga *Scenedesmus* (Bishop, Urbig & Senger, 1995) were identified which lack ε-cyclization but are unaffected in the synthesis of β-carotene.

Xanthophylls are enzymatically formed oxidation products of α- and β-carotene (Fig. 4.3). The most common oxygen groups found in plastidic xanthophylls are hydroxy at C-3 and epoxy at the 5, 6 position of the ionone ring. Lutein, the 3, 3′-dihydroxy α-carotene, is formed only by algae and higher plants which have the potential to form α-carotene with its ε-ring. Zeaxanthin, the corresponding 3, 3′-dihydroxy β-carotene, is synthesized by all organisms with oxygenic photosynthesis. As far as is known, all the oxygen groups originate from molecular oxygen. Epoxydation of zeaxanthin to violaxanthin is part of a reversible cycle operative in plant chloroplasts. Under light stress, de-epoxydation of violaxanthin occurs via antheraxanthin to zeaxanthin which is regarded to protect the chloroplast against high light conditions. The regulation and significance of the reversible epoxidation/de-epoxidation process for protection of the photosynthetic apparatus has been reviewed recently (Pfündel & Bilger, 1994).

In chloroplasts from algae and higher plants, the epoxy carotenoid neoxanthin with an allenic end group is present. It has been assumed that this allenic group originates from proton abstraction from C-7 neighbouring the 5, 6-epoxy-5, 6-dihydro-β-rings of violaxanthin followed by rearrangement of the epoxy group to 5-hydroxy.

## Genes

The first genes involved in the carotenoid biosynthetic pathway were identified, cloned and sequenced only a few years ago. They originated from prokaryotic organisms like *Rhodobacter capsulatus, Erwinia uredevora* and *Synechococcus*. A cDNA, pTOM5 was cloned from mRNA which accumulated during tomato fruit ripening by differential hybridization for genes specifically expressed during this process. It was later identified to code for phytoene synthase by functional complementation, *in vitro* enzymic analysis and by an antisense RNA approach. Several cloning strategies were followed to obtain other carotenogenic genes from plants. One technique involved was to use a DNA library from a *Synechococcus* mutant resistant against a bleaching herbicide and transfect a susceptible wild-type strain. After selection of the transformant for herbicide resistance, the *pds* gene was recovered by a special plasmid rescue technique. With a similar strategy, the gene for lycopene cyclase was cloned from *Synechococcus* (Cunningham *et al.*, 1994). Sequence information from the cyanobacterial genes was essential for the cloning of both cDNAs or genes mentioned above from plants. As the mechanisms for ζ-carotene desaturation is quite similar to phytoene desaturation, it was possible to clone the cDNA coding for the latter enzyme by screening databases for sequences which resemble known plant phytoene desaturase sequences but which are not identical to them. An *Arabidopsis thaliana* expressed sequence tag was found by this search which led to the cloning of a ζ-carotene desaturase cDNA from *Capsicum* (Albrecht *et al.*, 1995). A *Capsicum* cDNA for lycopene cyclase was cloned recently by a similar approach (Hugueney *et al.*, 1995). Other cDNAs of carotenogenic genes from *Capsicum* were isolated by the use of a biochemical approach. A cDNA library was screened with an antiserum raised against purified GGPP synthase as well as phytoene desaturase. Some of the carotenogenic genes have been

Table 4.2. *Complete sequenced cDNAs (c) or genes (g) encoding carotenogenic enzymes from plants*

| Enzyme | Plant |
|---|---|
| GGPP synthase | *Arabidopsis*[c,g] |
| | *Capsicum*[c] |
| Phytoene synthase | *Arabidopsis*[c] |
| | *Capsicum*[c] |
| | Tomato[c,g] |
| Phytoene desaturase | *Arabidopsis*[c] |
| | *Capsicum*[c] |
| | Soybean[c] |
| | Tomato[c,g] |
| ζ-Carotene desaturase | *Capsicum*[c] |
| Lycopene cyclase | *Arabidopsis*[c] |
| | *Capsicum*[c] |

Further literature on the cDNAs and genes mentioned here can be found in the review articles by Sandmann (1994) and Bartley and Scolnik (1995). Sequences of unpublished cDNAs for carotenogenic enzymes from other plants are deposited in the gene bank (see Bartley & Sconik, 1995 for accession numbers).

cloned directly by transposon tagging and subsequent rescue or localization with the transposon DNA. This powerful strategy may be the method of choice for future work to identify especially higher plant structural and also regulatory genes of the carotenoid biosynthetic pathway.

Very recently, the sequences of cDNAs for all steps of the pathway from synthesis of GGPP to β-carotene have been obtained from plants, including the genomic sequence of tomato phytoene synthase and desaturase as well as *Capsicum* GGPP synthase. A survey is given in Table 4.2.

## Enzymology

Many carotenogenic enzymes are associated with, or integrated into, membranes. Furthermore, their abundance is extremely low. This makes their solubilization in an active state and the purification extremely difficult. Therefore, only a few enzymes of the pathway are characterized biochemically in detail. This has not been achieved, to date, for enzymes involved in the formation of xanthophylls. Thus, the following paragraph is restricted to enzymes involved in the section of the carotenoid pathway from the synthesis of GGPP to the synthesis of β-carotene.

The enzymes catalysing the synthesis of GGPP have been shown to belong to a family of prenyl transferases with different allylic substrate and chain length specificities. Several types of this enzyme were isolated from carotenogenic and non-carotenogenic organisms. GGPP synthase was purified from *Capsicum* chromoplasts and characterized in detail. It has a dimeric structure with two subunits of 37 kDa. As indicated by the $K_m$ values, this enzyme from *Capsicum* shows no preference for a certain allylic substrate and converts the $C_5$ dimethylallyl pyrophospate (DMAPP), $C_{10}$ geranyl pyrophosphate (GPP), and $C_{15}$ farnesyl pyrophosphate (FPP) equally well. Therefore, DMAPP may be considered to be the genuine substrate for GGPP synthase in higher plant plastids.

*In vitro* studies involving phytoene synthesis were carried out with crude preparations from different sources. The results, however, were rather ambiguous. The only phytoene synthase purified to date originated from *Capsicum* chromoplasts. The main purification step was affinity chromatography using an analogue of GGPP. The protein obtained had a molecular weight of 47.5 kDa and can convert either GGPP or prephytoene pyrophosphate as substrates with high affinity. The enzyme is dependent on $Mn^{2+}$ and is inhibited by inorganic phosphate. Immunochemical localization of phytoene synthase revealed its presence in the stroma of chromoplasts as well as chloroplasts. There is evidence that the synthases in both compartments are distinct and originate from divergent genes (Fraser *et al.*, 1994).

Purification of integral membrane proteins like phytoene desaturase is very difficult to achieve from plant tissue. Nevertheless, this enzyme was isolated from *Capsicum* chromoplasts and purified in an active state. It had a size of 56 kDa and binding of FAD to the protein was suggested. Phytoene desaturase from *Synechococcus*, another plant-type enzyme was purified after overexpression of its gene in *E. coli*. A plasmid was used which mediated the expression of phytoene desaturase to a final concentration of more than 5% of total cellular protein. Under these conditions, the recombinant protein was sequestered in inclusion bodies, where it could be solubilized from by urea treatment which left the resulting enzyme poorly active. However, it was possible to regain its activity by removal of the denaturant and after lipid replenishment. By this procedure several milligrams of this 53 kDa membrane protein were obtained. Inhibition was observed by several bleaching herbicides. The cofactors for this cyanobacterial/algal/plant-type phytoene desaturase were either NAD or NADP whilst FAD was ineffective as an electron acceptor. The dependence of this purified phytoene desaturase on oxidized dinucleotides as electron and hydrogen mediators confirms a dehydrogenase-electron transferase mechanism for the desaturation reaction. Chloroplast import studies with the *pds* gene product from soybean showed that the expressed protein is processed after import into a smaller mature form. Very recently, also the ζ-carotene desaturase from *Anabaena* with a molecular mass of 53 kDa was purified and enzymologically characterized after expression of the *zds* gene in *E. coli*. In addition to ζ-carotene, the enzyme can also convert neurosporene to lycopene and β-zeacarotene to γ-carotene (Albrecht, Linden & Sandmann, 1996). However, it should be pointed out that this is not the plant-type ζ-carotene desaturase which is a completely different protein (Albrecht *et al.*, 1995).

Lycopene cyclase was solubilized from *Capsicum* chromoplasts by acetone treatment. This preparation was capable of converting lycopene into β-carotene. Cofactors were not required but the reaction was sensitive to sulfhydryl reagents. Furthermore, lycopene cyclization was inhibited by tertiary amines. Especially CPTA [2-(4-chlorophenylthio)-triethylamine HCl] has been used for several years as an inhibitor for lycopene cyclization *in vivo* or *in vitro* (see Chapter 25 on Herbicides). The only purified and well-characterized lycopene cyclase is the enzyme from the enterobacterium *Erwinia uredovora*. This enzyme shares a weak but significant similarity with the plant enzyme (Hugueney *et al.*, 1995). The lycopene cyclase gene from *Erwinia* was overexpressed in *E. coli* and an enzymatically active protein of 43 kDa was purified to homogeneity (Schnurr, Misawa & Sandmann, 1996). The cyclization reaction is dependent on NADPH or NADH. Conversion was observed for the substrates lycopene and γ-carotene to β-carotene, neurosporene to β-zeacarotene and 1-hydroxy-lycopene to 1'-hydroxy-γ-carotene.

## OUTLOOK

The most striking advances during the last years in the field of chloroplast pigments were made in the crystallization of pigment–protein complexes. The resulting establishment of atomic models revealed *in situ* details on the pigments, e.g. distortion from planarity or interactions with neighbouring molecules. The development of methods suitable for reconstituting or exchanging pigments into a variety of chlorophyll-carotenoid-proteins has resulted in an increasing demand of modified or selectively labelled pigments. These breakthroughs, together with the cloning of structural genes of the pigments' biosynthetic pathways, had an enormous stimulatory effect on biochemical investigations.

The high standard of the molecular biology techniques on the chlorophyll and carotenoid pathway makes it possible to adopt these procedures to analyse the interaction of both types of pigments. Genetic manipulations for quantitative and qualitative modification of thylakoid carotenoids has already been achieved in prokaryotes (Hunter *et al.*, 1994). Also, in higher plants a combination of molecular genetic techniques and biochemical approaches can be utilized to answer questions on pigment function. Future aspects, which can now be aimed at, are disruption of existing structural genes in plants and replacement by foreign genes for new carotenoids or chlorophylls followed by studies on their integration into the photosynthetic apparatus and their function in the photosynthetic pigment–protein complexes. The first attempt to modify the carotenoid content of plant thylakoloids was already successful (Misawa *et al.*, 1994*b*). Last, but not least, there is evidence that (bacterio) chlorophylls and carotenoids have a certain potential in medicine, e.g. in therapy of cancer. This may increase their demand and require commercial production on a larger scale.

## REFERENCES

Albrecht, M., Klein, A., Hugueney, P., Sandmann, G. & Kuntz, M. (1995). Molecular cloning and functional expression in *E. coli* of a novel plant enzyme mediating ζ-carotene desaturation. *FEBS Letters*, **372**, 199–202.

Albrecht, M., Linden, H. & Sandmann, G. (1996). Biochemical characterization of purified ζ-carotene desaturase from *Anabaena* PCC7120 after expression in

*Escherichia coli. European Journal of Biochemistry*, **236**, 115–20.

Aracri, B., Bartley, G. E., Scolnik P. A. & Giuliano, G. (1994). Sequence of the phytoene desaturase locus of tomato. Plant Gene Register. *Plant Physiology*, **106**, 789.

Armstrong, G. A. (1994). Eubacteria show their true colors: genetics of carotenoid pigment biosynthesis from microbes to plants. *Journal of Bacteriology*, **176**, 4795–802.

Badillo, A., Steppuhn, J., Deruere, J., Camara, B. & Kuntz, M. (1995). Structure of a functional geranylgeranyl pyrophosphate synthase gene from *Capsicum annuum*. *Plant Molecular Biology*, **27**, 425–8.

Bartley, G. E. & Scolnik, P. A. (1995). Plant carotenoids: pigments of photoprotection, visual attraction, and human health. *The Plant Cell*, **7**, 1027–38.

Bishop, N. I., Urbig, Th. & Senger, H. (1995). Complete separation of the β,ε- and β,β-carotenoid biosynthetic pathways by a unique mutation of the lycopene cyclase in the green alga, *Scenedesmus obliquus*. *FEBS Letters*, **367**, 158–62.

Blankenship, R., Madigan, M. T. & Bauer, C. E. (eds). (1995). *Anoxygenic Photosynthetic Bacteria*. Dordrecht: Kluwer Academic Publishers.

Bryant, D. A. (ed.) (1994). *The Molecular Biology of Cyanobacteria*. Dordrecht: Kluwer Academic Publishers.

Chadwick, D. J. & Ackrill, K. (eds). (1995). *Biosynthesis of the Tetrapyrroles*. Chichester: Wiley.

Cunningham, Jr, F. X., Sun, Z., Chamovitz, D., Hirschberg, J. & Gantt, E. (1994). Molecular structure and enzymatic function of lycopene cyclase from the cyanobacterium *Synechococcus* sp. strain PCC7942. *The Plant Cell*, **6**, 1107–21.

Deisenhofer, J. & Norris, J. R. (eds). (1993). *The Photosynthetic Reaction Center*. New York: Academic Press

Eichacker, L. A., Soll, J., Lauterbach, P., Rüdiger, W., Klein, R. R., and Mullet, J. E. (1990) *In vitro* synthesis of chlorophyll *a* in the dark triggers accumulation of chlorophyll *a* apoprotein in barley etioplasts. *Journal of Biological Chemistry*, **265**, 13566–71.

Engel, N., Curty, C. & Gossauer, A. (1996). Chlorophyll catabolism in *Chlorella protothecoides*. Part 8: Facts and artefacts. *Plant Physiology and Biochemistry*, **34**, 77–83.

Fraser, P. D., Truesdale, M. R., Bird, C. R., Schuch, W. & Bramley, P. M. (1994). Carotenoid biosynthesis during tomato fruit development. Evidence for tissue-specific gene expression. *Plant Physiology*, **105**, 405–13.

Helfrich, M. (1995). Chemische Modifikation von Chlorophyll-Vorstufen und deren Verwendung zur Charakterisierung von Enzymen der Chlorophyll-Biosynthese. Dissertation, München: Ludwig-Maximilians-Universität.

Hirschberg, J. & Chamovitz, D. (1994). Carotenoids in cyanobacteria. In *The Molecular Biology of Cyanobacteria*, ed. D. A. Bryant, pp. 559–79. Dordrecht: Kluwer Academic Publishers.

Holtorf, H., Reinbothe, S., Reinbothe, C., Bereza, B. & Apel, K. (1995). Two routes of chlorophyllide synthesis that are differentially regulated by light in barley (*Hordeum vulgare* L.). *Proceedings of National Academy of Sciences, USA*, **92**, 3254–8.

Hori, N., Kumar, A. M., Verkamp, E. & Söll, D. (1996). 5-Aminolevulinic acid formation in *Arabidopsis*. *Plant Physiology and Biochemistry*, **34**, 3–9.

Hugueney, P., Badillo, A., Chen, H. Ch., Klein, A., Hirschberg, J., Camara, B. & Kuntz, M. (1995). Metabolism of cyclic carotenoids: a model for the alteration of this biosynthetic pathway in *Capsicum annuum* chromoplasts. *The Plant Journal*, **8**, 417–24.

Hunter, C. N., Hundle, B. S., Hearst, J. E., Lang, H. P., Gardiner, A. T., Takaichi, S. & Cogdell, R. J. (1994). Introduction of new carotenoids into the bacterial photosynthetic apparatus by combining the carotenoid biosynthetic pathway of *Erwinia herbicola* and *Rhodobacter sphaeroides*. *Journal of Bacteriology*, **176**, 3692–7.

Ito, H. & Tanaka, A. (1996). Determination of the activity of chlorophyll *b* to chlorophyll *a* conversion during greening of etiolated cucumber cotyledons by using pyrochlorophyll *b*. *Plant Physiology and Biochemistry*, **34**, 35–40.

Jordan, P. M. (ed.) (1991). *Biosynthesis of Tetrapyrroles*. Amsterdam: Elsevier.

Krauss, N., Hinrichs, W., Witt, I., Fromme, P., Pritzkow, W., Dauter, Z., Betzel, C., Wilson, K. S., Witt, H. T. & Saenger, W. (1993). 3-Dimensional structure of system-I of photosynthesis at 6 angstrom resolution. *Nature*, **361**, 326–31.

Kühlbrandt, W., Wang, D. N. & Fujiyoshi, Y. (1994). Atomic model of plant light-harvesting complex by electron crystallography. *Nature*, **367**, 614–21.

Larkum, A. W. D., Scaramuzzi, C., Cox, G. C., Hiller, R. G. & Turner, A. G. (1994). Light-harvesting chlorophyll C-like pigment in *Prochloron*. *Proceedings of National Academy of Sciences USA*, **91**, 679–83.

Louie, G. V., Brownlie, P. D., Lambert, R., Cooper, J. B., Blundell, T. L., Wood, S. P., Warran, M. J., Woodcock, S. C. & Jordan, P. M. (1992). Structure of porphobilinogen deaminase reveals a flexible multidomain polymerase with a single active site. *Nature*, **359**, 33–9.

Matile, P. & Schellenberg, M. (1996). The cleavage of pheophorbide *a* is located in the envelope of barley gerontoplasts. *Plant Physiology and Biochemistry*, **34**, 55–9.

Misawa, N., Truesdale, Sandmann, G., Fraser, P. D. Bird, C. Schuch, W. & Bramley, P. M. (1994*a*). Expression of a tomato cDNA coding for phytoene synthase in *Escherichia coli*, phytoene formation *in vivo* and *in vitro* and functional analysis of the various truncated gene products. *Journal Biochemistry*, **116**, 980–5.

Misawa, N., Masomoto, K., Hori, T., Böger, P. & Sandmann, G. (1994*b*). Expression of an *Erwinia* phytoene desaturase gene not only confers multiple resistance to herbicides interfering with carotenoid biosynthesis but also alters xanthophyll metabolism in transgenic plants. *The Plant Journal*, **6**, 481–9.

Miyashita, H., Kurano, N., Ikemoto, H., Chihara, M. & Miyachi, S. (1995). A new photosynthetic prokaryote containing chlorophyll *d* as its major chlorophyll. *Chloroplast Development, Book of Abstracts*, ed. H. Senger, Marburg.

Mühlecker, W. & Kräutler, B. (1996). Breakdown of chlorophyll: constitution of nonfluorescent chlorophyll-catabolites from senescent cotyledons of the dicot rape. *Plant Physiology and Biochemistry*, **34**, 61–75.

Paulsen, H. (1995). Chlorophyll *a*/*b*-binding proteins. *Photochemistry and Photobiology*, **62**, 367–82.

Pfündel, E. & Bilger, W. (1994). Regulation and possible function of the violaxanthin cycle. *Photosynthesis Research*, **42**, 89–109.

Pogson, B., Norris, S., McDonald, K., Truong, M. & DellaPenna, D. (1995). Characterization of mutations disrupting carotenoid biosynthesis in *Arabidopsis thaliana*. In *Photosynthesis: from Light to Biosphere*. Vol. IV. ed. P. Mathis, pp. 75–8. Dordrecht: Kluwer Academic Publishers.

Porra, R. J., Schäfer, W., Cmiel, E., Katheder, I. & Scheer, H. (1994). The derivation of the formyl-group of chlorophyll *b* in higher plants from molecular oxygen. Achievement of high enrichment of the 7-formyl-oxygen from $^{18}O_2$ in greening maize leaves. *European Journal of Biochemistry*, **219**, 671–9.

Rüdiger, W. (1993). Esterification of chlorophyllide and its implications on thylakoid development, In *Pigment Complexes in Plastids*. Ed. C. Sundquist & M. Ryberg, pp. 219–40. San Diego: Academic Press.

Sandmann, G. (1994). Carotenoid biosynthesis in microorganisms and plants. *European Journal of Biochemistry*, **223**, 7–24.

Scheer, H. (ed.). (1991). *Chlorophylls*. Boca Raton: CRC-Press.

Schmitt, A., Frank, G., James, P., Staudenmann, W., Zuber, H. & Wilhelm, C. (1994). Polypeptide sequence of the chlorophyll *a*/*b*/*c*-binding protein of the prasinophycean alga *Mantoniella squamata*. *Photosynthesis Research*, **40**, 269–77.

Schnurr, G., Misawa, N. & Sandmann, G. (1996). Expression, purification and properties of lycopene cyclase from *Erwinia uredovoa*. *Biochemical Journal*, **315**, 869–74.

Senge, M. O. (1993). Recent advances in the biosynthesis and chemistry of the chlorophylls. *Photochemistry and Photobiology*, **57**, 189–206.

Shioi, Y., Tomita, N., Tsuchiya, T. & Takamiya, K. I. (1996). Conversion of chlorophyllide to pheophorbide by Mg-dechelating substance in extracts of *Chenopodium album*. *Plant Physiology and Biochemistry*, **34**, 41–7.

# 5 Import, assembly and degradation of chloroplast proteins

P. R. CHITNIS

## INTRODUCTION

Plastids are rich in diversity and amount of proteins. They contain complex machinery for many unique metabolic reactions, such as photosynthesis and biosynthesis of lipids, chlorophyll (Chl), and starch. The intricate architecture of the chloroplast presents unique and complex challenges in protein targeting and organelle biogenesis. Three distinct membrane systems divide chloroplasts into at least six suborganellar compartments (Fig. 5.1). The development and maintenance of this structure require an elaborate sorting system to ensure proper targeting and assembly of chloroplast proteins. The complexity is compounded by two sites of synthesis of chloroplast proteins, chloroplast stroma and cytoplasm. The chloroplast genes are expressed by a prokaryote-like expression system. Chloroplasts have their own translation system, consisting of 70S ribosomes, >30 tRNA species, initiation elongation factors and aminoacyl tRNA synthetases which are homologous to those in prokaryotes. Translation of several chloroplast RNAs, such as those encoding the PsaA-PsaB proteins of photosystem I (PSI) and the D1 protein of photosystem 11 (PSII), are regulated by light, the availability of cofactors, and other environmental and physiological regulatory signals. The proteins that are synthesized in chloroplasts may contain cleavable or internal targeting information.

Most chloroplast proteins are encoded in the nuclear genome and are translated on cytoplasmic ribosomes as precursors with a transit (leader) sequence at their amino-terminus. The precursors are imported into the chloroplasts in a post-translational, energy-dependent process, via a receptor in the envelope membranes. After import, the proteins are proteolytically processed to their mature forms, targeted to an appropriate subcompartment, associate with their cofactors and with other subunits to form active enzymes. During normal metabolism and under stress conditions the proteins are degraded and replaced with new proteins. Thus the life of a chloroplast protein involves synthesis, targeting, import, assembly, function and degradation. This chapter discusses our current understanding of the biogenesis of chloroplast proteins. Detailed descriptions of specific aspects and comprehensive lists of original references can be found in some recent reviews (Callis, 1995; Cohen, Yalovsky & Nechushtai, 1995; Keegstra *et al.*, 1995; Nechushtai, Cohen & Chitnis, 1995; Schnell, 1995).

## PROTEINS TARGETING TO THE CHLOROPLAST AND ITS COMPARTMENTS

Proteins synthesized in the plant cytoplasm can be targeted to many destinations; N-terminal presequences of mitochondrion- or chloroplast-specific proteins can target precursors to the respective organelles with variable efficiencies. The plant mitochondria generally do not import precursors of stromal or thylakoidal proteins. However, some chloroplast transit sequences contain sufficient information for specific interaction with mitochondrial import receptors (Brink *et al.*, 1994). Interestingly, a yeast mitochondrial leader peptide functions in transgenic plants as a dual targeting signal for both chloroplasts and mitochondria (Huang *et al.*, 1990). Therefore, additional control mechanisms in the cytosol that are independent of the presequence are required *in vivo* to achieve efficient sorting between chloroplasts and mitochondria.

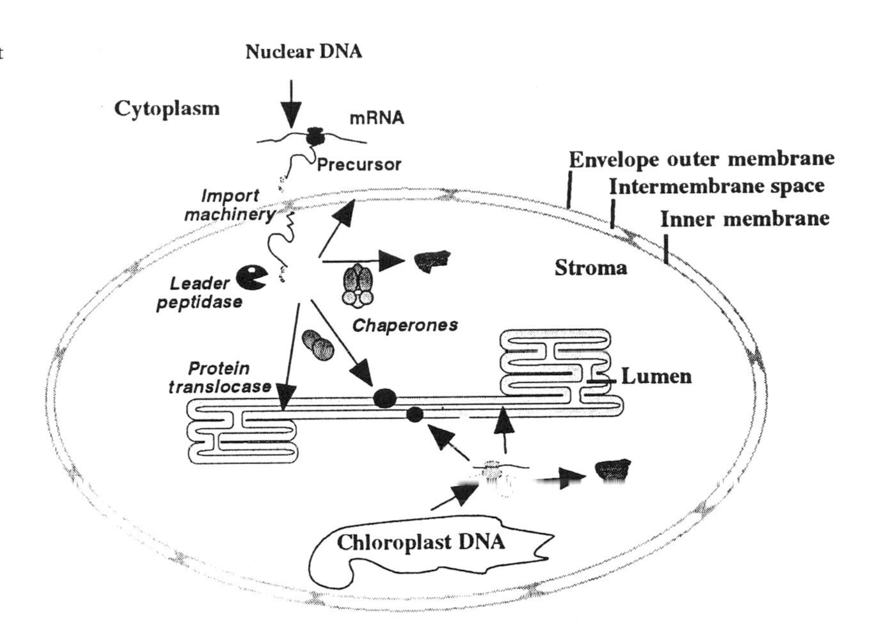


**Figure 5.1.** Biogenesis of chloroplast proteins. Chloroplast sub-compartments and targeting options for the chloroplast-localized proteins are shown.

Once directed to the chloroplast, precursor proteins need to be further targeted to the proper location within chloroplasts. The targeting information in precursor proteins resides in their leader sequences. Cytoplasmic chaperones assist many precursor proteins to maintain an import-competent form that is translocated across chloroplast envelope through a specialized machinery (Keegstra *et al.*, 1995; Schnell, 1995). Once the precursor is imported into stroma, the transit sequence is proteolytically removed by leader peptidases. These steps are common to most nuclear-encoded proteins that are targeted to the chloroplast.

## Targeting to the outer membrane of envelope

Many proteins targeted to the outer membrane of the chloroplast envelope use a mechanism that is distinct from targeting to the inner compartments (see Keegstra *et al.* (1995) for specific citations). These proteins do not contain a cleavable transit sequence. Studies involving deletion and chimeric constructs have shown that the targeting information for a cognate 70 kDa heat shock protein from the spinach chloroplast outer envelope (SCE70) resides in the N-terminal 42 amino acids. This targeting domain is required for directing SCE70 to a destination in the chloroplast outer envelope membrane, possibly through assembling the polypeptide into a protein complex. Another distinctive feature of the outer envelope proteins is that they do not require ATP hydrolysis for import. Treatment of chloroplasts with proteases or addition of a synthetic peptide corresponding to the central region of the transit peptide of the pSSU inhibits import of internal chloroplast proteins, but has no effect on the insertion of the outer membrane proteins (Keegstra *et al.*, 1995). Thus the translocation machinery that is responsible for recognition and import of precursor proteins destined for the inside of the chloroplasts is not involved in routing many proteins to the outer chloroplast envelope.

Recently, three unusual outer membrane proteins with different targeting and membrane-integration requirements have been described. Insertion of the omp24 protein into the membrane proceeds independently of surface receptors or targeting sequence, but is stimulated by ATP (Fischer *et al.*, 1994). Two other outer envelope proteins, OEP75 and OEP86, are synthesized as higher molecular weight precursors which are processed to the mature form. Binding and processing of these precursors requires ATP and one or more surface-exposed proteinaceous components. Integration of the OEP75 precursor can be competed by the precursor of the small subunit of Rubisco (pSSU) (Tranel *et al.*, 1995). In contrast, insertion of the OEP86 precursor cannot be competed by another precursor protein destined for the internal plastid compartments (Hirsch *et al.*, 1994). OEP75 and OEP86 are components of the protein import machinery of the chloroplast envelope. A complex route of insertion and processing of their precursors may exist to ensure targeting of these crucial components with high fidelity.

## Targeting to the inner membrane of envelope

The transport of cytoplasmically synthesized proteins to the inner envelope membrane is similar to the transport of stromal and thylakoid proteins. Import and targeting of several inner membrane proteins has been studied (see Keegstra *et al.*, 1995). These proteins are synthesized as higher molecular weight precursors that are imported into chloroplasts in an energy-dependent fashion. Proteinaceous components of the envelope are required for the import of the proteins of inner envelope membrane. Synthetic peptides that inhibit import of internal proteins also inhibit import and integration of inner membrane proteins (Keegstra *et al.*, 1995). Taken together, it can be concluded that targeting of proteins to the inner membrane of chloroplast envelope uses some of the same transport components used by internal proteins.

Which part of the precursor protein contains information for targeting to the inner envelope membrane? Chimeric precursor proteins were used to address this question (Keegstra *et al.*, 1995). The maize Bt1 gene product may serve as a metabolite translocator protein in the amyloplast membrane. A chimeric protein with the pSSU transit peptide fused to the mature region of the Bt1-encoded protein was targeted to the inner envelope membrane of chloroplasts. Moreover, a chimeric protein with the Bt1 transit peptide fused to the mature LHCP was targeted to the thylakoid. These results indicate that the Bt1 transit peptide functions primarily as a stromal targeting sequence. A similar chimeric construct strategy was used to study targeting of the spinach triose phosphate/phosphate translocator and the 37 kDa protein. N-terminal extensions of both envelope membrane proteins possess a stroma-targeting function and the information for the integration into the envelope membrane is contained in the mature parts of the proteins (Brink *et al.*, 1995). At least part of the integration signal is provided by hydrophobic domains in the mature sequences since the removal of such a hydrophobic segment from the 37 kDa protein leads to mis-sorting of the protein to the stroma and the thylakoid membrane. Thus the information for targeting to the chloroplastic inner envelope membrane is contained in the mature region of the protein.

The pathway of targeting to the inner envelope membrane is an unresolved problem. Based on analogy with mitochondrial proteins, the inner membrane proteins may be targeted and sorted by a conservative sorting or by a stop-transfer mechanism that involves halting in the translocation machinery and lateral migration in the lipid bilayer. It will be interesting to identify the mechanism that distinguishes between the inner envelope membrane and thylakoids.

## Targeting to the intrachloroplast compartments

### *Transit and transfer sequences*

Most nucleus-encoded chloroplast proteins are synthesized as precursor proteins with amino-terminal leader sequence. Early studies with the pSSU demonstrated that the leader peptide is essential and sufficient for targeting and import. Recent progress in cDNA cloning and sequencing has revealed deduced amino acid sequences of a few hundred precursors of chloroplast proteins; no striking homology is detected in their length or sequence or predicted secondary structure. In a broad sense, leader peptides can be divided into two classes (Fig. 5.2). The simple leader peptides contain a transit sequence which targets proteins to chloroplast stroma. The precursors destined to inner envelope membrane, stroma and thylakoids contain simple transit sequences. In contrast, some thylakoid proteins and the lumenal proteins have a compound leader peptide containing an amino terminal stroma-targeting transit sequence that is followed by a transfer peptide for thylakoid translocation (Keegstra *et al.*, 1995). The thylakoid transfer domain shows similarity to the bacterial presequence that targets proteins across bacterial plasma membrane. The simple transit peptides are cleaved by processing peptidases in

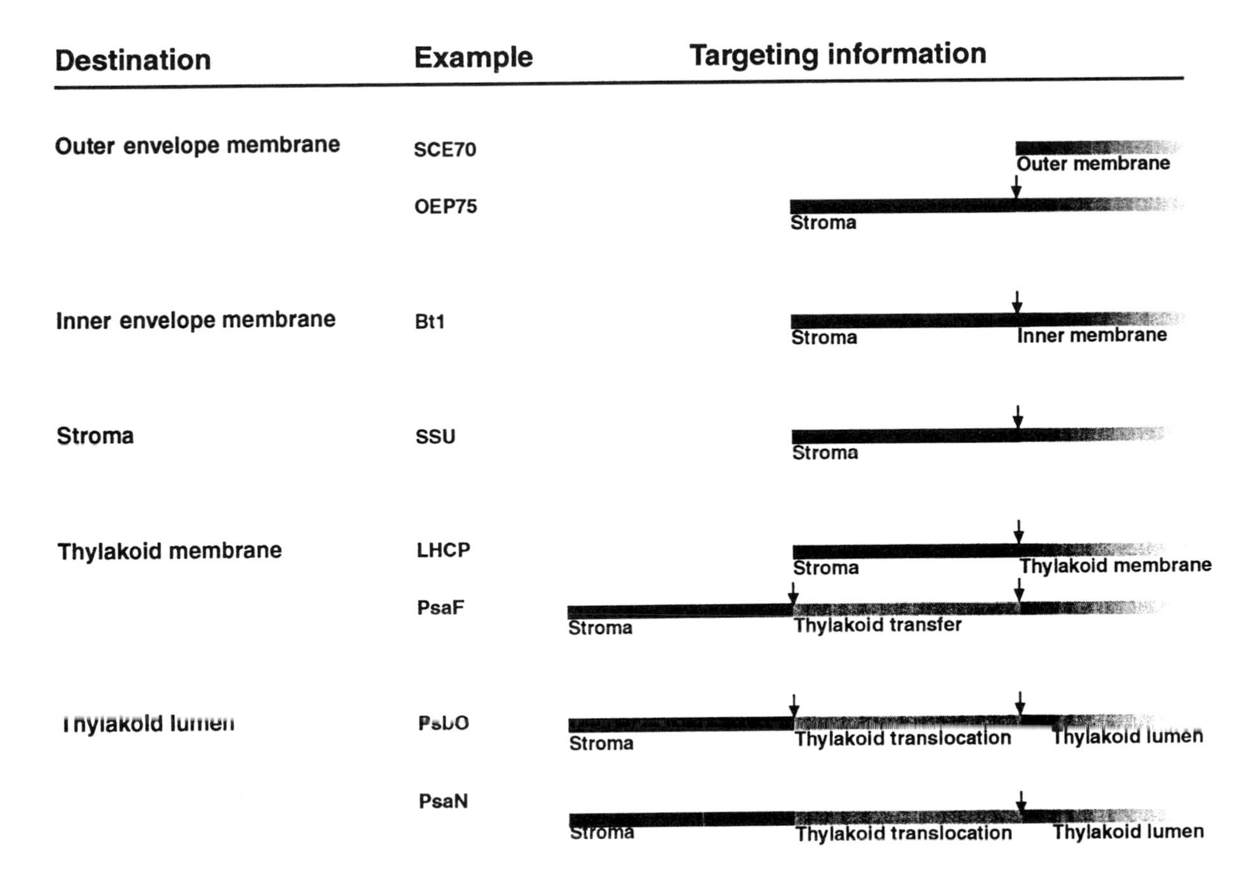


**Figure 5.2.** Transit and transfer sequences of chloroplast proteins. The diagram shows the targeting information in the domains in a chloroplast precursor. The arrow represents cleavage site for processing peptidase.

stroma. Most compound leader peptides are processed in two steps, the transit peptide is removed by stromal peptidases while the transfer sequence is cleaved by thylakoid leader peptidase. All of the transfer signals contain hydrophobic core sequences and a '-3,-1' motif reminiscent of those found in signal sequences, but the amino-terminal regions of the transfer signals of the 23 and 33 kDa PSII proteins are both longer and more highly charged. The net charge of each amino-terminal region of the transfer sequences is +1, including the amino-terminal amino group.

### *Structure–function relations in transit peptides*

Several approaches have been used to analyse the information content of a chloroplast transit sequence. Chemically synthesized transit peptides or its parts can be used in structural analysis and to study their ability to inhibit binding and import of the authentic precursor molecules. Twenty-residue synthetic peptides corresponding to regions of the transit peptide of the pSSU have been used as competitive inhibitors for the binding and translocation of precursor proteins into chloroplasts. These studies showed that the central regions of the transit peptide of pSSU mediate binding to the chloroplastic surface, whereas the ends of this transit peptide are more important for translocation across the envelope (Keegstra *et al.*, 1995). Analysis of substitution and deletion mutants of the ferredoxin transit sequence has revealed distinct functional domains (Pilon *et al.*, 1995). The N-terminal part and the C-terminal part are important for targeting. In addition, the C-terminal region is required for processing. The middle domain is important for translocation but not for the initial interaction with the envelope. In addition, the N-terminus is mainly involved in the insertion into mono-galactolipid-containing lipid surfaces whereas the C-terminus mediates the recognition of negatively charged lipids. These studies unravel a framework of a general structural design of transit sequences.

## IMPORT OF PROTEINS INTO CHLOROPLASTS

In the late 1970s, Chua and colleagues demonstrated post-translational import of the precursors for chloroplast proteins. They also pioneered *in vitro* reconstitution assay for studying protein import into chloroplasts. In this reaction, isolated intact chloroplasts are incubated with labeled precursor proteins that are synthesized by *in vitro* translation of polyA-RNA fraction. Subsequently, other researchers modified and improved the protein uptake protocols by using specifically labeled precursors that are derived from cloned genes. Recently many researchers have started to use pure, over-expressed precursor proteins. The *in vitro* uptake assays, coupled with biochemical and immunological methods, have led to characterization of physiological requirements for protein uptake and to identification of many components of the chloroplast import machinery (Schnell, 1995).

### Steps in protein import

Chloroplast protein import can be divided into two steps (Schnell, 1995). First, the cytosolic precursor specifically associates with the outer membrane in a high affinity interaction forming an early import intermediate. The establishment of this interaction is mediated by the transit sequence and requires hydrolysis of nucleoside triphosphates ($\sim$100 μM). In the second stage, the early intermediate is translocated into stroma across both envelope membranes at the contact sites. This process requires ATP (>1 mM), but a membrane potential is not involved.

### Machinery in envelope

Chloroplast envelope proteins play a major role in modulating the vectorial flow of proteins across the membrane. The import of proteins into the chloroplast requires the close collaboration of both the outer envelope and the inner envelope membranes. Competition experiments using different precursors or chemically synthesized partial transit peptides have shown the existence of a single common mechanism of envelope transport for all cytosolically synthesized precursors that are destined for the internal compartments of the chloroplast (see Keegstra *et al.*, 1995).

Many independent investigations have resulted in identification of four outer-membrane components of the import machinery (Table 5.1, Fig. 5.3; see Keegstra *et al.* (1995); Schnell, (1995) for specific citations). Schnell and colleagues used a tagged early intermediate and immunoaffinity purification to isolate a precursor-complex containing OEP43, OEP86, OEP76 and Hsp70. All four of these proteins also co-sediment with an early intermediate following detergent treatment of envelope membranes. OEP75 and OEP86 can also be cross-linked with the precursor proteins, indicating an intimate association. Thus the components of outer envelope membrane form a translocation complex that participates in precursor binding and the early stages of translocation. Deduced primary sequences of these outer membrane components have provided insight into their function and the mechanics of import. OEP34 and OEP86 are tightly anchored into the outer membrane with a GTP-binding domain exposed on the cytosolic side. These components may be involved in the recognition of precursors on the chloroplast surface. OEP75 is an integral membrane proteins, with no significant cytosolic domain. It may form extensive beta-barrel structures in the membrane and is proposed to serve as a component of the protein-conducting channel. Incidentally bacterial porins contain beta-barrel channels for ion transport. The Hsp70 IAP is a unique member of Hsp70 family; it is tightly anchored to the outer membrane. The nature of the membrane anchor is not known. Bulk of Hsp70 IAP lies in the intermembrane space, suggesting its role in protein transport from the outer membrane to the inner membrane. It may maintain the precursor in import competent form or 'pull' it during translocation. Besides the Hsp70 IAP, a second outer membrane Hsp70 homologue, Com70 may be involved in precursor import. This chaperone is peripherally associated with the outer sur-

Table 5.1. *Components of protein import machinery in the envelope*

| Protein | Mass (kDa) | Location | Activities | Proposed function |
|---|---|---|---|---|
| Com34/OEP34/IAP34 | 34 | OM/I | GTP binding | Precursor recognition |
| Com86/OEP86/IAP86 | 86 | OM/I | GTP binding | Precursor recognition |
| Com75/OEP75/IAP75 | 75 | OM/I | | Protein conducting channel |
| Cim44 | 44 | IM | Unknown | Unknown |
| Cim97/IAP100 | 100 | IM | Unknown | Unknown |
| IAP36 | 36 | Unknown | Unknown | |
| Hsp70 | 75 | OM/P | Hsp70 homologue | Chaperone |
| Com70 | 72 | OM/P | Hsp70 homologue | Chaperone |

OEP, outer envelope protein; IAP, import intermediate associated proteins; OM, outer membrane; IM, inner membrane; I, integral membrane protein; P, peripheral protein; Cim, chloroplast inner membrane; Com, chloroplast outer membrane.

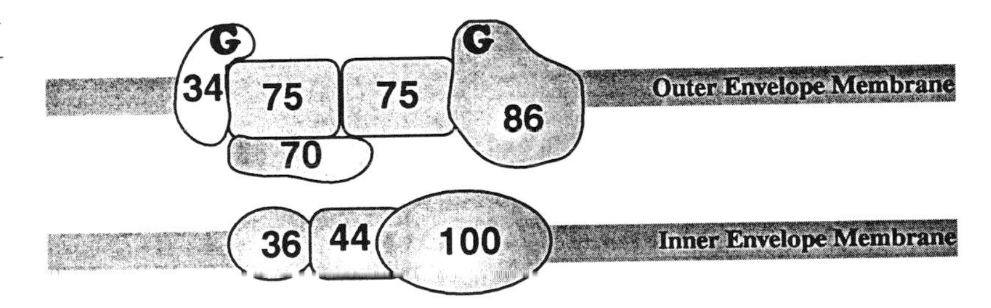


**Figure 5.3.** Components of the protein import machinery in the chloroplast envelope. The model shows locations of the proteins. Each protein is indicated by its size and the stoichiometry is not determined. G indicates GTP-binding domain of two translocase components. (Based on Schnell, 1995.)

face of chloroplasts and comes in physical proximity with the import intermediates.

Our understanding of the inner membrane components of the translocation machinery is less advanced than that of the outer membrane components. Inner membrane is inaccessible for biochemical manipulations such as cross-linking and the late intermediates that span inner membrane are difficult to isolate. At least three inner-membrane proteins, Cim97, Cim44 and IAP36, function in protein import. Their roles in protein import are not clear; they may be involved in the later stages of polypeptide translocation.

## PROTEOLYTIC PROCESSING OF CHLOROPLAST PROTEINS

The precursor proteins translocated across the chloroplast envelope are proteolytically processed to remove the leader peptide. The processing enzyme for removal of the stroma-targeting sequence is located in the stroma while a thylakoid-bound enzyme removes the presequence required for translocation across thylakoids (see Cohen *et al.*, (1995). A distinct C-terminal processing protease in thylakoid lumen removes a C-terminal extension of the D1 protein of PSII.

### Stromal processing proteases

Transit peptides of stromal and thylakoid proteins are removed after import by a stromal processing peptidase (SPP); the presequences of thylakoid lumen proteins are processed by SPP at intermediate sites prior to transport of these proteins across the thylakoid membrane. A loosely conserved consensus motif (Val/Ile)-X-(Ala/Cys) decreased Ala is found at the cleavage site of the transit peptides (von Heijne, 1995). SPP is a highly specific enzyme; the influence of the cleavage site residues on the SPP reaction mechanism has been examined by introducing mutations within the presequence of the lumenal 33 kDa PSII protein. These studies indicated that the identity of the -1 residue, within the context of a given presequence, is important in determining processing efficiency, but that the site of cleavage is specified by other determinants near the cleavage site (Bassham *et al.*, 1994). Structurally related proteins of 145 and 143 kDa copurify with a soluble chloroplast processing enzyme that cleaves the pLHCP, pSSU and acyl carrier protein. The 145 and 143 kDa proteins have not been found as a heterodimer and thus may represent functionally independent isoforms encoded by separate genes. Antibodies raised against the 145/143 kDa doublet were used to isolate a cDNA clone coding for a 140 kDa polypeptide that contains a transit peptide (VanderVere *et al.*, 1995). This protein also contains a His-Xaa-Xaa-Glu-His zinc-binding motif that is conserved in a recently recognized family of metalloendopeptidases, which includes *Escherichia coli* protease III, insulin-degrading enzyme, and subunit beta of the mitochondrial processing peptidase (VanderVere *et al.*, 1995).

## Thylakoid processing protease

In addition to the stromal processing protease, a thylakoid processing protease (TPP) is required for complete maturation of the lumenal proteins. The reaction specificity of this enzyme is very similar to those of signal peptidases located in the endoplasmic reticulum and bacterial plasma membrane. The reaction mechanism of the thylakoidal peptidase has been investigated by substituting a variety of amino acids for the alanine residues at the -3 and -1 positions of a thylakoid lumen protein precursor (Shackleton & Robinson, 1991). Small neutral side chains play a critical role in defining the thylakoidal processing peptidase cleavage site. However, the requirements of the thylakoidal enzyme at these sites are significantly more restrictive than those of the bacterial or endoplasmic reticulum peptidases.

## Carboxyl terminal processing peptidase

The D1 subunit of PSII is synthesized by thylakoid-bound ribosomes on the stromal surface of membranes as a precursor protein, which has a short C-terminal extension. This part of the protein is translocated immediately after synthesis into the lumenal space and then it is excised enzymatically. As shown by previous studies of the LF-1 mutant of *Scenedesmus obliquues* and of genetically engineered mutants of *Synechocystis* sp. PCC 6803, this C-terminal cleavage process is essential for the evolution of oxygen in photosynthesis, although the biological function(s) of the carboxyl-terminal extension of the D1 protein is still obscure. Without the cleavage and removal of the extension, PSII remains inactive due to the inability to incorporate Mn. The amino acids upstream of the cleavage site are mostly conserved, whereas the carboxyl-terminal extension, which is cleaved off by the processing protease, is variable both in terms of its amino acid sequence and its length.

The protease involved in the cleavage of the carboxyl-terminal extension of pD1 has been extracted and purified from spinach, pea, and *Scenedesmus*, and its molecular and enzymatic properties have been analysed (Bowyer *et al.*, 1994). A gene (*ctpA*) that seems to correspond to this enzyme has recently been identified in a cyanobacterium and sequenced (Shestakov *et al.*, 1994). However, details of the characteristics and the mode of action of this enzyme have not been fully elucidated. For example, the inhibitors of this protease are very different from those of most proteases and the enzyme is now believed to be a new type of protease.

# ASSEMBLY OF SOLUBLE PROTEINS: INVOLVEMENT OF CHAPERONES

Molecular chaperones, prolyl and disulfide isomerases, and cofactor addition processes are involved in the correct folding of soluble proteins in chloroplast stroma and lumen. Distinct classes of peptidylprolyl *cis,trans*-isomerase are present in the cytosol, mitochondria, and chloroplast (Breiman *et al.*, 1992). Cyclosporin A-sensitive prolyl isomerase in the chloroplast is also localized to the thylakoids, which is suggestive of its function in the folding of membrane proteins.

Folding of newly synthesized polypeptides in the crowded cellular environment requires the assistance of molecular chaperone proteins (Schatz & Dobberstein, 1996). Chaperones of the Hsp70 class and their partner proteins interact with nascent polypeptide chains on ribosomes and prevent their premature (mis)folding at least until a domain capable of forming a stable structure is synthesized. For many proteins, completion of folding requires the subsequent interaction with one of the large oligomeric ring-shaped proteins of the chaperonin family, which is composed of the GroEL-like proteins in eubacteria, mitochondria, and chloroplasts (Gatenby & Viitanen, 1994). These proteins bind partially folded polypeptide and promote folding by ATP-dependent cycles of release and rebinding. In these reactions, molecular chaperones interact predominantly with the hydrophobic surfaces exposed by non-native polypeptides, thereby preventing incorrect folding and aggregation.

Proteins in stroma need molecular chaperones for correct assembly (see Gatenby & Viitanen (1994) for specific citations). Prior to its recognition as the chloroplast chaperonin (ch-Cpn60), 'the large subunit binding protein' was implicated in the assembly of the higher plant Rubisco. Assembly occurs in the chloroplast stroma, following post-translational import of the small subunits. However, numerous studies have shown that the holoenzyme does not assemble spontaneously. Indeed, the nascent large subunits initially form a stable complex with the ch-Cpn60. Then, in a complicated and poorly understood set of reactions, the bound large subunits are discharged in an ATP-dependent manner and are subsequently incorporated into the Rubisco holoenzyme. The chloroplast chaperonins play a prominent role in the folding of many other plastid proteins. However, a detailed characterization of the assisted-folding reaction mediated by these proteins is currently lacking.

Chaperonin 60 and chaperonin 10 (cpn 10) are key cellular components in numerous folding pathways leading to biologically active proteins (Gatenby & Viitanen, 1994). Some features of the higher plant chloroplast chaperonins clearly distinguish them from their bacterial and mitochondrial counterparts. First, the ch-cpn60 consists of roughly equal amounts of two different subunits, that are no more similar to each other than they are to GroEL (Gatenby & Viitanen, 1994). It is not known whether different subunits reside in the same or different tetradecamers. Secondly, the ch-cpn10 subunit is nearly twice the size of other GroES homologues (Baneyx *et al.*, 1995). It consists of two complete cpn10 molecules that are fused 'head-to-tail' to form a single protein. Thirdly, in contrast to GroEL and its mitochondrial counterparts, the assisted-folding reaction mediated by the ch-cpn60 does not require K ions (Viitanen *et al.*, 1995).

Besides chaperonins, the chloroplast stroma also contain HSP70 homologues, but their role in the biogenesis of chloroplast proteins is not characterized. Another chaperone found in chloroplast is homologous to the 54 kDa protein of signal recognition particle; this protein is involved in integration of thylakoid proteins (Li *et al.*, 1995).

## PROTEIN ASSEMBLY IN THYLAKOIDS

A large proportion of chloroplast proteins are found in thylakoids. Most nuclear-encoded thylakoid proteins are synthesized as precursors with stromal-targeting transit sequences and are processed to mature form by stromal processing peptidases. Upon integration into thylakoids, the thylakoid proteins may be assembled into multiprotein complexes and may bind cofactors before they function in photosynthesis, translocation or other thylakoid-associated functions. Due to their abundance, integration and assembly of the photosynthetic thylakoid proteins have been extensively studied and are discussed here.

### Membrane integration

As in any biological membrane, the proteins of the thylakoids can be either integral or peripheral membrane proteins. PsaD and PsaE, two peripheral subunits of PSI incorporate spontaneously into the membranes and assemble into the PSI complex, without needing energy input, assistance from molecular chaperones or membrane potential. Thus the primary sequence information in the protein and the topological information in the previously assembled membrane proteins is sufficient for incorporation of these peripheral proteins. In contrast, the intrinsic proteins require energy, chaperones and precise structural information for stable integration into thylakoids.

#### *Information in the protein sequence and structure*

Determination of the structure of the bacterial photosynthetic reaction centre, bacterial porins, the light harvesting complexes of PSII and of purple bacteria, channel-forming colicins, and bacteriorhodopsin allows a comparison of the basic structural features of integral membrane proteins. Helix–helix recognition, hydrophobic and electrostatic interactions, and distribution of charged residues have major roles in determining the specificity and mechanism of membrane integration. In addition, lipid–protein interactions, that are yet to be elucidated for thylakoid proteins, are also crucial in determining stable integration of transmembrane domains into lipid bilayer. The localization of polar and electrostatic residues in the polypeptide chain have a major role in determining topography of membrane proteins (von Heijne, 1995). The topology of hydrophobic intramembrane proteins is characterized by a statistical asymmetry in the distribution of positively charged residues on the two sides of the membrane, the 'inside- or *cis*-positive rule'. Domains facing the cytoplasm are generally enriched in basic residues, whereas regions of proteins that are translocated through the membrane and face the periplasm are largely devoid of basic residues. Thus, segments containing an excess of positively charged residues tend not to cross the lipid bilayer. This correlation was found to be true for proteins of the endoplasmic reticulum, the bacterial plasma membrane, the inner mitochondrial membrane and the chloroplast thylakoid membrane. The positively charged domains of thylakoid protein do not cross lipid bilayer into the thylakoid lumen and remain exposed on the stromal side.

### *Physiological requirements*

Most studies on membrane integration of thylakoid proteins have been performed with pLHCP, the apo-protein of the major light harvesting antenna of PSII (see Nechushtai *et al.* (1995) for specific citations). The presence of Mg-ATP is essential for the insertion of the protein into the thylakoid membranes. When the import of pLHCP into intact chloroplasts was performed in the presence of uncouplers, significant amounts of the imported pLHCP accumulated in the stromal fraction and did not insert into the membrane. The accumulation of LHCP in the stroma was enhanced by reducing the amounts of ATP in the reaction mixture. Dissipation of the proton-motive force during chloroplast import reactions also partially inhibited the integration of LHCP into the thylakoids. Thus the integration of LHCP into the thylakoids requires ATP hydrolysis and is enhanced by a trans-thylakoid pH gradient.

### *Molecular chaperone assists thylakoid integration*

The integration into the membrane is often assisted by molecular chaperones. A protein factor from the chloroplast stroma was found to be essential for the integration of pLHCP into isolated thylakoids. The apoprotein is assembled into a soluble 'transit complex' before its integration into the thylakoid membrane. The identity and function of the stromal molecular chaperone has received extensive attention in recent years (Nechushtai *et al.*, 1995). A chloroplast homologue of the 54 kDa subunit (54CP) of the mammalian signal recognition particle is essential for transit complex formation, is present in the complex, and is required for LHCP integration into the thylakoid membrane (Li *et al.*, 1995). 54CP functions post-translationally as a molecular chaperone and potentially pilots LHCP to the thylakoids. Thus one of several pathways for protein routing to the thylakoids is homologous to the SRP pathway and points to a common evolutionary origin for the protein transport systems of the endoplasmic reticulum and the thylakoid membrane (Li *et al.*, 1995).

## Assembly of multi-protein complexes

Many thylakoid proteins exist as multi-heteromeric protein complexes. Mode of assembly of these complexes can be deduced from *in vitro* reconstitution, from the sequence of appearance of proteins during chloroplast development, and from the phenotype of subunit-deficient algal and cyanobacterial mutants. These studies have shown that the thylakoid protein complexes assemble by different modes. PSI is assembled in a *step-by-step* fashion (Nechushtai *et al.*, 1995). Initially, PsaA and PsaB form a heterodimer; its stability is dependent on the correctly formed $F_X$ cluster. Similarly, cubane iron–sulphur clusters at both $F_A$ and $F_B$ are needed for assembly of PsaC into PSI. Assembly of PsaD and PsaE is dependent on the presence of PsaC in the complex. The sequence in which the hydrophobic proteins and cofactors are added, is not known. Assembly of PSII and LHCII involves *lateral migration* of proteins. *In vitro* reconstitution experiments have shown that LHCP is integrated into the stromal lamellae in a 'free' form which unidirectionally migrates to the granal lamellae and assembles into the pigmented LHCII complex.

## Addition of cofactors

Many thylakoid proteins associate with cofactors. Being the major pigment in thylakoids, Chl binding has been extensively studied by *in vitro* reconstitutions and in mutants lacking specific pigments. Several studies indicate that the pigments bind to the apo-protein after their integration into membrane. LHCP is capable of inserting into the thylakoid in the apo-protein form, i.e. without attached pigments. When LHCP is expressed in *E. coli* as a fusion protein with a bacterial signal peptide, LHCP is inserted into the inner bacterial envelope membrane without attached pigments.

Binding to Chl and other photosynthetic pigments stabilizes the apo-proteins (see (Nechushtai *et al.*, 1995). Mutants of *Chlamydomonas, Arabidopsis*, maize, and barley that are blocked in specific stages of Chl and carotenoids biosynthesis do not contain the apo-proteins of the LHC complexes (of both PSI and PSII), suggesting a stabilizing effect of both the carotenoids as well as the Chl molecules on the apo-protein of the Chl-binding proteins. Incubation of over-expressed LHCP with isolated pigments results in the formation of pigment–protein complexes that have spectroscopic properties of LHCII. The presence of Chl*a*, Chl*b*, as well as two out of three xanthophylls of LHCII, are essential for reconstitution. Moreover, the Chl*a* to Chl*b* ratio significantly affects the reconstitution efficiency. Pigment-binding leads to change in conformation of LHCP, from about 20% to about 60% α-helical content. In the recently published LHCII structure, two lutein molecules form cross-braces in the centre of the protein molecule. This binding may explain why they are essential for *in vitro* reconstitution of LHCII. The pigments probably orientate the protein helices, and form a compact protein–pigment structure in LHCII.

## TRANSLOCATION OF PROTEINS ACROSS THYLAKOIDS

Thylakoid lumen occupies a relatively minor fraction of chloroplast volume, yet is complex and rich in its protein complement. Most lumenal proteins are synthesized in the cytosol and targeted across all three chloroplast membranes into the thylakoid lumen. This complex import pathway can be broadly divided into two phases, the first of which involves the transport of a cytosolically synthesized precursor protein into the stroma (see p. 69), after which the stromal form is transported across the thylakoid membrane into the lumenal space. The two translocation events are directed by distinct envelope transit and thylakoid transfer signals in the presequences of lumenal proteins.

### Multiple pathways with distinct requirements

The available evidence suggests that most proteins are transported across the envelope membranes by a common mechanism, but recent studies have pointed to the operation of at least three completely different mechanisms for protein transport across the thylakoid membrane (Cline *et al.*, 1993; Robinson & Klosgen, 1994). A subset, including plastocyanin (PC) and the extrinsic 33 kDa PSII protein (PsbO) and the PsaF subunit of PSI require the presence of a stromal protein factor and nucleoside triphosphates for their transport across the thylakoid membrane. In contrast, neither of these elements is required for transport of the extrinsic 23 and 16 kDa PSII proteins, PSII subunit PsbT, and PSI subunit PsaN; these proteins appear to be dependent only on the transthylakoidal pH. In contrast, both 33K and PC can be transported across the thylakoid membrane of intact chloroplasts in the complete absence of a pH gradient. A third set of requirements are found for the CFoII intrinsic subunit of the ATP synthase. This protein is synthesized with a bipartite, lumen-targeting presequence that is shown to possess a terminal cleavage site for the thylakoidal processing peptidase, but no intermediate site for the stromal processing peptidase (Michl *et al.*, 1994). The integration of CFoII into the pea thylakoid membranes does not require the presence of either stromal extracts or nucleoside triphosphates. The thylakoidal delta pH does not play a significant role in the integration mechanism. In each of these respects, the requirements for CFoII integration differ notably from those determined for integration of LHCP. The integration mechanism also differs significantly from the two mechanisms involved in the translocation of lumenal proteins across the thylakoid membrane. The role of thylakoidal delta pH depends on the identity of the passenger protein and the concentration of ATP. The delta pH may be obligatory when the passenger protein is abnormally difficult to translocate, possibly due to the folding of the polypeptide chain (Mant *et al.*, 1995).

These differing requirements reflect the operation of separate thylakoidal protein translocation systems. Chimeric proteins were used to show that the presequences of

23K and 16K are able to direct translocation of mature PC solely by the 23K/16K-type pathway, indicating that the presequences of 23K and PC contain different types of targeting signal which almost certainly specify translocation by distinct translocases (Robinson & Klosgen, 1994). PsbO and PC compete for transport across the thylakoid membrane, and 23K competes with 16K, but the two groups do not compete with each other. Recent studies have shown that transfer signals for the pH-driven system contain a twin arginine motif which is critical for translocation across the thylakoid membrane (Chaddock *et al.*, 1995).

The origins of the pH-driven translocation mechanism are presently unclear, but the ATP-dependent, azide-sensitive mechanism may have been inherited from a sec-related mechanism in the cyanobacterial-type progenitor of the chloroplast. 33K and PC are present in cyanobacteria, where they are synthesized with presequences resembling signal peptides and transfer signals. Furthermore, the transport of these proteins across the chloroplast thylakoid membrane is blocked by azide (Knott & Robinson, 1994), a known inhibitor of bacterial SecA proteins, and a SecA homologue has been shown to participate in the translocation process (Nakai *et al.*, 1994; Yuan *et al.*, 1994). None of the proteins targeted by the other, ATP-independent pathway has been found in cyanobacteria, and it has therefore been suggested that their appearance in chloroplasts may have been accompanied by the emergence of a novel mechanism for their translocation across the thylakoid membrane.

### Machinery

Genetic and biochemical studies have identified several proteins, encoded by *sec* genes, that are required for the translocation of proteins across the plasma membrane in bacteria (Table 5.2). These proteins include soluble/peripheral components (SecB and SecA) and a membrane-bound translocase complex that includes SecY and SecE. Other integral membrane proteins, SecD and SecF, appear to participate in the later stages of the translocation process. Several lines of evidence have indicated the existence of a Sec-type machinery in the thylakoid membrane (Table 5.2). The *secA* and *secY* homologues have also been found in the plastid genomes of red algae and cyanophytes. Similarly, cDNA clones for SecA and SecY homologues from plant plastids have also been cloned. Other components of the thylakoidal Sec machinery still need to be characterized.

## DEGRADATION OF CHLOROPLAST PROTEINS

Protein degradation is an essential component of normal cellular homeostasis. During metabolism in a cell, proteins are damaged by metabolic byproducts or abiotic stresses and are degraded. Protein degradation is also essential to remove regulatory proteins or enzymes that are no longer necessary. Thus protein degradation is essential for the survival and death of a cell and for the biogenesis and degeneration of its membrane-bound compartments. The major sites of intracellular proteolysis are lysosomes and cytosol. Proteins in semi-autonomous organelles are presumably degraded inside those organelles.

Protein degradation is a major factor regulating levels of many chloroplast proteins (Callis, 1995). The substrates include the unassembled subunits of multimeric proteins, multimeric complexes lacking or containing a single mutated polypeptide and the apoprotein of unassembled holoproteins. Environmental stresses, nutrient deficiency and senescence causes rapid degradation of chloroplast proteins. Oxidative stress causes rapid degradation of Rubisco, while high temperature stress stimulates protein degradation in chloroplasts. Many researchers have extensively studied damage, degradation, and replacement of the D1 protein of PSII. This protein is rapidly turned over following damage to PSII (Aro, Virgin & Andersson, 1993). A light-dependent charge separation or UV-B radiation may damage D1. The damage may cause conformational changes which in turn may lead to the degradation of damaged D1 protein. Degradation of the D1- and D2 proteins of PSII in higher plants is regulated by reversible phosphorylation. Despite the knowledge of the regulation of D1 turnover, the mechanism of its fragmentation is still not clear. Degradation has been reported to be enzymatic, as serine-type protease inhibitors can block D1 degradation. Proteolysis of damaged D1 protein allows incorporation of newly synthesized D1 protein into PSII.

Proteolytic systems in chloroplasts are not well understood. ATP is required for protein degradation in

Table 5.2. *Components of translocation machinery in thylakoids*

| Component | In *E. coli* membranes: Location | Function | Evidence for the presence and role in thylakoids |
|---|---|---|---|
| SecA | Peripheral | (i) Membrane receptor and (ii) protein motor functions | Gene cloned from plants (GenBank # X82404) and algae (Z21642, Z35718, X65961)<br>Biochemical evidence (Nakai *et al.*, 1994; Yuan *et al.*, 1994) |
| SecY<br>SecE<br>SecG | Intrinsic | (i) translocation ATPase activity, (ii) preprotein translocation, (iii) capacity for SecA binding, and (iv) formation of the membrane-inserted form of SecA | Sec Y gene cloned from plant (GenBank # U37247) and algae (X64732, S38893, X74773, Z50047, Z36235, X62348, Z67753)<br>SecE and Sec G not demonstrated |
| SecD<br>SecF | Intrinsic | Late steps in translocation | Not demonstrated |

chloroplasts. Specific proteases that are involved in the degradation of chloroplast proteins are just being identified. Since Sec-type prokaryotic protein translocation apparatus is conserved in chloroplasts, it is reasonable to assume that many of the major prokaryotic proteases may also function in plastids. Among the proteases found in prokaryotes, Clp and Lon proteases require ATP for their function. Clp protease universally present in prokaryotes and in eukaryotic organelles. It contains the 21 kDa ClpP subunit with proteolytic activity and another regulatory subunit (ClpA, ClpB, ClpC or ClpX) with ATPase activity (Chung, 1993). ClpP contains the serine active site. It functions as a tetradecamer with an axial pore (Flanagan *et al.*, 1995). By itself, it has a peptidase activity, but its transient association with the regulatory subunit enables it to hydrolyse proteins (Chung, 1993). In plants, the *clpP* gene is present in both the chloroplast genome and the nucleus (GenBank L38581, L09547). ClpA, a 84 kDa protein, is a conserved molecular chaperone and is implicated in protein degradation under normal conditions. ClpB, a 94 kDa protein, is also conserved among prokaryotes and eukaryotes, where it is located in the cytoplasm and is known as HSP104. Plant homologues of ClpA and ClpB are known (GenBank X75328, D17582, U13949), but they are presumably present in the cytoplasm. ClpC is the chloroplast-localized regulatory subunit of the Clp protease.

The stroma of higher plant plastids contain ClpP and ClpC (Shanklin, DeWitt & Flanagan, 1995). In addition, many other proteolytic activities have been demonstrated in the chloroplast, but these proteases have not been purified and identified. A stromal zinc protease from pea chloroplast can hydrolyse the large subunit of native Rubisco (Bushnell, Bushnell & Jagendorf, 1993). Similarly, membrane-bound proteases that may degrade incompletely folded light-harvesting complex apoprotein, are partially characterized from chloroplasts of *C. reinhardtii* (Hoober & Hughes, 1992) and bean (Anastassiou & Argyroudiakoyunoglou, 1995). The role of Clp or any other prokaryotic protease in degradation of membrane proteins has not been demonstrated.

## ACKNOWLEDGMENTS

Research in the author's laboratory is supported by grants from NSF (MCB9405325), USDA-NRICGP (92–37306–7661), NIH (GM53104-R01) and the US-Israel BARD Fund (IS-2191–92RC).

## REFERENCES

Anastassiou, R. & Argyroudiakoyunoglou, J. H. (1995). Thylakoid-bound proteolytic activity against LHC II apoprotein in bean. *Photosynthesis Research*, **43**, 241–50.

Aro, E. M., Virgin, I. & Andersson, B. (1993). Photoinhibition of Photosystem II. Inactivation, protein damage and turnover. *Biochimica et Biophysica Acta*, **1143**, 113–34.

Baneyx, F., Bertsch, U., Kalbach, C. E., van der Vies, S. M., Soll, J. & Gatenby, A. A. (1995). Spinach chloroplast cpn21 co-chaperonin possesses two functional domains fused together in a toroidal structure and exhibits nucleotide-dependent binding to plastid chaperonin 60. *Journal of Biological Chemistry*, **270**, 10695–702.

Bassham, D. C., Creighton, A. M., Karnauchov, I., Herrmann, R. G., Klosgen, R. B. & Robinson, C. (1994). Mutations at the stromal processing peptidase cleavage site of a thylakoid lumen protein precursor affect the rate of processing but not the fidelity. *Journal of Biological Chemistry*, **269**, 16062–6.

Bowyer, J. R., Packer, J. C. L., McCormack, B. A., Whitelegge, J. P., Robison, C. & Taylor, M. A. (1994). Carboxyl-terminal processing of the D1 protein and photoactivation of water-splitting in photosystem II. Partial purification and characterization of the processing enzyme from *Scenedesmus obliquus* and *Pisum sativum*. *Journal of Biological Chemistry*, **267**, 5424–33.

Breiman, A., Fawcett, T. W., Ghirardi, M. L. & Mattoo, A. K. (1992). Plant organelles contain distinct peptidylprolyl *cis,trans*-isomerases. *Journal of Biological Chemistry*, **267**, 21293–6.

Brink, S., Flugge, U. I., Chaumont, F., Boutry, M., Emmermann, M., Schmitz, U., Becker, K. & Pfanner, N. (1994). Preproteins of chloroplast envelope inner membrane contain targeting information for receptor-dependent import into fungal mitochondria. *Journal of Biological Chemistry*, **269**, 16478–85.

Brink, S., Fischer, K., Klosgen, R. B. & Flugge, U. I. (1995). Sorting of nuclear-encoded chloroplast membrane proteins to the envelope and the thylakoid membrane. *Journal of Biological Chemistry*, **270**, 20808–15.

Bushnell, T. P., Bushnell, D. & Jagendorf, A. T. (1993). A purified zinc protease of pea chloroplasts, EP1, degrades the large subunit of ribulose-1,5-bisphosphate carboxylase/oxygenase. *Plant Physiology*, **103**, 585–91.

Callis, J. (1995). Regulation of protein degradation. *Plant Cell*, **7**, 845–57.

Chaddock, A. M., Mant, A., Karnauchov, I., Brink, S., Herrmann, R. G., Klosgen, R. B. & Robinson, C. (1995). A new type of signal peptide: central role of a twin-arginine motif in transfer signals for the Delta pH-dependent thylakoidal protein translocase. *EMBO Journal*, **14**, 2715–22.

Chung, C. H. (1993). Proteases in E. coli. *Nature*, **262**, 372–4.

Cline, K., Henry, R., Li, C. & Yuan, J. (1993). Multiple pathways for protein transport into or across the thylakoid membrane. *EMBO Journal*, **12**, 4105–14.

Cohen, Y., Yalovsky, S. & Nechushtai, R. (1995). Integration and assembly of photosynthetic protein complexes in chloroplast thylakoid membranes. *Biochimica et Biophysica Acta*, **1241**, 1–30.

Fischer, K., Weber, A., Arbinger, B., Brink, S., Eckerskorn, C. & Flugge, U. I. (1994). The 24 kDa outer envelope membrane protein from spinach chloroplasts: molecular cloning, *in vivo* expression and import pathway of a protein with unusual properties. *Plant Molecular Biology*, **25**, 167–77.

Flanagan, J. M., Wall, J. S., Capel, M. S., Schneider, D. K. & Shanklin, J. (1995). Scanning transmission electron microscopy and small-angle scattering provide evidence that native *Escherichia coli* ClpP is a tetradecamer with an axial pore. *Biochemistry*, **34**, 10910–17.

Gatenby, A. A. & Viitanen, P. V. (1994). Structural and functional aspects of chaperonin mediated protein folding. *Annual Review of Plant Physiology and Plant Molecular Biology*, **45**, 469–91.

Hirsch, S., Muckel, E., Heemeyer, F., von Heijne, G. & Soll, J. (1994). A receptor component of the chloroplast protein translocation machinery. *Science*, **266**, 1989–92.

Hoober, J. K. & Hughes, M. J. (1992). Purification and characterization of a membrane-bound protease from *Chlamydomonas reinhardtii. Plant Physiology*, **99**, 932–7.

Huang, J., Hack, E., Thornburg, R. W. & Myers, A. M. (1990). A yeast mitochondrial leader peptide functions *in vivo* as a dual targeting signal for both chloroplasts and mitochondria. *Plant Cell*, **2**, 1249–60.

Keegstra, K., Bruce, B., Hurley, M., Li, H. M. & Perry, S. E. (1995). Targeting of proteins into chloroplasts. *Physiologia Plantarum*, **93**, 157–62.

Knott, T. G. & Robinson, C. (1994). The secA inhibitor, azide, reversibly blocks the translocation of a subset of proteins across the chloroplast thylakoid membrane. *Journal of Biological Chemistry*, **269**, 7843–6.

Li, X., Henry, R., Yuan, J., Cline, K. & Hoffman, N. E. (1995). A chloroplast homologue of the signal recognition particle subunit SRP54 is involved in the posttranslational integration of a protein into thylakoid membranes. *Proceedings of the National Academy of Sciences, USA*, **92**, 3789–93.

Mant, A., Schmidt, I., Herrmann, R. G., Robinson, C. & Klosgen, R. B. (1995). Sec-dependent thylakoid protein translocation. Delta pH requirement is dictated by passenger protein and ATP concentration. *Journal of Biological Chemistry*, **270**, 23275–81.

Michl, D., Robinson, C., Shackleton, J. B., Herrmann, R. G. & Klosgen, R. B. (1994). Targeting of proteins to the thylakoids by bipartite presequences: CFoII is imported by a novel, third pathway. *EMBO Journal*, **13**, 1310–17.

Nakai, M., Goto, A., Nohara, T., Sugita, D. & Endo, T.

(1994). Identification of the SecA protein homolog in pea chloroplasts and its possible involvement in thylakoidal protein transport. *Journal of Biological Chemistry*, **269**, 31338–41.

Nechushtai, R., Cohen, Y. & Chitnis, P. R. (1995). Assembly of the chlorophyll–protein complexes. *Photosynthesis Research*, **44**, 165–81.

Pilon, M., Wienk, H., Sips, W., de Swaaf, M., Talboom, I., van 't Hof, R., de Korte-Kool, G., Demel, R., Weisbeek, P. & de Kruijff, B. (1995). Functional domains of the ferredoxin transit sequence involved in chloroplast import. *Journal of Biological Chemistry*, **270**, 3882–93.

Robinson, C. & Klosgen, R. B. (1994). Targeting of proteins into and across the thylakoid membrane – a multitude of mechanisms. *Plant Molecular Biology*, **2624**, 15–24.

Schatz, G. & Dobberstein, B. (1996). Common principles of protein translocation across membranes. *Science*, **271**, 1519–26.

Schnell, D. J. (1995). Shedding light on the chloroplast protein import machinery. *Cell*, **83**, 521–4.

Shackleton, J. B. & Robinson, C. (1991). Transport of proteins into chloroplasts. The thylakoidal processing peptidase is a signal-type peptidase with stringent substrate requirements at the -3 and -1 positions. *Journal of Biological Chemistry*, **266**, 12152–6.

Shanklin, J., DeWitt, N. D. & Flanagan, J. M. (1995). The stroma of higher plant plastids contain ClpP and ClpC, functional homologs of *Escherichia coli* ClpP and ClpA: an archetypal two-component ATP-dependent protease. *Plant Cell*, **7**, 1713–22.

Shestakov, S. V., Anbudurai, P. R., Stanbekova, G. E., Gadzhiev, A., Lind, L. K. & Pakrasi, H. B. (1994). Molecular cloning and characterization of the ctpA gene encoding a carboxyl-terminal processing protease. *Journal of Biological Chemistry*, **269**, 19354–9.

Tranel, P. J., Froehlich, J., Goyal, A. & Keegstra, K. (1995). A component of the chloroplastic protein import apparatus is targeted to the outer envelope membrane via a novel pathway. *EMBO Journal*, **14**, 2436–46.

VanderVere, P. S., Bennett, T. M., Oblong, J. E. & Lamppa, G. K. (1995). A chloroplast processing enzyme involved in precursor maturation shares a zinc-binding motif with a recently recognized family of metalloendopeptidases. *Proceedings of the National Academy of Sciences, USA*, **92**, 7177–81.

Viitanen, P. V., Schmidt, M., Buchner, J., Suzuki, T., Vierling, E., Dickson, R., Lorimer, G. H., Gatenby, A. & Soll, J. (1995). Functional characterization of the higher plant chloroplast chaperonins. *Journal of Biological Chemistry*, **270**, 18158–64.

von Heijne, G. (1995). Membrane protein assembly: rules of the game. *Bioessays*, **17**, 25–30.

Yuan, J., Henry, R., McCaffery, M. & Cline, K. (1994). SecA homolog in protein transport within chloroplasts: evidence for endosymbiont-derived sorting. *Science*, **266**, 796–8.

# 6 Expression and regulation of plastid genes

S. KAPOOR AND M. SUGIURA

## INTRODUCTION

Plastids are plant-specific, semi-autonomous organelles that possess their own genome. The studies on chloroplast genome started with physical mapping in mid-1970s and reached their pinnacle with the sequencing of entire chloroplast genomes of tobacco and a liverwort, *Marchantia polymorpha*. Since then several plastid genomes representing taxonomically diverse groups have been analysed and sequenced. The following is our attempt to summarize the enormous information thus generated on the structural, functional and evolutionary aspects of plastid genomes (see Sugiura, 1992, 1996).

Another area dealt with in this review is the regulation of plastid gene expression. Various aspects of plastid gene expression have been reviewed in detail elsewhere (Gillham, Boynton & Hauser, 1994; Mayfield *et al.*, 1995; Mullet, 1993). In this chapter we focus on the newly emerging ideas about the transcriptional and post-transcriptional molecular mechanisms of plastid gene regulation. These developments have been made possible by the much awaited breakthroughs like, development of *in vitro* translation and reliable transformation system in the field of plastid molecular biology. Plastids being semi-autonomous, endosymbiotic organelles depend, to a large extent on nuclear genes for several key structural and regulatory elements. This requires coordinated expression from both the genomes at all the stages of plastid development and differentiation. Therefore, the nuclear mutants impaired in plastid functions have been extremely useful in the delineation of plastid regulatory mechanisms. Hence, wherever necessary, the data on these mutants has also been included. Due to the vast scope of this subject we have limited our emphasis only on the factors/processes that are, or might be, involved in regulating the plastid gene expression, in higher plants. The references only to the major contributions or reviews published since 1991 are incorporated.

## PLASTID GENOMES AND GENES

The plastid genetic material is a circular DNA molecule with an average size of 135 kb, present in multiple copies in each organelle (Palmer, 1991). The most extreme variations in the sizes of ctDNA have been observed among algal species, i.e. from 89 kb in *Codium fragile* to 292 kb in *Chlamydomonas moewusii*. Preliminary studies have indicated that plastome size of the giant green alga *Acetabularia* could be as large as 400 kb. The non-photosynthetic plants, on the other hand, possess relatively small plastomes due to the loss of various photosynthesis related genes. These plants however, retain most of the genes for the genetic expression machinery (Wolfe *et al.*, 1992).

### Sequencing efforts

Extensive characterization of ctDNAs from diverse plant groups has helped a great deal in understanding the structure, function and evolution of plastid genomes. Entire plastome sequences are now available from dicot (*Nicotiana tabacum, Epifagus virginiana*), monocot (*Oryza sativa* and *Zea mays*), gymnosperm (*Pinus thunbergii*), bryophyte (*Marchantia polymorpha*) and algal (*Euglena gracilis, Porphyra purpurea, Odontella sinesis, Cyanophora paradoxa* and *Chlorella vulgaris*) species (see Sugiura, 1996). Most of the ctDNAs are characterized by presence of a large inverted repeat (IR) ranging from 5 to 76 kb in length (Palmer, 1991). A few dicot (*Pisum sativum, Vicia faba*) and algal (*Chlorella vulgaris, Porphyra purpurea* and *Codium fragile*) species, however, do not contain the IRs (Palmer, 1991; Wakasugi *et al.*, 1997). Recently, complete sequencing of black pine ctDNA revealed presence of an unusually small (495 bp) IR comprising 3 portions of *psb*A and tRNA-Ile gene (Wakasugi *et al.*, 1994). The IR segments divide the rest of the DNA into a large (LSC, 65–86 kb) and a small single copy (SSC, 12–53 kb) region. With our present-day knowledge of chloroplast gene expression it is difficult to comment upon the significance of the IR regions which result in duplication of the *rrn* as well as other genes included within these regions. However, existence of IRs in all the major plant groups is suggestive of their inheritance from a common ancestor of land plants and successive loss of one segment in some legumes, conifers and algal species.

## Gene classes

During the late 1970s and early 1980s, by using techniques like DNA–RNA hybridization and 2D PAGE of soluble and membrane proteins coupled with *in vivo* labeling and immunological techniques it was possible to identify major RNA and protein species in chloroplasts. The chloroplast polypeptide count was estimated to be about 500, out of which up to 80 might be products of plastidic translation as they could be labelled in isolated plastids. Later work using coupled *in vitro* transcription translation in *E. coli* cell-free extracts and cloned ctDNA sponsored hybrid select translation of specific mRNAs *in vitro*, followed by immunological detection, provided sufficient basis for identification of their genes. However, after complete sequences of plastid genomes were available, partial polypeptide sequencing followed by comparisons with the deduced amino acid sequences of unidentified ORFs became the favourite strategy for relating the structural and functional aspects of the chloroplast genome (see Palmer 1991; Sugiura 1992, 1996).

Depending on their function, plastid genes can be grouped into three categories, i.e. (i) genes for genetic system, (ii) genes for photosynthesis and (iii) a few protein coding genes, presumably involved in biosynthesis or other diverse functions. The information regarding the genes and their products from fully sequenced plastomes has been collated in Table 6.1. For gene nomenclature, CPGN recommendations have been followed.

The group of photosynthetic genes codes for about half of the thylakoid polypetides involved in the light reaction of photosynthesis. These polypeptides, together with equal number of nuclear-encoded ones, form four supra-molecular complexes, i.e. photosystem I (PSI), photosystem II (PSII), cytochrome b6/f complex and ATP synthatase. In land plants, 5–6 components of PSI, 12–13 of PSII, 4 of Cytb6/f complex and 6 subunits of the ATP synthatase complex are encoded by chloroplast *psa, psb, pet* and *atp* genes, respectively (see Table 6.1). These genes also include those for the reaction centre proteins for both the photosystems. Among the components of dark reaction only one polypeptide, i.e. large subunit (LSU) of ribulose-1,5-bisphosphate carboxylase (Rubisco), is encoded in the plastid, while the rest are nuclear encoded.

The plastome codes for a complete set of rRNAs and 27–35 species of tRNAs. In higher plants four rRNA genes are present in the following order: 16S, 23S, 4.5S and 5S. The 4.5S rRNA species is unique to land plants whereas two small rRNAs, 3S and 7S, were detected only in *Chlamydomonas reinhardtii* (see Sugiura, 1996). Of an estimated 60 polypeptides in plastid ribosomes, genes for about 21 have been found to be encoded by ctDNA in land plants (see Subramanan, 1993). These genes were mainly deduced owing to their respective homologies to *E. coli* counterparts. The *rps16, rpl21* and *rpl5* genes are unique to the angiosperms, liverwort and *Euglena*, respectively. Black pine lacks *rpl21* and *rps16*, whereas, *rps15* and *rps16* are missing in *Euglena*. Legume ctDNAs lack *rpl22*. The cyanelle and non-green algae, *Porphyra paradoxa* and *Odontella sinensis*, plastid DNAs contain the maximum number of ribosomal protein genes, i.e. 37, 46 and 44, respectively (Kowallik *et al.*, 1995; Reith & Munholland, 1995; Stirewalt *et al.*, 1995).

The nucleotide sequencing of several ctDNAs has revealed the presence of 27–35 tRNA genes. Thirty tRNA genes exist in angiosperms (see Sugiura, 1996) whereas *Euglena* and *Porphyra* possess the maximum and the minimum number of tRNA genes, respectively (Hallick *et al.*, 1993; Reith & Munholland, 1995). According to normal wobble base pairing, at least 32 tRNA species are required to translate all of the 61 codons utilized in plastids. However, no tRNA that

Table 6.1. *The genes and their products in the plastids*

| Mnemonic | Gene product | *Cyanelle* | *Porphyra* | *Odontella* | *Chlorella* | *Euglena* | *Marchantia* | *Pinus* | *Epiphagus* | *Nicotiana* | *Oryza sativa* | *Zea mays* |
|---|---|---|---|---|---|---|---|---|---|---|---|---|
| **Genes for the genetic system** | | | | | | | | | | | | |
| *rrn 4.5* | ribosomal RNA 4.5S | — | — | — | — | — | O(X2) | O | O(X2) | O(X2) | O(X2) | O(X2) |
| *rrn 5* | ribosomal RNA 5S | O(X2) | O | O(X2) | O | O(X2) | O(X2) | O | O(X2) | O(X2) | O(X2) | O(X2) |
| *rrn 16* | ribosomal RNA 16S | O(X2) | O(X2) | O(X2) | O | O(X4) | O(X2) | O | O(X2) | O(X2) | O (X2) | O(X2) |
| *rrn 23* | ribosomal RNA 23S | O(X2) | O(X2) | O(X2) | O(S) | O(X3) | O(X2) | O | O(X2) | O(X2) | O(X2) | O(X2) |
| *rps1* | ribosomal protein, 30S, S1 | — | O | — | — | — | — | — | — | — | — | — |
| *rps2* | ribosomal protein, 30S, S2 | O | O | O | O | O(S) | O | O | O | O | O | O |
| *rps3* | ribosomal protein, 30S, S3 | O | O | O | O | O(S) | O | O | O | O | O | O |
| *rps4* | ribosomal protein, 30S, S4 | O | O | O | O | O(S) | O | O | O | O | O | O |
| *rps5* | ribosomal protein, 30S, S5 | O | O | O | — | — | — | — | — | — | — | — |
| *rps6* | ribosomal protein, 30S, S6 | O | O | O | — | — | — | — | — | — | — | — |
| *rps7* | ribosomal protein, 30S, S7 | O | O | O | O | O(S) | O | — | O(X2) | O(X2) | O(X2) | O(X2) |
| *rps8* | ribosomal protein, 30S, S8 | O | O | O | O | O(S) | O | O | O | O | O | O |
| *rps9* | ribosomal protein, 30S, S9 | O | O | O | O | O | — | — | — | — | — | — |
| *rps10* | ribosomal protein, 30S, S10 | O | O | O | — | — | — | — | — | — | — | — |
| *rps11* | ribosomal protein, 30S, S11 | O | O | O | O | O(S) | O | O | O | O | O | O |
| *rps12* | ribosomal protein, 30S, S12 | O | O | O | O | O | O | O3′(S) | O3′(X2) | O3′(X2) | O3′(X2) | O3′(X2) |
| | | | | | | | | O5′ | O5′ | O5′ | O5′ | O5′ |
| *rps13* | ribosomal protein, 30S, S13 | O | O | O | — | — | — | — | — | — | — | — |
| *rps14* | ribosomal protein, 30S, S14 | O | O | O | O | O(S) | O | O | O | O | O | O |
| *rps15* | ribosomal protein, 30S, S15 | — | — | — | — | — | O | O | — | O | O(X2) | O(X2) |
| *rps16* | ribosomal protein, 30S, S16 | O | O | O | — | — | — | — | — | O | O(S) | O |
| *rps17* | ribosomal protein, 30S, S17 | O | O | O | — | — | — | — | — | — | — | — |
| *rps18* | ribosomal protein, 30S, S18 | O | O | O | O | O(S) | O | O | O | O | O | O |
| *rps19* | ribosomal protein, 30S, S19 | O | O | O | O | O(S) | O | O | O | O | O(X2) | O(X2) |
| *rps20* | ribosomal protein, 30S, S20 | O | O | O | — | — | — | — | — | — | — | — |
| *rpl1* | ribosomal protein, 50S, L1 | O | O | O | — | — | — | — | — | — | — | — |
| *rpl2* | ribosomal protein, 50S, L2 | O | O | O | O | O | O | O(S) | O(X2) | O(S)(X2) | O(S)(X2) | O(S)(X2) |
| *rpl3* | ribosomal protein, 50S, L3 | O | O | O | — | — | — | — | — | — | — | — |
| *rpl4* | ribosomal protein, 50S, L4 | — | O | O | — | — | — | — | — | — | — | — |
| *rpl5* | ribosomal protein, 50S, L5 | O | O | O | O | O | — | — | — | — | — | — |
| *rpl6* | ribosomal protein, 50S, L6 | O | — | O | — | — | — | — | — | — | — | — |
| *rpl7* | ribosomal protein, 50S, L7 | O | — | — | — | — | — | — | — | — | — | — |
| *rpl9* | ribosomal protein, 50S, L9 | — | O | — | — | — | — | — | — | — | — | — |
| *rpl11* | ribosomal protein, 50S, L11 | O | O | O | — | — | — | — | — | — | — | — |
| *rpl12* | ribosomal protein, 50S, L12 | — | O | O | O | O(S) | — | — | — | — | — | — |
| *rpl13* | ribosomal protein, 50S, L13 | — | O | O | — | — | — | — | — | — | — | — |
| *rpl14* | ribosomal protein, 50S, L14 | O | O | O | O | O(S) | O | O | ψO | O | O | O |
| *rpl16* | ribosomal protein, 50S, L16 | O | O | O | O | O(S) | O | O(S) | O | O | O(S) | O |
| *rpl18* | ribosomal protein, 50S, L18 | O | O | O | — | — | — | — | — | — | — | — |
| *rpl19* | ribosomal protein, 50S, L19 | O | O | O | O | — | — | — | — | — | — | — |
| *rpl20* | ribosomal protein, 50S, L21 | O | O | O | O | O | O | O | O | O | O | O |
| *rpl21* | ribosomal protein, 50S, L21 | O | O | O | — | — | O | — | — | — | — | — |
| *rpl22* | ribosomal protein, 50S, L22 | O | O | O | | O(S) | O | O | | O | O | O |
| *rpl23* | ribosomal protein, 50S, L23 | — | O | O | O | O(S) | O | O | ψO(X2) | O(X2) | O(X2) | O(X2) |
| *rpl24* | ribosomal protein, 50S, L24 | — | O | O | — | — | — | — | — | — | — | — |
| *rpl27* | ribosomal protein, 50S, L27 | — | O | O | — | — | — | — | — | — | — | — |
| *rpl28* | ribosomal protein, 50S, L28 | — | O | O | — | — | — | — | — | — | — | — |
| *rpl29* | ribosomal protein, 50S, L29 | — | O | O | — | — | — | — | — | — | — | — |
| *rpl31* | ribosomal protein, 50S, L31 | — | O | O | — | — | — | — | — | — | — | — |
| *rpl32* | ribosomal protein, 50S, L32 | — | O | O | O | O | — | O | — | — | — | O |
| *rpl33* | ribosomal protein, 50S, L34 | O | O | O | — | — | O | O | O | O | O | O |
| *rpl34* | ribosomal protein, 50S, L34 | O | O | O | — | — | — | — | — | — | — | — |
| *rpl35* | ribosomal protein, 50S, L35 | O | O | O | — | — | — | — | — | — | — | — |
| *rpl36* | ribosomal protein, 50S, L36 | O | O | O | O | O | — | O | O | O | O | O |
| *trnA(ggc)* | transfer RNA ala | — | O | — | — | — | — | — | — | — | — | — |
| *trnA(ugc)* | transfer RNA ala | O(X2) | O(X2) | O(X2) | O | O(X3) | O(X2) | O(S) | ψO(X2) | O(S)(X2) | O(S)(X2) | O(S)(X2) |
| *trnC(gca)* | transfer RNA cys | O(X2) | O | O | O | O | O | O | ψO | O | O | O |
| *trnD(guc)* | transfer RNA asp | O | O | O | O | O | O | O | O | O | O | O |
| *trnE(uuc)* | transfer RNA asp | O | O | O | O | O | O | O | O | O | O | O |
| *trnF(gaa)* | transfer RNA phe | O | O | O | O | O | O | O | O | O | O | O |

Table 6.1. (*cont.*)

| Mnemonic | Gene product | *Cyanelle* | *Porphyra* | *Odontella* | *Chlorella* | *Euglena* | *Marchantia* | *Pinus* | *Epiphagus* | *Nicotiana* | *Oryza sativa* | *Zea mays* |
|---|---|---|---|---|---|---|---|---|---|---|---|---|
| *trnG(gcc)* | transfer RNA gly | O | O | O | O(X2) | O | O | O | — | O | O | O |
| *trnG(guu)* | transfer RNA gly | — | O | — | — | — | — | — | — | — | — | — |
| *trnG(ucc)* | transfer RNA gly | O | O | O | O | O | — | O(S) | — | O | O(S) | O(S) |
| *trnH(gug)* | transfer RNA his | O | O | O | O | O | O | O(X2) | O | O | O(X2) | O(X2) |
| *trnI(cau)* | transfer RNA ile | O | O | — | O | O | O | O | O(X2) | O(X2) | O(X2) | O(X2) |
| *trnI(gau)* | transfer RNA ile | O(X2) | O(X2) | O(X2) | O | O(X3) | O(X2) | O(S) | ψO(X2) | (S)(X2) | O(X2) | O(X2) |
| *trnK(uuu)* | transfer RNA lys | O | O | O | O | O | O | O(S) | — | O | O(S) | O(S) |
| *trnL(caa)* | transfer RNA leu | O | O | — | O | O | O | — | O(X2) | O(X2) | O(X2) | O(X2) |
| *trnL(gag)* | transfer RNA leu | O | O | — | O | — | — | — | — | — | — | — |
| *trnL(uaa)* | transfer RNA leu | O | O | O | O(S) | O | O | O(S) | — | O | O(S) | O(S) |
| *trnL(uag)* | transfer RNA leu | O | O | O | O | O | O(X2) | O | O(X2) | O | O | O |
| *trnfM(cau)* | transfer RNA fmet | O | O | O | O | O | O | O | O | O | O | O |
| *trnM(cau)* | transfer RNA met | O | O | O | O | O | O | O | O | O | O | O |
| *trnN(guu)* | transfer RNA asn | O | — | O | O | O | O(X2) | O | O(X2) | O(X2) | O(X2) | O(X2) |
| *trnP(ggg)* | transfer RNA pro | — | — | — | — | — | ψO | O | — | — | — | — |
| *trnP(ugg)* | transfer RNA pro | O | O | O(X2) | O | O | O | O | O | O | O | O |
| *trnQ(uug)* | transfer RNA gln | O | O | O | O | O | O | O | O | O | O | O |
| *trnR(acg)* | transfer RNA arg | O | O | O | O | O | O(X2) | O | O(X2) | O(X2) | O(X2) | O(X2) |
| *trnR(ccg)* | transfer RNA arg | O | O | — | O | — | O | O | — | — | — | — |
| *trnR(ccu)* | transfer RNA arg | — | O | — | — | — | — | — | — | — | — | — |
| *trnR(ucu)* | transfer RNA arg | O | O | O | O | O | O | O | ψO | O | O | O |
| *trnS(cga)* | transfer RNA ser | O | O | — | — | — | — | — | — | — | — | — |
| *trnS(gct)* | transfer RNA ser | O | — | — | — | — | — | — | — | — | — | — |
| *trnS(gaa)* | transfer RNA ser | — | — | — | O | — | — | — | — | — | — | — |
| *trnS(gcu)* | transfer RNA ser | O | O | O | O | O | O | O(X2) | O | O | O | O |
| *trnS(gga)* | transfer RNA ser | — | O | — | — | — | O | O | ψO | O | O | O |
| *trnS(uga)* | transfer RNA ser | O | O | O | O | O | O | O | O | O | O | O |
| *trnT(ggu)* | transfer RNA thr | O | O | — | O | — | O | O(X2) | — | O | O | O |
| *trnT(ugu)* | transfer RNA thr | O | O | O | O | O | O | O | — | O | O | O |
| *trnV(gac)* | transfer RNA val | O | O | — | — | — | O | O | — | O(X2) | O(X2) | O(X2) |
| *trnV(uac)* | transfer RNA val | O | O | O | O(X2) | O | O | O(S) | — | O(S) | O(S) | O(S) |
| *trnW(cca)* | transfer RNA trp | O | O | O | O | O+5ψ | O | O | O | O | O | O |
| *trnY(gua)* | transfer RNA tyr | O | O | O | O | O | O | O | O | O | O | O |
| *rpoA* | RNA polymerase, α chain | O | O | O | O | — | O | O | ψO | O | O | |
| *rpoB* | RNA polymerase, β chain | O | O | O | O | O(S) | O | O | — | O | O | O |
| *rpoC1* | RNA polymerase, β′ chain | O | O | O | O | O(S) | O | O(S) | — | O(S) | O | O |
| *rpoC2* | RNA polymerase, β″ chain | O | O | O | O | O(S) | O | O | — | O | O | O |
| *dnaB* | replication helicase subunit | — | O | O | — | — | — | — | — | — | — | — |
| *dnaK* | hsp-70-type chaperone | O(X2) | O | O | — | — | — | — | — | — | — | — |
| *infA* | initiation factor 1 | — | — | — | O | | O | O | O | O | O | O |
| *infB* | initiation factor 2 | — | O | — | — | — | — | — | — | — | — | — |
| *infC* | initiation factor 3 | — | O | — | — | — | — | — | — | — | — | — |
| *matK* | RNA-maturase K | — | — | — | — | — | O | O | O | O | O | O |
| *minD* | putative organelle division protein | — | — | — | O | — | — | — | — | — | — | — |
| *rne* | RNase E | — | O | — | — | — | — | — | — | — | — | — |
| *rnpB* | RNA component of RNase P | O | O | — | — | — | — | — | — | — | — | — |
| *syfB* | phenylalanine tRNA synthetase | — | O | — | — | — | — | — | — | — | — | — |
| *syh* | histidine tRNA synthetase | — | O | — | — | — | — | — | — | — | — | — |
| *trpA* | tryptophan synthase α subunit | — | O | — | — | — | — | — | — | — | — | — |
| *trpG* | anthranilate synthase component | O | O | — | — | — | — | — | — | — | — | — |
| *tsf* | elongation factor Ts | | O | — | — | — | — | — | — | — | — | — |
| *tufA* | elongation factor Tu | O | O | O | O | O(S) | — | — | — | — | — | — |
| **Genes involved in photosynthesis** | | | | | | | | | | | | |
| *rbcL* | ribulose bisphosphate carboxylase, small subunit | O | O | O | O | O(S) | O | O | ψO | O | O | O |
| *rbcS* | ribulose bisphosphate carboxylase, large subunit | O | O | O | — | — | — | — | — | — | — | — |

Table 6.1. (*cont.*)

| Mnemonic | Gene product | *Cyanelle* | *Porphyra* | *Odontella* | *Chlorella* | *Euglena* | *Marchantia* | *Pinus* | *Epiphagus* | *Nicotiana* | *Oryza sativa* | *Zea mays* |
|---|---|---|---|---|---|---|---|---|---|---|---|---|
| *psaA* | PSI, P700 apoprotein, 1a | O | O | O | O | O(S) | O | O | — | O | O | O |
| *psaB* | PSI, P700 apoprotein, 1b | O | O | O | O | O(S) | O | O | — | O | O | O |
| *psaC* | PSI, Fe-S polypeptide, 9-kDa | O | O | O | O | O(S) | | O | — | O | O | O |
| *psaD* | PSI, ferredoxin binding protein, subunit II | O | O | O | — | — | — | — | — | — | — | — |
| *psaE* | PSI, subunit IV, 18—20 kDa | O | O | O | — | — | — | — | — | — | — | — |
| *psaF* | PSI, plasticyanin-binding protein, subunit III | — | O | O | — | — | — | — | — | — | — | — |
| *psaI* | PSI, PS1-I polypeptide | O | O | O | — | — | — | O | — | O | O | O |
| *psaJ* | PSI, PS1-J polypeptide | O | O | O | — | O | — | O | — | O | O | O |
| *psaK* | PSI, PS1-K polypeptide (P37) | — | O | — | — | — | — | — | — | — | — | — |
| *psaL* | PSI, reaction centre subunit XI | — | O | O | — | — | — | — | — | — | — | — |
| *psaM* | PSI, reaction centre subunit M | O | O | O | O | O | — | O | — | — | — | — |
| *psbA* | PSII, D1 rxn-centre polypeptide | O | O | O | O | O(S) | O | O | ψO | O | O | O |
| *psbB* | PSII, CP47 apoprotein | O | O | O | O | O(S) | O | O | ψO | O | O | O |
| *psbC* | PSII, CP43 apoprotein | O | O | O | O | O(S) | O | O | — | O | O | O |
| *psbD* | PSII, D2 rxn-centre polypeptide | O | O | O | O | O(S) | O | O | — | O | O | O |
| *psbE* | PSII, cytochrome b-559 α-subunit (8 kDa) | O | O | O | — | O(S) | O | O | — | O | O | O |
| *psbF* | PSII, cytochrome b-559 β-subunit (4 kDa) | O | O | O | — | O(S) | O | O | — | O | O | O |
| *psbH* | PSII, 10-kDa phosphoprotein | O | O | O | O | O | O | O | — | O | O | O |
| *psbI* | PSII, protein I (4.8 kDa) | O | O | O | — | — | — | O | — | O | O | O |
| *psbJ* | PSII, protein J | O | O | O | — | O | — | O | — | — | — | O |
| *psbK* | PSII, protein K | O | O | O | O | O(S) | — | O | — | O | O | O |
| *psbL* | PSII, protein L | O | O | O | — | O | — | O | — | O | O | O |
| *psbM* | PSII, protein M | O | — | — | O | — | — | O | — | O | O | O |
| *psbN* | PSII, protein N | O | O | O | O | O | — | O | — | O | O | O |
| *psbT* | PSII, protein T | O | O | O | O | O(S) | O | O | — | O | O | O |
| *psbV* | cytochrome C 550 (oxygen evolving complex component) | O | O | O | — | — | — | — | — | — | — | — |
| *psbW* | PSII, protein W (13 kDa) | O | O | O | — | — | — | — | — | — | — | — |
| *psbX* | PSII, protein X (4.1 kDa) | O | O | O | — | — | — | — | — | — | — | — |
| *petA* | cytochrome F | O | O | O | O | — | O | O | — | O | O | O |
| *petB* | cytochrome b | O | O | O | O | O(S) | O | O(S) | — | O(S) | O(S) | O(S) |
| *petD* | Rieske Fe–S polypeptide, subunit IV | O | O | O | O | — | O | O(S) | — | O(S) | O(S) | O(S) |
| *petF* | ferredoxin | O | O | O | — | — | — | — | — | — | — | O |
| *petG* | subunit V | O | O | O | O | O(S) | — | O | — | O | O | — |
| *petJ* | cytochrome C 553 | — | O | — | — | — | — | — | — | — | — | — |
| *petL(ycf7)* | cytochrome b/f complex 3.5 kDa subunit | O | O | O | O | — | O | O | — | O | O | O |
| *atpA* | ATP synthase CF1 subunit α | O | O | O | O | O(S) | O | O | ψO | O | O | O |
| *atpB* | ATP synthase CF1 subunit β | O | O | O | O | O(S) | O | O | ψO | O | O | O |
| *atpD* | ATP synthase CF1 subunit δ | O | O | O | — | — | — | — | — | — | — | — |
| *atpE* | ATP synthase CF1 subunit ε | O | O | O | O | O(S) | O | O | — | O | O | O |
| *atpF* | ATP synthase CFo subunit I | O | O | O | O | O(S) | O | O(S) | — | O(S) | O(S) | O(S) |
| *atpG* | ATP synthase CF1 subunit γ | O | O | O | — | — | — | — | — | — | — | — |
| *atpH* | ATP synthase CFo subunit III | O | O | O | O | O | O | O | — | O | O | O |
| *atpI* | ATP synthase CFo subunit IV | — | O | O | O | O(S) | O | O | — | O | O | O |

Table 6.1. (*cont.*)

| Mnemonic | Gene product | *Cyanelle* | *Porphyra* | *Odontella* | *Chlorella* | *Euglena* | *Marchantia* | *Pinus* | *Epiphagus* | *Nicotiana* | *Oryza sativa* | *Zea mays* |
|---|---|---|---|---|---|---|---|---|---|---|---|---|
| **Genes involved in other functions** | | | | | | | | | | | | |
| *ndhA* | NADH-PQ oxidoreductase, chain 1 | — | — | — | — | — | O | — | — | O(S) | O(S) | O |
| *ndhB* | NADH-PQ oxidoreductase, chain 2 | — | — | — | — | — | O | — | ψO(X2) | O(S)(X2) | O(S)(X2) | O(S)(X2) |
| *ndhC* | NADH-PQ oxidoreductase, chain 3 | — | — | — | — | — | O | ψO | — | O | O | O |
| *ndhD* | NADH-PQ oxidoreductase, chain 4 | — | — | — | — | — | O | ψO | — | O | O | O |
| *ndhE* | NADH-PQ oxidoreductase, chain 4L | — | — | — | — | — | O | ψO | — | O | O | O |
| *ndhF* | NADH-PQ oxidoreductase, chain 5 | — | — | — | — | — | O | — | — | O | O | O |
| *ndhG* | NADH-PQ oxidoreductase, chain 6 | — | — | — | — | — | O | — | — | O | O | O |
| *ndhH* | NADH-PQ oxidoreductase, 49 kDa | — | — | — | — | — | O | ψO | — | O | O | O |
| *ndhI* | NADH-PQ oxidoreductase, subunit I | — | — | — | — | — | O | ψO | — | O | O | O |
| *ndhJ* | NADH-PQ oxidoreductase, subunit J | — | — | — | — | — | O | — | — | O | O | O |
| *ndhK* | NADH-PQ oxidoreductase, subunit K | — | — | — | — | — | O | ψO | — | O | O | O |
| *accA* | acetyl-CoA carboxylase carboxytransferase, α subunit | — | O | — | — | — | — | — | — | — | — | — |
| *accB* | acetyl-CoA carboxylase biotin carboxyl carrier protein subunit | — | O | — | — | — | — | — | — | — | — | — |
| *accD* | acetyl-CoA carboxylase carboxytransferase, β subunit | — | O | — | O | — | O | O | O | O | ψO | — |
| *acpP* | acyl carrier protein | O | O | O | — | — | — | — | — | — | — | — |
| *fabH* | β-ketoacyl acyl carrier protein synthase III | — | O | — | — | — | — | — | — | — | — | — |
| *apcA* | allophycocynin α subunit | O | O | — | — | — | — | — | — | — | — | — |
| *apcB* | allophycocyanin β subunit | O | O | — | — | — | — | — | — | — | — | — |
| *apcD* | allophycocyanin γ subunit | O | O | — | — | — | — | — | — | — | — | — |
| *apcE* | phycobilisome core linker polypeptide | O | O | — | — | — | — | — | — | — | — | — |
| *apcF* | allophycocyanin B18 subunit | O | O | — | — | — | — | — | — | — | — | — |
| *argB* | acetyl glutamate kinase | — | O | — | — | — | — | — | — | — | — | — |
| *carA* | carbamoyl phosphate synthase small subunit | — | O | — | — | — | — | — | — | — | — | — |
| *cemA* | chloroplast envelop protein (hbp: putative haem binding protein) | — | O | — | — | — | O | O | — | O | O | O |
| *cfxQ* | | — | O | O | — | — | — | — | — | — | — | — |
| *chlB* | protochlorophyllide reductase chlB chain | O | O | — | O | — | — | O | — | — | — | — |
| *chlI* | magnesium chelatase subunit | O | O | O | O | O | — | — | — | — | — | — |
| *chlL* | protochlorophyllide reductase iron–sulphur ATP-binding protein | O | O | — | O(S) | — | O | O | — | — | — | — |
| *chlN* | protochlorophyllide reductase chlN chain | O | O | — | O | — | — | O | — | — | — | — |
| *clpC* | clp protease ATP-binding subunit | — | O | O | — | — | — | — | — | — | — | — |
| *clpP* | clp protease catalytic subunit | O(X2) | — | — | O | — | — | O | O | O | O | O |
| *cpcA* | phycocyanin α-subunit | O | O | — | — | — | — | — | — | — | — | — |
| *cpcB* | phycocyanin β-subunit | O | O | — | — | — | — | — | — | — | — | — |
| *cpcG* | phycobilisome rod-core linker polypeptide | — | O | — | — | — | — | — | — | — | — | — |

Table 6.1. (*cont.*)

| Mnemonic | Gene product | *Cyanelle* | *Porphyra* | *Odontella* | *Chlorella* | *Euglena* | *Marchantia* | *Pinus* | *Epiphagus* | *Nicotiana* | *Oryza sativa* | *Zea mays* |
|---|---|---|---|---|---|---|---|---|---|---|---|---|
| *cpeA* | phycoerethrin $\alpha$-subunit | — | O | — | — | — | — | — | — | — | — | — |
| *cpeB* | phycoerethrin $\beta$-subunit | — | O | — | — | — | — | — | — | — | — | — |
| *crtE* | geranylgeranyl pyrophosphate synthase | O | O | — | — | — | — | — | — | — | — | — |
| *cysA* | transporter (sulfate) (mbpX) | — | — | — | O | — | O | — | — | — | — | — |
| *cysT* | transporter (sulfate) (mbpY) | — | — | — | O | — | O | — | — | — | — | — |
| *ftrB* | ferredoxin–thioredoxin reductase $\beta$ subunit | — | O | — | — | — | O | — | — | — | — | — |
| *ftsW* | putative cell (organelle) division protein | O | | — | — | — | — | — | — | — | — | — |
| *glnB* | nitrogen regulatory protein PII | — | O | — | — | — | — | — | — | — | — | — |
| *gltB* | glutamate synthase (GOGAT) | — | O | — | — | — | — | — | — | — | — | — |
| *groEL* | 60 kDa chaperonin | O(X2) | O | O | — | — | — | — | — | — | — | — |
| *groES* | 10 kDa chaperonin | O(X2) | — | — | — | — | — | — | — | — | — | — |
| *hemA* | 5-aminolevulnic acid synthase | O | — | — | — | — | — | — | — | — | — | — |
| *hisH* | histidinol-phosphate aminotransferase | O | — | — | — | — | — | — | — | — | — | — |
| *hisP* | histidine transport ATP-binding protein | O | — | — | — | — | — | — | — | — | — | — |
| *ilvB* | acedohydroxyacid synthase large subunit | — | O | — | — | — | — | — | — | — | — | — |
| *ilvH* | acedohydroxyacid synthase small subunit | — | O | — | — | — | — | — | — | — | — | — |
| *nadA* | quinolinate synthetase | O | — | — | — | — | — | — | — | — | — | — |
| *odpA* | pyruvate dehydrogenase E1 component, $\alpha$ subunit | — | O | — | — | — | — | — | — | — | — | — |
| *odpB* | pyruvate dehydrogenase E1 component, $\beta$ subunit | — | O | — | — | — | — | — | — | — | — | — |
| *pbsA* | heme oxygenase | — | O | — | — | — | — | — | — | — | — | — |
| *pgmA* | phosphoglycerate mutase | — | O | — | — | — | — | — | — | — | — | — |
| *preA* | prenyl transferase | O | — | — | — | — | — | — | — | — | — | — |
| *secA* | preprotein translocase subunit | — | O | O | — | — | — | — | — | — | — | — |
| *secX* | preprotein translocase subunit | — | — | — | — | — | O | — | — | — | — | — |
| *secY* | preprotein translocase subunit | O | O | O | — | — | — | — | — | — | — | — |
| *thiG* | thiG protein, thiamine biosynthesis | — | O | O | — | — | — | — | — | — | — | — |
| *trxA* | thioredoxin | O | O | — | — | — | — | — | — | — | — | — |
| *ycf1* | | — | — | — | — | — | O | O | O | O | — | — |
| *ycf2* | | — | — | — | — | — | O | O | O | O | — | — |
| *ycf3* | | O | O | O | O | — | O | O | — | O | O | O |
| *ycf4* | | O | O | O | O | O(S) | O | O | — | O | O | O |
| *ycf5* | | O | O | O | O | — | O | O | — | O | O | O |
| *ycf6* | | O | O | O | — | — | O | O | — | O | O | O |
| *ycf9* | | O | O | O | O | O | O | O | — | O | O | O |
| *ycf12* | | 0 | O | O | O | O(S) | O | O | — | — | — | — |

O Present; - Absent; (X2) Number of copies (e.g. 2); (S) Split genes; $\psi$ Pseudogenes.

recognizes several codons according to normal wobble base pairing has been found. However, if both 'two-out-of-three' and 'U:N wobble' mechanisms operate, the plastid encoded tRNA species might just suffice the needs of translation. In *Epifagus*, absence of 13 of the tRNA species that are present in tobacco, indirectly raises another possibility of import of nuclear-encoded tRNA species to effect translation (Wolfe *et al.*, 1992). But, so far, there has been no direct evidence to support RNA transport into plastids.

Apart from *rrn* and *trn* genes, homologues of bacterial RNA polymerase subunits α, β and β′ are also encoded in the plastids by *rpo* genes. The plastid *rpoA and rpoB* genes are homologous to their bacterial counterparts (see Igloi & Kossel, 1992). However, two genes *rpoC1* and *rpoC2* were found to show homologies to different regions of β′ subunit of *E. coli*. The precise matching of maize 38 kDa, 120 kDa, 78 kDa and 180 kDa amino terminal peptide sequence to the *rpoA, rpoB, rpoC1* and *rpoC2* deduced sequences also vindicates that, at least part of, the RNA polymerase is plastid encoded. In monocots like rice and maize an extra, non-homologous stretch of 380–450 bp has also been observed in the *rpoC2* gene. Interestingly, all of the male sterile lines of sorghum (a monocot) lack this extra sequence suggesting its involvement in cytoplasmic male sterility.

## Polycistronic transcription units

Presence of polycistronic transcription units is a common feature of plastids. More than 60% of the plastid genes are transcribed as multicistronic transcripts, which constitute 16 and 19 multicistronic transcription units in rice and tobacco, respectively (Kanno & Hirai, 1993 and unpublished data). To some extent, the grouping of genes in these units seems to have a functional bias as well. For example, all the *rrn* genes are clustered in an operon along with a few *trn* genes, in the IR region. Even in the species like black pine which do not possess IRs, the arrangement of *rrn* genes is unaltered (Wakasugi *et al.*, 1994). Similarly, *atp* genes are grouped in two operons i.e. *atpB/E* and *atpI/H/F/A* (*see* Fig. 6.1). The *psaA/B, psbD/C* and *rpoB/C* coding for PSI, PSII and plastid RNA polymerase subunits, respectively, have also been found grouped irrespective of the plant species (see Herrmann *et al.*, 1991). Ten ribosomal protein genes are transcribed together with *infA* and *rpoA* genes to make a large *rpl23* operon. The arrangement of *rpl23* operon corresponds to that of homologous genes in *E. coli* S10, *spc* and α operons, indicating towards their development from a common ancestral gene set (see Sugiura, 1996).

## Split genes

Introns have been found in plastid genes belonging to all functional categories. In tobacco plastids, 18 genes have been found to contain introns that range in size from 503–2526 bp. The *Euglena* and *Porphyra* genomes exhibit extreme situations as the former contains 149 introns as opposed to about 13 introns in land plants and the latter does not contain any intron (Hallick *et al.*, 1993; Reith & Munholland, 1995). Hence, considering these two as special cases, individual genes in these two species have not been discussed.

The *rrn23* was the first plastid gene found to contain an intron in *Chlamydomonas reinhardtii* (see Rochaix, 1992). This gene in *C. eugametos* contains six introns of which one codes for a sequence-specific endonuclease that could mediate transposition of introns. Complete sequencing *Chlorella vulgaris* IAM C-27 ctDNA has also revealed an intron in the *rrn23* gene that shows high degree of homology to that of *C. reinhardtii* (Kapoor *et al.*, 1997). This gene, however, does not contain any intron in land plants. Among ribosomal protein genes, *rpl2* was the first one to be identified in *Nicotiana debneyi*. The *rpl2* of rice and pine is also split but no intron was found in this gene in spinach and some other related species. Introns have been detected also in *rps16* and *rpl16* of land plants (see Sugiura, 1992). Most unique, however, is land plant *rps12* gene whose 5 coding region is present in the LSC region whereas two coding regions for the 3 exon have been localized in the IRs. These separately transcribed RNAs consequently require *trans*-splicing for expression. Six plastid tRNAs from land plants are known to possess introns ranging in size from 0.5 to 2.5 kb. The introns in tRNA genes were first reported in maize *trnI*(GAU) and *trnA*(UGC). Intron in all [e.g. *trnA*(UGC), *trnI*(GAU), *trnL*(UAA), *trnK*(UUU) and *trnV*(UAC)] but one (*trnG*(UUC)) have been found in the anticodon loop. The *trnG*(UUC) intron is located in the D-stem region, a feature unique to chloroplasts. Intron of *trnG*(UCC), *trnK*(UUU) and

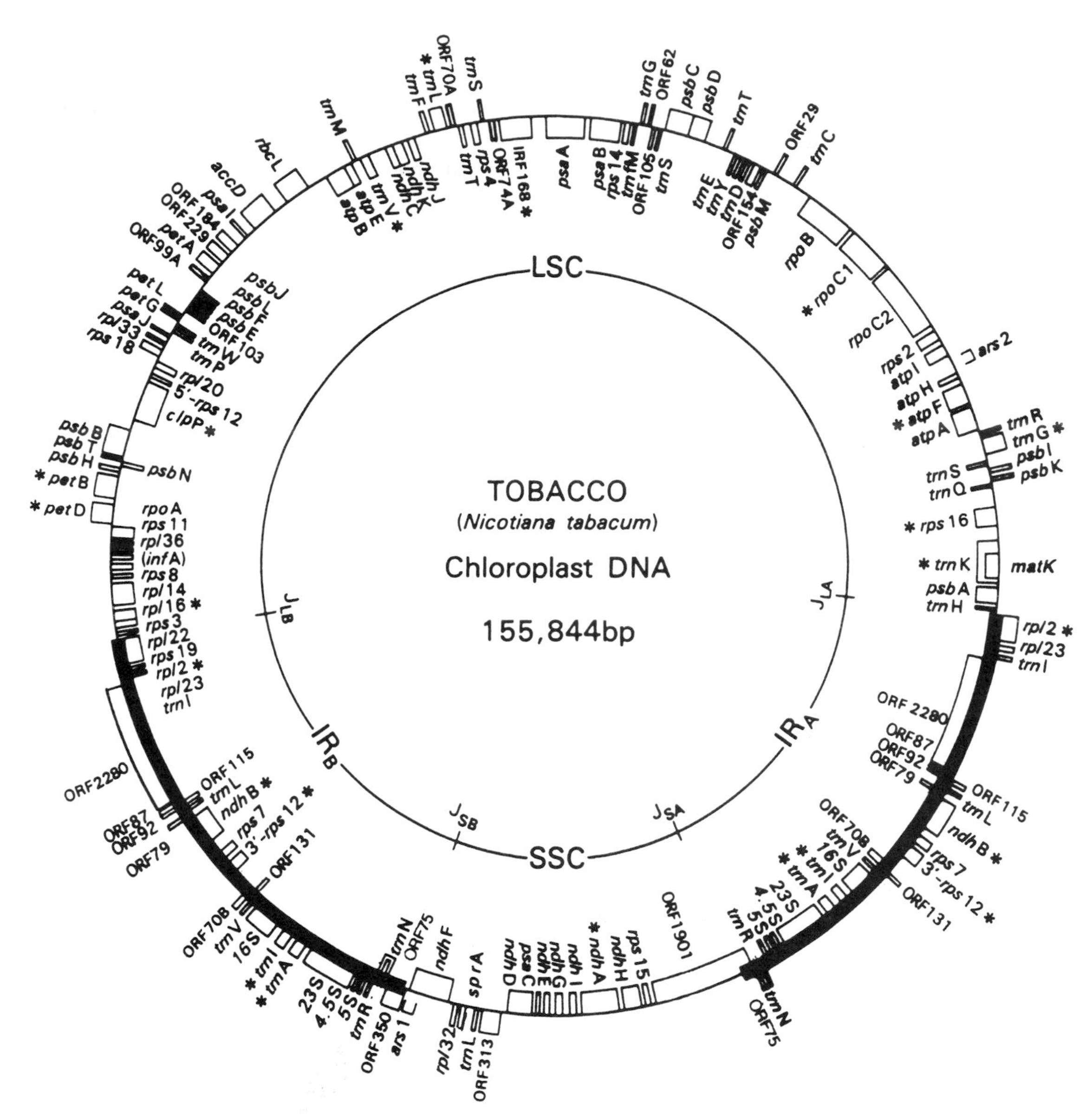


**Figure 6.1.** Genetic map of tobacco chloroplast DNA. Genes shown inside the circle are transcribed clockwise, and those on the outside are transcribed counter-clockwise. Asterisks denote split genes.

*trnV*(UAC) have conserved boundaries as found in the protein coding genes suggesting a similar splicing pathway being operational in these two RNA categories. Interestingly, no split tRNA or rRNA gene have been found in *Euglena*, which otherwise contains maximum number of introns. However, presence of introns in the tRNA genes of some of the members of the *Charophyceae* is suggestive of this group being part of the lineage that gave rise to land plants (see Sugiura, 1996).

Existence of introns has been consistent in photosynthetic genes such as *petB*, and *petD* in the land plant chloroplasts. The *psbA* gene in *Euglena* and *Chlamydo-*

*monas* contains 4 introns each. In *Euglena* the photosynthesis-related genes have been detected to have an unusually high number of introns, in all 31 in 10 genes. However, *petD* that usually is an intron-containing gene in land plants, does not contain any intron in *Euglena*. As a unique example, three exons of *Chlamydomonas*, *psaA* genes have been found transcribed separately and subsequently joined by a *trans*-splicing event that happens to involve a product of another plastid gene, *tscA* (Goldschmidt-Clermont *et al.*, 1991). Among RNA polymerase genes only *rpoC1* has been found to be split only in dicots.

On the basis of intron boundary sequences and deduced secondary structures, chloroplast introns can be classified into four groups. The group I (*trnL*, *rrn23* and *C. moewusii psbA*) and group II (*trnI and trnA* in the *rrn16-rrn23* spacer region) intron can be folded into a secondary structure typical of fungal mitochondrial group I and group II introns, respectively. The structure of chloroplast group III introns (in *trnV*(UAC), *trnG*(UCC) and *trnK*(UUU) genes) is similar to that of group II, however, they differ in the conserved 5′ and 3′ boundary sequences. The group III conserved intron boundary sequences are similar to that of nuclear protein-coding genes, suggesting existence of a similar splicing mechanism in two compartments of the plant cell. In *Euglena* 46 introns constitute the fourth group with distinctive features such as uniform, relatively smaller size (91–119 bp) and sharing of common features among themselves (Hallick *et al.*, 1993).

## REGULATION OF PLASTID GENE EXPRESSION

As in the nucleus, one would expect the plastid gene expression to be regulated at transcriptional, post-transcriptional and translational levels. Although ctDNA codes for several basic components for gene expression, it must rely on the nucleo-cytoplasm for a major portion of the components involved in its biogenesis (which are largely unknown) and photosynthetic machinery. Moreover, number of plastids in a cell and copies of the ctDNA in each plastid vary depending on the external and internal cues. For some unknown reasons, faithful *in vitro* reproduction of the molecular processes involved in the plastid gene expression has been difficult. For example, plastid *in vitro* transcription system is available for only a few selected species and there is no *in vitro* system that can reproduce plastid RNA splicing. The systems for plastid transformation and *in vitro* translation have been optimized very recently using the tobacco plants (Hirose & Sugiura, 1996; Svab & Maliga, 1993). These complexities have, so far, made it difficult to ascertain the degree of control at each level of plastid gene expression.

### Controls at the level of transcription

Contradicting with the view prevailing in the late 1980s that most of the chloroplast genes are constitutively expressed and major controls for plastid gene expression lie mainly at the post-transcriptional level, several examples of modulation of chloroplast transcripts at the level of transcriptional initiation have been reported recently (see Igloi & Kössel 1992, Mayfield *et al.*, 1995). Although the *cis*- and *trans*-acting regulatory elements involved in this regulation remain to be identified, efforts are already under way to dissect this so far obscure means of regulation in the cell organelle. The following is our current understanding of various parameters involved in the process of chloroplast transcription.

#### *RNA polymerase-dependent differential usage of plastid promoters*

Due to the presence of 5′ triphosphates, chloroplast primary transcripts can be specifically labeled *in vitro* using [α $^{32}$P]GTP and guanylyltransferase, making it possible to distinguish between primary and processed 5′ ends. This technique, in combination with northern hybridization, primer extension and SI nuclease/ribonuclease protection analyses, has been used extensively for unambiguous localization of chloroplast promoter regions. Systematic studies in tobacco led to the identification of ~60 transcription initiation sites (unpublished data). Most of the chloroplast promoters (including those for tRNA, rRNA and protein coding genes) thus identified resemble that of prokaryotes at least in terms of −10 and −35 consensus sequences. These studies, and others, have also revealed existence of multiple promoters within polycistronic transcription units as well as for individual genes such as *rbcL* gene from maize, and *rrn16* and *atpB* from spinach and tobacco (see Igloi & Kössel 1992). Both spinach and tobacco *atpB* genes contain four different promoters, however, their locations are species specific (see Gruissem & Tonkyn, 1993 and Kapoor *et al.*, 1997).

Besides the $\sigma^{70}$ type promoters, some examples of non-consensus type promoters have also been found in chloroplasts that show no similarity to the typical −10 and −35 promoter elements. The *trnS* and *trnR* promoters were first such examples (Gruissem & Tonkyn, 1993). However, even in the absence of these consensus elements, correct transcription from these promoters was observed *in vitro*. The *in vitro* utilized promoter (Pc, initiation site −141) and one of the two promoters (P2, initiation site −64) of spinach and tobacco *rrn16* gene, respectively, also lack both the elements (Baeza *et al.*, 1991; Iratni *et al.*, 1994; Vera & Sugiura, 1995). A promoter containing only −10 but no −35 region has been reported for the *rps16* gene of the mustard plastome (see Igloi & Kössel, 1992).

Recently, we have found that at least some of the non-consensus type promoters identified in the 5′ region of the genes, which are not directly involved in photosynthesis, express in a functionally distinctive manner from the previously described promoter elements (Kapoor *et al.*, 1997). These have been named as non-consensus type II (NCII) promoters. The abundance of the transcripts initiating from these promoter regions remains unaltered in different light conditions and plastid types. *In vivo* inhibition of plastid protein synthesis by spectinomycin/streptomycin has no inhibitory effect on the abundance of the NCII promoter-derived transcripts. However, transcripts from Consensus-Type promoters – referred to as CT promoters – are considerably reduced when either the light-grown seedlings are shifted to dark or plastid translation is inhibited. *In vivo* tagetitoxin treatment greatly reduces the levels of CT promoter-derived transcripts whereas no such inhibitory effect on the accumulation of NCII promoter-derived transcripts is observed. We hypothesize that the two plastid promoter types might be utilized by different RNA polymerase activities, thus providing a switch during varying developmental and environmental conditions. A similar theory has also been proposed by Alison *et al.* (1996), by creating *rpoB*-minus tobacco mutant plants using a particle gun. As these mutants are deficient in plastid-encoded RNA polymerase, any active transcription would result from the non-plastid-encoded RNA polymerase. Their data also show that, in the absence of plastid-encoded RNA polymerase, different transcription initiation sites are utilized in the plastids.

A similar promoter switching mechanism has also been proposed for the *psbA* promoter in mustard chloroplasts (Eisermann, Tiller & Link, 1990). This promoter region is characterized by a TATA-like element sandwiched between canonical −10 and −35 elements. Alterations in the TATA-box like region negatively affect *psbA* transcription *in vitro* using plastid extracts from either chloroplasts or etioplasts. Positions within the −35 element, however, play an important role in chloroplast but not in etioplast, thereby, raising a possibility of development-dependent switching between the two promoters through recognition by two different RNA polymerases (see Igloi & Kössel, 1992).

## Regulation at post-transcriptional level

Post-transcriptional processes involve processing, stability, splicing and editing of pre-RNAs. However, to what extent these processes influence plastid gene expression is debatable.

### *Intron splicing – limited evidence for direct involvement in the regulation of gene expression*

Presence of split genes is a common feature of plastid DNA. In tobacco plastome alone 18 genes contain introns that vary in size from 503–2526 bp. As a consequence, in addition to 5′ and 3′ processing, the intron-containing plastid transcripts have to undergo correct excision of the intervening sequences before they can become translationally functional. The mechanism of plastid splicing in higher plants is largely unknown. Most of the introns found in higher plant plastids are of group II type and the splicing intermediates are similar to that found in the nucleus. Therefore, involvement of RNA–protein complexes has also been envisaged in plastid processing and splicing. In a preliminary observation, spliced products of three photosynthesis-related transcript species (*atpF*, *petB* and *petD*) have been shown to be specifically under-represented in roots and proplastids, compared with green tissue of maize, whereas splicing of *rpl16* transcript is not affected to the same magnitude (Barkan, 1989). Later, a transposon-induced nuclear mutation was shown to involved in the processing of *petB* and *petD* mRNAs (Barkan *et al.*, 1994). Tissue-specific differential splicing and/or processing was also reported for rice *psbB* operon (Kapoor, Maheshwari & Tyagi, 1993). However, due to lack of any plastid *in vitro* splicing system delineation of *cis* or *trans* factors involved in these processes has still not been possible. Several complexities, e.g. introns within introns (twintrons, Copertino & Hallick, 1991), ORFs (including those coding for endonucleases or RNA splicing maturases) within introns (Saldanha *et al.*, 1993) have also been found in plastid introns. Taken together, these observations suggest that plastid splicing is an important post-transcriptional process; but more evidence is needed to ascertain its role in the regulation of plastid gene expression.

### *Editing might affect gene expression*

The first evidence for the modification of plastid RNAs by editing came from the maize *rpl2* transcripts in which an ACG codon changes to a functional start codon – AUG (Hoch *et al.*, 1992). Later, similar modifications in the initiation codons of *psbL* mRNA of tobacco and spinach were also reported. Editing of the *psbF-psbL* transcripts is differentially down-regulated in leucoplasts and proplastids, whereas only fully edited transcripts exist in etioplasts and chloroplasts. The editing of *petB* transcript, however, is unaffected in proplastids and chloroplasts. As the unedited *psbL* transcripts are unavailable for translation, regulation of editing can affect plastid gene expression in a tissue-specific manner. In the coding regions, editing has been suggested to restore conserved amino acids and in the introns it perhaps helps in maintaining the structure and function of introns. Editing has been proposed to precede both splicing and cleavage and must therefore, be independent of other processing events. It has also been suggested to be unaffected by plastid transcription, indicating that protein components involved in editing machinery must be nuclear encoded (see Maier *et al.*, 1995). Although mitochondrial and plastidic editing sites exhibit some similarity, none of the seven sites in petunia mitochondrial *coxII* transcript is edited when transcribed in transgenic tobacco chloroplasts indicating organelle specificity of the two editing machineries (Sutton *et al.*, 1995).

### *Stability of transcripts*

Plastid mRNA half-lives have been shown to differ dramatically depending on the plastid developmental stage, type and growth conditions (see Mullet, 1993). Investigation regarding the possible involvement of untranslated regions (UTRs) have indicated that both 5′ and 3′ UTRs may contribute to mRNA stability. The inverted repeat (IR) regions within the 5′ and 3′ UTRs of many plastid transcripts have been the target of investigations, as the stem–loop structure thus formed might act as a signal for RNA processing or as a preventive structure against the nucleolytic degradation (see Gruissem & Tonkyn, 1993; Rochaix, 1992).

The analysis of 3′ UTRs of several *in vitro* synthesized *psbA*, *rbcL*, *petD* and atpE transcripts using homologous *in vitro* transcription systems revealed that IRs are involved in 3′ end formation and the stability of these transcripts. Gene-specific interaction of IR regions with some RNA-binding proteins was also demonstrated. *In vivo* experiments with transformed *C. reinhardtii* also indicated towards a possible role of 3′ IRs in determining the stability of plastid transcripts. Several RNA binding proteins have also been shown to interact with plastid transcript 3′ UTRs. A nuclear-encoded 28 kDa protein that accumulates in a development-dependent manner was reported to be essential for correct 3′ end processing of several plastid RNAs. Due to correlation between accumulation of this protein and plastid RNA during development, its role in stability of transcripts was inferred (see Gruissem & Tonkyn, 1993). In another study, the *petD* 3′ IR region was specifically found to interact with a complex consisting of 55, 41, and 29 kDa RNA-binding proteins. In this study an AU-rich sequence downstream of the IR was found to be important for this interaction (Chen *et al.* 1995).

The evidence regarding the involvement of 5′ UTR in transcript stability has mainly come from the *in vivo* experiments in *Chlamydomonas*. Analysis of sequentially deleted *petD* 5′ UTR shows that entire 5′ UTR is essential for the stability of the chimeric transcript (Sakamoto *et al.* 1993). A light-dependent specific degradation of the chimeric message containing *rbcL* 5′ UTR also indicates its involvement in the regulation of transcript stability (Salvador, Klein & Bogorad, 1993). Analysis of the *psbD* 5′ UTR in the background of a nuclear mutant *nac2-26* indicates that a 74 nt region that interacts with a 47 kDa polypeptide in wild-type chloroplast lysates is essential for the stability of the chimeric transcript (Nickelson *et al.*, 1994). The *psbD* 3′ UTR, however, had no effect on the stability of a chimeric transcript in this mutant.

In summary, these data are suggestive of the involvement of 3′ and 5′ UTRs in the regulation of plastid RNA stability, but molecular components involved in these processes have not been discretely identified. The 3′ IR seems more to be a signal for the formation of the 3′ end rather than being involved in control of transcript stability. On the other hand, some direct correlation has been found between the 5′ UTRs and RNA stability. But it is still early to comment whether these regions have a direct role in controlling the RNA stability or these data represent a pleiotropic effect of some other function related to these UTRs.

## Translation

Most of the structural components involved in the plastid translation machinery, e.g. 70S ribosomes, no capping of the transcripts, presence of Shine-Dalgarno(SD)-like sequence, are similar to that of prokaryotes. Factors equivalent to prokaryotic initiation factors IF-2 and IF-3 have also been identified in *Euglena* chloroplast and a chloroplast gene with sequence similarity to *infA* encoding IF-1 has been found in land plants (see Mayfield *et al.*, 1995; Gillham *et al.*, 1994; Sugiura, 1992). Despite these remarkable similarities in structural aspects, plastid translation has several distinct features. These include lack of apparent coupling between transcription and translation and less stringent spacing between the SD-like sequences and the initiator codon than in prokaryotes. Moreover, in higher plant chloroplast genomes, about one-third of protein coding genes do not contain SD-like sequences within the 25 nt upstream region from the initiation codon.

Of all the processes involved in the regulation of gene expression, the step of translation seems to be affected most severely by light and developmental signals. The accumulation of some of the proteins may increase from 100 to 10 000 fold during light-induced greening of plastids (see Gillham *et al.*, 1994; Mayfield *et al.*, 1995; Rochaix, 1992). The most actively synthesized protein D1, in spite of the presence of abundant respective message, are not synthesized in dark growth conditions and in non-photosynthetic plastids in higher plants. In these instances, although the transcripts were present, they were not recruited into plastid polysomes. These results suggested the presence of a regulatory switch at the step of translation initiation. In barley, translation of two 65–70 kDa PSI chlorophyll *a*-apoproteins was found to be arrested on membrane-bound polysomes at the level of polypeptide chain elongation.

Barring a few reports in higher plants, most of the evidence for the involvement of nuclear factors in the process of translation has come from genetic and molecular analyses of nuclear mutants of *Chlamydomonas* (reviewed in Gillham *et al.* 1994; Mayfield *et al.*, 1995; Rochaix, 1992). These results indicated that 5′ UTRs play an important role in regulation of plastid translation by interacting with specific and/or general nuclear factors. Characterization of virdis-115, a barley nuclear mutant lacking PSII activity, also indicate that some nuclear factor(s) specifically affects translation and stability of plastid-encoded D1 and CP47 after the onset of light-induced chloroplast development. In *Chlamydomonas* binding activity of the RNA binding proteins specific for *psbA* mRNA 5′ UTR was shown to be modulated by light through ADP-dependent phosphorylation and redox potential. Later, using site-specific mutations in the 5′ UTR, the SD-like sequences and a region upstream were identified to be essential for high levels of *psbA* translation *in vivo*. Recently, using a newly developed *in vitro* translation system for tobacco plastids, *cis*-elements essential for tobacco *psbA* mRNA translation have been identified (Hirose & Sugiura, 1996). Two of them are complementary to the 3′ terminus of chloroplast 16S rRNA and probably necessary for the association of 30S ribosomal subunit (termed as RBS1 and RBS2). The other is an AU-rich sequence (UAAAUAAA) located between RBS1 and RBS2 that is another critical element for translation and termed as AU-box. As an important development, this system holds promises for the future for the elucidation of *cis*/*trans*-elements involved in the translation of plastid transcripts.

REFERENCES

Alison, L. A., Simon, L. D. & Maliga, P. (1996). Deletion of *rpoB* reveals a second distinct transcription system in plastids of higher plants. *EMBO Journal* **15**, 2802–9.

Baeza, L., Bertrand, A., Mache, R. & Lerbs-Mache, S. (1991). Characterization of a protein binding sequence in the promoter region of the 16S rRNA gene of the spinach chloroplast genome. *Nucleic Acids Research*, **19**, 3577–81.

Barkan, A. (1989). Tissue-dependent plastid RNA splicing in maize: transcripts from four plastid genes are predominantly unspliced in leaf meristems and roots. *Plant Cell*, **1**, 437–45.

Barkan, A., Walker, M., Nolasco, M. & Johnson, D. (1994). A nuclear mutation in maize blocks processing and translation of several chloroplast mRNAs and provides evidence for the differential translation of alternative mRNA forms. *EMBO Journal*, **13**, 3170–81.

Blowers, A. D., Klein, U., Ellmore, G. S. & Bogorad, L. (1993). Functional in vivo analyses of the 3′ flanking sequences of the *Chlamydomonas* chloroplast *rbcL* and *psaB* genes. *Molecular & General Genetics* **238**, 339–49.

Chen, Z. J., Muthukrishnan, S., Liang, G. H., Schertz, K. F. & Hart, G. E. (1993). A chloroplast DNA deletion located in RNA polymerase gene *rpoC2* in CMS lines of sorghum. *Molecular & General Genetics* **236**, 251–9.

Chen, Q., Adams, C. C., Usack, L., Yang, J., Monde, R. A. & Stern, D. B. (1995). An AU-rich element in the 3 untranslated region of the spinach chloroplast *petD* gene participates in sequence-specific RNA protein complex formation. *Journal of Molecular Biology*, **15**, 2010–18.

Copertino, D. W. & Hallick, R. B. (1991). Group II twintron: an intron within an intron in a chloroplast cytochrome b-559 gene. *EMBO Journal*, **10**, 433–42.

Eisermann, A., Tiller, K. & Link, G. (1990). *In vitro* transcription and DNA binding characteristics of chloroplast and etioplast extracts from mustard (*Sinapis alba*) indicate differential usage of the *psbA* promoter. *EMBO Journal*, **9**, 3981–7.

Gillham, N. W., Boynton, J. E. & Hauser, C. R. (1994). Translational regulation of gene expression in chloroplasts and mitochondria. *Annual Review of Genetics*, **28**, 71–93.

Goldschmidt-Clermont, M., Choquet, Y., Girard-Bascou, J., Michel, F., Schirmer-Rahire, M. & Rochaix, J. D. (1991). A small chloroplast RNA may be required for *trans-splicing in Chlamydomonas reinhardtii. Cell*, **65**, 135–43.

Gruissem, W. & Tonkyn, J. C. (1993). Control mechanisms of plastid gene expression. *Critical Review in Plant Science*, 12: 19–55.

Hallick, R. B., Hong, L., Drager, R. G., Favreau, M. R., Monfort, A., Orsat, B., Spielmann, A. & Stutz, E. (1993). Complete sequence of *Euglena gracilis* chloroplast DNA. *Nucleic Acids Research*, **21**, 3537–44.

Herrmann, R. G., Oelmüller, R., Bichler, J., Scheiderbauer, A., Steppuhn, J., Wedel, N., Tyagi, A. K. & Westhoff, P. (1991). The thylakoid membrane of higher plants: genes, their expression and interaction. In *Plant Molecular Biology* 2 ed R. G. Herrmann & B. Larkins, pp. 411–27. New York: Plenum Press.

Hirose, T. & Sugiura, M. (1996). *Cis*-acting elements and *trans*-acting factors for accurate translation of chloroplast *psbA* mRNA: development of an *in vitro* translation system from tobacco chloroplasts. *EMBO Journal*, **15**, 1687–95.

Hoch, B., Maier, R., Appel, K., Igloi, G. L. & Kossel, H. (1992). Editing of a chloroplast mRNA by creation of an initiation codon. *Nature*, **353**, 178–80.

Igloi, G. L. & Kössel, H. (1992). The transcription apparatus of chloroplasts. *Critical Reviews of Plant Science*, **10**, 525–58.

Iratni, R., Baeza, L., Andreeva, A., Mache, R., & Lerbs-Mache, S. (1994). Regulation of rDNA transcription in chloroplast: promoter exclusion by constitutive repression. *Genes & Development*, **8**, 2928–38.

Kanno, A. & Hirai, A. (1993). A transcription map of the chloroplast genome from rice (*Oryza sativa*). *Current Genetics*, **23**, 166–74.

Kapoor, S., Maheshwari, S. C. & Tyagi, A. K. (1993). Organ-specific expression of plastid-encoded genes in rice involves both quantitative and qualitative changes in messenger RNAs. *Plant Cell Physiology*, **34**, 943–7.

Kapoor, M., Nagai, T., Wakasugi, T., Yosginaga, K. & Sugiura, M. (1977). Organization of chloroplast ribosomal RNA genes and in vitro self-splicing of the large subunit rRNA intron from the green alga *Chlorella vulgaris* C-27. *Current Genetics* (in press).

Kapoor, S., Suzuki, J. Y. & Sugiura, M. (1997). Identification and functional significance of a new class of non-consensus-type plastid promoters. *Plant Journal*, **11**, 327–37.

Kowallik, K. V., Stoebe, B., Schaffran, I. & Freier, U. (1995). The chloroplast genome of a chlorophyll *a+c*-containing alga, *Odontella sinensis. Plant Molecular Biology Reporter*, 13: 336–342.

Maier, R. M., Neckermann, K., Igloi, G. L. & Kossel, H. (1995). Complete sequence of the maize chloroplast genome: gene content, hotspots of divergence and fine tuning of genetic information by transcript editing. *Journal of Molecular Biology*, **251**, 614–28.

Mayfield, S. P., Yohn, C. B., Cohen, A. & Danon, A. (1995). Regulation of chloroplast gene expression. *Annual Review of Plant Physiology and Plant Molecular Biology*, 46: 167–188.

Mullet, J. E. (1993). Dynamic regulation of chloroplast transcription. *Plant Physiology*, **103**, 309–13.

Nickelson, J., Dillewijn, J. V., Rahire, M. & Rochaix, J. D. (1994). Determinants for stability of the chloroplast *psbD* RNA are located within its short leader region in *Chlamydomonas reinhardtii*. *EMBO Journal*, **13**, 3182–91.

Palmer (1991). Plastid chromosomes: structure and evolution. In ed. *The Molecular Biology of Plastids* L. Bogorad & I. K. Vasil pp. 5–53. San Diego: Academic Press.

Reith, M. & Munholland, J. (1995). Complete nucleotide sequence of the *Porphyra purpurea* chloroplast genome. *Plant Molecular Biology Reporter*, **13**, 333–5.

Rochaix, J. D. (1992). Post-transcriptional steps in the expression of chloroplast genes. *Annual Review of Cell Biology*, **8**, 1–28.

Sakamoto, W., Kindle, K. L. and Stern, D. B. (1993). *In vivo* analysis of *Chlamydomonas* chloroplast *petD* gene expression using stable transformation of beta-glucuronidase translational fusions. *Proceedings of the National Academy of Sciences*, USA, **90**, 497–501.

Saldanha, R., Mohr, G., Belfort, M. & Kambowitz, A. M. (1993). Group I and Group II introns. *FASEB Journal*, **7**, 15–24.

Salvador, M. L., Klein, U. & Bogorad, L. (1993). Light-regulated and endogenous fluctuations of chloroplast transcript levels in *Chlamydomonas* regulation by transcription and RNA degradation. *The Plant Journal*, **3**, 213–19.

Stirewalt, V. L., Michalowski, C. B., Loffelhardt, W., Bohnert, H. J. & Bryant, D. A. (1995). Nucleotide sequence of the cyanelle genome from *Cyanophora paradoxa*. Plant Molecular Biology Reporter, **13**, 327–32.

Subramanian, A. R. (1993). Molecular genetics of chloroplast ribosomal proteins. *Trends in Biochemical Sciences*, **18**, 177–81.

Sugiura, M. (1992). The chloroplast genome. *Plant Molecular Biology*, **19**, 149–68.

Sugiura, M. (1996). Structure and replication of chloroplast DNA. In *Molecular Genetics of Photosynthesis* ed. B. Anderson, & J. Barber pp. 58–74. Oxford: Oxford Press.

Sutton, C. A., Zoubenko, O. V., Hanson, M. R. & Maliga, P. (1995). A plant mitochondrial sequence transcribed in transgenic tobacco chloroplast is not edited. *Molecular and Cellular Biology*, **15**, 1377–81.

Svab, Z. & Maliga, P. (1993). High-frequency plastid transformation in tobacco by selection of a chimeric *aadA* gene. *Proceedings of the National Academy of Sciences* USA, 90: 913–917.

Vera, A. & Sugiura, M. (1995). Chloroplast rRNA transcription from structurally different tandem promoters: an additional novel-type promoter. *Current Genetics*, **27**, 280–4.

Wakasugi, T., Nagai, T., Kapoor, M., Sugita, M., Ito, M., Ito, S., Tsudzuki, J., Nakashima, K., Tsudzuhi, T., Suzuki, Y., Hamada, A., Ohta, T., Inamura, A., Yoshinaga, K. & Sugiura, M. (1997). Complete nucleotide sequence of the chloroplast genome from the green alga *Chlorella vulgaris*: The existence of genes possibly involved in chloroplast division. *Proceedings of the National Academy of Sciences, USA*, **94**, (in press).

Wakasugi, T., Tsudzuki, J., Ito, S., Nakashima, K., Tsudzuki, T. & Sugiura, M. (1994). Loss of all *ndh* genes as determined by sequencing the entire chloroplast genome of the black pine *Pinus thunbergii*. *Proceedings of the National Academy of Sciences*, USA **91**, 9794–8.

Wolfe, K. H., Morden, C. W., Ems, S. C. & Palmer, J. D. (1992). Rapid evolution of the plastid translational apparatus in a nonphotosynthetic plant-loss or accelerated sequence evolution of tRNA and ribosomal protein genes. *Journal of Molecular Evolution*, **35**, 304–17.

# 7 Electron transport and energy transduction

J. WHITMARSH

## INTRODUCTION

Photosynthetic organisms use light energy to synthesize organic compounds. The primary product of photosynthesis is reduced carbon, which serves as both the energy source and building block for other organic compounds. In oxygenic photosynthesis, which occurs in plants, algae and some types of bacteria, electrons for the reduction of carbon dioxide are taken from water in a reaction that results in the release of molecular oxygen. In anoxygenic photosynthesis, purple bacteria and other types of bacteria use light energy to create organic compounds, but do not release oxygen.

Here we focus on oxygenic photosynthesis in plants, a process that is traditionally divided into two stages: (i) the 'light reactions', which include light absorption, transfer of exciton energy to reaction centres, followed by electron and proton transfer reactions that produce NADPH, ATP and $O_2$; and (ii) the 'dark reactions', which include the reduction of carbon dioxide and the synthesis of carbohydrates using the NADPH and ATP produced by the light reactions. This chapter describes electron and proton transfer reactions in oxygenic photosynthesis and their role in the transduction of light energy to a form of chemical energy that can be readily used for biosynthesis.

## PHOTOSYNTHETIC ELECTRON TRANSPORT

In plants, the photosynthetic process occurs inside chloroplasts, which are small organelles (5–10 microns across) found inside specialized cells. Most chloroplasts consist of three membranes, an outer envelope membrane, an inner envelope membrane, and an internal membrane system, known as the photosynthetic or thylakoid membrane. The photosynthetic membrane absorbs light, transfers electrons and protons, and produces ATP (see Chapter 1 for a more complete description of the structure of the chloroplast). The photosynthetic membrane is composed mostly of glyceral lipids in the form of a bilayer, which is heavily embedded with the protein complexes that make up the photosynthetic apparatus. To transduce light energy, the photosynthetic membrane functions as a vesicle, with an inner (lumen) and outer (stromal) water phase. The protein complexes are asymmetrically arranged in the photosynthetic membrane, enabling electron transport to create a proton gradient and an electric potential across the membrane. The energy stored in the proton electrochemical gradient drives a membrane-bound ATP synthase that produces ATP from ADP and inorganic phosphate. The enzymes required for the fixation and reduction of $CO_2$ are located outside the photosynthetic membrane in the surrounding aqueous phase of the chloroplast.

The photosynthetic membrane is arranged in circular stacks (grana) that are interconnected by non-stacked membranes (stromal membranes) (see Chapter 1). To further complicate this picture, the protein complexes in the membrane are deployed unevenly between grana and stroma membranes. Fortunately, to understand the fundamentals of photosynthetic electron transport and energy transduction, we can ignore the complexity of the design and treat the photosynthetic membrane as a simple vesicle with an inner and outer aqueous space. The structure of the photosynthetic membrane and the consequences of separating electron transport complexes between grana and stroma membranes are discussed on pp. 100–2.

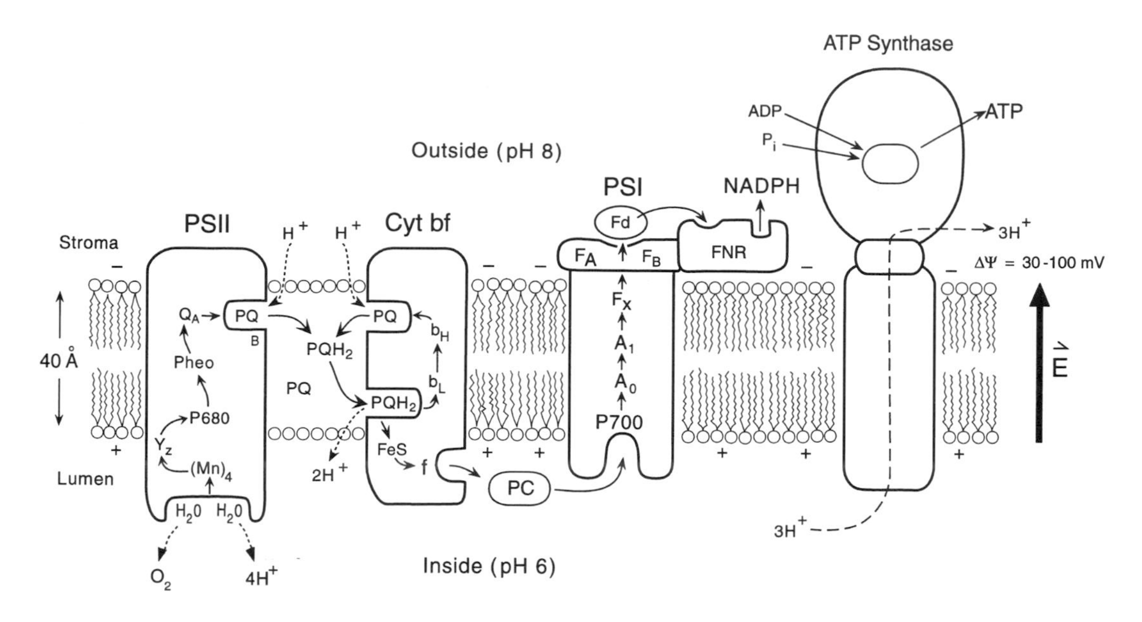


**Figure 7.1.** Model of the photosynthetic membrane of plants showing the electron transport components and the ATP synthase enzyme (cross-sectional view). The complete membrane forms a vesicle. The pathways of electrons are shown by solid arrows. The membrane-bound electron transport protein complexes involved in transferring electrons are the PSII and PSI reaction centres and the cytochrome *bf* complex (Cyt *bf*). Abbreviations: Tyr, a specific tyrosine on the D1 protein; P680 and P700, the reaction centre chlorophyll of PSII and PSI, respectively; Pheo, pheophytin; $Q_A$, and $Q_B$ bound plastoquinones; $QH_2$, reduced plastoquinone; Cyt $b_L$ and Cyt $b_H$, low and high forms of cytochrome $b_6$; FeS, Rieske iron–sulphur centre; *f*, cytochome *f*; PC, plastocyanin; $A_0$, chlorophyll; $A_1$, phylloquinone; $F_X$, $F_A$ and $F_B$, iron–sulphur centres; Fd, ferredoxin; FNR, ferredoxin/$NADP^+$ oxidoreductase; NADPH, nicotinamide adenine dinucleotide phosphate (reduced form); ADP, adenosine diphosphate; ATP, adenosine triphosphate; $P_i$, inorganic phosphate; $H^+$, protons; ΔΨ, the light-induced electrical potential across the membrane. The light-harvesting protein complexes are not shown.

Electron transfer from water to $NADP^+$ involves three integral membrane protein complexes operating in series: the PSII reaction centre, the cytochrome *bf* complex, and the PSI reaction centre (Fig. 7.1). The two reaction centres are the site of primary charge separation, in which light energy or exciton energy is transformed into redox-free energy. Photosynthetic electron transport consists of a series of electron transfer steps from one electron carrier to another over relatively short distances (typically less than 20 Å). Most of the electron carriers are metal ion complexes bound within proteins, although a few of the carriers are aromatic groups. The only non-protein carriers involved in photosynthetic electron transport are plastoquinone and NADPH. The transfer of an electron from one site to another is known as an oxidation/reduction or redox reaction (for a discussion of the thermodynamics of redox reactions see Cramer & Knaff, 1991). In photosynthetic electron transport an oxidation reaction is always coupled to a reduction reaction, so that each component acts as an electron acceptor and as an electron donor. Furthermore, some redox reactions involve the loss or gain of protons along with electrons. This coupling of proton transfer to electron transport reactions is essential for energy transduction by the photosynthetic membrane.

The energy for this uphill electron transfer reaction is provided by light that is either absorbed by the reaction centre directly or transferred to it as an exciton from the light harvesting antenna system (light absorption and the antenna system are described in Chapter 2). The primary photochemical reaction of photosynthesis is charge transfer from an excited electronic state of a specialized electron donor to an electron acceptor. This reaction occurs in each reaction centre. In the case of PSII, the primary electron donor is reduced by electrons removed from water. The oxidation of two water molecules by PSII results in the release of molecular oxygen into the atmosphere. In this section we will ignore the pathway of electrons and protons within PSII and view it as a protein complex that extracts electrons from water molecules and transfers them to plastoquinone. The internal workings of PSII and the other protein complexes will be discussed on pp. 92–4.

As shown in Fig. 7.1, electrons are transferred from PSII to the cytochrome *bf* complex by plastoquinone, which functions as a mobile electron carrier within the hydrophobic core of the photosynthetic membrane. Plastoquinone is a key player in energy transduction because it links electron transport to proton transfer across the photosynthetic membrane. The reduction of plastoquinone by PSII requires two electrons and two protons, creating $PQH_2$. The reduced plastoquinone molecule unbinds from PSII and diffuses in the photosynthetic membrane until it encounters a specific binding site on the cytochrome *bf* complex, which is a membrane-bound protein complex containing four electron carriers. In a complicated reaction sequence that is not fully understood, the cytochrome *bf* complex removes two electrons from $PQH_2$ and releases protons into the inner aqueous space. The electrons extracted from plastoquinone are transferred to plastocyanin, a small copper-containing protein that operates in the inner aqueous space of the photosynthetic membrane.

The next step in electron transfer is driven by light and occurs in the PSI reaction centre. As in PSII, the primary photochemical reaction is charge separation between the primary donor of PSI, P700, and the primary acceptor, a chlorophyll *a* molecule. Plastocyanin serves as the electron donor to $P700^+$. Ferredoxin, a small FeS protein located in the stromal aqueous space of the chloroplast, serves as the electron acceptor. Electrons are transferred from ferredoxin to $NADP^+$ by ferredoxin–NADP oxidoreductase (FNR), a peripheral flavoprotein located on the stromal surface of the photosynthetic membrane. The final product of light-driven linear electron transport is NADPH, a small mobile electron carrier operating in the stromal phase of the chloroplast. The transfer of a single electron from water to $NADP^+$ involves about 29 metal ions, including Fe, Mg, Mn and Cu, and approximately seven non-metal carriers, including quinones, pheophytin, NADPH, tyrosine, and flavin.

## PHOTOSYNTHETIC ENERGY TRANSDUCTION

To convert the transient energy of a photon into stable chemical bond energy, the photosynthetic apparatus performs a series of energy transducing reactions (Fig. 7.2). The process is initiated by absorption of a photon by an antenna molecule, which converts light energy to an excited electronic state known as an exciton. The antenna system for each reaction centre contains 200–300 antenna molecules (mainly chlorophyll and carotenoids) that are anchored to light harvesting proteins within the photosynthetic membrane (see Chapter 2 for a discussion of the light harvesting proteins and Chapter 4 for a discussion of the antenna pigments). The fate of the exciton is determined by the structure of the light harvesting protein. Because of the proximity of other antenna molecules with the same or similar energy states, the exciton is transferred over the antenna system. During this process some excitons are converted back into photons and emitted as fluorescence, some are converted to heat, and some are trapped by a reaction centre. Under optimum conditions over 90% of the absorbed quanta are transferred within a few hundred picoseconds from the antenna system to a reaction centre. The excited state energy trapped by PSII and PSI provides the energy for primary charge separation at the reaction centre.

The relative affinity of the photosynthetic redox components for electrons is shown by their equilibrium midpoint potentials in what is traditionally known as the Z-scheme (Fig. 7.3). The electron transfer reactions are energetically downhill, from a lower (more negative) to a higher (more positive) redox potential. Primary charge separation differs from other types of electron transfer reactions in that the equilibrium midpoint potential of the primary donor (e.g. P680 in PSII) is lower in energy than the primary electron acceptor (pheophytin). Electron transfer from P680 to pheophytin can only occur if P680 is in an excited electronic state, which is created

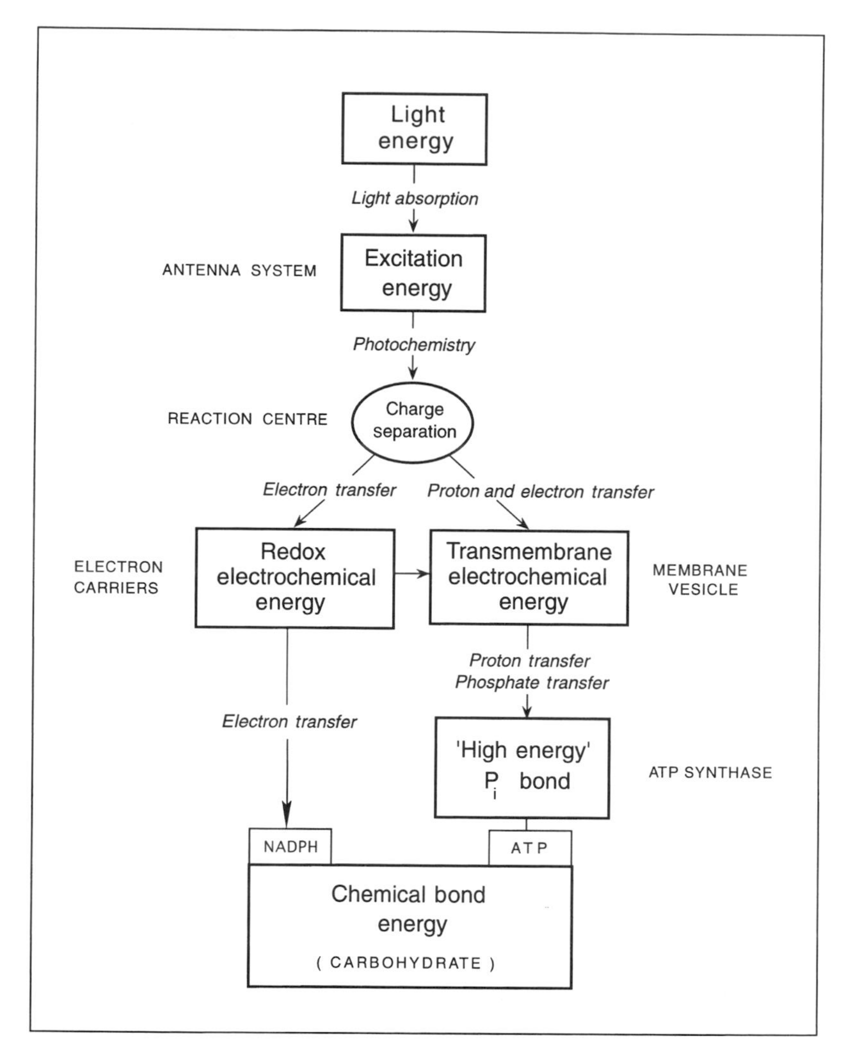


**Figure 7.2.** Photosynthesis is shown as a series of reactions that transform energy from one form to another. The different forms of energy are shown in boxes and the direction of energy transformation is shown by the arrows. The energy-transforming reactions are shown in italics. The physical site at which the energy is located is shown in capital letters outside the boxes. The primary photochemical reaction, charge separation, is shown in the oval.

either by direct absorption of a photon or by transfer of an exciton from the antenna system. It is the downhill flow of electrons, from one carrier to another, that provides the free energy for the creation of a proton electrochemical gradient and the reduction of $NADP^+$.

In addition to producing NADPH, the electron transfer reactions concentrate protons inside the membrane vesicle and create an electric potential across the membrane. In this process the redox free energy is converted into two forms of energy, both of which depend on the photosynthetic vesicle: (i) A chemical potential of protons in the form of a pH gradient across the membrane. Electron transport increases the concentration of protons in the inner aqueous space of the photosynthetic vesicle by: (a) the release of protons during the oxidation of water by PSII, and (b) the translocation of protons from the outer aqueous phase to the inner aqueous phase by the coupled reactions of PSII and the cytochrome *bf* complex. This process depends on plastoquinone reduction by PSII taking up the necessary protons from the outer aqueous space, and plastoquinol oxidation by the cytochrome *bf* complex releasing protons into the inner aqueous space (Fig. 7.1). The result of these reactions is a concentration difference of protons across the

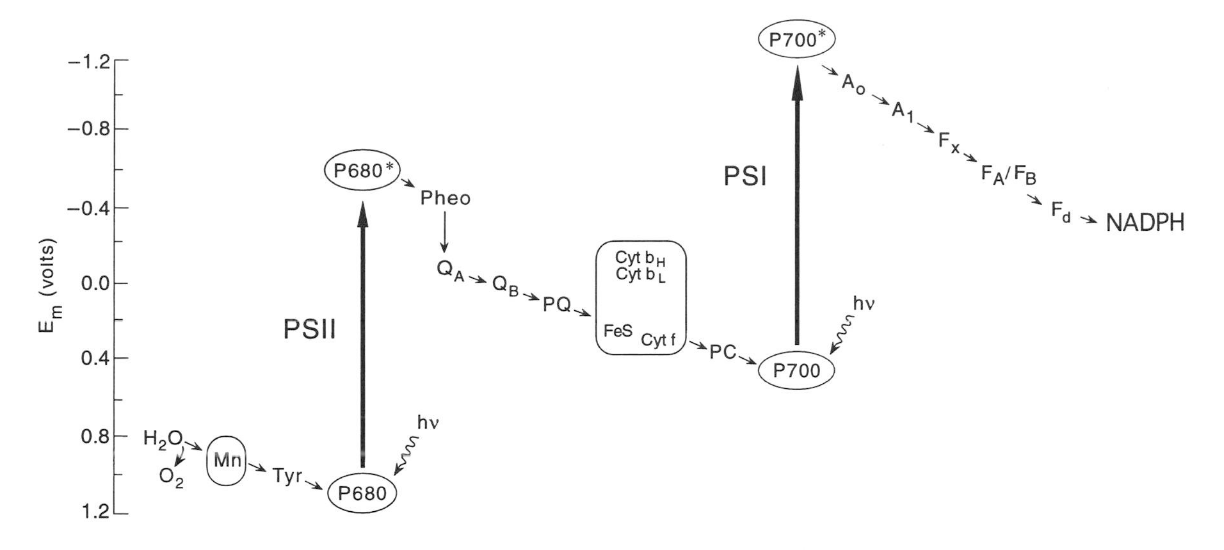


**Figure 7.3.** Z-scheme showing equilibrium midpoint potential of the photosynthetic electron transport components.

membrane ($\Delta pH = pH_{in} - pH_{out}$). (ii) An electric potential across the membrane that is created by: (a) primary charge separation at the reaction centre, and (b) charge transfer by the cytochrome *bf* complex. These reactions transfer uncompensated charge across the photosynthetic membrane (positive charge inside), producing a charge difference that creates a transmembrane electric potential ($\Delta\Psi = \Psi_{in} - \Psi_{out}$). Together, these two forms of energy make up the proton electrochemical potential ($\Delta\mu_{H+}$), which is related to the pH difference across the membrane and the electrical potential difference across the membrane by the following equation:

$$\Delta\mu_{H+} = F\ \Delta\Psi - 2.3\ RT\ \Delta pH$$

where F is the Faraday constant, R is the gas constant, and T is the temperature in Kelvin. Although the value of $\Delta\Psi$ across the photosynthetic membrane in chloroplasts can be as large as 100 mV, under steady-state conditions it appears to be below 30 mV, in which case the proton gradient dominates. For example, during photosynthesis the outer pH is typically near 8, while the inner pH may be around 6, giving a pH difference of 2 across the membrane (equivalent to about 120 mV/proton transferred across the membrane). Under these conditions the free energy for proton transfer from the inner to the outer aqueous phase due to the pH difference is −12 kJ/mol.

The next step in energy transduction is the conversion of the proton electrochemical energy stored across the photosynthetic membrane into phosphate group transfer energy in the form of ATP. This reaction depends on a single enzyme, ATP synthase (also known as coupling factor, or $CF_1$–$CF_0$). ATP synthase can be separated into two protein complexes: $CF_0$, which crosses the membrane, and; $CF_1$, which is attached to $CF_0$ on the stromal side of the membrane. Proton flow through the $CF_0$ portion of the ATP synthase provides the energy for the $CF_1$ portion to covalently attach a phosphate group to ADP, thereby forming ATP.

The net product of the light reactions is redox free energy in the form of NADPH and phosphate group transfer free energy in the form of ATP. These two compounds drive the reduction of $CO_2$ by the Calvin Cycle. Under optimum conditions, each electron extracted from water is transferred to $NADP^+$, so that for every four electrons transferred, one molecule of oxygen is released and two NADPH molecules are produced. Although the number of ATP molecules produced for each oxygen molecule released varies, values of three have been measured. Because the reduction of $CO_2$ requires two NADPH and three ATP, the theoretical minimum number of photons required for reduction of a single carbon is eight quanta (four quanta required by PSII and four by PSI). These stoichiometries can be used to calculate the theoretical energy conversion efficiency of photosynthesis (the free energy stored as reduced carbon divided by the light energy absorbed). If eight red quanta are absorbed (8 mol of red photons are equivalent to 1400 kJ) for each $CO_2$ molecule reduced (480 kJ/mol), the theoretical maximum energy efficiency for carbon reduction is 34%. During brief periods, photosynthesis in plants can achieve energy conversion efficiencies within 90% of the theoretical maximum. However, under normal growing conditions the actual performance of the plant is far below these theoretical values. The factors that conspire to lower the quantum yield include limitations imposed by biophysical and biochemical reactions and environmental conditions that limit photosynthetic performance.

## ELECTRON TRANSPORT COMPONENTS

### Photosystem II

Photosystem II is composed of over 20 polypeptides and contains at least nine different redox components (chlorophyll, pheophytin, plastoquinone, tyrosine, Mn, Fe, cytochrome b559, carotenoid, and histidine) that have been shown to undergo light-induced electron transfer. However, only five of these redox components are known to be involved in transferring electrons from $H_2O$ to the plastoquinone pool: the water oxidizing manganese cluster $(Mn)_4$, the amino acid tyrosine, the reaction centre chlorophyll (P680), pheophytin, and two plastoquinone molecules, $Q_A$ and $Q_B$. Of these essential redox components, tyrosine, P680, pheophytin, $Q_A$, and $Q_B$ have been shown to be bound to two key polypeptides (D1 and D2) that form the reaction centre core of PSII. There is mounting evidence that the D1 and D2 polypeptides also provide ligands for the $(Mn)_4$ cluster.

Despite a great deal of effort, the three-dimensional structure of PSII has not been determined. Structural models have been developed that are based on the atomic structure of the reaction centre in purple bacteria (Deisenhofer *et al.*, 1985) and biochemical and spectroscopic data for PSII (*e.g.* Ruffle *et al.*, 1992; Svensson *et al.*, 1996; Xiong, Subramaniam & Govindjee, 1996). In Chapter 3, He and Malkin describe the biochemistry, molecular biology, and function of PSII (see Fig. 3.2 for a schematic view of PSII in the membrane). In this chapter, we focus on electron and proton transfer. (For detailed discussions of the structure, function, and composition of PSII, see Bricker & Ghanotakis, 1996; Britt, 1996; Haumann & Junge, 1996; Satoh, 1996; Diner & Babcock, 1996.)

Photosystem II uses light energy to drive two chemical reactions: the oxidation of water and the reduction of plastoquinone. Photochemistry in PSII is initiated by charge separation between P680 and pheophytin, creating $P680^+/Pheo^-$. Although P680 is known to be chlorophyll *a*, it is not known whether it functions as a dimer or monomer. The primary charge separation reaction takes only a few picoseconds (Fig. 7.4). Subsequent electron transfer steps prevent the primary charge separation from recombining by transferring the electron within 200 picoseconds from pheophytin to a plastoquinone molecule ($Q_A$) that is permanently bound to PSII. Although plastoquinone normally acts as a two-electron acceptor, it works as a one-electron acceptor at the $Q_A^-$ site. The electron on $Q_A^-$ is then transferred to another plastoquinone molecule that is loosely bound at the $Q_B^-$ site. Plastoquinone at the $Q_B^-$ site differs from $Q_A$ in that it works as a two-electron acceptor, and becomes fully reduced and protonated after two photochemical turnovers of the reaction center. The full reduction of $Q_B$ requires the addition of two electrons and two protons. The reduced plastoquinone then unbinds from the reaction centre and diffuses in the hydrophobic core of the membrane, after which an oxidized plastoquinone molecule finds its way to the $Q_B^-$ binding site and the process is repeated. Because the $Q_B^-$ site is near the outer aqueous phase, the protons added to plastoquinone during its reduction are taken from the outside of the membrane. There are several compounds that inhibit electron transport by binding at or near the $Q_B^-$ site, preventing access to plastoquinone. These $Q_B^-$ site inhibitors have been extremely useful for investigating PSII function, and some are important commercial herbicides (for review see Oettmeier, 1992; see Chapter 25 for discussion of herbicides).

As stated earlier, some of the exciton energy in the antenna system is emitted as fluorescence from excited chlorophyll. Because PSII competes for exciton energy, the level of fluorescence is controlled in part by the redox state of components within the reaction centre. Measurements of chlorophyll fluorescence, then, provide a

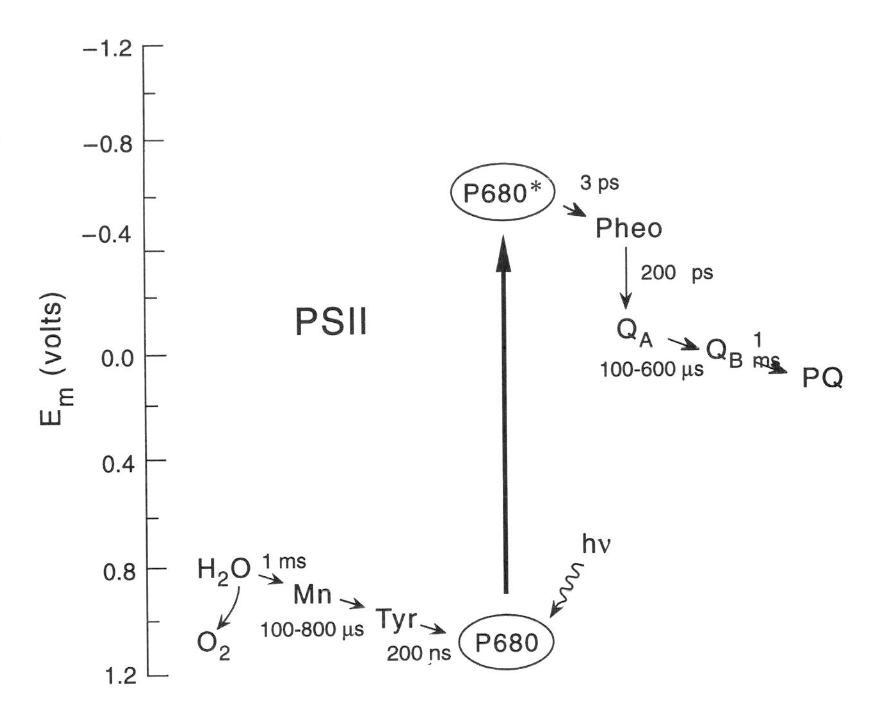


**Figure 7.4.** Photosystem II electron transport pathway showing half-times of electron transfer reactions. The vertical axis shows the midpoint potential of the electron carriers. The heavy vertical arrow indicates light absorption. P680* is the electronically excited state of P680. The abbreviations are given in the legend of Fig. 7.1.

powerful, non-invasive method for studying PSII function and its interaction with the antenna system in isolated PSII complexes, isolated membranes, and leaves (see Chapter 24 for a discussion of chlorophyll fluorescence as a diagnostic tool of plant performance).

Photosystem II is the only known protein complex that can oxidize water, which results in the release of $O_2$ into the atmosphere. Water is a poor electron donor with an oxidation–reduction midpoint potential of +0.82 V (pH 7). Although we know that oxidation is driven by the oxidized primary donor, $P680^+$ (the midpoint potential of $P680/P680^+$ is estimated to be +1.2 V at pH 7), it is unclear how electrons are transferred from water to $P680^+$. We do know that $P680^+$ oxidizes a tyrosine on the D1 protein and that four Mn ions are at the core of the oxygen evolving complex; and that water oxidation requires two molecules of water and involves four sequential turnovers of the reaction centre. In addition, X-ray spectroscopy shows that Mn in the cluster undergoes light-induced oxidation. Other components, including $Y_z$ and histidine, may be directly involved in the oxidation of water. Although calcium ions and chloride ions are required for water oxidation, their roles are unclear.

One of the classic experiments in photosynthesis is the demonstration that oxygen release by PSII occurs with a four flash dependence (for review see Joliot & Kok, 1975). Using the results of this experiment, Kok and coworkers developed a kinetic model of oxygen evolution based on five S-states (for review see Britt, 1996). Each photochemical reaction creates an oxidant that removes one electron, driving the oxygen-evolving complex to the next higher S-state. The result is the creation of four oxidizing equivalents in the oxygen-evolving complex. It is not known whether the oxidation of two water molecules leading to oxygen evolution occurs in sequential steps during the S-state transitions, or occurs in a concerted reaction in the $S_3 \rightarrow S_4 \rightarrow S_0$ transition. The net reaction results in the release of one $O_2$ molecule, the deposition of four protons into the inner water phase, and the sequential transfer of four electrons through the reaction centre to the plastoquinone pool.

Photosystem II reaction centres contain a number of redox components with no clear function. For example, cytochrome b559, a haem protein, is an essential component of all PSII reaction centres (for review see Whitmarsh & Pakrasi, 1996). If the cytochrome is not present in the membrane, a stable PSII reaction centre cannot be

Q-cycle are shown in Fig. 7.5. Initially, plastoquinol binds to the $Q_O$-site and plastoquinone binds to the $Q_R$-site. In the dark, the Rieske FeS centre and cytochrome *f* is reduced, and cytochrome $b_L$ and cytochrome $b_H$ are oxidized. The oxidation of plastoquinol at the $Q_O$-site is initiated by oxidation of the Rieske FeS centre by PSI turnover. The oxidized FeS prompts a concerted reaction, in which two electrons are transferred from plastoquinol, the first electron going to the Rieske centre, and the second electron to cytochrome $b_L$. The removal of electrons from plastoquinol leads to the release of two protons into the inner aqueous phase. The electron transferred to the Rieske centre is subsequently transferred to cytochrome *f*. The electron transferred to cytochrome $b_L$ is subsequently transferred to cytochrome $b_H$, which then transfers the electron to plastoquinone at the $Q_R$-site. Once plastoquinol is oxidized, it unbinds from the complex, and is replaced by $PQH_2$. The complete Q-cycle requires another turnover of the cytochrome *bf* complex. The sequence of events is the same as for the first half of the cycle, except plastoquinone at the $Q_R$-site is present in its semiquinone form ($PQ^-$ or PQH). As before, the oxidation of plastoquinol at the $Q_O$-site is driven by oxidation of the Rieske FeS centre, with the two electrons following the same two pathways. The second half of the cycle leads to the full reduction of plastoquinone at the $Q_R$-site, including the uptake of two protons from the outer aqueous phase. $PQH_2$ then unbinds from the $Q_R$-site, and enters the hydrophobic core of the membrane. Much of the experimental evidence supporting two quinone binding sites on the cytochrome *bf* complex and the operation of a Q-cycle depends on the action of electron transport inhibitors (Kallas, 1994).

Overall, the Q-cycle requires two photochemical turnovers of PSI and leads to: (i) transfer of two electrons from the plastoquinone pool to PSI; (ii) oxidation of two $PQH_2$; (iii) reduction of one PQ; (iv) release of four protons into the inner aqueous space; (v) uptake of two protons from the outer aqueous space; and (vi) creation of an electric potential across the membrane ($\Delta\Psi$) due to transfer of an electron from cytochrome $b_L$ to cytochrome $b_H$ and possibly proton uptake associated with plastoquinone reduction.

There is experimental evidence for a Q-cycle operating in isolated photosynthetic membranes and in leaves. However, under some conditions (steady-state electron transport at high light intensities, for example), it is not clear if the Q-cycle is operating; the predicted additional proton translocation is not observed, nor is the additional ATP. These observations have led to the suggestion that the Q-cycle may be switched off under some conditions, and that plastoquinol oxidation may proceed without the translocation of an extra proton.

## Plastocyanin

Plastocyanin operates in the inner aqueous phase of the photosynthetic vesicle, transferring electrons from cytochrome *f* to P700, the primary donor of PSI. It is a small protein (ca. 10 kDa) composed of a single polypeptide that is coded for in the nuclear genome. Plastocyanin contains a copper ion that is ligated to four residues of the polypeptide. The copper ion serves as a one-electron carrier with a midpoint potential (370 mV) near that of cytochrome *f*. Plastocyanin shuttles electrons from the cytochrome *bf* complex to PSI by diffusion (and may be involved in shuttling electrons from grana to stromal membranes (see pp. 101–2). Although all plants depend on plastocyanin for linear electron transport, some algae and cyanobacteria can replace it with cytochrome $c_6$. Because algae and cyanobacteria can use either protein for photosynthesis, they provide powerful systems for using directed mutagenesis to study genetic and physiological processes.

Plastocyanin is one of the best characterized proteins in the electron transport chain (for review see Gross, 1996). Amino acid sequences from numerous species are known, as is the atomic structure of plastocyanin from several species. Kinetic measurements of electron transfer show that reduced plastocyanin can bind to PSI, donate an electron to $P700^+$, unbind from the complex, diffuse through the inner aqueous phase, bind to reduced cytochrome *f*, and accept an electron in a few hundred microseconds. Under steady-state conditions this process can be repeated at rates in excess of 1000 times per second. Biochemical and kinetic studies show that charges of the surface of cytochrome *f* and plastocyanin are important in determining both binding and the proper orientation of the two proteins for electron transfer. Attention is currently focused on the effect of alter-

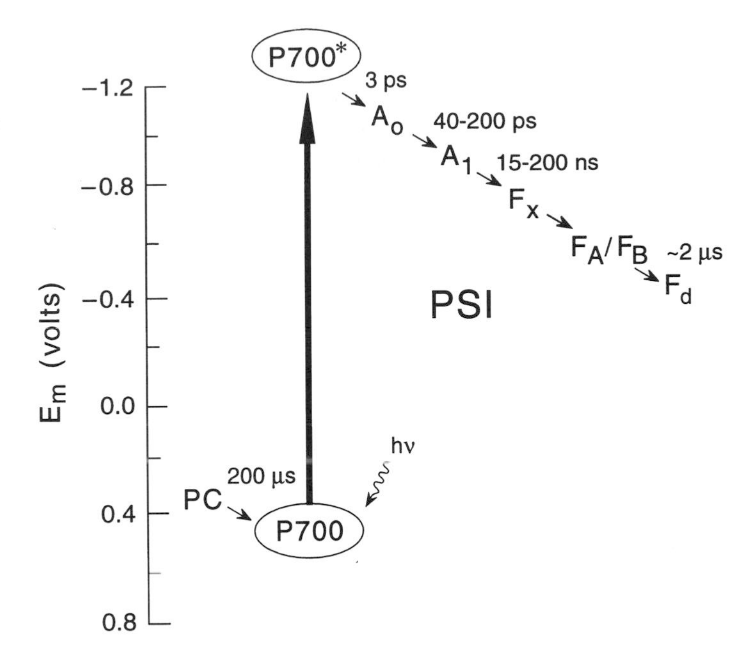


**Figure 7.6.** Photosystem I electron transport pathway showing half-times of electron transfer reactions. The vertical axis shows the midpoint potential of the electron carriers. The heavy vertical arrow indicates light absorption. P700* is the electronically excited state of P700. The abbreviations are given in the legend of Fig. 7.1.

ing charged residues of cytochrome *f* and plastocyanin on binding and electron transfer rates.

## Photosystem I

In plants, the PSI complex catalyses the oxidation of plastocyanin and the reduction of ferredoxin. The reaction centre is composed of more than ten polypeptides and shows considerable similarity to its counterpart, the PSII reaction centre (discussed in detail in Chapter 3). Two polypeptides, *psaA* and *psaB*, form a heterodimer that bind the primary electron donor and acceptor and the transmembrane electron carriers (see Fig. 3.3 for a schematic view of PSI in the membrane). The three-dimensional structure of PSI from a cyanobacteria has been resolved to 4.0 Å (Krauss *et al.*, 1996). The primary donor, P700, appears to be a chlorophyll dimer, and the primary acceptor, $A_O$, a chlorophyll monomer. As in PSII, a quinone known as phylloquinone (vitamin $K_1$), operates as a single electron acceptor ($A_1$) (Fig. 7.6). Electrons are transferred from $A_1$ to a 4Fe4S cluster ($F_X$) that is ligated to both polypeptides of the heterodimer. An extrinsic protein, PsaC, containing two 4Fe4S clusters, $F_A$ and $F_B$, is located on the stromal side of the reaction centre close to $F_X$. There is clear evidence that electron transfer from $F_X$ to ferredoxin involves at least one of these centres, but it is unknown whether $F_A$ and $F_B$ operate in series or in parallel, or if just one of the centres is involved. A complete description of the organization and function of PSI is given in Chapter 3. More detailed treatments are given in recent reviews by Nechustai *et al.*, 1996 and Malkin, 1996.

While there are similarities between PSI and PSII, it is worthwhile to keep in mind some of the differences that distinguish PSI. In terms of electron transport, PSI produces a strong reductant that transfers electrons to ferredoxin, while PSII produces a strong oxidant that removes electrons from water. These two reactions are at the opposite extremes of redox chemistry in biological systems. The reducing side of PSI depends on FeS clusters, which are not present in PSII. In contrast to PSII, electron transfer within PSI does not involve protons, so the only direct contribution of PSI to the proton electrochemical potential is through the electric potential created by electron transfer across the membrane, at least in linear electron transport. The reaction centre heterodimer of PSI contains about 100 chlorophyll molecules, whereas the reaction centre of PSII contains less than ten chlorophyll molecules. Lastly, PSI operates in cyclic electron transport, contributing to the conversion of redox free energy into the proton electrochemical potential (see p. 104).

## Ferredoxin

Ferredoxin operates in the stromal aqueous phase of the chloroplast, transferring electrons from PSI to a membrane-associated flavoprotein, known as FNR. A 2Fe2S cluster, ligated by four cysteine residues, serves as a one-electron carrier. Ferredoxin is a small protein (ca. 11 kDa), and has the distinction of being one of the strongest soluble reductants found in cells ($E_m \simeq -420$ mV). The amino acid sequence for ferredoxin from different species is known as well as the three-dimensional structure. Plants contain different forms of ferredoxin, all of which are encoded in the nuclear genome. In some algae and cyanobacteria, ferredoxin can be replaced by a flavoprotein.

In linear electron transport essentially all electron flow from water follows the pathway shown in Fig. 7.1, at least up to ferredoxin. Once an electron reaches ferredoxin, however, the electron pathway becomes branched, enabling redox free energy to enter other metabolic pathways in the chloroplast. For example, ferredoxin can transfer electrons to nitrite reductase, glutamate synthase, and thioredoxin reductase (for review see Knaff, 1996). In addition, ferredoxin may be a participant in PSI cyclic electron transport (p. 104), and it may be involved in a Mehler reaction that leads to the reduction of molecular oxygen (Asada & Takahashi, 1987).

## FNR

Ferredoxin-$NADP^+$ oxidoreductase (FNR) links the one-electron donor, ferredoxin, to the two-electron acceptor $NADP^+$ (for review see Knaff, 1996). The electron carrier in FNR is flavin adenine dinucleotide (FAD) which, under equilibrium conditions, is a two-electron acceptor with a midpoint potential near −360 mV at pH 7. For catalytic activity, two molecules of ferredoxin must donate electrons to FNR, which in turn reduces $NADP^+$ to NADPH. FNR is a single polypeptide (ca. 35 kDa) which is encoded by a nuclear gene. It operates as a peripheral protein, bound to the stromal side of the photosynthetic membrane near PSI. The FNR amino acid sequence is known for several species and the three-dimensional structure of spinach FNR has been determined to 1.7 Å (Bruns & Karplus, 1995). It appears that ferredoxin and $NADP^+$ bind to FNR at two different sites.

## NADP

The final electron acceptor in the photosynthetic electron transport chain is $NADP^+$, which is fully reduced by two electrons (and one proton) to form NADPH. NADPH is a strong reductant ($E_m = -320$ mV at pH 7), and serves as a mobile electron carrier in the stomal aqueous phase of the chloroplast. Although NADPH is a powerful reductant, it reacts slowly with oxygen, which enables it to serve a stable source of electrons for the reduction of carbon dioxide (Chapter 8), as well as in other reductive biosynthetic pathways.

### ATP SYNTHASE

The conversion of proton electrochemical energy into chemical free energy is accomplished by a single protein complex known as ATP synthase (for review see McCarty, 1996). This enzyme catalyses a phosphorylation reaction, which is the formation of ATP by the addition of inorganic phosphate ($P_i$) to ADP:

$$ADP^{-3} + P_i^{-2} + H^+ \rightarrow ATP^{-4} + H_2O$$

The reaction is energetically uphill ($\Delta G = +32$ kJ/mol) and is driven by the transmembrane proton electrochemical gradient, $\Delta\mu_{H^+}$. ATP synthase is a membrane-bound enzyme that is composed of two oligomeric subunits, $CF_0$ and $CF_1$ (Fig. 7.7). $CF_0$ is hydrophobic and spans the membrane, forming a proton channel through the membrane. $CF_0$ is made from four different polypeptides known as I (17 kDa), II (16.5 kDa), III (8 kDa), IV (27 kDa). The stoichiometries of the polypeptide subunits are not known, but appear to be in the range of 1:1:(6–12):1, respectively, giving a molecular mass of 110–160 kDa. $CF_1$ is hydrophilic and is attached to the top of the $CF_0$ on the stromal side of the membrane. $CF_1$ has a molecular mass near 400 kDa and is composed of five different protein subunits, $\alpha$ (55 kDa), $\beta$ (54 kDa), $\gamma$ (36 kDa), $\delta$ (20 kDa) and $\varepsilon$ (15 kDa), at stoichiometries of 3:3:1:1:1, respectively. Although the atomic structure of chloroplast ATP synthase is not known, biochemical and

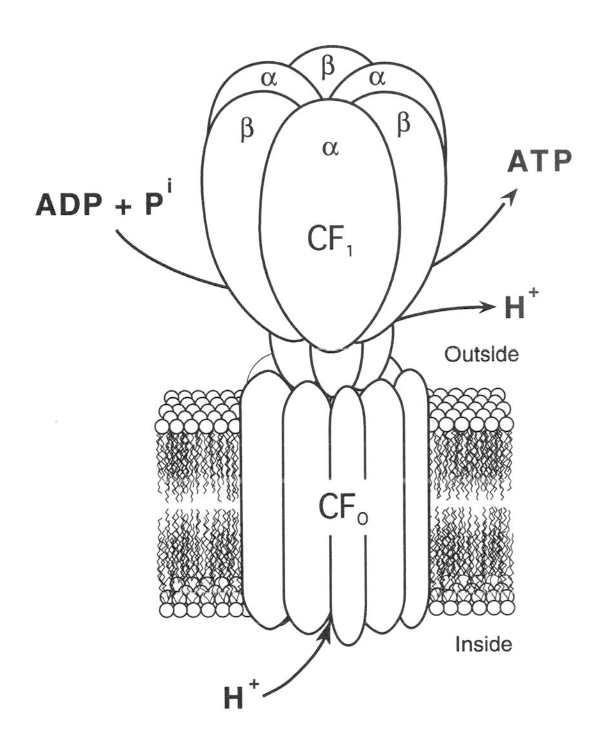


**Figure 7.7.** Model of ATP synthase embedded in the membrane. Proton transfer through the ATP synthase provides the energy for the creation of ATP from ADP and $P_i$. Abbreviations are given in the legend of Fig. 7.1.

biophysical data for the $CF_1$ part of the enzyme plus the structure of mitochondrial ATP synthase from beef heart (Abrahams *et al.*, 1994), which has been resolved to 2.8 Å, provides the basis for a structural model (for review see Boekema and Lücken 1996). $CF_0$ is embedded in the membrane and is thought to have a stalk that extends into the stroma aqueous phase and acts as an anchor for $CF_1$. If $CF_1$ is removed from the photosynthetic membrane, protons can pass rapidly through $CF_0$, preventing the build-up of a proton gradient. Rebinding of $CF_1$ to $CF_0$ blocks the free flow of electrons. Observations such as these provide a clear demonstration that $CF_0$ acts as a proton channel and that the flux of protons is regulated by $CF_1$. The α- and β-subunits of $CF_1$ alternate to form a ring structure with a diameter near 110 Å. The outside surface of the αβ structure contains the catalytic sites for ATP synthesis. Although, when viewed from above the membrane, the six αβ-subunits appear symmetrical, there are significant differences in the environments of the subunits due to their interaction with the γ, δ, and ε subunits. This asymmetry plays a key role in models attempting to explain the mechanism of ATP synthesis. Structural and biochemical data reveal six nucleotide binding sites on each $CF_1$. However, which sites are catalytic and which are non-catalytic remains to be determined. Located on the membrane side of the αβ structure are the γ, δ, and ε subunits, which serve to attach $CF_1$ to $CF_0$ and play a role in regulation of the catalytic activity and controlling proton flux through $CF_0$.

Protons moving through the ATP synthase protein (from inside to outside the vesicle) provide the energy for ATP synthesis. The molecular processes that couple proton transfer through the protein to the chemical addition of phosphate to ADP are not clear. It is known that phosphorylation is driven by a pH gradient, a transmembrane electric field, or a combination of the two. Evidence for this process depends, in part, on compounds that collapse the transmembrane proton electrochemical gradient (uncouplers). Uncouplers prevent the formation of ATP, but do not inhibit electron flow (see Cramer & Knaff, 1991). Experiments indicate that at least three protons must pass through the ATP synthase complex for the synthesis of one molecule of ATP. However, the protons are not involved in the chemistry of adding phosphate to ADP at the catalytic site. Current models envision that the flux of protons through the enzyme causes a conformational change of $CF_1$ that alters binding affinities at the catalytic sites for substrates and ATP (for review see Boyer 1993). In other words, energy in the form of a proton electrochemical gradient is used to cause a protein conformational change that leads to synthesis of ATP, resulting in free energy in the form of a group transfer potential. Support for this idea is provided by compounds, known as energy transfer inhibitors (e.g. DCCD), that block phosphorylation by binding to ATP synthase (McCarty, 1996).

These models envision two or three catalytic sites on each $CF_1$. The sites alternate between different states that have different binding affinities for nucleotides. One model proposes that there are three catalytic sites on each $CF_1$ that cycle among different states. The states differ in their affinity for ADP, $P_i$ and ATP. At any one time each site is in a different state. This model is supported by the structure of mitochondrial ATP synthase (Abrahams *et al.*, 1994). Initially, one catalytic site on

$CF_1$ binds one ADP and one inorganic phosphate molecule relatively loosely. Due to a conformational change of the protein, the site becomes a tight binding site that stabilizes ATP. Next, proton transfer induces an alteration in protein conformation that causes the site to release the ATP molecule into the aqueous phase. In this model, the energy from the proton electrochemical gradient is used to lower the affinity of the site for ATP, allowing its release to the water phase. The three sites on $CF_1$ act co-operatively (i.e. the conformational states of the sites are linked). It has been suggested that protons cause the conformational change in $CF_1$ by driving the rotation of the $\alpha\beta$-subunits relative to the $\gamma$, $\delta$, and $\varepsilon$ subunits (see Boyer, 1993). Support for the revolving site mechanism has been provided by work showing the rotation of the $\gamma$-subunit relative to the $\alpha\beta$ subunits by Sabbert, Engelbrecht and Junge (1996). It is worth noting that the revolving site mechanism would require rotation rates as high as 100 revolutions per second to account for *in vivo* rates of phosphorylation. Modelling studies based on control theory indicate that ATP synthase turnover does not limit the rate of photosynthetic electron transport (see Mills, 1996). This is not surprising, considering that ATP synthase can turnover 400 times per second, whereas rates of phosphorylation *in vivo* requires less than half this turnover number.

In the dark chloroplast ATP synthase is down regulated (for review see Mills, 1996). This regulation is essential to prevent ATP synthase activity, the exergonic hydrolysis of ATP, which would deplete the chloroplast of ATP. Activation of ATP synthase is controlled chiefly by the transmembrane electrochemical gradient ($\Delta\mu_{H+}$). Once $\Delta\mu_{H+}$ goes above a threshold value, ATP synthase becomes active. Activation by $\Delta\mu_{H+}$ depends in part on the concentrations of substrates (ADP and $P_i$) as well as ATP, indicating a role for substrate or product binding in controlling activation of the enzyme. Although the molecular process by which $\Delta\mu_{H+}$ switches on the ATP synthase is not known, it is presumed to involve protonation and deprotonation reactions of residues on $CF_0$ and $CF_1$. In addition to control by $\Delta\mu_{H+}$, an oxidation–reduction reaction plays a secondary, but significant, role in controlling activity. The reduction reaction is driven by thioredoxin, a soluble protein that is also involved in redox control of enzymes in the carbon reduction cycle. Thioredoxin is reduced by ferredoxin via thioredoxin reductase (Knaff, 1996), which represents one of the branches of electron transport discussed on p. 98. Thioredoxin contributes to the activation of ATP synthase by reducing a disulphide bridge in the $\gamma$-subunit formed by two cysteine residues. When the $\gamma$-subunit is reduced, activation by $\Delta\mu_{H+}$ occurs at a much lower energy threshold. Only chloroplast ATP synthase appears to be under redox control. ATP synthase in cyanobacteria and other photosynthetic bacteria operates like the reduced form of chloroplast ATP synthase. This difference is probably due to the observation that photosynthetic bacteria can use the photosynthetic membrane ATP synthase for oxidative phosphorylation.

## ORGANIZATION OF THE COMPONENTS IN THE PHOTOSYNTHETIC MEMBRANE

### Distribution of protein complexes

The photosynthetic membrane of chloroplasts exhibits a remarkable degree of lateral asymmetry, both in its architecture and in the distribution of protein complexes between stacked and unstacked membranes (for review Staehelin & van der Staay, 1996). Despite decades of research directed at understanding the structural organization of the membrane and the forces involved in holding it together, the functional significance of the asymmetry is still not clearly delineated. The heterogeneous distribution of protein complexes between grana and stroma membranes for a typical sun adapted plant is shown in Fig. 7.8. Also shown in Fig. 7.8 are the relative amounts of various components of the photosynthetic apparatus on a chlorophyll basis. (Using a membrane area per chlorophyll molecule of 200 Å, the density of the photosynthetic components can be calculated in grana and stroma membranes). The most distinct features of the asymmetry are the exclusion of ATP synthase from grana appressed membranes and the separation of PSII and PSI between grana and stroma membranes. Why deploy electron transport complexes (that operate in series) so far from one another that significant demands are placed on the mobile electron carriers? In the mitochondria inner membrane, the electron transport components are arranged conveniently close to one another, and the

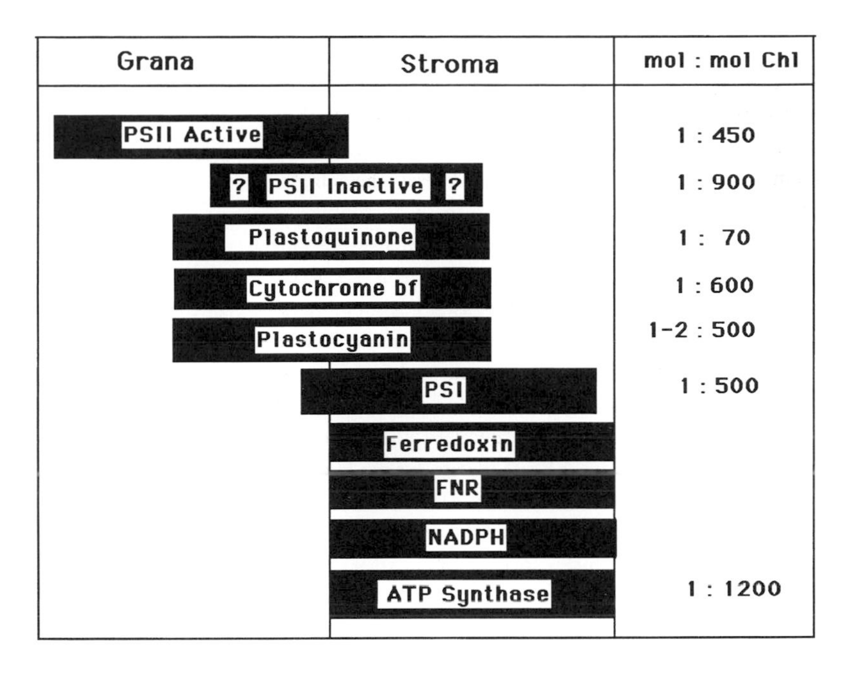


**Figure 7.8.** Diagram showing the distribution of photosynthetic components between grana and stroma membranes. The ratio of the component to chlorophyll is shown on the left. This Figure is representative of a sun-adapted plant (e.g. pea or spinach) grown in the field. References for the distribution of complexes between grana and stroma membranes and a discussion of variations that occur in different species can be found in Staehelin and van der Staay (1996). The chlorophyll to component ratios are taken from Lee and Whitmarsh (1989) for the electron transport components and Hangarter et al. (1987) for ATP synthase. Photosystem II active refers to reaction centres that contribute to energy transduction and Photosystem II inactive refers to reaction centres impaired in the oxidation of $Q_A$ (Chylla & Whitmarsh, 1989).

photosynthetic membranes of cyanobacteria do not form grana, yet they can transport electrons at reasonably robust rates. In addition, mutants of barley that lack chlorophyll *b* and do not form grana stacks, grow photosynthetically, which is an indication that membrane stacking is not essential for photosynthetic activity in chloroplasts.

Over the years several different ideas have been developed with the aim of revealing an advantage for the chloroplast in membrane stacking. The observation that PSII light-harvesting complexes can move from grana to stromal membranes prompted the suggestion that membrane stacking allows rearrangement of antenna complexes between PSII and PSI in response to different light environments. The notion is that light energy can be balanced between the two photosystems for maximum quantum efficiency. While this remains a viable explanation, most experiments reveal a relatively small (or even negligible) change in the antenna sizes of the two photosystems. Trissl and Wilhelm (1993) have developed the idea that separation of PSII and PSI is necessary because the reaction centre of PSI traps exciton energy more rapidly than does PSII. In addition, the average energy level of an exciton in the PSI antenna system is lower than in the PSII antenna system, an imbalance that would again favour energy trapping by PSI.

## Electron transport by diffusion

Photosynthetic electron transport depends on a few small electron carriers that transport electrons or hydrogen atoms over relatively long distances (hundreds or even thousands of Å). Within the photosynthetic membrane, PSII and PSI are linked by two of these mobile carriers: plastoquinone and plastocyanin (for review see Whitmarsh, 1986). In the stromal aqueous phase, ferredoxin and NADPH carry electrons from PSI to other redox components in the chloroplast. This transport is driven by diffusion, which is the movement of molecules due to random thermal motion. For molecules like plastoquinone, the motion is limited to the two-dimensional region defined by the liquid hydrophobic core of the membrane. For plastocyanin, the motion is limited to the inner water phase of the photosynthetic membrane. The net movement would be from a region of higher concentration to a region of lower concentration, as described by Fick's laws of Diffusion.

The estimated average distance between the major protein complexes in the photosynthetic membrane is based on a number of different experimental results and assumptions (e.g. Whitmarsh, 1986; Staehelin & van der Staay, 1996). For example, within grana membranes PSII complexes are estimated to be a few hundred Å from one another (centre to centre distance). Assuming the cytochrome *bf* complexes are evenly distributed between grana and stroma membranes, the maximum edge to edge distance from PS II to cytochrome *bf* would be less than 200 Å. This means that plastoquinone must shuttle electrons over hundreds of Å within a few milliseconds. To accomplish this, plastoquinone operates at a stoichiometry of 6–8 molecules per PSII. Because the membrane is densely packed with protein complexes (approximately 70% protein and 30% lipid by weight), the pathway of plastoquinone is likely tortuous. Measurements of plastoquinone mobility in lipid membranes containing various amounts of protein indicate that diffusional motion is significantly slowed by integral proteins (Blackwell & Whitmarsh 1990). Nevertheless, the rates of diffusion of plastoquinone are adequate to account for even the highest rates of photosynthetic electron transport. Plastocyanin, like plastoquinone, appears to be evenly distributed between grana and stromal membranes, and measurements of plastocyanin mobility indicate that its rate of diffusion in the stromal membranes is rapid and easily can account for high rates of electron transport.

The separation between PSII and PSI puts increased strain on the diffusional motion necessary to account for the electron transfer between grana and stromal membranes. It has been estimated that electron transfer from grana to stromal membranes would require plastoquinone and/or plastocyanin to diffuse over an average distance of at least 1000 Å within a few milliseconds (Whitmarsh, 1986). The experimental evidence indicates that both plastoquinone and plastocyanin can diffuse rapidly enough to do the job. However, the experiments do not measure diffusion directly, and the question of how extensive and rapid electron transfer is between the two membrane regions remains unanswered. Factors like dimerization of membrane protein complexes (Huang *et al.*, 1994), or formation of domains in the membrane that limit plastoquinone mobility (Lavergne & Briantais 1996), increase the constraints put on the diffusional motion of plastoquinone and plastocyanin.

## CONTROL OF ELECTRON TRANSPORT

In the complex system of photosynthesis, numerous factors limit and regulate the rate and efficiency of electron transport and energy transduction. Plants have evolved intricate molecular systems for regulation that include changes in light harvesting, exciton usage, PSII activity, cyclic electron transport, ATP synthase activity, and other protective and repair processes. In addition to these genetically based regulatory mechanisms, powerful environmental controls, such as light and temperature, dictate rates and efficiencies. At physiological extremes, environmental factors can damage the photosynthetic apparatus, impairing performance and inducing a number of repair and protective processes. This section will provide a brief description of some of the major factors involved in regulating and controlling photosynthetic electron transport, as well as references for detailed discussions (control of ATP synthesis was discussed on p. 100). It is worth keeping in mind, however, that an integrated understanding of the factors that control photosynthesis is still beyond us.

### Electron transport within proteins

Electron transport proteins can be thought of as polypeptide chains that provide a scaffolding for metal ions and aromatic groups. In the case of PSII, an electron is extracted from a water molecule bound to a specific site, after which it is transferred within the protein from one electron carrier to another. The pathway of electrons within the protein is controlled by the location and environment of the redox components. By setting the distance between electron carriers and controlling the electronic environment surrounding a redox component, the protein can control the rate and direction of the electron transfer reactions. The key factors determining the rate of electron transfer between two redox components are the distance between the donor and acceptor and the free energy of activation. For a general description of electron transfer mechanisms, see Krishtalik and Cramer (1996).

One of the current issues in understanding intraprotein electron transport is the importance of specific amino acids on the rate of electron transfer between two redox components. Based on an analysis of electron transfer reactions in biological and chemical systems in terms of electron tunneling theory developed by R. Marcus and others (for review see Marcus & Sutin 1985), it has been argued that the specific amino acid residues between an electron transfer pair is generally of less importance than the distance between them in determining the rate of pairwise electron transfer (Moser *et al.*, 1992). Protein controls the rate of electron transfer mainly through the distance between the donor and acceptor molecules, the free energy, and the reorganization energy of the reaction. The importance of distance is demonstrated by electron transfer data from biological and synthetic systems showing that the dependence of the electron transport rate on distance (edge to edge) is exponential over 12-orders of magnitude when the free energy is optimized (Moser *et al.*, 1992). Increasing the distance between two carriers by 1.7 Å slows the rate of electron transfer ten-fold. The extent to which this view is generally applicable for intraprotein transfer remains to be established.

## Rate limiting reactions and temperature

Measurements of the rate of photosynthetic electron transfer and the quantum requirement for oxygen evolution at low light intensities reveal that the rate is linearly dependent on the light intensity and that the quantum requirement is within 10–20% of the theoretical minimum. These observations show that light intensity controls the rate of electron transport at subsaturating levels. This is not surprising, because reaction centre turnover is limited by the rate that excitons are delivered by the light-harvesting system. The situation is different in saturating light, at which point enzyme turnover becomes limiting. For example, under a wide range of physiological conditions the rate-limiting reaction in electron transport is the oxidation of plastoquinol by the cytochrome *bf* complex. This conclusion is based in part on the observation that in high light plastoquinol is mainly reduced, while the electron acceptors in the cytochrome *bf* complex (Rieske FeS protein and cytochrome *f*) are oxidized. Measurements in leaves and isolated photosynthetic membranes from a variety of species show that maximum turnover-rate of the cytochrome *bf* complex is about 200 electrons/second around 20 °C (e.g., Lee & Whitmarsh 1989). The rate-limiting step in photosynthetic electron transport is strongly dependent on temperature. For example, from 5–25 °C the maximum rate of electron transport increases two fold for every 10 °C increase in temperature for isolated photosynthetic membranes from spinach (e.g. Whitmarsh & Cramer, 1979). This temperature dependence applies to light saturation response of electron transport also. For every 10 °C increase in temperature, the amount of light required to saturate electron transport increases two-fold (Lee & Whitmarsh, 1989).

The observation that the oxidation of plastoquinol is often the slowest step in photosynthetic electron transport does not exclude other reactions from also limiting the maximum rate (Kacser & Burns, 1973). One can ask, what is the effect on the rate of electron transport of increasing or decreasing the concentration of a given redox component (Kacser & Burns, 1973)? For example, what is the impact on the rate of electron transport of decreasing the concentration of plastocyanin by 20%, 50%, and 80% on the rate of electron transport. Experiments using isolated membranes show that a 20% decrease has virtually no impact on the rate, while an 80% decrease lowers the rate 30% (Paulson, 1986), showing that plastocyanin exerts relatively little control until it is reduced to quite a low level, and furthermore, that increasing the concentration of plastocyanin would not increase the rate of electron transport. It must be emphasized that demonstrating that plastocyanin (or any other component) is not rate-limiting under conditions optimized for electron transport does not exclude the possibility that it may become rate-limiting under suboptimal conditions, such as low temperature where the diffusional motion of plastocyanin could become a limiting factor.

## Regulation of exciton energy

Plants have evolved molecular machinery that controls the delivery of exciton energy to the reaction centres. Because each reaction centre is served by 200–300 antenna molecules, the intensity of light required to saturate electron transport is often much less than full sunlight (e.g. Lee & Whitmarsh, 1989). Consequently, it is not uncommon for PSII to operate at light intensities that far exceed its capacity for electron transport, conditions that can lead to irreversible damage to the reaction centre. Plants can avoid this damage by diverting excess excitons to a trap within the antenna system that transforms the energy into heat (Demmig-Adams & Adams, 1992; Yamamoto & Bassi, 1996). Although the details of this process are not fully understood, it is clear that down-regulation of PSII plays an important role in plant photosynthesis (for review see Demmig-Adams & Adams, 1992; Yamamoto & Bassi, 1996).

Other types of control at the light-harvesting level, known as state transitions, have been proposed based on the observation that a change in the spectral distribution of the incident light can trigger the phosphorylation of light-harvesting proteins serving PSII (for review see Allan, 1992). The process is reversible and leads to a decrease in the antenna size of PSII and a small increase in the antenna size of PSI. However, because the observed changes in antenna size are small, it is not clear that the purpose of the phosphorylation reactions is to redirect exciton energy.

## PSI Cyclic Electron Transport

Since the observation that PSI alone can drive photophosphorylation in isolated photosynthetic membranes (see Arnon & Chain, 1975), the notion of PSI cyclic electron transport has had a strong hold on the photosynthetic literature (Heber & Walker, 1992; Bendall & Manasse, 1995; Malkin, 1996). The evidence for cyclic electron transport occurring in chloroplasts, however, is sketchy and the proposed pathway for electrons has significant gaps. Compared to linear electron transport, where the carriers are known, PSI cyclic electron transport is enigmatic and its role under normal physiological conditions uncertain. The idea is that electrons from PSI are transferred from the stromal side of the membrane (via ferredoxin, FNR, or NADPH) to the inner side, presumably by plastoquinone and/or the cytochrome *bf* complex in a process that releases protons that are used to drive ATP synthesis. Despite intriguing reports of electron flow from ferredoxin (or NADPH) to plastoquinone, a clear pathway has yet to emerge (*e.g.* Cleland & Bendall, 1992). Nevertheless, circumstantial evidence for PSI cyclic electron transport in chloroplasts is strong. For example, PSI cyclic electron transport is the likely driving force for ATP synthesis in chloroplasts from bundle sheath cells, which contain very little PSII (see Chapter 9). In addition, controlling the ratio of cyclic to linear electron transport would provide a mechanism for regulating the ATP to NADPH ratio in chloroplasts, which is necessary for efficient metabolic activity.

## Environmental stress

Both field and laboratory measurements show that the photosynthetic apparatus can perform efficiently in a wide range of environmental conditions. However, plants must survive under conditions that push their molecular and physiological machinery to the limit and often beyond (see the section in this book on 'Agronomy and environmental factors'). At the molecular level, extreme environmental conditions can lead to irreversible damage, as well as the induction of protective and repair mechanisms, which can lower photosynthetic performance. For example, excess light can inhibit PSII, leading to protein degradation and repair (see Chapter 20). In the past decade our understanding of the molecular process of photosynthesis has advanced to the stage where we can begin to understand the effect of environmental stress on the photosynthetic apparatus (Baker, 1997). This knowledge, coupled with the ability to alter photosynthetic components genetically, provides a foundation for designing plants that can perform well over a broad range of environmental conditions.

### ACKNOWLEDGEMENTS

I thank Jason Whitmarsh and Himadri Pakrasi for editorial comments. Portions of this text are revised, with permission from the publisher, from an article entitled 'Photosynthesis' by J. Whitmarsh and Govindjee (1995), published in the *Encyclopedia of Applied Physics* (Vol. 13) by VCH Publishers, Inc.

### REFERENCES

Abrahams, J., Leslie, A. G. W., Lutter, R. and Walker, J. E. (1994). Structure at 2.8 A resolution of F1-ATPase from bovine heart mitochondria. *Nature*, **370**, 621–8.

Allan, J. F. (1992). Protein phosphorylation in regulation of photosynthesis. *Biochimica et Biophysica Acta*, **1098**, 275–335.

Arnon, D. & Chain, R. K. (1975). Regulation of ferredoxin-catalyzed photosynthetic phosphorylation. *Proceedings of the National Academy of Sciences, USA*, **72**, 4961–5.

Asada, K. & Takahashi, M. (1987). Production and scavenging of active oxygen in photosynthesis. In *Photoinhibition*, ed. D. J. Kyle, B. Osmond & C. A. Arntzen, pp. 227–88. Amsterdam: Elsevier.

Baker, N. (ed.) (1997). *Photosynthesis and the Environment*. The Netherlands: Kluwer Academic Press.

Bendall, D. S. & Manasse, R. S. (1995). Cyclic photophosphorylation and electron transport, *Biochimica et Biophysica Acta*, **1229**; 23–38.

Blackwell, M. F. & Whitmarsh, J. (1990). The effect of integral membrane proteins on the lateral mobility of plastoquinone in phosphotidylcholine proteoliposomes. *Biophysical Journal*, **58**, 1259–71.

Boekema, E. J. & Lücken, U. (1996). The structure of the CF1 part of the ATP-synthase complex from chloroplasts. In *Advances in Photosynthesis, vol. 4, Oxygenic Photosynthesis: The Light Reactions*, ed. D. R. Ort & C. F. Yocum, pp. 487–92. The Netherlands: Kluwer Academic Publishers.

Boyer, P. D. (1993). The binding change mechanism of ATP synthase – some probabilities and possibilities. *Biochimica et Biophysica Acta*, **1140**, 215–50.

Bricker, T. M. & Ghanotakis, D. F. (1996). Introduction to oxygen evolution and the oxygen-evolving complex. In *Advances in Photosynthesis, vol. 4, Oxygenic Photosynthesis: The Light Reactions*, ed. D. R. Ort & C. F. Yocum, pp. 113–36. The Netherlands: Kluwer Academic Publishers.

Britt, R. D. (1996). Oxygen evolution. In *Advances in Photosynthesis, vol. 4, Oxygenic Photosynthesis: The Light Reactions*, ed. D. R. Ort & C. F. Yocum, pp. 137–64. The Netherlands: Kluwer Academic Publishers.

Bruns, C. M. & Karplus, P. A. (1995). Refined crystal structure of spinach ferredoxin reductase at 1.7 Å resolution: oxidize, reduced and 2-phospho-S-AMP bound states. *Journal of Molecular Biology*, **247**, 125–45.

Chylla, R. A. & Whitmarsh, J. (1989). Inactive PS II complexes in leaves: turnover rate and quantitation. *Plant Physiology*, **90**, 765–72.

Cleland, R. E. & Bendall, D. S. (1992). Photosystem I cyclic electron transport: measurement of ferredoxin-plastoquinone reductase activity. *Photosynthesis Research*, **34**, 409–18.

Cramer, W. A., Black, M. T., Widger, W. R. & Girvin, M. E. (1987). Structure and function of photosynthetic cytochrome bc1 and b6f complexes. In *The Light Reactions*, ed. J. Barber, pp. 447–93. Amsterdam: Elsevier.

Cramer, W. A. & Knaff, D. B. (1991). *Energy Transduction in Biological Membranes*. Berlin: Springer-Verlag.

Deisenhofer, J., Epp, O. Miki, K. Huber, R. & Michel, H. (1985). Structure of the protein subunits in the photosynthetic reaction centre of *Rhodopseudomonas viridis* at 3 Å resolution. *Nature*, **318**, 618–24.

Demmig-Adams, B. & Adams, W. W. (1992). Photoprotection and other responses of plants to high light stress. *Annual Review of Plant Physiology and Plant Molecular Biology*, **43**, 599–626.

Diner, B. A. & Babcock, G. T. (1996). Structure, dynamics, and energy conversion efficiency in PS II. In *Advances in Photosynthesis, vol. 4, Oxygenic Photosynthesis: The Light Reactions*, ed. D. R. Ort & C. F. Yocum, pp. 213–47. The Netherlands: Kluwer Academic Publishers.

Gross, E. L. (1996). Plastocyanin: structure, location, diffusion and electron transfer mechanisms. *Advances in Photosynthesis, vol. 4, Oxygenic Photosynthesis: The Light Reactions*, ed. D. R. Ort & C. F. Yocum, pp. 413–29. The Netherlands: Kluwer Academic Publishers.

Hangarter, R. P., Grandoni, P. & Ort, D. R. (1987). The effects of chloroplast coupling factor reduction on the energetics of activation and on the energetics and efficiency of ATP formation. *Journal of Biologial Chemistry*, **262**, 13513–19.

Haumann, M. & Junge, W. (1996). Protons and charge indicators in oxygen evolution. In *Advances in Photosynthesis, vol. 4, Oxygenic Photosynthesis: The Light Reactions*, ed. D. R. Ort & C. F. Yocum, pp. 165–92. The Netherlands: Kluwer Academic Publishers.

Hauska, G., Schütz, M. & Büttner, M. (1996). The cytochrome $b_6f$ complex – composition, structure and function. In *Advances in Photosynthesis, vol. 4, Oxygenic Photosynthesis: The Light Reactions*, ed. D. R. Ort & C. F. Yocum, pp. 377–98. The Netherlands: Kluwer Academic Publishers.

Heber, U. & Walker, D. (1992). Concerning the dual function of coupled cyclic electron transport in leaves. *Plant Physiology*, **100**, 1621–6.

Huang, D. Everly, R. M., Cheng, R. H., Heymann, J. B., Schägge, H., Sled, V., Ohnishi, T. Baker, T. S. & Cramer, W. A. 1994). Characterization of the cytochrome b6f complex as a structural and functional dimer. *Biochemistry*, **33**, 4401–9.

Iwata, S., Saynovits, M., Link, T. A. & Michel, H. (1996). Structure of a water soluble fragment of the 'Rieske' iron-sulphur protein of the bovine heart mitochondrial cytochrome $bc_1$ complex determined by MAD phasing at 1.5 Å resolution. *Structure*, **4**, 567–79.

Joliot, P. & Kok, B. (1975). Oxygen evolution in photosynthesis. In *Bioenergetics of Photosynthesis*, ed. Govindjee pp. 387–412. NY: Academic Press.

Kacser, H. & Burns, J. A. (1973). The control of flux. In *Rate Control of Biological Processes*, ed. D. D. Davies, pp. 65–104. Cambridge, UK: Cambridge University Press.

Kallas, T. (1994). The cytochrome $b_6f$ complex. In ed. D. Bryant *The Molecular Biology of Cyanobacteria*, pp. 259–317. The Netherlands: Kluwer Academic.

Knaff, D. B. (1996). Ferredoxin and ferredoxin-dependent enzymes. In *Advances in Photosynthesis, vol. 4, Oxygenic Photosynthesis: The Light Reactions*, ed. D. R. Ort & C. F. Yocum, pp. 333–61. The Netherlands: Kluwer Academic Publishers.

Krauss, N., Schubert, W. D., Klukas, O., Fromme, P., Witt, H. T. & Saenger, W. (1996). Photosystem I at 4.0 Å resolution represents the first structural model of a joint photosynthetic reaction center and core antenna system. *Nature Structural Biology*, **3**, 965–73.

Krishtalik, L. I. & Cramer, W. A. (1996). Basic aspects of electron and proton trasnfer reactions with applications to photosynthesis. In *Advances in Photosynthesis, vol. 4, Oxygenic Photosynthesis: The Light Reactions*, ed. D. R. Ort & C. F. Yocum, pp. 399–411. The Netherlands: Klumer Academic Publishers.

Lavergne, J. & Briantais, J-M. (1996). Photosystem II heterogeneity. In *Advances in Photosynthesis, vol. 4, Oxygenic Photosynthesis: The Light Reactions*, ed. D. R. Ort & C. F. Yocum, The Netherlands: Kluwer Academic Publishers, pp. 265–287.

Lee, W. J. & Whitmarsh, J. (1989). The photosynthetic apparatus of pea thylakoid membranes: Response to growth light intensity. Plant Physiology, **89**, 932–40.

Malkin, R. (1996). Photosystem I electron transfer reactions – components and kinetics. In *Advances in Photosynthesis, vol. 4, Oxygenic Photosynthesis: The Light Reactions*, ed. D. R. Ort & C. F. Yocum, pp. 313–32, The Netherlands: Kluwer Academic Publishers.

Marcus, R. A. & Sutin, N. (1985). Electron transfers in chemistry and biology. *Biochimica et Biophysica Acta*, **811**, 265–322.

Martinez, S. E., Huang, D. Szczepaniak, A. Cramer, W. A. & Smith, J. L. (1994). Crystal structure of chloroplast cytochrome *f* reveals a novel cytochrome fold and unexpected heme ligation. *Structure (Current Biology Ltd)*, **2**, 95–105.

McCarty, R. E. (1996). An overview of the function, composition and structure of the chloroplast ATP synthase. In *Advances in Photosynthesis, vol. 4, Oxygenic Photosynthesis: The Light Reactions*, ed. D. R. Ort & C. F. Yocum, pp. 439–451. The Netherlands: Kluwer Academic Publishers.

Metzger, S., Cramer, W. A. & Whitmarsh, J. (1997). Critical analysis of the extinction coefficient of chloroplast cytochrome *f*. *Biochimica et Biophysica Acta*, In press.

Mills, J. D. (1996). The regulation of chloroplast ATP synthase, $CF_0CF_1$. In *Advances in Photosynthesis, vol. 4, Oxygenic Photosynthesis: The Light Reactions*, ed. D. R. Ort & C. F. Yocum, pp. 469–85, The Netherlands: Kluwer Academic Publishers.

Moser, C.C., Keske, J. M. Warncke, K. Farid, R. S. & Dutton, P. L. (1992). Nature of biological electron transfer. *Nature*, **355**, 796–802.

Nechushtai, R., Eden, A. Cohen, Y. & Klein, J. (1996). Introduction to photosystem I: reaction center, function, composition and structure. In *Advances in Photosynthesis, vol. 4, Oxygenic Photosynthesis: The Light Reactions*, ed. D. R. Ort & C. F. Yocum, pp. 289–311. The Netherlands: Kluwer Academic Publishers.

Oettmeier, W. (1992). Herbicides of photosystem II. In *The Photosystems: Structure, Function and Molecular Biology*, ed. J. Barber. Amsterdam: Elsevier.

Oxborough, K., Nedbal, L., Chylla, R. A. & Whitmarsh, J. (1996). Light-dependent modification of photosystem II in spinach leaves. *Photosynthesis Research*, **48**, 247–54.

Paulson, S. E. (1986). The mobility and function of plastocyanin in chloroplast photosynthetic membranes. Master's Thesis, University of Illinois, Champaign-Urbana, IL.

Ruffle, S. V., Donnelly, D., Bundell, T. L. & Nugent, J. H. A. (1992) A three-dimensional model of the photosystem II reaction centre of *Pisum sativum*. *Photosynthesis Research*, **34**, 287–300.

Sabbert, D., Engelbrecht, S. & Junge, W. (1996). Intersubunit rotation in active F-ATPase. *Nature*. **381**, 623–5.

Satoh, K. (1996). Introduction to the photosystem II reaction center–isolation and biochemical and biophysical characterization. In *Advances in Photosynthesis, vol. 4, Oxygenic Photosynthesis: The Light Reactions*, ed. D. R. Ort & C. F. Yocum, pp. 193–211. The Netherlands: Kluwer Academic Publishers.

Staehelin, L. A. & van der Staay, G. W. M. (1996). Structure, composition, functional organization and dynamic properties of thylakoid membranes. In: *Advances in Photosynthesis, vol. 4, Oxygenic Photosynthesis: The Light Reactions*, ed. D. R. Ort & C. F. Yocum, pp. 11–30. The Netherlands: Kluwer Academic Publishers.

Svensson, B., Etchebest, C., Tuffery, P., van Kan, P., Smith, J. & Styring, S. (1996). A model for the photosystem II reaction center core including the structure of the primary donor P680. *Biochemistry*, **35**, 14486–502.

Trissl, H. W. & Wilhelm, C. (1993). Why do thylakoid membranes from higher plants form grana stacks? *Trends in Biochemical Science*, **18**, 415–19.

Whitmarsh, J. (1986). Mobile electron carriers in thylakoids. *Encyclopedia of Plant Physiology* New Series, **19**, 508–27.

Whitmarsh, J. & Cramer, W. A. (1979). Cytochrome *f* function in photosynthetic electron transport. *Biophysical Journal*, **26**, 223–34.

Whitmarsh, J. & Pakrasi, H.B. (1996). Form and function of cytochrome b559. In *Advances in Photosynthesis, vol. 4, Oxygenic Photosynthesis: The Light Reactions*, ed. D.R. Ort & C.F. Yocum, pp. 137–64. The Netherlands: Kluwer Academic Publishers.

Xiong, J., Subramaniam, S. & Govindjee (1996). Modeling of the D1/D2 proteins and cofactors of the photosystem II reaction center: Implications for herbicide and bicarbonate binding. *Protein Science*, **5**, 2054–73.

Yamamoto, H.Y. & Bassi, R. (1996). Carotenoids: localization and function. In *Advances in Photosynthesis, vol. 4, Oxygenic Photosynthesis: The Light Reactions*, ed. D.R. Ort & C.F. Yocum, pp. 539–563. The Netherlands: Kluwer Academic Publishers.

Yu, C-A., Xia, J-Z., Kachurin, A.M., Yu, L., Xia, D., Kim, H. & Deisenhofer, J. (1996). Crystallization and preliminary structure of beef heart mitochondrial cytochrome-bc1 complex. *Biochimica et Biophysica Acta*, **1275**, 47–53.

Zhang, J., Carrell, C.J., Huang, D., Sled, V., Ohnishi, T., Smith, J.L. & Cramer, W.A. (1996). Characterization and crystallization of the lumen side domain of the chloroplast Rieske Iron–Sulfur protein. *Journal Biological Chemistry*, **271** 31360–6.

PART II

# Physiology and biochemistry

# 8 Photosynthetic carbon reduction

T. D. SHARKEY

## INTRODUCTION

The energy captured in photosynthetic electron transport is used several ways; the most common is the reduction of carbon dioxide to sugars. Two ATP and two NADPH are used to reduce one carbon dioxide to the level of a sugar ($CH_2O$) and one ATP is used to activate the substrate in preparation for carboxylation. The reactions are linked together in a cycle called the Calvin cycle, the Benson–Calvin cycle, or the photosynthetic carbon reduction cycle (PCRC) (Fig. 8.1).

The PCRC is autocatalytic; the PCRC makes its own substrates. For this reason there is often a lag in the capacity for the PCRC when photosynthetic material is first exposed to light. In addition, all of the reactions of the cycle must proceed at the same rate, making analysis of limitations and controls difficult. Each metabolite of the cycle is both a substrate and a product. This affects schemes for analysing metabolic pathways which rely on changes in substrate and product concentrations to identify critical control points.

Control issues are also important because there is no capacitance in the PCRC. The amount of triose-phosphate or ATP is often sufficient to last for only 0.1 s; the amount of NADPH is even less. Thus, excess capacity in one set of reactions cannot compensate for slow reactions in another part of the cycle. A limited carboxylation capacity cannot be offset by a higher capacity for regeneration of the $CO_2$ acceptor ribulose 1,5-bisphosphate (RuBP). The low light availability early in the morning cannot be offset by high light later in the day. Matching the PCRC to the availability of ATP and NADPH is one of the critical functions of regulation within the PCRC. Matching the rate of photosynthetic electron transport to the capacity of the PCRC is also important for the same reason. The issue of control and regulation will be emphasized here.

The PCRC is often divided into three phases: reduction, regeneration, and carboxylation. This division can be modified slightly to reflect the underlying biochemistry. There are (i) gluconeogenic reactions, (ii) pentose phosphate reactions, and (iii) reactions unique to photosynthesis. This is how the PCRC reactions will be discussed.

In eukaryotes, the PCRC occurs in the stroma of the chloroplast. In this way photosynthetic carbon metabolism is similar to respiratory carbon metabolism. The cytosol is the site of the primary sugar metabolism, glycolysis in respiration, and primarily sucrose synthesis in photosynthesis. A three-carbon compound crosses from the cytosol to the organelle or vice versa; pyruvate enters the mitochondrion in respiration and triose-phosphate leaves the chloroplast in photosynthesis. The reactions producing or consuming carbon dioxide occur inside the organelle in both processes while electron transport occurs on an internal membrane system in both organelles. Many facts about the PCRC make sense in light of the endosymbiotic theory that chloroplasts and mitochondria were once free-living bacteria that began a symbiotic existence and eventually became totally dependent on the host cell.

In this chapter the reactions of the PCRC and the consequences of the properties of some of the enzymes and regulatory mechanisms involved will be described.

## GLUCONEOGENIC REACTIONS

The reactions in gluconeogenesis from phosphoglyceric acid (PGA) to fructose 6-phosphate (F6P) occur as one major section of the PCRC. The reactions from PGA to

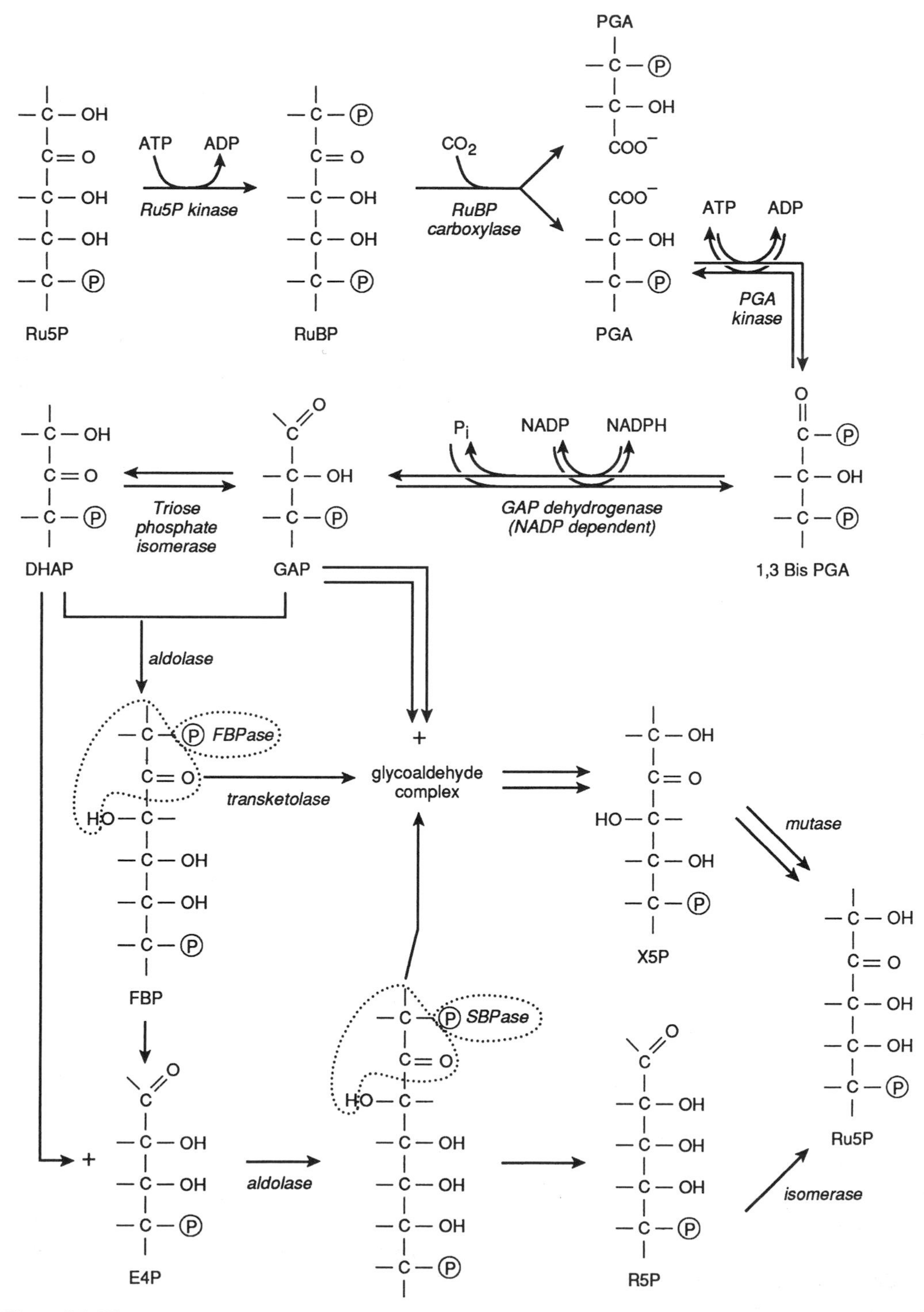


Figure 8.1. The reactions and enzymes of the photosynthetic carbon reduction cycle.

fructose 1,6-bisphosphate (FBP) are all reversible. The pool of carbon potentially in equilibrium extends outside of the chloroplast to FBP in the cytosol. The PGA kinase in the stroma is similar to that in respiration. The concentration of 1,3 bisPGA is low in the PCRC as in respiration.

## Energy-consuming reactions

Stromal glyceraldehyde-3-phosphate dehydrogenase (GAPDH) has some similarities to the respiratory enzyme but it also has two important differences. First, the photosynthetic enzyme can use both NADH and NADPH while the respiratory enzyme uses exclusively NAD. This is consistent with the rule that NAD is involved in degradation while NADPH is involved in synthesis. However, the electron transport reactions impose the use of NADPH in photosynthesis since NAD is not reduced in photosynthetic electron transport. Photosynthetic GAPDH is sometimes called NADP dependent, but this is incorrect. It is NADP competent.

The second way that photosynthetic GAPDH differs from the respiratory enzyme is that it can exist in an active or inactive form, depending on the redox status of thiol groups on the enzyme. When the thiol groups of cysteine residues become oxidized, they link together to form a cystine residue. This alters the conformation of GAPDH, making it inactive. When electron transport occurs, some electrons reduce thioredoxin f, which in turn reduces GAPDH to break the thiol bridges. In this way, the enzyme becomes active in the light and inactive in the dark. Some investigators prefer to reserve the term 'light activation' to this process which occurs with GAPDH, stromal FBPase, and phosphoribulokinase among others. There are other important changes in the stroma that occur as a result of photosynthetic electron transport, which are needed for the enzymes of the PCRC to operate efficiently, especially increases in pH and $Mg^{2+}$ concentration. For these three reasons (i) thiol reduction, (ii) increased $Mg^{2+}$, and (iii) increased stromal pH the light, what are sometimes called the dark, reactions will not proceed in darkness even when supplied with NADPH and ATP (except for very brief periods of darkness).

The PGA kinase and GAPDH reactions consume most of the energy needed for the reduction of carbon. In aerobic respiration, the reducing power generated by GAPDH is converted to ATP in oxidative phosphorylation and so ATP is the only energy currency in respiration. In photosynthesis, both ATP and NADPH are required to run the reactions and they are needed in strict stoichiometry. The ratio of ATP to NADPH required depends upon what other processes are occurring, such as photorespiration and nitrite reduction. This requires the photosynthetic electron transport reactions to be flexible in the ratio of ATP to NADPH produced, which is accomplished through cyclic or pseudocyclic phosphorylation.

Since the gluconeogenic reactions consume energy, many investigators have looked to this part of the cycle to understand what limits the rate of photosynthesis. The ratios of ATP to ADP and NADPH to NADP, or more sophisticated measures that take into account phosphorylation potential and pH, have been used and are often called assimilatory power. Sometimes, investigators have debated whether ATP or NADPH was the primary driver of carbon reduction. However, both ATP and NADPH are required, they are required in strict stoichiometries, and there is no reason to suspect that the rate of the gluconeogenic reactions go at anything less than that allowed by either the rate at which ATP and NADPH become available (i.e. light-limited photosynthesis) or the rate at which PGA becomes available (i.e. $CO_2$-limited photosynthesis).

## Importance of the triose-phosphates

In the gluconeogenic reactions, the triose-phosphates are key intermediates. The GAP made initially is converted by triose-phosphate isomerase to dihydroxyacetone phosphate (DHAP) (perhaps this compound is more properly a bis hydroxy, since the hydroxyls are attached to different carbons but this term is never used). The carbon in the triose-phosphates comes mostly from the $CO_2$ acceptor molecule. Only one-sixth of the carbon comes from $CO_2$. In order to keep the cycle going, the five-sixths of the carbon which came from RuBP must be reformed into RuBP, but one-sixth of the triose-phosphates can be used to make sugars, amino acids, fatty acids, etc. In fact, it is critical that this one-sixth, but only one-sixth, be used to make end products that are not phosphorylated, otherwise the phosphate concentration inside the chloroplast will fall too low.

During photosynthesis, triose-phosphate leaves the chloroplast, making triose-phosphate the end product of photosynthesis inside the chloroplast in the light. Some of the triose-phosphate made can be stored temporarily as starch inside the chloroplast, but generally this accounts for half or less of the carbon fixed in photosynthesis. Triose-phosphate exchanges across the chloroplast envelope on the phosphate translocator. The translocator is an abundant protein in chloroplast membranes which exchanges one triose-phosphate for one inorganic phosphate much of the time. The phosphate translocator can also transport PGA in the -2 ionization state and can exchange any of the transported compounds in either direction for any other transported compound. When starch breaks down, the carbon may leave the chloroplast as hexose or hexose phosphate but this is under active investigation.

From the standpoint of the chloroplast, the purpose of sucrose and starch synthesis is to release phosphate. All of the phosphate added to the organic phosphate pool by the PGA kinase reaction is released in the GAPDH reaction. Of the phosphate added by phosphoribulokinase, one-third is released in the FBPase reaction, one-third in SBPase, and the rest must be released in end product synthesis. Even though end product synthesis is responsible for only one-ninth of the phosphate turnover, photosynthesis begins to oscillate within seconds of changes that alter the balance between triose-phosphate production and triose-phosphate utilization. This is a clear indication of the importance of balancing the rates of all steps involved in photosynthesis.

## The FBPases

The reactions from PGA to FBP and the phosphate translocator are reversible. Therefore, all of the carbon from PGA to FBP whether inside or out of the chloroplast is effectively one pool. The fate of the carbon is determined by the activity of stromal and cytosolic FBPase. If dephosphorylated by the stromal enzyme, the carbon is committed to starch synthesis or RuBP regeneration. If dephosphorylated by the cytosolic FBPase, the carbon will be used in sucrose synthesis. Thus regulation of the two FBPases is critical. Regulation of the cytosolic FBPase responds to both conditions in the chloroplast and conditions in the cytosol Cytosolic considerations include whether there is an excess of hexose phosphate which indicates that sucrose synthesis is not keeping up with the cytosolic FBPase.

Cytosolic FBPase must respond to signals from the stroma that indicate, when there is sufficient triose-phosphate, that some can be removed from the cycle. At low rates of photosynthesis, the cytosolic FBPase must go very slowly to avoid pulling too much carbon from the PCRC, while at high rates of photosynthesis the cytosolic FBPase must go fast to avoid feedback resulting from insufficient end product synthesis capacity. The primary regulation of cytosolic FBPase is through the regulatory metabolite fructose 2,6-bisphosphate (F26BP). A specific F6P 2-kinase is stimulated by high levels of phosphate, while a F26BP 2-phosphatase is stimulated by triose-phosphate. As photosynthesis proceeds, the concentration of phosphate falls as triose-phosphate is produced. This turns off the kinase and stimulates the phosphatase. The presence of F26BP strongly inhibits cytosolic FBPase and this inhibition is removed as the F26BP is metabolized by the phosphatase. This system is a type of feedforward. A high concentration of triose-phosphate causes an activation of the enzyme needed to convert that triose-phosphate to end products. The level of F26BP is nearly always high at night. If most, or all, of the carbon from starch breakdown at night leaves the chloroplast as hexose or hexose phosphate, the cytosolic FBPase is not required at night. This pathway for carbon export also could save one ATP per six carbons leaving the chloroplast at night. The hexose transporter would bypass the regulation by cytosolic FBPase and so is potentially disadvantageous in the light when the amount of carbon left in the PCRC must be regulated.

The stromal FBPase does not use the F26BP system of regulation. Instead, it is regulated by dithiol reduction-type light activation, pH, and $Mg^{2+}$. The activation of stromal FBPase also depends upon the presence of FBP. If one-half of the carbon fixed becomes starch and one-half becomes sucrose, the cytosolic FBPase needs to catalyse 11 reactions for every one reaction catalysed by the cytosolic FBPase. Generally, when measured under optimal conditions for each enzyme, there is about ten times more stromal than cytosolic FBPase activity.

### PENTOSE PHOSPHATE REACTIONS

Regeneration of the five-carbon acceptor of the PCRC from the three-carbon triose-phosphates involves reactions of the pentose phosphate pathway. These reactions are not unique to photosynthesis. In fact, they are widely distributed throughout the biological world as are the gluconeogenic reactions. The pentose phosphate reactions often make learning the PCRC appear to be a formidable task. However, presenting the reactions in terms of a simple game seems to help students learn the pentose phosphate reactions. The goal is to convert three-carbon compounds into five-carbon compounds by combining them in different ways. The number of carbon atoms in each molecule must be kept as small as possible except that one-carbon compounds are not allowed. The simplest way to solve this puzzle is also a description of the regeneration reactions. Also, if the reactions are shown without crossing lines, it makes them easier to learn.

The transfer of the two carbon glycoaldehyde top of FBP and SBP to GAP by transketolase is central to the pentose phosphate reactions. Transketolase requires thiamine pyrophosphate as a cofactor for this reaction. Showing the glycoaldehyde–thiamine pyrophosphate complex as a separate metabolite aids understanding the simplicity of the pentose phosphate reactions.

The only irreversible reaction in the pentose phosphate pathway reactions is the sedoheptulose bisphosphatase. This enzyme is regulated by pH, $Mg^{2+}$, and dithiol reduction. While the stromal FBPase commits carbon to either regeneration or starch synthesis, the SBPase commits carbon only to regeneration.

## UNIQUE REACTIONS OF THE PCRC

While most of the steps of the PCRC are distributed widely in the biological world, two steps are unique to photosynthesis: the phosphorylation of ribulose 5-phosphate to ribulose 1,5-bisphosphate (RuBP) and the carboxylation of RuBP to produce two molecules of PGA by ribulose-1,5-bisphosphate carboxylase/oxygenase (rubisco).

### Phosphoribulokinase

Phosphoribulokinase uses one ATP, bringing the total ATP required per $CO_2$ to three and competing with PGA kinase for ATP. Phosphoribulokinase is highly regulated by pH, $Mg^{2+}$, and dithiol reduction. In addition, it is inhibited by PGA and 6-phosphogluconate. The PGA sensitivity allows phosphoribulokinase to be slowed if there is insufficient ATP and NADPH to drive the gluconeogenic reactions. This regulation is necessary because phosphoribulokinase has a higher affinity for ATP than does PGA kinase, so if ATP availability is low, RuBP synthesis continues while PGA builds up to high levels. This regulation is also necessary because RuBP carboxylation is irreversible, so a high ratio of PGA to RuBP would not slow the rate of RuBP carboxylation. PGA to RuBP ratios are sometimes reported but there seems to be no significance of this ratio.

The sensitivity of phosphoribulokinase to changes that occur in the chloroplast upon darkening make this the step that stops PCRC activity in the dark. It is important to prevent the PCRC before the ATP and NADPH status of the chloroplast becomes too low since this could damage other processes occurring inside chloroplasts, not least of which is synthesis of new proteins from genes encoded in the chloroplast. While it is true that the PCRC does not occur in darkened chloroplasts, the mechanism is not low ATP/ADP ratios or low NADPH/NADP ratios, but regulation of phosphoribulokinase. In this way ATP/ADP ratios can be kept high at night to allow all of the normal metabolism required in cells.

### Rubisco-basic considerations

The other unique reaction of the PCRC is carboxylation of RuBP by RuBP carboxylase. This enzyme will also catalyse the oxygenation of RuBP, the first step in photorespiration, discussed in Chapter 13. The enzyme name was shortened to rubisco, a play on the name of the National Biscuit Company, Nabisco. The capitalization of rubisco is quite variable, sometimes it is written as RuBisCO. There seems no reason to capitalize rubisco differently from any other enzyme name. Rubisco is an exceedingly complex enzyme and the mechanism and regulation of rubisco has been studied at length.

The basic catalytic unit of rubisco is a homodimer. The active site of rubisco is made up of the top of an 8-staved $\alpha$-$\beta$ barrel formed from the carboxy terminus of one subunit together with the N-terminus of the other subunit. The $\alpha$-$\beta$ barrel is a common tertiary structure in enzymes even though there is little or no sequence homology among enzymes with this structure.

In many cyanobacteria, rubisco exists as the dimer, but in some bacteria and all higher plants, four of the catalytic dimers aggregate and eight copies of a smaller subunit are added to give the $L_8S_8$ form of rubisco. Some bacteria have both the single dimer (called $L_2$ form) and the $L_8S_8$ form. In bacteria, the large and small subunits are coded for in the same region of the chromosome (i.e. dicistronic coding) but in higher plants, small subunit genes are in the nucleus, while the large subunit gene is in the chloroplast. The small subunit is made on cytosolic ribosomes and targeted to the chloroplast by a leader sequence which is cleaved after the small subunit is imported into the chloroplast. It may be that rubisco genes have moved among different organisms making rubisco sequences less reliable for predicting evolutionary relationships than other genes.

The import of small subunits and assembly of rubisco in the stroma requires another protein. This protein belongs to a class now called chaperonins. It was originally discovered as a 70 kDa heat shock protein and so is called HSP 70. ATP is required and, at least in some conditions, an HSP 10 is also required. Without HSP 70, rubisco does not assemble, although exactly how HSP 70 causes the assembly of catalytically competent rubisco is unknown.

## Rubisco – post-translational modifications

As with many enzymes, rubisco requires post-translational modification to become catalytically competent. A lysine (Lys 201 in the spinach enzyme) is located in the barrel of the active site. This lysine must be deprotonated, then carbamylated (addition of $CO_2$ to an amino group). The $CO_2$ added to the lysine for carbamylation is not the same $CO_2$ involved in catalysis, as can be demonstrated by using radiolabelled $CO_2$ for the carbamylation and showing that the label stays on rubisco and does not move off with the products of carboxylation. Carbamylation turns the positive lysine residue into a negatively charged region. This slow process is followed by a rapid co-ordination of $Mg^{2+}$ onto the active site. Once carbamylated, and with $Mg^{2+}$ present, rubisco is ready for catalysis. RuBP binds to the carbamylated rubisco and undergoes five steps of catalysis. If RuBP binds before rubisco is carbamylated, it is very difficult to remove.

Rubisco is usually less carbamylated at night than during the day because, among other things, the low pH and $Mg^{2+}$ that occurs at night is not conducive to carbamylation. For this reason, carbamylation is often referred to as light activation, but this is misleading. Rubisco carbamylation changes in ways that are easily predicted, based on the capacity for electron transport and sucrose synthesis. Large changes in carbamylation can be demonstrated with no change in light level. Also, the decarbamylation is not the reason that the PCRC stops at night. At night, there is no detectable RuBP in photosynthetic tissue, confirming that it is phosphoribulokinase that disrupts the PCRC at night, not rubisco. This, plus the common usage of 'light activatable', to mean dithiol reduction activation, makes it desirable to avoid calling changes in carbamylation, light activation of rubisco.

The degree of carbamylation can be estimated by measuring the activity of rubisco in extracts as quickly as possible after extraction, and comparing that with the activity that can be measured after incubating with $CO_2$ and $Mg^{2+}$ to fully carbamylate the enzyme. The ratio of the initial activity to total activity is an accurate reflection of the number of rubisco sites carbamylated.

The term 'activated' was applied to the state of being carbamylated, but it was discovered that rubisco could be carbamylated but inactive. It was found that carboxyarabinitol 1-phosphate occurs in many plants at night and that it can bind to carbamylated rubisco and prevent catalysis. In fact, many phosphorylated molecules will bind to rubisco but few as tightly as carboxyarabinitol 1-phosphate. Carboxyarabinitol 1-phosphate is similar in structure to the reaction intermediate.

A compound that binds even more tightly to rubisco is carboxyarabinitol 1,5-bisphosphate (CABP). This compound does not occur naturally but is easily synthesized chemically from RuBP. It has often been used to determine the amount of rubisco present in plant extracts. Radioactive CABP is added to the plant extract whereupon it binds essentially irreversibly to rubisco active sites. The protein is then precipitated using antibodies or polyethylene glycol. Other methods of precipitation such as ammonium sulphate will remove CABP, but antibody or PEG precipitation do not. Then the precipitated protein is washed to remove any unbound CABP and, finally, the precipitated protein is counted by scintillation counting. The amount of label in the precipitated protein is a direct measure of the total amount of rubisco in the extract.

The very tight binding of CABP and other properties of rubisco are believed to result from a movable section of rubisco. One of the loops of amino acids linking α-helix to β-strand, called loop 6, can move and there may be other parts of the enzyme that move upon binding phosphate-containing molecules. It has been hypothesized that loop six moves and locks CABP into place. CABP must bind to rubisco for a few seconds before this occurs and, during this time, it is easily displaced from the enzyme. This explains why CABP added to a rubisco reaction, when RuBP is already present, does not stop the rubisco reaction.

When extracted and held in conditions believed to exist in chloroplasts, rubisco is mostly uncarbamylated and so inactive. It turns out that there is another enzyme, called rubisco activase, that leads to high levels of carbamylation. The enzyme requires hydrolysis of ATP but the mechanism of rubisco activase is not yet known. It is known that activase can remove RuBP bound to uncarbamylated rubisco and can remove carboxyarabinitol 1-phosphate. Some investigators speculate that rubisco activase works by catalysing the removal of anything bound to the active site of rubisco.

## Rubisco – mechanism of carboxylation

The carboxylation of RuBP occurs in five steps (Fig. 8.2). In the first step, a hydrogen is abstracted to give an ene–diol intermediate. Carbon dioxide attacks this intermediate to give 3-keto 2-carboxy arabinitol bisphosphate. The tight binding inhibitor CABP lacks only the 3-keto group when compared to this reaction intermediate. The third step is hydration and this is followed by carbon bond cleavage. The products are PGA from the bottom three carbon atoms and a carbanion of PGA from the top three atoms. Finally, the carbanion is protonated to give the second PGA. It is this PGA that has the newly fixed carbon atom.

## Rubisco – catalytic anomalies

The carboxylation of RuBP is a complex reaction, and alternative reactions are possible at several steps. The best known of these is the substitution of oxygen for carbon dioxide leading to photorespiration. Another reaction that can occur is the isomerization of RuBP to make xylulose 1,5-bisphosphate. This occurs when the abstraction of a hydrogen from RuBP to form the enediolate intermediate (Fig. 8.2) is reversed but the hydrogen is added to the wrong side of the molecule. Xylulose 1,5-bisphosphate binds tightly to rubisco but does not undergo catalysis. It is estimated that substrate isomerization occurs about once for every 400 carboxylations. While this is not too often, each time it occurs one active site will become bound with xylulose 1,5-bisphosphate and so no longer be active. In *in vitro* assays, this gives rise to a slow decline in the rate of the rubisco reaction. Rubisco assays are typically conducted for just 30 s in order to avoid this fall-off in activity. In the plant, rubisco activase may be responsible for removing the xylulose 1,5-bisphosphate from rubisco active sites so that it can be metabolized.

At the other end of the rubisco reactions, the top three carbons become a carbanion of PGA which is then protonated to PGA. However, about 1% of the time, the carbanion undergoes β-elimination of the phosphate to produce pyruvate. This phenomenon could be important for the supply of acetyl CoA inside the chloroplast. In some studies, little or no phosphoglyceromutase was detectable in chloroplasts preventing the metabolism of PGA to pyruvate. The formation of pyruvate by rubisco is sufficient to satisfy some of the needs of the chloroplast for acetyl CoA for such things as fatty acid synthesis and possibly carotenoid synthesis, both of which are restricted to plastids.

## Rubisco – catalytic efficiency

Rubisco is a relatively inefficient enzyme. The $k_{cat}$ is around 3 $s^{-1}$ per site. In other words, each site can catalyse three reactions per second. Typical $k_{cat}$s for enzymes are $10^2$ to $10^4$ $s^{-1}$. Because carbon reduction is the major sink for energy captured in photosynthetic electron transport and rubisco is a very slow enzyme, plants

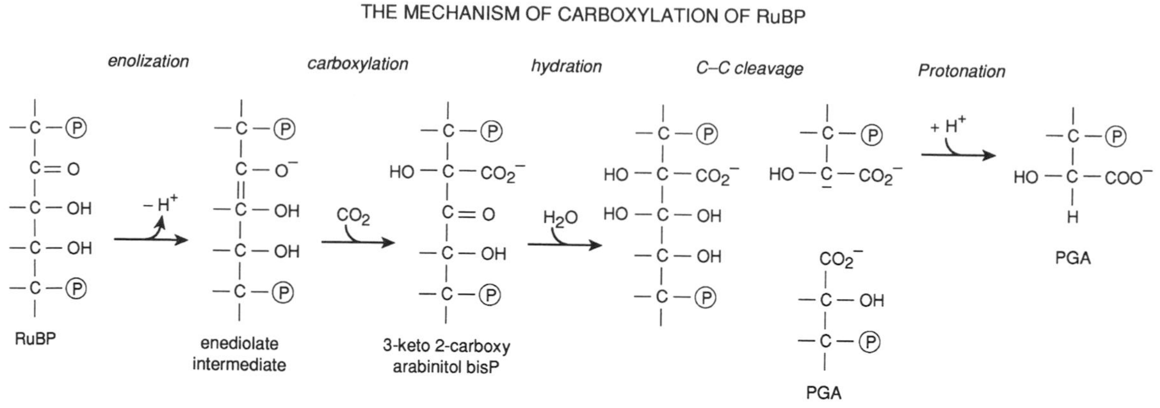

THE MECHANISM OF CARBOXYLATION OF RuBP

**Figure 8.2.** The five steps of carboxylation.

invest a tremendous amount of nitrogen in rubisco. About 20% of the nitrogen in leaves is in just one enzyme, rubisco. Rubisco is believed to be the most abundant enzyme in the world.

Rubisco is also inefficient because its affinity for $CO_2$ is relatively low. The $K_m$ for $CO_2$ is roughly equal to the level of $CO_2$ in the air. However, the effective $K_m$ is increased by the competitive inhibition by oxygen. This nearly doubles the effective $K_m$ for $CO_2$. In addition, in higher plants $CO_2$ must diffuse through the stomata and then the liquid phase of the mesophyll before it gets to rubisco. These diffusion resistances lower the effective $CO_2$ level by about half. So, instead of working near the $K_m$, rubisco is working at just one-quarter of its effective $K_m$ for $CO_2$. In many plants, a substantial amount of carbonic anhydrase is present in the chloroplast, which could enhance diffusion by converting $CO_2$ to bicarbonate and allowing bicarbonate to diffuse through the stroma. Since the pH in the stroma is high, the gradient for bicarbonate diffusion is nearly two orders of magnitude greater than the gradient in $CO_2$ if there is equilibrium between $CO_2$ and bicarbonate. Carbonic anhydrase is also present in many aquatic algae and bacteria and seems to be involved in energy-requiring pumps of $CO_2$ or bicarbonate.

The low affinity of rubisco for $CO_2$ also force the plant to keep stomata open, allowing substantial water loss. Thus the properties of rubisco are behind the large requirement for water and nitrogen in agricultural crops.

## REGULATION OF, AND BY, RUBISCO

Rubisco is the entry point for $CO_2$ into photosynthesis. The properties of rubisco make it central to issues of nitrogen and water use by photosynthetic organisms. Rubisco, then, is a central point from which to understand what limits photosynthesis. Generally, photosynthesis is considered limited by the availability of light or of $CO_2$. From the vantage point of rubisco, this appears as a limitation of the rate of regeneration of RuBP (light limited) or of $CO_2$, in which case rubisco would be RuBP saturated. If the ability to regenerate RuBP were in excess, then more rubisco could be added for greater rates of the PCRC. These concepts were formalized in the Farquhar model of photosynthesis. A third limitation was recognized: the limitation on the PCRC when the capacity to use triose-phosphates to make end products is less than the capacity of the chloroplast to produce triose-phosphates. In this case, photosynthesis is not limited by either light or $CO_2$. This condition occurs in natural conditions but is rare. More important is the insight into photosynthesis and its regulation provided by this phenomenon.

Many of the reactions of the PCRC are highly regulated, but few ever limit the overall rate of photosynthesis. Ideally, an enzyme that can be regulated should be regulated to take maximum advantage of the availability of light and $CO_2$. In an ideal situation, a process becomes limiting only after all of the capacity for regulation has been exhausted. Furthermore, once a process becomes limiting, the characteristics of that process should become apparent in the characteristics of photosynthesis of the whole system. Thus, when rubisco limits photosynthesis (i.e. the capacity to regenerate RuBP is in excess of what can be used by rubisco), photosynthesis takes on characteristics of rubisco including a strong sensitivity to $CO_2$ and low sensitivity to temperature.

In many respiratory pathways, control of the overall flux of carbon through the pathway is distributed among a number of reactions. The rate of, for example, glycolysis is determined by the rate at which the products of glycolysis are needed. This is very different from photosynthesis where resources such as light are available at one instant, and resources not captured during that instant are lost forever. Regulation in the PCRC is critical to keep all of the reactions in balance in order to keep reactions from limiting the overall flux. As enzymes are activated, a point will come where a process is fully activated but no further increase in rate is possible. This process then limits the overall rate of photosynthesis, and photosynthesis will have the characteristics of this process. Thus, in photosynthesis, regulation should prevent limits on the overall rate while, in respiration, regulation should set limits on the overall rate.

In the PCRC many important steps are regulated by enzyme activation rather than by changes in metabolite concentration (mass action ratios). Many times it has been observed that the concentration of RuBP does not change when photosynthetic rate changes over more than an order of magnitude. Enzyme activation and deactivation appear to be the primary mechanism by which the rates of PCRC reactions are regulated.

Currently, the limits to the rate of photosynthesis are divided among three broad classes in the Farquhar model of photosynthesis. These are (i) $CO_2$ limited or rubisco limited (RuBP saturated), (ii) light limited or RuBP regeneration limited, and (iii) end product synthesis limited or triose-phosphate use limited. The third limitation will limit the rate of RuBP regeneration but it has very different characteristics. It is hard to label each limitation but each is well defined mathematically. The mathematical treatment of these limitations have been presented many times but will not be presented here. A non-mathematical way to gain insight into these limitations is to view them from the standpoint of the ATP synthase. ATP synthesis can be limited by the availability of ADP, phosphate, or energy gradient. The rubisco limitation described above causes an ADP limitation, the RuBP regeneration limitation causes an energy gradient limitation, and the end product synthesis limitation causes a phosphate limitation. There are some ways that RuBP regeneration could be limited that would result in an ADP rather than energy limitation of ATP synthase but this situation would seem to be rare.

One of the most controversial aspects of the Farquhar model is the idea that photosynthesis appears limited by only one process at a time; in other words there is very little colimitation. Yet another way of expressing this is that metabolic control rests primarily in one or another enzyme over a large range of conditions, but the control coefficients can change quickly from nearly one to nearly zero and vice versa. When we think of the spatial scale of the whole plant and temporal scale of growth (days to weeks), it is obvious that both $CO_2$ and light are limiting in most situations. Even within one leaf, since some cells are in very different environments than others, we expect variation in the relative degree to which light or $CO_2$ limits photosynthesis. However, the underlying biochemistry does exhibit sharp transitions from one limitation to another, and one cause for this is known. The affinity of rubisco for RuBP is very high. The $K_m$ is approximately two orders of magnitude lower than the concentration of active sites. If the affinity were infinite, it would be easy to imagine each RuBP that becomes available becoming bound to rubisco and increasing the rate. Increasing RuBP would increase the amount of enzyme ready for catalysis linearly. However, once all of the rubisco was bound with RuBP, any additional RuBP would have no effect since there would be nowhere to bind. This was the original explanation for the abrupt transition from rubisco limited to RuBP regeneration-limited photosynthesis that could be observed in thin leaves.

Now it is known that rubisco is regulated to match the rate required. If the light is low and so RuBP regeneration is slow, rubisco is decarbamylated. This results in the pool of RuBP remaining high. Uncarbamylated rubisco binds RuBP and the RuBP concentration remains at about 1.6 times the number of rubisco active sites. The decarbamylation is even easier to demonstrate when end product synthesis limits the rate of photosynthesis. The sharp transition from little to substantial overall control by rubisco occurs when rubisco becomes fully carbamylated (or as carbamylated as possible under prevailing conditions). As light increases, the amount of RuBP remains constant and rubisco carbamylation increases. Rubisco exerts little or no control. At some point, carbamylation can increase no further and rubisco begins to exert substantial control. Thus, it is the highly regulated nature of rubisco and other PCRC enzymes that leads to sharp transitions in control from one enzyme to another. The idea of sharp transitions among limitations is often called the Blackman view of photosynthesis.

## Elevated $CO_2$

When limited by rubisco, photosynthesis can be stimulated by elevated $CO_2$ for two different reasons. First, $CO_2$ is a substrate normally well below the effective $K_m$, and so increased concentration increases the rate of reactions. Secondly, since increasing $CO_2$ suppresses photorespiration, the $CO_2$ release in photorespiration is reduced in elevated $CO_2$. When limited by RuBP regeneration, photosynthesis is still stimulated by elevated $CO_2$ as a result of reduced photorespiration. The effects of photorespiration are somewhat different. When RuBP regeneration limits, suppressing photorespiration allows more of the energy captured by photosynthetic electron transport to go to carboxylation reactions, an effect not relevant when rubisco limits photosynthesis. The suppression of $CO_2$ release in photorespiration is relevant, regardless of whether rubisco or RuBP regeneration limit photosynthesis.

Since photosynthesis is stimulated by elevated $CO_2$ even when it is effectively light limited, it is sometimes difficult to see the abrupt transition from rubisco-limited to RuBP regeneration-limited photosynthesis. The change is much more apparent in the rate of electron transport. In rubisco-limited photosynthesis, RuBP use and so electron transport, increases with increasing $CO_2$ since more RuBP can be used. In RuBP regeneration-limited photosynthesis, increasing $CO_2$ directs more RuBP and electron transport to carboxylation but does not alter the limiting process, the rate at which RuBP can be made. Therefore, electron transport is independent of $CO_2$ when photosynthesis is limited by RuBP regeneration. It is easier to detect the change from rubisco-limited to RuBP regeneration-limited photosynthesis using photosynthetic electron transport. The advent of fluorescence methods for measuring the rate of electron transport in intact leaves has allowed a confirmation of the predictions of the Farquhar model.

Rubisco, other PCRC enzymes, electron transport components, etc. are most efficient when they are present in balanced amounts. When plants are grown in elevated $CO_2$, rubisco becomes more efficient and so there is, in relative terms, more rubisco than needed. The carbamylation state of rubisco often falls under elevated $CO_2$ and so decarbamylation helps restore the balance among the various components of photosynthesis. In longer-term experiments the total extractable rubisco activity falls when plants are grown in elevated $CO_2$. This effect is most pronounced when plants are grown in small pots and in older plants. It has been suggested that the signal that results in reduced rubisco levels in plants grown in elevated $CO_2$ is glucose or some process associated with glucose such as hexokinase activity. There remain a number of important discoveries to be made in this area.

The promoter of the rubisco small subunit appears to be particularly sensitive to regulation. In addition to some mechanism of hexose regulation, expression of rubisco small subunit genes is affected by phytochrome.

## Molecular biology studies

Studies of metabolic control have been aided by the recent ability to selectively change expression of enzymes of the PCRC. To study how rubisco controls photosynthesis, the rubisco small subunit gene has been cloned and reinserted in the antisense direction in a number of plants by several investigators. While there are many effects, one of the more direct effects is reduced photosynthetic rate. However, this is only true if plants are grown in high light. Plants grown in low light can lose up to half of their rubisco before the rate of photosynthesis declines. Plants grown and assayed in higher light exhibit reduced rates of photosynthesis when any rubisco is lost. In low light-grown tobacco plants, the control coefficient for rubisco was essentially zero until one-half of the rubisco was lost, then the control coefficient was nearly one as rubisco was reduced further.

Does this indicate that plants have excess rubisco? Probably not. Instead, this finding can be interpreted as indicating that plants adapted to high light will make a lot of rubisco since, at high light, rubisco often limits photosynthesis. When that plant is in low light, acclimation does not reduce the activity of rubisco as much as it could be reduced. This may indicate the limits of plant adaptability or it may be advantageous for a plants in some ecological niches to always allow for high light by making high levels of rubisco.

Antisense NADP GAPDH plants have also been made. It was predicted that this manipulation would shift the balance of control toward RuBP regeneration just as the antisense rubisco plants had control shifted toward rubisco. This was found to be true by metabolite analysis and fluorescence quenching analysis. These observations add to the evidence that the Farquhar model is a good description of photosynthesis.

Another enzyme that has been targeted by antisense is rubisco activase. Again, several investigators have reported such plants. When measured early in the plant's life, little effect of reduced activase has yet been demonstrated. However, later in the life of the plant the total level of rubisco is higher in transgenic plants than in control while carbamylation is much lower. The net effect is that the rate of photosynthesis is reduced in antisense rubisco activase plants. Whether the rate or extent of short-term changes in carbamylation rate will be observed in the antisense rubisco activase is not yet known.

The gene for carbonic anhydrase has also been put into plants in the antisense direction. Carbonic anhydrase could aid diffusion of $CO_2$ within the chloroplasts of $C_3$ plants. Little effect on photosynthesis was seen, but calculations indicate that the effect of removing all carbonic anhydrase from the chloroplast should be small enough to be difficult to measure. Another parameter that is affected by the loss of mesophyll conductance is the discrimination against $^{13}C$. Plants lacking carbonic anhydrase in their leaves had less $^{13}CO_2$ confirming that carbonic anhydrase can aid $CO_2$ diffusion.

Another well-known genetic manipulation has been the introduction of an antisense stromal FBPase gene. These plants do not grow as well as untransformed plants but the effects are not easily interpreted since the stromal FBPase is needed for RuBP regeneration and starch synthesis.

The triose-phosphate transporter has been reduced by antisense expression of the gene. In these plants, reductions in carbon export from the chloroplast during the day were compensated by increased carbon export at night. This is consistent with a hexose transporter in the chloroplast membrane that is inhibited by light and is important for export of carbon.

## FUTURE OUTLOOK

The path of carbon in photosynthesis has been generally accepted for over 30 years. While some additional details may yet come to light, it is unlikely that there will be much reported in the future that changes the currently accepted pathway. One part of the pathway which may change in the near future is the path of carbon out of the chloroplast at night. It may be proven that, in darkness, carbon comes out of the chloroplast through a hexose transporter and that this transporter is active only at night.

Major progress is likely to come in understanding the regulation of the PCRC now that specific enzymes can be targeted and changed. Most experiments have involved antisense gene constructs; in the future it would be nice to see more overexpression of genes. It is suspected that many genes would appear limiting if their expression were reduced, but in many fewer cases would photosynthesis be increased by increasing gene expression.

Finally, most progress is likely to come in understanding how the PCRC and its regulation are related to other processes in the plant. It is often assumed that sink strength (growth of roots, new leaves, and fruits) affects the rate of photosynthesis but there is little data to support this assumption. Tracing information from sinks back to the reactions of the PCRC will continue to be an active area of research.

### FURTHER READING

Andrews, T. J. & Kane, H. J. (1991). Pyruvate is a by-product of catalysis by ribulosebisphosphate carboxylase/oxygenase. *Journal of Biological Chemistry*, **266**, 9447–52.

Andrews, T. J., Hudson, G. S., Mate, C. J., Von Caemmerer, S., Evans, J. R. & Arvidsson, Y. B. C. (1995). Rubisco:

the consequences of altering its expression and activation in transgenic plants. *Journal of Experimental Botany*, **46**, 1293–300.

Buchanan, B. B. (1991). Regulation of $CO_2$ assimilation in oxygenic photosynthesis: the ferredoxin/thioredoxin system: perspective on its discovery, present status, and future development. *Archives of Biochemistry and Biophysics*, **288**, 1–9.

Dietz, K-J. & Heber, U. (1986). Light and $CO_2$ limitations of photosynthesis and states of the reactions regenerating ribulose 1,5-bisphosphate or reducing 3-phosphoglycerate. *Biochimica et Biophysica Acta*, **848**, 392–401.

Edmondson, D. L., Badger, M. R. & Andrews, T. J. (1990). Slow inactivation of ribulosebisphosphate carboxylase during catalysis is caused by accumulation of a slow, tight-binding inhibitor at the catalytic site. *Plant Physiology*, **93**, 1390–7.

Edwards, G. and Walker, D. (1983). $C_3$, $C_4$: Mechanisms, and Cellular and Environmental Regulation, of Photosynthesis, pp. 1–542. Berkeley: University of California Press.

Farquhar, G. D., Von Caemmerer, S. & Berry, J. A. (1980). A biochemical model of photosynthetic $CO_2$ assimilation in leaves of $C_3$ species. *Planta*, **149**, 78–90.

Leegood, R. C. (1989). Biochemical studies of photosynthesis: from $CO_2$ to sucrose. In *Photosynthesis Proceedings of the C. S. French Symposium* held in Stanford California, ed. W. R. Briggs, pp. 457–73, New York: Alan R. Liss Inc.

Mate, C. J., Hudson, G. S., Von Caemmerer, S., Evans, J. R. & Andrews, T. J. (1993). Reduction of ribulose bisphosphate carboxylase activase levels in tobacco (*Nicotiana tabacum*) by antisense RNA reduces ribulose bisphosphate carboxylase carbamylation and impairs photosynthesis. *Plant Physiology*, **102**, 1119–28.

Price, G. D., Evans, J. R., Von Caemmerer, S., Yu, J.-W. & Badger, M. R. (1995). Specific reduction of chloroplast glyceraldehyde-3-phosphate dehydrogenase activity by antisense RNA reduces $CO_2$ assimilation via a reduction in ribulose bisphosphate regeneration in transgenic tobacco plants. *Planta*, **195**, 369–78.

Robinson, S. P. & Walker, D. A. (1981). Photosynthetic carbon reduction cycle. In *The Biochemistry of Plants. A Comprehensive Treatise*, ed. M. D. Hatch & N. K. Boardman, pp. 193–236, New York: Academic Press.

Sage, R. F. (1994). Acclimation of photosynthesis to increasing atmospheric $CO_2$: the gas exchange perspective. *Photosynthesis Research*, **39**, 351–68.

Sharkey, T. D. (1990). Feedback limitation of photosynthesis and the physiological role of ribulose bisphosphate carboxylase carbamylation. *Botanical Magazine Tokyo Special Issue*, **2**, 87–105.

Sheen, J. (1994). Feedback control of gene expression. *Photosynthesis Research*, **39**, 427–38.

Walker, D. A. & Herold, A. (1977). Can the chloroplast support photosynthesis unaided. *Plant and Cell Physiology* Supplement SI, 295–310.

# 9 $C_4$ pathway

R. T. FURBANK

## INTRODUCTION

A large proportion of the plant kingdom carries out $C_3$ photosynthesis (commercially relevant examples are wheat, rice, barley, rape-seed, rye-grass, and tree species); however, an important group of plants carries out photosynthesis via the $C_4$ pathway (see Hatch, 1987). Plants in this group include maize, sugar-cane, sorghum, a wide variety of tropical pasture grasses and some of the world's worst weeds (such as crab-grass, *Digitaria sanguinalis* and nut-grass, *Cyperus rotunda*). $C_4$ plants are generally characterized by high rates of photosynthesis and growth, particularly in subtropical/tropical environments. Based on annual biomass accumulation, 11 out of the 12 most productive higher plant species on the planet are $C_4$ (see Hatch, 1992). $C_4$ plants are particularly successful in environments where temperature and light intensities are high. One of the reasons for the success of $C_4$ plants under these conditions is that these plants have a $CO_2$ concentrating mechanism which reduces the rate of photorespiration and increases the rate of $CO_2$ fixation in air levels of $CO_2$. The result is an increase in photosynthetic efficiency in air, the details of which are discussed later.

## THE PATH OF CARBON THROUGH THE $C_4$ PATHWAY

The $C_4$ pathway (Fig. 9.1) is basically composed of the photosynthetic carbon reduction (PCRC) cycle, located within the bundle sheath cells, with an added biochemical $CO_2$ 'pump' in the form of specialized mesophyll cells which concentrate $CO_2$ at the site of rubisco. For this mechanism to evolve, a complex combination of cell specialization and differential expression of photosynthetic genes was necessary and these aspects were recently reviewed by Furbank and Taylor (1995). Figure 9.1A shows a low magnification electron micrograph of a $C_4$ leaf related to a generalized scheme for the $C_4$ pathway. Mesophyll cells carry out the initial steps of $CO_2$ fixation utilizing the enzyme phosphoenolpyruvate (PEP) carboxylase and the product of $CO_2$ fixation is the 4-carbon organic acid, oxaloacetate (rather than the 3-C sugar phosphate formed in $C_3$ plants), hence the designation '$C_4$'. Oxaloacetate is converted to either malate or aspartate, which diffuses to the bundle sheath cells. In the bundle sheath cells, the $C_4$ acid is decarboxylated and the resulting $CO_2$ fixed by ribulose 1,5-bisphosphate carboxylase/oxygenase (rubisco). A $C_3$ product returns to the mesophyll to be recycled to PEP for the carboxylation reaction. $C_4$ plants have evolved three distinct mechanisms for decarboxylating $C_4$ acids in the bundle sheath cells (Fig. 9.1B). (i) Chloroplast localized NADP-malic enzyme which decarboxylates malate to produce pyruvate (NADP-ME-type species), (ii) Mitochondrial localized NAD-malic enzyme (NAD-ME-type), or (iii) Cytosolic localized PEP carboxykinase which produces PEP from oxaloacetate (PCK-type species) (see Hatch, 1987). In the case of the NADP-ME types, the predominant translocated $C_4$ acid is malate while, in the other two types, mainly aspartate is translocated to the bundle sheath cells, whereupon it is converted by transamination to oxaloacetate. In PCK types, mitochondrial respiration is closely linked to the energy requirements of the PEP carboxykinase reaction and it is thought that some malate is transported in these species, supplying both ATP from respiration and $CO_2$ from NAD-malic enzyme (supported by the presence of NAD-malic enzyme in

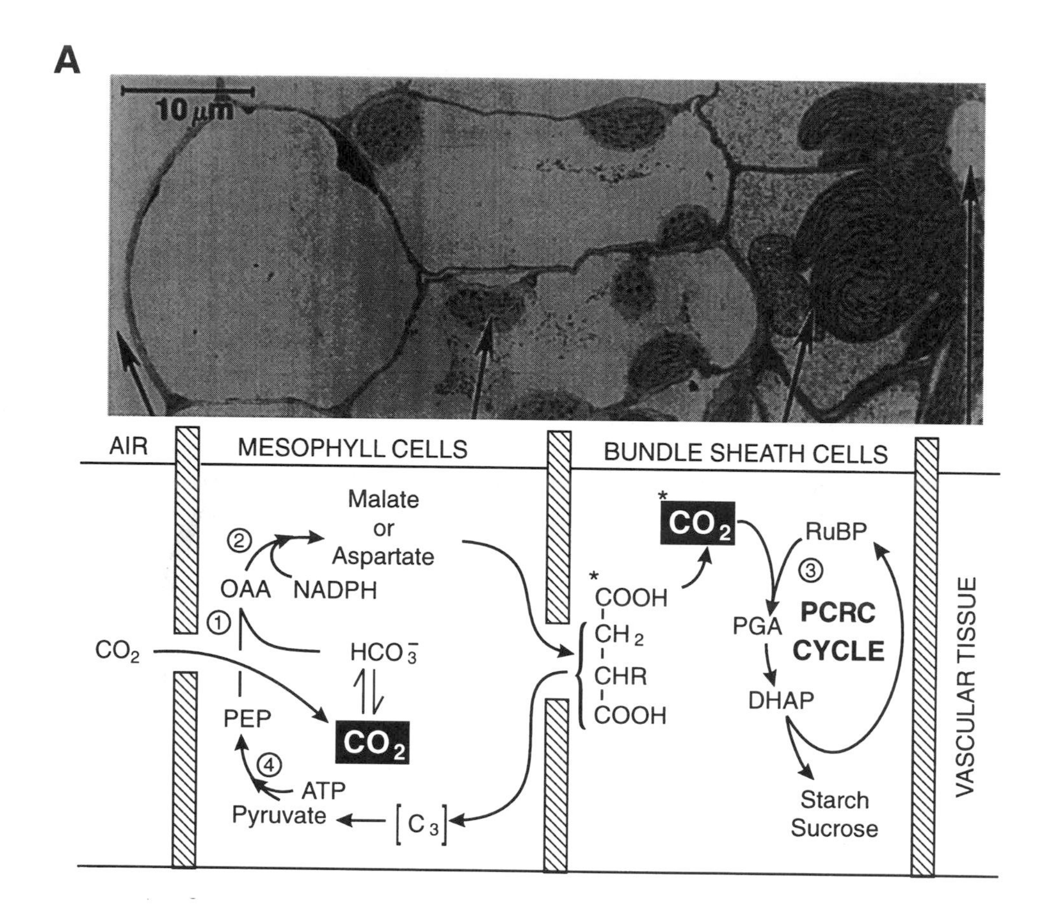
A
10 μm
AIR
MESOPHYLL CELLS
BUNDLE SHEATH CELLS
VASCULAR TISSUE
Malate
or
Aspartate
OAA
NADPH
$CO_2$
$HCO_3^-$
PEP
ATP
Pyruvate
$C_3$
COOH
$CH_2$
CHR
COOH
RuBP
PGA
PCRC
CYCLE
DHAP
Starch
Sucrose

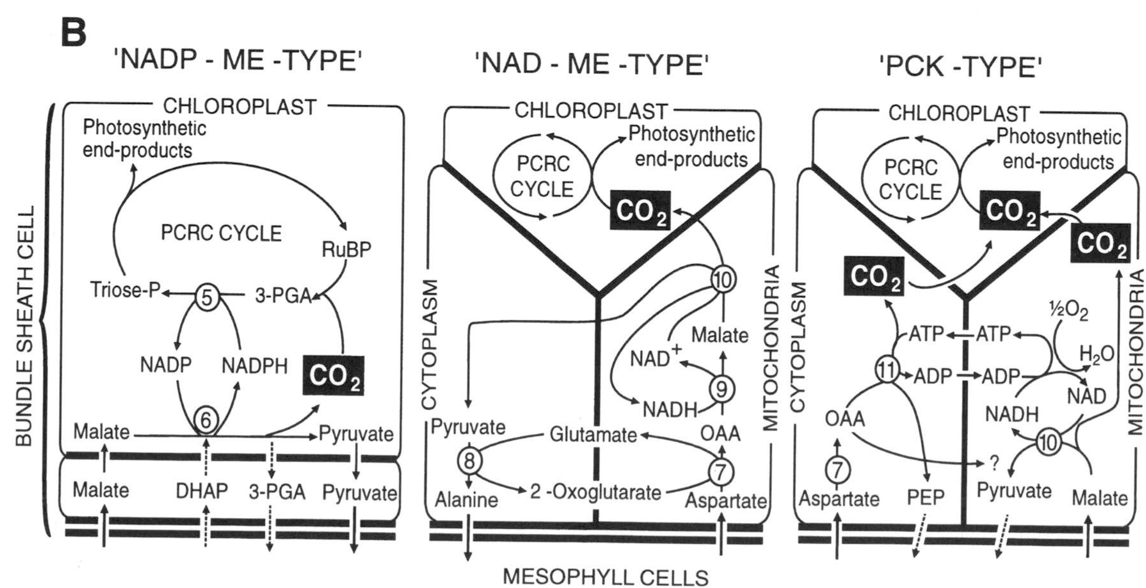
B
'NADP - ME -TYPE'
'NAD - ME -TYPE'
'PCK -TYPE'
BUNDLE SHEATH CELL
CHLOROPLAST
Photosynthetic end-products
PCRC CYCLE
RuBP
Triose-P
3-PGA
NADP
NADPH
$CO_2$
Malate
Pyruvate
DHAP
CYTOPLASM
MITOCHONDRIA
Malate
$NAD^+$
NADH
OAA
Glutamate
Alanine
2 -Oxoglutarate
Aspartate
ATP
ADP
½$O_2$
$H_2O$
NAD
PEP
MESOPHYLL CELLS

these plants and also by labelling data; Hatch (1987)). It is now widely established that the function of these diverse biochemical mechanisms is to concentrate $CO_2$ in the bundle sheath cells at the site of rubisco. To contain this pool of $CO_2$, a physical barrier to $CO_2$ diffusion exists in the cell walls of the bundle sheath compartment of many species, preventing $CO_2$ diffusion back to the mesophyll and allowing $CO_2$ to build up to levels in excess of 10 × atmospheric concentrations. The biophysical properties of this diffusion barrier and the consequences for the efficiency of photosynthesis will be dealt with in the following section. In summary, $C_4$ plants have evolved a mechanism to concentrate $CO_2$ at the site of rubisco, reducing photorespiration and resulting in a higher 'affinity' for $CO_2$ than $C_3$ plants, so that photosynthesis in these plants is almost saturated at current air levels of $CO_2$ (Hatch, 1992).

## PHOTOSYNTHETIC EFFICIENCY AND EVOLUTION OF THE $C_4$ PATHWAY

When considering the evolution and physiological significance of the $C_4$ pathway, it is instructive to consider the energetics of the $C_3$ photosynthetic mechanism. A major inefficiency in photosynthesis is the competition between $CO_2$ and $O_2$ which occurs at the catalytic site of rubisco (for a review, see Hatch, 1987; see also chapters 8 and 13). It is thought that this oxygenase function of rubisco is an inevitable consequence of its reaction mechanism. The oxygenase reaction and the biochemical recycling of the phosphoglycollate formed (photorespiration; see Chapter 13) is an energetically expensive process, the magnitude of which, relative to the rate of $CO_2$ fixation, varies with environmental parameters such as temperature and $CO_2$ concentration. The rate of the oxygenase reaction increases relative to the rate of carboxylation as temperature is increased and likewise when atmospheric $CO_2$ concentration is reduced. In the absence of photorespiration (for example, at high $CO_2$), the energetic requirements of $C_3$ photosynthesis are well established: 3ATP and 2NADPH per $CO_2$ fixed, inferred from the biochemical requirements of the PCRC cycle. Under current atmospheric levels of $CO_2$, this requirement increases to approximately 5ATP and 3.2 NADPH per $CO_2$ fixed, due to the cost of the oxygenase reaction (see Hatch, 1987). As the ratio of photorespiration to photosynthesis increases with temperature, this cost will also rise. Due to this energetic cost of photorespiration, there would be constant selective pressure, at least under current atmospheric $CO_2$ levels, for the evolution of either a more efficient rubisco or a more efficient alternative carbon fixation mechanism.

The occurrence of $C_4$ photosynthesis is limited to the more recently evolved genera of angiosperms and it appears that the $C_4$ pathway may have arisen as recently as 7 million years ago (see Ehleringer *et al.*, 1991). In contrast, it is thought that $C_3$ angiosperms may have evolved as early as 300 million years ago. On the basis of the energetics of photorespiration discussed above, Ehleringer *et al.* (1991) suggested that a critical factor in the evolution of $C_4$ photosynthesis was a sustained fall in global atmospheric $CO_2$ concentration, particularly during the Oligocence and Meiocene periods (fig 9.2). Atmospheric $CO_2$ concentration was approximately ten times current levels 100 million years ago, falling steeply during and following the Cretaceous, potentially providing a strong selective pressure for the evolution of more efficient carboxylation mechanisms. The earliest fossil evidence of plants with well-developed bundle sheath anatomy dates from the late Meiocene (see Ehleringer *et al.*, 1991), correlating well with sustained $CO_2$ concentrations of around 300 $\mu l\ l^{-1}$. Hatch (1992) calculated

**Figure 9.1.** A shows a simplified scheme of $C_4$ photosynthesis below an electron micrograph of a transverse section of a $C_4$ leaf, showing the relationship between anatomy and biochemistry in these plants. B shows the three decarboxylation mechanisms present in the bundle sheath cells of $C_4$ plants. Key enzymes are labelled as follows: 1. Phosphoenolpyruvate (PEP) carboxylase, 2. NADP-malate dehydrogenase, 3. Ribulose 1,5-bisphosphate carboxylase/oxygenase (rubisco), 4. Pyruvate-Pi dikinase (PPdK), 5. Glyceraldehyde 3-phosphate dehydrogenase, 6. NADP-malic enzyme, 7. Aspartate aminotransferase, 8. Alanine aminotransferase, 9. NAD-malate dehydrogenase, 10. NAD-malic enzyme, 11. Phosphoenolpyruvate carboxykinase. Regulation of enzymes 1, 2, 3 and 4 is dealt with in more detail on pp. 129–31. (Adapted from Hatch, 1987 and Hatch, 1992.)

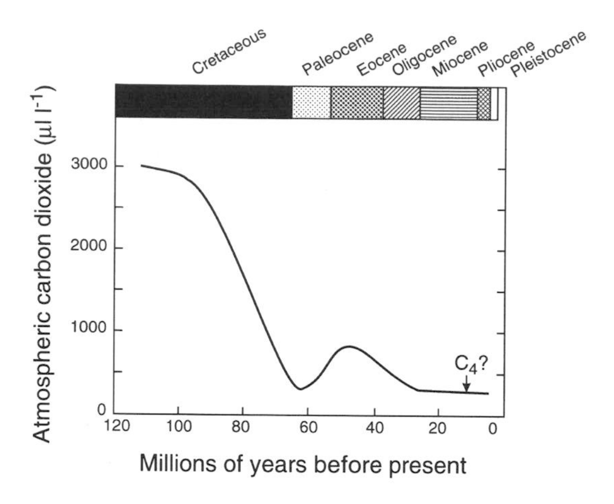


**Figure 9.2.** Modelled change in atmospheric $CO_2$ concentration during the past million years (adapted from Ehleringer *et al.*, 1991). An arrow indicates the likely appearance of $C_4$ type anatomy in the fossil record.

that, during the early Cretaeous period, the rate of Rubisco carboxylation relative to oxygenation at 30 °C in a $C_3$ plant would have been around 31 falling to around 2.2 at current $CO_2$ levels. A useful way to quantify this effect on photosynthetic efficiency in a physiological context is to calculate (or measure) the the quantum yield of photosynthesis (i.e. the mol $CO_2$ fixed per mol quanta absorbed by the leaf, determined at limiting irradiance). At an atmospheric $CO_2$ concentration of 3000 $\mu l\ l^{-1}$, the quantum yield of $C_3$ photosynthesis is around 0.082 while, under current $CO_2$ levels, this efficiency falls to 0.052 (i.e. approximately 12 quanta are required per $CO_2$ fixed in the former case, 19 in the latter). Such a loss in photosynthetic efficiency in response to global change would provide a powerful selective pressure for $C_3$ plants to evolve mechanisms to concentrate $CO_2$ at the active site of rubisco, resulting in the variants of the $C_4$ pathway in existence today.

## BIOPHYSICAL ASPECTS OF THE $CO_2$ CONCENTRATING MECHANISM

Although the biochemical details of the $C_4$ pathway have been resolved for many years (see Hatch, 1987), it is only relatively recently that researchers have studied the biophysical aspects of the diffusion of $CO_2$ and metabolites central to the operation of this mechanism (for review see Hatch, 1992). The importance of these factors becomes apparent when we consider that the diffusion characteristics of the bundle sheath cells, and how well $CO_2$ 'leakage' out of the bundle sheath is contained, will largely determine both the efficiency of the mesophyll $CO_2$ pump and the rate of photorespiration in $C_4$ plants. In the last decade, both the size of the $CO_2$ pool in the bundle sheath cells of $C_4$ plants and the diffusion characteristics of the bundle sheath/mesophyll cell interface have been determined (see Hatch, 1992).

From radioisotope experiments in intact leaves, the total inorganic carbon pool ($CO_2$ +$HCO_3$) in a range of $C_4$ species is between 30 and 100 nmol $mg^{-1}$ Chl (see Hatch, 1987). This pool is up to 20 times that found in a $C_3$ leaf in the light. From a model of the cellular distribution of inorganic carbon between $CO_2$ and $HCO_3$ in the bundle sheath (see Hatch, 1992) we can calculate that this pool would be equivalent to around 75 $\mu M$ $CO_2$, more than ten times normal ambient $CO_2$ levels. This $CO_2$ concentration at the active site of rubisco would be sufficient to reduce photorespiration to less than 2% of gross photosynthesis (see Hatch, 1992). These estimates also fall within the range of photorespiration rates determined in $C_4$ leaves by stable $O_2$ isotope mass spectrometry (Furbank & Badger, 1982).

The additional 'cost' of the mesophyll reactions in an NADP-ME type $C_4$ plant is 2ATP per $CO_2$ fixed by PEP carboxylase (see Fig 9.1B and Hatch, 1987), as the NADPH required to reduce oxaloacetate is regenerated in the decarboxylation reaction. Because the ATP and NADPH requirements of the PCRC cycle are the same as in $C_3$ plants, the efficiency of $C_4$ carbon assimilation is largely determined by the rate at which the mesophyll 'pump' must run to maintain the $CO_2$ levels described above. A key feature in this equation is the rate at which $CO_2$ leaks back to the mesophyll cells without being fixed following decarboxylation in the bundle sheath. Without any leak of $CO_2$, 5ATP and 2NADPH are required for every $CO_2$ fixed by rubisco. If $CO_2$ were to leak out of the bundle sheath at an appreciable rate, the energetic cost of running the $CO_2$ pump could become greater than the cost of photorespiration, affording no selective advantage to $C_4$ plants over their $C_3$ counterparts. However, if the bundle sheath cells were too resistant to the penetration of small molecules, the flux of metabolites between the bundle sheath and mesophyll compartments necessary for the operation of $C_4$ photosynthesis (see Fig 9.1) could not occur. The permeability coefficient of the bundle sheath / mesophyll interface to $CO_2$ has been examined in isolated bundle sheath cells and in intact $C_4$ leaves in which PEP carboxylase was inhibited (Furbank, Jenkins & Hatch, 1989; Jenkins, Furbank & Hatch, 1989). Both methods indicated that permeability of bundle sheath cells is around two orders of magnitude lower than an equivalent $C_3$ cell (values of 6 to 30 μmol $min^{-1}$ $mg^{-1}$ chl $mM^{-1}$ were obtained for bundle sheath cells and up to 3000 μmol $min^{-1}$ $mg^{-1}$ Chl $mM^{-1}$ for $C_3$ cells). Notably, bundle sheath cells are still two orders of magnitude more permeable to $CO_2$ than the cyanobacteria, which also concentrate $CO_2$ within their cells, indicative of an optimization of bundle sheath permeability during evolution to cater for both high rates of metabolite diffusion and low permeability to $CO_2$.

Quantification of the $CO_2$ leak *in vivo* is technically difficult but both indirect estimations, and direct measurements by radioisotope techniques have been made. Indirect estimations of $CO_2$ leakage have been made by modelling, using either the $CO_2$ pool-size measurements and permeability coefficients described above, or from measurements of $^{13}C$ isotope discrimination (Farquhar, 1983; Henderson, von Caemmerer & Farquhar, 1992). A direct measure of $CO_2$ leakage has recently been made by following $CO_2$ release from intact leaves following a $^{14}C$ pulse (Hatch, Agostino & Jenkins, 1995). Estimates from modelling of $CO_2$ pools and permeability coefficients generally give the lowest values for leakage (around 0.14, expressed as $CO_2$ lost from the bundle sheath as a fraction of $CO_2$ produced in the decarboxylation reaction). $^{13}C$ isotope discrimination gives values at the higher end of the range (up to 0.5) while the radioisotope determinations give values which generally agree well with the lower estimates of 0.1 to 0.3. There is still some controversy concerning estimation of $CO_2$ leakage in $C_4$ plants, compounded by the fact that none of the methods used is free from assumptions implicit in their calculations. Sampling gas which has passed over a leaf ('on-line' measurements) and determining resultant $^{13}CO_2$ discrimination gives leakage values quite close to those obtained by the radioisotope method. However, if the NADP–ME dicots are excluded from the data of Hatch *et al.* (1995), leakage values from this method average out around 0.1 for 11 different species with a standard deviation of only 0.02. The reason NADP-ME dicots show high leak rates is unknown, and there appears to be no correlation between the presence or absence of a suberin layer between bundle sheath and mesophyll cells and leakiness. Apparently, an asymetric cellular arrangement of chloroplasts within bundle sheath cells can provide a large resistance to diffusion (see Hatch, 1992).

The discrepancies between different methods used to determine leakage are yet to be resolved, but it seems likely that between 10 and 30% of the $CO_2$ released in the bundle sheath leaks back to the mesophyll, creating 'slippage' in the $CO_2$ pump and an inherent inefficiency in $C_4$ photosynthesis. The implications of this inefficiency on photosynthetic performance are dealt with in the following section.

## PHOTOSYNTHETIC ELECTRON TRANSPORT AND QUANTUM YIELD IN $C_4$ PLANTS

Quantum yield is a sensitive measure of the efficiency of light interception and utilization in plants widely used as a non-intrusive probe of photorespiratory activity and photoinhibitory damage in $C_3$ and $C_4$ plants. Quantum yield in unstressed $C_3$ plants is incredibly invariant across a wide range of species when measured under standardized temperature, $CO_2$ and $O_2$ concentrations (see Hatch, 1987). Values of 0.1 to 0.11 are commonly measured for $C_3$ plants under non-photorespiratory conditions, approaching the theoretical maximum calculated from the ATP and NADPH requirements of the PCRC cycle. Of course, if the $CO_2$ concentration in the gas phase is reduced, the energetic cost of photorespiration is reflected in a reduction in quantum yield in $C_3$ plants to around 0.065 at 20 °C in air (see Furbank, Jenkins & Hatch, 1990). Due to the increase in photorespiration with increasing temperature, a typical quantum yield for a $C_3$ plant in air drops still further to around 0.052 at 30 °C. At high $CO_2$, quantum yield in $C_3$ plants is insensitive to temperature because photorespiration is inhibited. Likewise, the quantum yield of $C_4$ photosynthesis is insensitive to temperature because of reduced photorespiration due to the $CO_2$ concentrating mechanism. The cost of the $CO_2$ concentrating mechanism means that at lower temperatures (below about 25 °C),

$C_3$ plants possess the more efficient photosynthetic process. This observation in part explains the absence of $C_4$ plants from cooler habitats (see Hatch, 1992) where they would have no great competitive advantage over $C_3$ plants.

As discussed above, the amount of leakage of $CO_2$ from the bundle sheath cells which occurs *in vivo* could have a profound impact on the efficiency of $C_4$ photosynthesis and hence the quantum yield of $C_4$ plants. Unlike $C_3$ plants, the quantum yield of photosynthesis in $C_4$ plants shows considerable interspecific variation (see Hatch, 1987), ranging from 0.050 to 0.069 across 32 $C_4$ species (this represents a variation of 5 quanta per $CO_2$ fixed). It has been suggested that this large range of quantum yields might reflect interspecific variation in the leakiness of the bundle sheath compartment and this hypothesis has been discussed at length in subsequent papers (Farquhar, 1983; Furbank *et al.*, 1990). Considering that there are three diverse mechanisms responsible for the $C_4$ photosynthetic mechanism, one might expect some trends in the energy requirements and hence quantum yields within decarboxylation types. There is some evidence for such a trend, although the most striking differences are between NAD-ME dicot types and the remainder of the $C_4$ monocots and dicots, where the former have average quantum yields of around 0.053 compared to 0.0625 for the latter (see Hatch, 1987, 1992). It is interesting to note that there is, in fact, no consistent correlation between measured quantum yields and leak rates estimated either by carbon isotope discrimination or radioisotope methods (see Hatch *et al.*, 1995).

It is possible to model the effect of leakiness to $CO_2$ on quantum yield in $C_4$ plants and this has been done on at least two occasions (Farquhar, 1983; Furbank *et al.*, 1990). Furbank *et al.* (1990) constructed a model which predicts the quantum requirement of photosynthesis from a given leak rate, based on the ATP and NADPH requirements of the mesophyll reactions. It is assumed that, if a plant has a high leak rate, flux through the mesophyll reactions must increase to maintain the bundle sheath $CO_2$ concentration, which will be reflected in an increased quantum requirement. In these calculations, it became apparent that a key assumption was the number of protons partitioned per electron transported in the chloroplast electron transport chain, affecting the number of ATP produced per quanta. Currently,

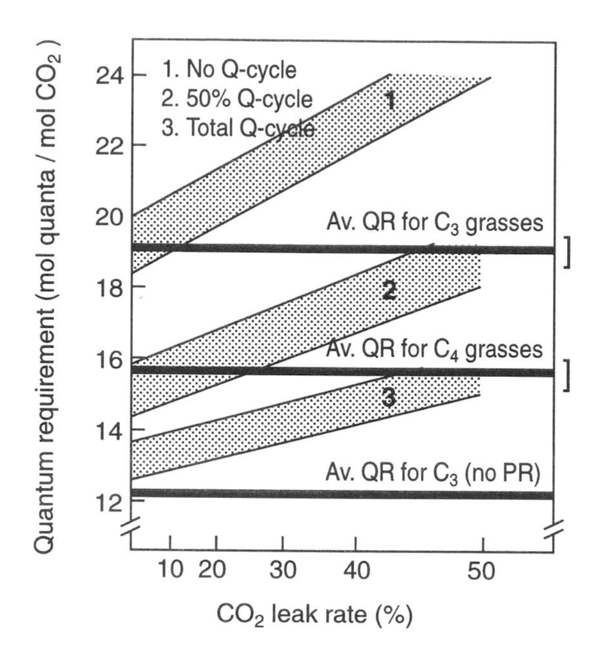


**Figure 9.3.** Calculated quantum requirements (1/quantum yield) for $C_4$ plants with different rates of $CO_2$ leakage from the bundle sheath cells, assuming either no engagement of a Q-cycle in thylakoid electron transport or with full or partial engagement of a Q-cycle. Average measured quantum requirements for $C_3$ and $C_4$ grasses are also shown. (Adapted from Furbank *et al.*, 1990 and Hatch, 1992.)

opinion is divided as to the mechanism of proton partitioning during electron transport through the thylakoid cytochrome $b_6f$ complex (see Chapter 7). One mechanism (which now has considerable support from *in vitro* experiments) involves a cycle around cytochrome $b_6f$ (the 'Q-cycle') and essentially doubles the efficiency of proton pumping through this complex per electron transported when compared to a simple linear model. Figure 9.3 summarizes the model of Furbank *et al.* (1990) by comparing the relationship between calculated quantum requirement and $CO_2$ leak rate with measured quantum yields for $C_3$ and $C_4$ plants (adapted from Hatch, 1992). Three calculated relationships are shown using the following assumptions: (i) no Q-cycle operation in proton partitioning; (ii) partial Q-cycle engagement and (iii) electron transport operating with full Q-cycle efficiency. The shaded areas represent the variability of calculated values obtained for each assumption, depending on the other uncertain factors implicit in the calculation. The horizontal lines show the average measured quantum requirements for $C_4$ and $C_3$ grasses measured at 30 °C. Without the operation of the Q-cycle, $C_4$ plants would be no more efficient than $C_3$ plants, even at 30 °C and, assuming even low to moderate leakage rates, there would be no energetic advantage in the operation of the $C_4$ pathway. In fact, the only way we can account for the low measured values for quantum requirement in $C_4$ monocots is to assume the operation of the Q-cycle, and in its absence, it is doubtful if the $C_4$ mechanism could have evolved (see Furbank *et al.*, 1990; Hatch, 1992).

Apart from the unique additional energy requirements of $C_4$ photosynthesis and their implications for chloroplast thylakoid processes, there are several other aspects of the $C_4$ mechanism which require a high degree of specialization of the light-harvesting apparatus. First, due to the spatial separation of the PCRC cycle from the $CO_2$ pump, the balance between the ATP and NADPH requirements in the mesophyll is quite different to that in the bundle sheath (see Fig. 9.1). In the mesophyll chloroplast of an NAD–ME or NADP–ME-type $C_4$, for example, 2ATP and 1 NADPH is required for each $CO_2$ fixed by PEP carboxylase while, in a PCK type, the ATP requirement in the mesophyll compartment is reduced as PEP is returned to the mesophyll directly (see Fig. 9.1b). An added complication is that, in many NADP–ME grasses, like maize and sorghum, the decarboxylation of malate in the bundle sheath chloroplast provides NADPH directly to the PCRC cycle and consequently, these species have lost most of their PSII activity and the ability to carry out whole-chain electron transport (see Hatch, 1987). The bundle sheath chloroplasts of these species are highly specialized in that they have a high capacity for cyclic photophosphorylation but almost no capacity to generate NADPH. In these plants, a proportion of the 3-phosphoglycerate (3-PGA) produced in the bundle sheath is shuttled to the mesophyll chloroplast for reduction and isolated maize chloroplasts, for example, can support high rates of 3-PGA reduction but lack any other enzymes of the PCRC cycle (see Fig. 9.1 and Hatch, 1987). Despite the differences in specialized electron transport pathways required in the different decarboxylation types, it has recently been shown that the rate of photosynthesis can be predicted quite accurately in a range of $C_4$ plants using chlorophyll fluorescence techniques (e.g. Krall & Edwards, 1990; Edwards & Baker, 1993), which assess electron flux through PSII by measuring the quantum efficiency of light harvesting. These fluorescence measurements also support the hypothesis that little or no photorespiration occurs in $C_4$ plants in air or, in fact, at low $CO_2$ (Krall & Edwards, 1990).

## ENZYME REGULATION IN $C_4$ PHOTOSYNTHESIS

Although the intricacies of the $C_4$ photosynthetic mechanism have been challenging to unravel, the regulation of flux through the $C_4$ pathway in response to environmental parameters is equally complex. To date, the most common approach to understanding biochemical regulation has been to look for potential regulatory properties of isolated enzymes and attempt to extrapolate these properties to the cellular environment *in vivo*. In many cases, for example in the case of rubisco, this approach has proven to be very successful, allowing us to construct mechanistic models of $C_3$ photosynthesis which predict the response of photosynthesis to light and $CO_2$ concentration quite accurately. For this approach to succeed, reliable measurements of likely enzyme activity *in vivo* and cellular concentrations of substrates and effectors are required. In the case of $C_4$ plants, these criteria are difficult to meet due primarily to the technical demands of making such measurements in a highly compartmentalized system (not only the organelle/cytosol compartmentation must be considered but also the bundle sheath/mesophyll compartmentation). Thus, in the $C_4$ context there is a vast number of 'potentially' rate-limiting and highly regulated enzymes which, in addition to the PCRC cycle enzymes, could provide a regulatory interface between $C_4$ photosynthesis and the environment. In the space available, it is not possible to discuss all of these. In the subsequent sections, with the exception of rubisco, discussions on enzyme regulation will be limited to those enzymes unique to the $C_4$ pathway which show clear changes in activity in response to environmental cues. There will be a particular emphasis on recent advances in the application of molecular biology to $C_4$ photosynthesis (see also Furbank & Taylor, 1995). The properties of the $C_4$ enzymes not dealt with here are reviewed in Ashton *et al.* (1990).

**A** Thioredoxin mediated activation of MDH

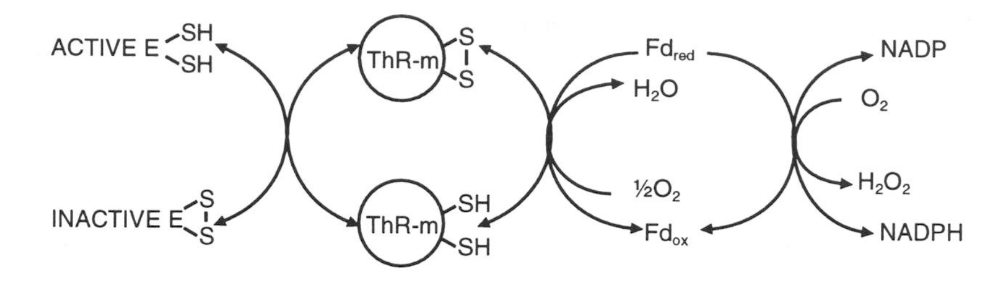


**B** Phosphorylation/dephosphorylation regulation of PPdK

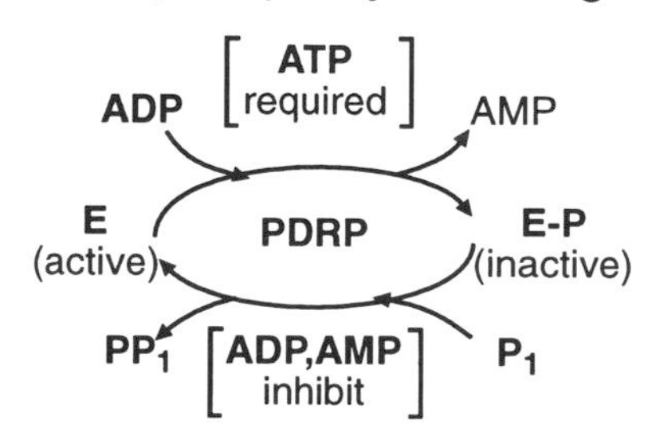


**C** Phosphorylation/dephosphorylation regulation of PEP carboxylase

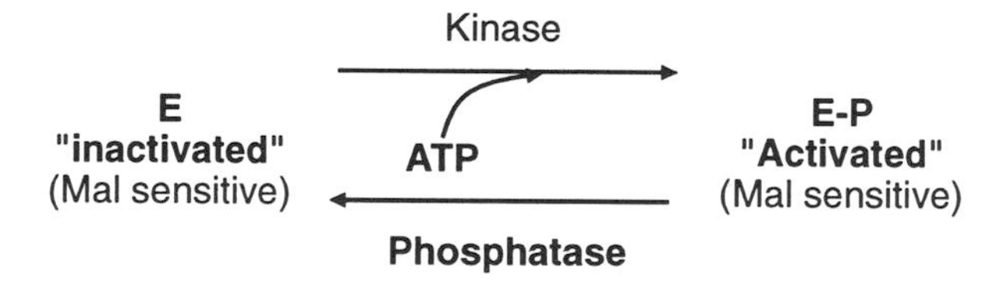


**Figure 9.4.** Three examples of enzyme regulation by covalent modification. A Regulation of MDH by the thioredoxin system. ThR-m, thioredoxin m; Fd, ferredoxin; B Phosphorylation regulation of PPdK. E, un-phosphorylated enzyme, E-P, phosphorylated enzyme; PDRP, PPdK regulatory protein; C Phosphorylation regulation of PEP-carboxylase. (Adapted from Furbank & Taylor, 1995.)

The key regulatory enzymes unique to the $C_4$ pathway, which will be the focus of the following section, are indicated on Fig. 9.4 and their regulation summarized in Fig. 9.4. Pyruvate Pi-dikinase (PPdK), phosphoenolpyruvate carboxylase (PEPC) and NADP-dependent malate dehydrogenase (NADP-MDH) are all localized in mesophyll cells, with PPdK and NADP-MDH located in the chloroplast and PEPC in the cytosol. The extractable activity of PPdK is only just sufficient to account for observed rates of photosynthesis and the activity of all three enzymes *in vivo* is increased markedly by illumination (Hatch, 1987).

In the past two decades it has become apparent that a major form of enzyme regulation in plants is brought about by covalent modification of specific amino acid residues either by the formation of disulphide bridges or phosphorylation by protein kinases (see Fig. 9.4). This type of regulation can be directly responsive to light (via the thioredoxin system) or can respond to cellular energy reserves such as ATP. With gene sequences of more photosynthetic enzymes appearing in genetic databases almost daily, this area of enzyme regulation is particularly amenable to study by expression of recombinant protein *in vitro* and *in vivo*. NADP-MDH for example, is a light-activated enzyme in the $C_4$ pathway (Fig. 9.1) which undergoes reductive activation in the light via photosynthetic electron transfer and the thioredoxin system (Buchanan, 1991). Reduction of NADP-MDH by reduced thioredoxin breaks a disulfide bridge between two cysteine residues, activating the enzyme (see Fig. 9.4). The amino acid sequence of NADP-MDH is very similar to the NAD-dependent form, but also has C- and N-terminal extensions which have been implicated in conferring the unique regulatory properties of the photosynthetic enzyme. Using site-directed mutagenesis, the pairs of cysteine residues responsible for redox activation at both the N and C termini of the sorghum enzyme have been identified (see Issakidis *et al.*, 1994). Mutant forms of this enzyme, which are not inactivated by oxidation and thus should not be inactivated in the dark, have been produced.

Two other key enzymes in $C_4$ photosynthesis are regulated by covalent modification: PPdK and PEPC (Fig. 9.1 and 9.4). PEPC, the primary $CO_2$ fixing enzyme of the $C_4$ pathway, is 'activated' when phosphorylated by a specific protein kinase (for a review on phosphorylation regulation of photosynthetic proteins see Budde & Chollet, 1988; PEPC regulation was recently reviewed by Rajagopalan, Devi & Raghavendra, 1994 and Lepiniec *et al.*, 1994). Although the maximum extractable activity of this enzyme far exceeds measured photosynthetic flux, it is thought that this enzyme never achieves its $V_{max}$ *in vivo* (see Hatch, 1987) and that phosphorylation status, combined with levels of the effector malate, play a pivotal role in providing 'feedback' regulation of this enzyme in leaves. 'Activation' of PEP carboxylase does not result in an increase in the $V_{max}$ of the enzyme, but manifests itself through a decrease in sensitivity to the inhibitor malate. In the sorghum protein, phosphorylation of serine 8 is responsible for this activation. Substitution of this serine in a recombinant enzyme expressed in *E. coli* prevents this activation (Duff *et al.*, 1993). This region of the amino acid sequence appears to be highly conserved among PEPC from many sources, suggesting that this regulatory mechanism may be universal. So far, no transformation experiments have been performed with a recombinant enzyme in higher plants. The signal transduction chain which controls the activity of the protein-serine kinase remains somewhat of a mystery as does the identity of the kinase itself.

PPdK, responsible for regenerating PEP, the acceptor for $CO_2$ fixation in the mesophyll chloroplast, has long been recognized as a light-activated enzyme (see Hatch, 1987). Its activity in intact leaves changes rapidly in response to changing light intensity and it is largely inactive in the dark. It is now known that, unlike PEPC, this enzyme is inactivated by phosphorylation, and the residue involved is threonine 527 (see Furbank & Taylor, 1995). This threonine is distinct from the phosphorylation which occurs during catalysis (at histidine residue 529). The enzyme that catalyses regulatory phosphorylation of PPdK ('regulatory protein') is unique in that ADP rather than ATP is the phosphate donor and this same enzyme catalyses the removal of the phosphate, reactivating PPdK by phosphorolysis rather than by hydrolysis (Fig. 9.4). It is not clear how the activity of the regulatory protein is itself controlled, although phosphorylation is sensitive to adenylate energy charge and high pyruvate levels appear to block inactivation (see Hatch, 1987 and Furbank & Taylor, 1995). The regulatory protein is not abundant in leaves and neither purification to homogeneity nor cloning of this gene has been possible. However, a partial purification was recently reported (Smith, Duff & Chollet, 1994).

## TRANSGENIC PLANTS

Recently, molecular genetics and plant genetic transformation have allowed biochemists to manipulate the amounts and properties of enzymes *in vivo* with great precision. We are now able to examine the role an enzyme plays in controlling metabolic flux both by titrat-

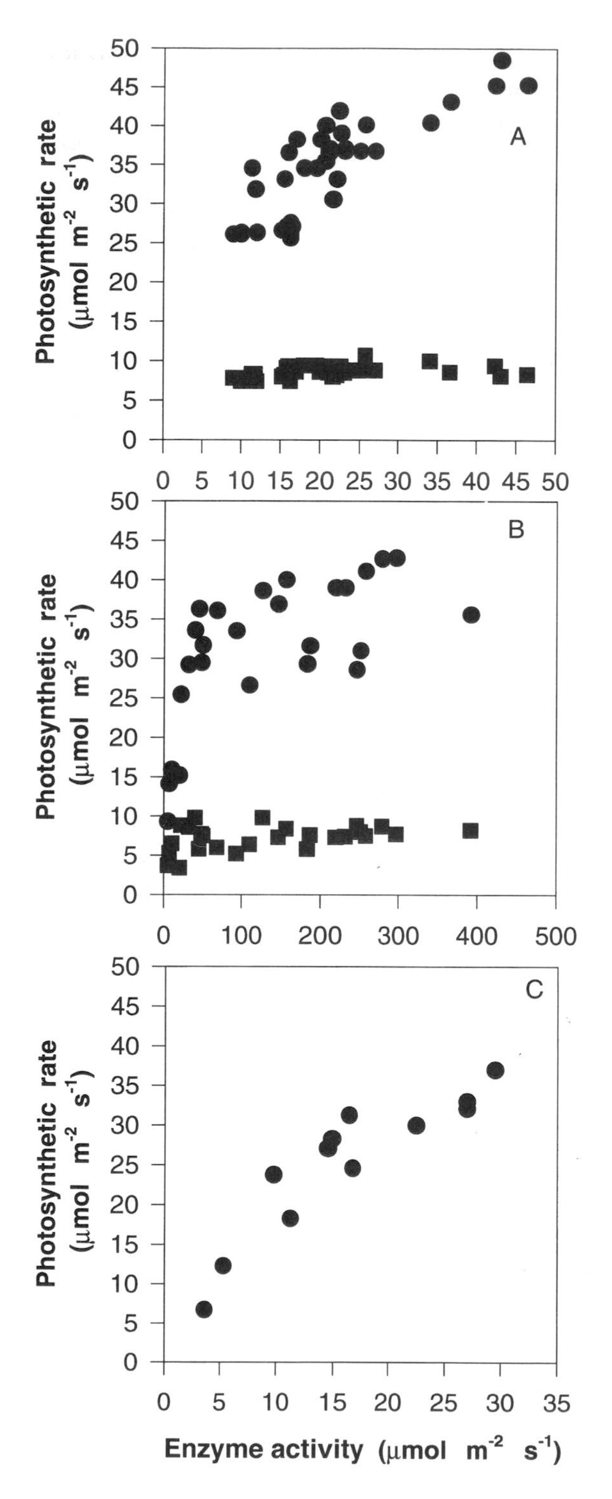


ing down its activity in transgenic plants with gene suppression techniques and by increasing its activity by over-expression experiments (recently reviewed by Furbank & Taylor, 1995). To examine the finer points of enzyme regulation, it is possible to alter regulatory properties of enzymes by site-directed mutagenesis and determine the consequences in bacterial systems as described above and then *in planta*. The most commonly used technique in metabolic engineering of photosynthesis has been gene suppression, principally antisense RNA technology (see Furbank & Taylor, 1995). Transgenic $C_3$ plants containing gene suppression constructs have proven useful in furthering our understanding of the role of key enzymes in determining rates of photosynthesis and growth under varying environmental conditions (for review see Furbank & Taylor, 1995) and in the control of carbon partitioning (Sonnewald *et al.*, 1994). The advantage of this technique is the precision with which the activity of a single enzyme can be decreased *in vivo* without appreciably altering levels of other enzymes. By producing a number of transgenic plants, each genetically identical but with a wide range of levels of the target enzyme, it is possible to accurately assess the role this enzyme plays in controlling photosynthetic flux *in vivo* and under different environmental conditions. The use of the antisense approach in studying $C_4$ photosynthesis has so far been impeded by the lack of efficient and reliable genetic transformation systems. It is only recently, following the advent of an efficient *Agrobacterium* mediated transformation system for the $C_4$ NADP-ME dicot *Flaveria bidentis* (Chitty *et al.*, 1994), that attempts to use this approach to study $C_4$ photosynthesis have been reported.

Recently, transgenic *Flaveria bidentis* plants containing antisense constructs targeted to the enzymes rubisco, NADP-MDH, PPdK and the mesophyll enzyme involved in conversion of $CO_2$ to $HCO_3$, carbonic anhydrase, have been generated (Furbank *et al.*, 1996, 1997; Trevanion et al., 1997; Ludwig, Price, Burnell and

**Figure 9.5.** A plot of extractable enzyme activity against photosynthetic rate in air, at high light (●) or limiting light (■) in transgenic *Flaveria*. A, with reduced PPdK activity; B, NADP-MDH activity; C, rubisco activity.

Furbank, pers. comm.). Plants with reduced rubisco, NADP-MDH and PPdK have been fully characterized and the results are summarized in a plot of photosynthetic rate against enzyme activity in a range of transformants (Fig. 9.5). This plot (along with the data in Furbank *et al.*, 1996) indicates that none of these three enzymes is particularly important in controlling photosynthetic flux at low light intensity. By analogy with the $C_3$ photosynthetic models, photosynthetic electron transport probably limits photosynthetic flux at the light intensity used (175 $\mu$mol $m^{-2}$ $s^{-1}$, or one-tenth of the growth irradiance). However, at high light intensities, it appears that there is only just sufficient rubisco present in $C_4$ leaves to support photosynthesis, as is the case in $C_3$ plants (the four points on the extreme right of Figs 9.5A and C are wild-type individuals). Regulation of this enzyme could provide effective fine control of photosynthetic flux. In contrast, NADP-MDH activity can be reduced more than 80% with only a marginal effect on photosynthetic rate. PPdK appears to fall mid-way between these two extremes and transgenic plants with only one-quarter to one-third of wild-type enzyme activity are still capable of photosynthetic rates half to two-thirds of wild-type. This is somewhat surprising, as maximum extractable activities from wild-type plants are only just sufficient to account for photosynthetic flux, suggesting that this enzyme, of all those in the $C_4$ pathway, would be most likely to be rate limiting (see Hatch, 1987). From these data, we must conclude that maximum extractable PPdK activity seriously underestimates the potential activity of this enzyme *in vivo*. These data also suggest that, in the case of NADP-MDH, 'light regulation' may be more important in switching the enzyme off in the dark rather than providing fine control of photosynthetic flux in response to changing light intensity. Down-regulation of this enzyme by more than 90% in wild-type individuals would be necessary before this enzyme restricted the rate of photosynthesis.

## MUTANTS OF THE $C_4$ PATHWAY

Another approach to obtaining plant material with reduced levels of key photosynthetic enzymes is by mutagenesis and screening individuals for lack of photosynthetic competence in air. Recently, mutants of the NAD-ME-type $C_4$ dicot *Amaranthus edulis* with greatly reduced levels of PEP-carboxylase or NAD-malic enzyme were obtained by this method (Dever *et al.*, 1995). PEP carboxylase mutants contained less than 10% wild-type activities and when crossed to wild-type material gave progeny with around 50% wild-type activity. These heterozygous progeny grew normally in air and despite the reduction in PEP carboxylase activity, there was less than a 30% decrease in maximum photosynthetic rate. NAD-malic enzyme mutants accumulated malate, possessed less than 10% of wild-type enzyme activity and appeared to show an increased $K_m$ for malate. Although this is a powerful approach, there are potential limitations to this technique when compared to the antisense system. A major shortcoming is that, because of the reliance on gene dosage to affect protein level, it is not possible to produce plants with a wide range of enzyme activities from a single gene mutation.

## FUTURE RESEARCH DIRECTIONS

One of the greatest challenges in the area of photosynthesis is to utilize the large body of information available in the literature on the mechanism and regulation of enzymes *in vitro* to understand and perhaps improve photosynthetic performance in whole plants. As discussed earlier, the major barrier to this research has been the difficulty of extrapolating data obtained *in vitro* to the whole leaf, particularly with $C_4$ plants. Early indications are that metabolic engineering, by allowing us to perform biochemical manipulations *in vivo* in a precise and interpretable fashion, coupled with more traditional biochemical and physiological approaches, may provide the means to achieve this aim. So far, only gene suppression has been used successfully in the $C_4$ system and attempts at heterologous over-expression led to co-suppression of the target gene with a decrease in enzyme levels (Trevanion, Furbank & Ashton, 1997). Future experiments will, no doubt, include over-expression of both heterologous and mutant forms of highly regulated enzymes like PPdK and PEPC in an effort to better understand the regulatory properties of these enzymes *in vivo*. Another area in which major advances could be made is in the non-destructive estimation of metabolite concentrations within metabolic compartments in intact leaves; a field which, after a flurry of activity involving nuclear magnetic resonance techniques, has moved quite slowly in recent years. Undoubtedly, the complexities of $C_4$ photosynthesis will retain a fascination for plant biochemists, molecular geneticists and physiologists for many years to come.

## REFERENCES


Ashton, A. R., Burnell, J. N., Furbank, R. T., Jenkins, C. L. D. & Hatch, M. D. (1990). Enzymes of $C_4$ photosynthesis. *In Methods in Plant Biochemistry*, vol. 3, ed. P. J. Lea, pp. 39–72. London: Academic Press.

Buchanan, B. B. (1991). Regulation of $CO_2$ assimilation in oxygenic photosynthesis: The ferredoxin/thioredoxin system. *Archives of Biochemestry and Biophysics*, **288**, 1–8.

Budde, R. J. A. & Chollet, R. (1988). Regulation of enzyme activity in plants by reversible phosphorylation. *Physiologia Plantanum*, **72**, 435–39.

Chitty, J. A., Furbank, R. T., Marshall, J. S., Chen, Z. & Taylor, W. C. (1994). Genetic transformation of the $C_4$ plant *Flaveria bidentis. Plant Journal*, **6**, 949–56.

Dever, L. V., Blackwell, R. D., Fullwood, N. J., Lacuesta, M., Leegood, R. C., Onek, L. A., Pearson, M. & Lea, P. J. (1995). The isolation and characterisation of mutants of the $C_4$ pathway. *Journal of Experimental Botany* **46**, 1363–76.

Duff, S. M. G., Lepiniec, L., Cretin, C., Andreo, C., Condon, S. A., Sarath, G., Vidal, J., Gadal, P. & Chollet, R. (1993). An engineered change in the L-malate sensitivity of a site directed mutant of sorghum phosphoenolpyruvate carboxylase: The effect of sequential mutagenesis and s-carboxymethylation at position 8. *Archives of Biochemistry and Biophysics*, **306**, 272–6.

Edwards, G. E. & Baker, N. R. (1993). Can $CO_2$ assimilation in maize leaves be predicted accurately from chlorophyll fluorescence analysis? *Photosynthesis Research*, **37**, 89–102.

Ehleringer, J. R., Sage, R. W., Flanagan, B. & Pearcy, R. W. (1991). Climate change and the evolution of $C_4$ photosynthesis. *Tree*, **6**, 95–9.

Farquhar, G. D. (1983). On the nature of carbon isotope discrimination in $C_4$ species. *Australian Journal of Plant Physiology*, **10**, 205–26.

Furbank, R. T. & Badger, M. R. (1982). Photosynthetic oxygen exchange in attached leaves of $C_4$ monocotyledons. *Australian Journal of Plant Physiology*, **9**, 553–8.

Furbank, R. T. & Hatch M. D. (1987). Mechanism of $C_4$ photosynthesis. The size and composition of the inorganic carbon pool in bundle sheath cells. *Plant Physiology*, **85**, 958–64.

Furbank, R. T. & Taylor, W. C. (1995). Regulation of photosynthesis in $C_3$ and $C_4$ plants: a molecular approach. *The Plant Cell*, **7**, 797–807.

Furbank, R. T., Jenkins, C. L. D. & Hatch, M. D. (1989). $CO_2$ concentrating mechanism of $C_4$ photosynthesis: permeability of isolated bundle sheath cells to inorganic carbon. *Plant Physiology*, **91**, 1364–71.

Furbank, R. T., Jenkins, C. L. D. & Hatch, M. D. (1990). $C_4$ photosynthesis: quantum requirement, $C_4$ acid overcycling and Q-cycle involvement. *Australian Journal of Plant Physiology*, **17**, 1–7.

Furbank, R. T., Chitty, J. A., von Caemmerer, S. & Jenkins, C. L. D. (1996). Antisense RNA inhibition of RbcS gene expression reduced rubisco level and photosynthesis in the $C_4$ plant *Flaveria bidentis. Plant Physiology*, **111**, 725–34.

Furbank, R. T., Chitty, J. A., Jenkins, C. L. D., Taylor, W. C., Trevanion, S. J., Von Caemmerer, S. & Ashton, A. R. (1997). Genetic manipulation of key photosynthetic enzymes in the $C_4$ plant, *Flaveria bidentis. Australian Journal of Plant Physiology*, in press.

Hatch, M. D. (1987). $C_4$ photosynthesis: a unique blend of modified biochemistry, anatomy and ultrastructure. *Biochimica et Biophysica Acta*, **895**, 81–106.

Hatch, M. D. (1992). The making of the $C_4$ pathway. In *Research in Photosynthesis*, vol. III, ed. N. Murata, pp. 747–56. Amsterdam: Kluwer Academic Publishers.

Hatch, M. D., Agostino, A. & Jenkins, C. L. D. (1995). Measurement of the leakage of $CO_2$ from bundle sheath cells of leaves during $C_4$ photosynthesis. *Plant Physiology*, **108**, 173–81.

Henderson, S. A., von Caemmerer, S. & Farquhar, G. D. (1992). Short-term measurements of carbon isotope discrimination in several $C_4$ species. *Australian Journal of Plant Physiology*, **19**, 263–85.

Issakidis, E., Saarinen, M., Decottignies, P., Jacquot, J-P., Cretin, C., Gadal, P. & Miginiac-Maslow, M. (1994). Identification and characterisation of the second regulatory disulphide bridge of recombinant sorghum leaf NADP-malate dehydrogenase. *Journal of Biological Chemistry*, **269**, 3511–17.

Jenkins, C. L. D., Furbank, R. T. & Hatch, M. D. (1989). Inorganic carbon diffusion between $C_4$ mesophyll and bundle sheath cells: direct bundle sheath $CO_2$ assimilation in intact leaves in the presence of an inhibitor of the $C_4$ pathway. *Plant Physiology*, **91**, 1356–63.

Krall, J. P. & Edwards, G. E. (1990). Quantum yield of photosystem II electron transport and carbon dioxide fixation in $C_4$ plants. *Australian Journal of Plant Physiology*, **17**, 579–88.

Lepiniec, L., Vidal, J., Chollet, R., Gadal, P. & Cretin, C. (1994). Phosphoenolpyruvate carboxylase: structure, regulation and evolution. *Plant Science*, **99**, 111–24.

Rajagopalan, A. V., Devi, M. T & Raghavendra, A. S. (1994). Molecular biology of $C_4$ phosphoenolpyruvate carboxylase: structure, regulation and genetic engineering. *Photosynthesis Research*, **39**, 115–35.

Smith, C. M., Duff, S. M. G., & Chollet, R. (1994). Partial purification and characterisation of maize-leaf pyruvate, Pi dikinase regulatory protein: a low abundance mesophyll chloroplast stromal protein. *Archives of Biochemistry and Biophysics*, **308**, 200–6.

Sonnewald, U., Lerchl, J., Zrenner, R. & Frommer, W. (1994). Manipulation of sink-source relations in plants. *Plant, Cell and Environment*, **17**, 649–58.

Trevanion, S. J., Furbank, R. T. & Ashton, A. (1997). Inhibition of NADP-linked malate dehydrogenase gene expression in the $C_4$ plant *Flaveria bidentis. Plant Physiology*, in press.

von Caemmerer, S. & Farquhar, G. D. (1981). Some relationships between the biochemistry of photosynthesis and the gas exchange of leaves. *Planta*, **53**, 376–87.

Woodrow, I. E., & Berry, J. A. (1988). Enzymatic regulation of photosynthetic $CO_2$ fixation in $C_3$ plants. *Annual Review of Plant Physiology and Plant Molecular Biology*, **39**, 533–94.

# 10 Crassulacean acid metabolism

U. LÜTTGE

## THE CYTOLOGICAL AND BIOCHEMICAL REACTION SCHEME OF CAM

The essence of crassulacean acid metabolism (CAM) is that it allows massive uptake of $CO_2$ in the darkness (Fig. 10.1), which is stored over the night in the form of organic acids. This is brought about by phosphoenolpyruvate carboxylase (PEPC) followed by malate dehydrogenase. The $CO_2$ acceptor PEP may be derived from nocturnal carbohydrate metabolism starting mainly from starch or soluble glucans but in many cases also from free hexose stored in the vacuole. Triosephosphate may be generated from starch breakdown in the plastids, glycolysis in the cytoplasm or the oxidative pentose-phosphate cycle which circumvents phosphofructokinase.

For two reasons, nocturnally synthesized malic acid must be removed from the site of PEPC and the cytoplasm. First, malate exerts a feedback inhibition on PEPC. Secondly, malate synthesis is associated with the formation of acid equivalents which tend to acidify the cytoplasm. Therefore, malic acid is sequestered in the vacuole. This is an active mechanism. The energy for the transport of malic acid into the vacuole is derived from ATP via an $H^+$-transporting ATPase at the tonoplast (V-ATPase), which establishes an electrochemical gradient of protons, $\Delta\bar{\mu}_{H^+}$, across the tonoplast membrane:

$$\Delta\bar{\mu}_{H} = RT \ln \frac{10^{-pH_{vac}}}{10^{-pH_{cyt}}} + F\Delta E \qquad (1)$$

where R is the gas constant; T, absolute temperature; vac, vacuole; cyt, cytoplasm; F, the Faraday constant and $\Delta E$ the transmembrane electrical potential.

In some CAM plants there may be also smaller or larger oscillations of citric acid in addition to malic acid. Massive night/day changes of citric acid levels are observed under certain conditions in species of the tropical tree genus *Clusia*. Mitochondrial reactions must be involved in citrate synthesis (Fig. 10.1). However, unlike accumulation of malic acid citric acid does not lead to a net gain of carbon since as much $CO_2$ as is fixed by PEPC to generate OAA is lost by pyruvate decarboxylase, where the acetyl-CoA needed for citrate synthase is formed. This has led to questions regarding the physiological significance of citrate accumulation (see below).

Nocturnal accumulation of acid ($\Delta H^+$), i.e. day–night changes of titratable protons, reaches 400 to 500 mmol titratable protons $l^{-1}$ in the cell sap of species of the Crassulaceae genus *Kalanchoë* which are standard objects of CAM research. The highest acid oscillations observed are 1410 mmol titratable protons $l^{-1}$ in species of *Clusia*. Nocturnal accumulation of malate ($\Delta$mal) is up to 200–250 mmol $l^{-1}$ in *Kalanchoë* and 360 mmol $l^{-1}$ in *Clusia* which may also accumulate citric acid (up to 200 mmol $l^{-1}$) over night (Lüttge 1996; see p. 142).

During the day, the organic acids are diffusing out of the vacuole again (Fig. 10.2). Mostly malate is decarboxylated by NADP-dependent decarboxylating malate dehydrogenase in the cytoplasm, although other decarboxylation reactions also occur among CAM plants (Winter & Smith 1996a). The $CO_2$ regenerated is assimilated in the chloroplasts, where carbohydrates $(CH_2O)_n$ are synthesized, which may be exported from the assimilating leaves for growth and the formation of new

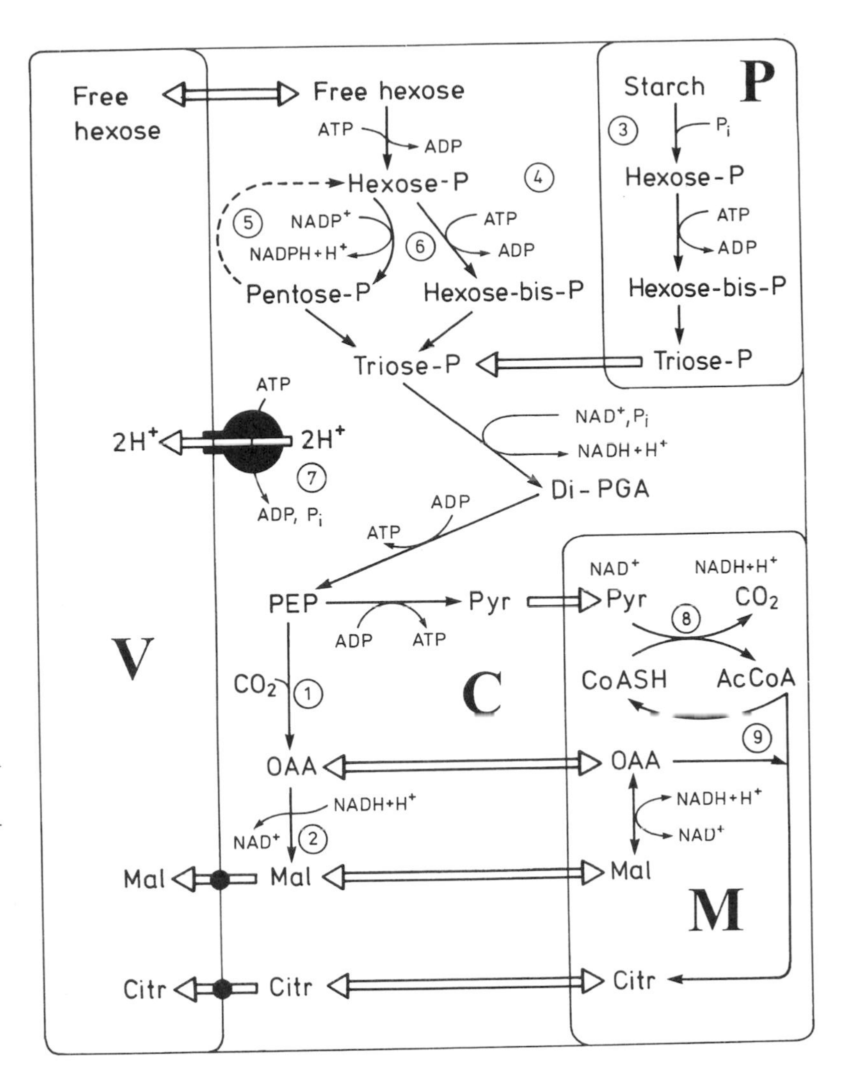


**Figure 10.1.** The dark-period of the loop of CAM: a simplified metabolic scheme. Stoichiometries have not been included in the scheme because they may vary due to the different possible combinations of reaction sequences. (For more detailed schemes of the flow of carbon in CAM see Winter & Smith 1996*a*.) *Abbreviations:* AcCoA = acetyl coenzyme A; Citr = citrate; CoASH = coenzyme A; Di-PGA = 2,3-di-phosphoglyceric acid; Mal = malate, OAA = oxaloacetate; $P_i$ = inorganic phosphate; PEP = phosphoenolypyruvate; Pyr = pyruvate. Compartments are C = cytoplasm; M = mitochondria; P = plastids; V = vacuole. Numbers represent enzymes or pathways: 1. PEPC; 2. NAD-malate dehydrogenase; 3. Phosphorlysis of starch to hexose-P; 4. Glycolytic enzymes; 5. Oxidative pentose phosphate pathway; 6. Phosphofructokinase; 7. V-ATPase; 8. Pyruvate decarboxylase; 9. Citrate synthase.

biomass, and starch is regenerated as a precursor for PEP formation in the next dark period. The pyruvate is funnelled into gluconeogenesis via pyruvate-$P_i$-dikinase (PPDK,) and may also contribute to triose-phosphate in the chloroplasts or lead to hexose formation in the cytoplasm.

Citrate can be broken down to $CO_2$ and malate in the mitochondria. As a first step, it may also be decarboxylated to α-ketoglutarate (αKGA) by cytoplasmic isocitrate dehydrogenase. Thus, potentially day-time breakdown of citrate may lead to more $CO_2$ than that of malate. This may be one of the keys to understand the physiological function of citric acid oscillations in CAM (see p. 142).

## GAS EXCHANGE AND THE FOUR PHASES OF CAM

The biochemical loop of the path of carbon in CAM (see above) brings about a very characteristic pattern of gas exchange (Fig. 10.3). There is $CO_2$ uptake and some transpirational loss of $H_2O$ during the nocturnal part of

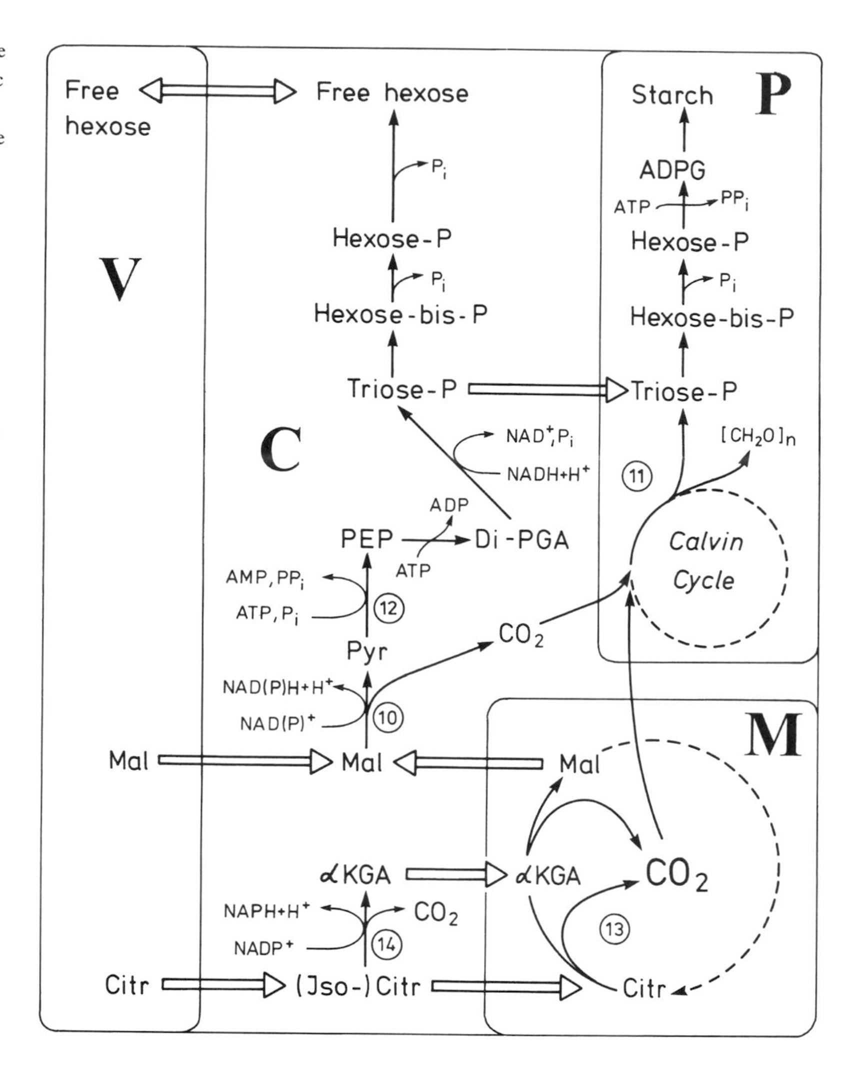


**Figure 10.2.** The light-period of the loop of CAM: a simplified metabolic scheme. Stoichiometries have not been included in the scheme because they may vary due to the different possible combinations of reaction sequences. (For more detailed schemes of the flow of carbon in CAM see Winter & Smith 1996*a*.) For abbreviations see Fig. 10.1. *Additional abbreviations:* ADPG = adenosine-diphosphate-glucose; $[CH_2O]n$ = carbohydrate; αKGA = α-keto-glutaric acid; $PP_i$ = inorganic pyrophosphate. Numbers represent enzymes or pathways:
10. NAD(P)-dependent decarboxylating malate dehydrogenase.
11. Calvin cycle; 12. Pyruvate Pi dikinase; 13. TCA cycle; 14. Cytoplasmic isocitrate dehydrogenase.

the loop in the dark period with $CO_2$ acquisition and storage as organic acids. There is no apparent gas exchange during the day-time part of the loop when acids are remobilized, decarboxylated and the $CO_2$ regained is reduced and assimilated behind closed stomata. Between these two phases of the loop there are transitions. One transition occurs in the morning when stomata remain open for a period of up to a few hours and both PEPC and RuBPC (ribulose-bisphosphate carboxylase) participate in primary $CO_2$ fixation. The other transition is observed sooner or later in the afternoon or evening. When the organic acid stored overnight is consumed and water relations permit, stomata may open in the later part of the light period, and $CO_2$ is fixed via RuBPC and carbon flows directly into the Calvin cycle. The major phases of the loop have been numbered I and III, respectively, i.e.

phase I, nocturnal $CO_2$ acquisition and storage of organic acid;

phase III, day-time remobilization of $CO_2$ with photochemical assimilation.

The transition phases were numbered II and IV, respectively, i.e.

phase II, early morning;

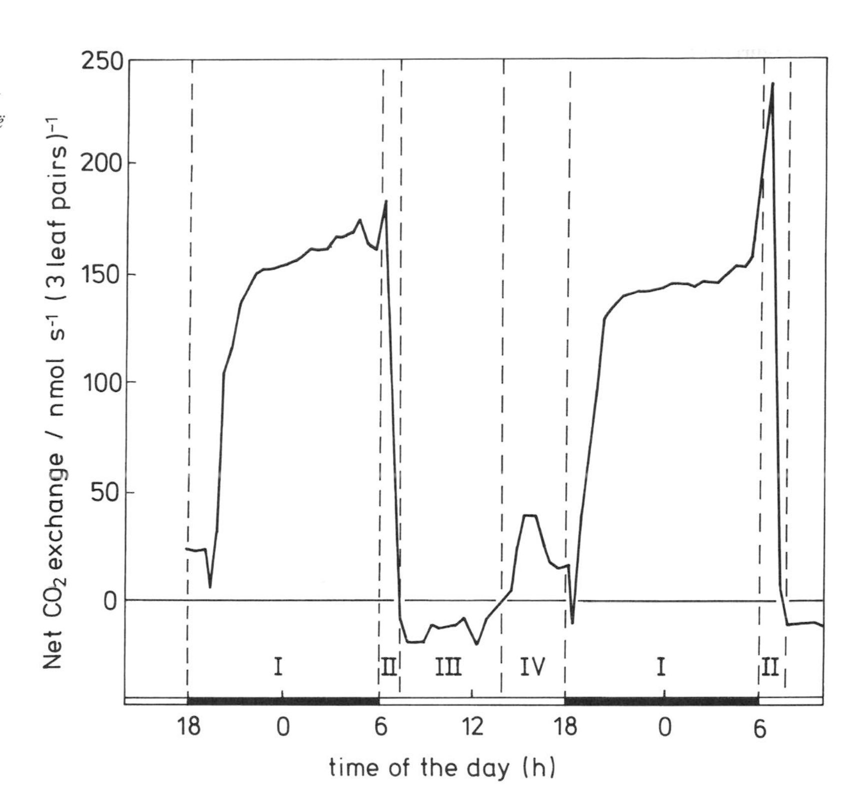


**Figure 10.3.** The four phases of $CO_2$ gas exchange after an experiment with the CAM plant *Kalanchoë daigremontiana*. For explanation of phases I–IV see pp. 138–139. High internal $CO_2$ concentration in phase III even leads to a small loss of $CO_2$ from the leaves in this phase.

phase IV afternoon and evening stomatal opening

(Osmond, 1978).

## THE EVOLUTION OF CAM AND CAM AS AN ECOPHYSIOLOGICAL ADAPTATION TO ENVIRONMENTAL STRESS

Basically, there is nothing special about the metabolic machinery of CAM as depicted in Fig. 10.1 and 10.2. Even the key enzymes PEPC and PPDK are found in plants ubiquitously, the former is important for anaplerotic reactions and the latter a key step in gluconeogenesis. The tonoplast $H^+$-ATPase is observed in all plant cells. Although we shall see later (p. 144) that these enzymes may be expressed in CAM in the form of specific isoenzymes and that they are highly regulated, *prima facie* it is nothing but a trick to use this rather general and simple enzymatic machinery for massive nocturnal acquisition of $CO_2$.

With the rather basic enzymic complement that is required for CAM performance, it may not be surprising that CAM has arisen polyphyletically, i. e. independently in very different taxa as suggested by its wide distribution on the various branches of the phylogenetic tree of vascular plants (Lüttge, 1987). CAM occurs in almost all divisions, classes and subclasses of vascular plants (Lüttge, 1987; Smith & Winter, 1996). This includes the pteridophyta with CAM plants among the Polypodiaceae in the Filicatae but also in the Lycopodiatae in genera of the family of the Isoëtaceae. The consideration of CAM as a special mechanism for $CO_2$ acquisition as presented on p. 136 gives the key for understanding the selective pressures that may have led to the evolution of CAM.

In the genus *Isoëtes* and among comparable life forms of angiosperms, which are submerged fresh-water macrophytes, we find CAM species (Griffiths, 1989; Keeley, 1996). They perform CAM with their submerged leaves and, to the extent that they are amphibic, they even lose the capacity for CAM in their emerged leaves. In view of the reduced diffusivity of $CO_2$ in water as compared to air and the strong competition among photosynthezising organisms for the $CO_2$ (or bicarbonate) in the aqueous habitat, CAM can be regarded as an adaptation to limited availability of $CO_2$, as it is also a $CO_2$ concentrating mechanism (Griffiths, 1989). $CO_2$ is more abundantly available during the night, when microorganisms, animals and plants produce $CO_2$ in respiration and there is no competition by other photosynthesizing organisms for CAM plants which can fix $CO_2$ in the dark.

In hot and dry climates or under osmotic stress like salinity in the root medium, land plants often suffer the danger of desiccation, especially under strong solar irradiation on clear and hot days. Loss of water from these plants can be much attenuated by stomatal closure. A typical phenomenon is the so-called midday depression of photosynthetic gas exchange when stomata close during the hottest time of the day when a high water-vapour deficit of the atmosphere and hence high vapour-pressure gradient between the leaf internal air spaces and the atmosphere (VPD) cause the greatest evaporative demand. However, this is a dangerous compromise. Closed stomata also prevent $CO_2$ entry into the leaves and, in the long run, midday depressions may severely curtail productivity. However, the even greater problem is that the photosynthetic electron transport systems remain energized by solar irradiation during the midday depression. In the absence of sufficient $CO_2$ for dissipation of this excitation energy via photosynthetic $CO_2$ assimilation, and when other mechanisms of energy dissipation become insufficient, this may not only lead to temporal and reversible photoinhibition but also to severe and irreversible photodamage. Nocturnal $CO_2$ acquisition, with stomata open in the night when VPD is much lower, not only means that much less water is lost by transpiration for each $CO_2$ taken up, i.e. water use efficiency (WUE), given by:

$$WUE = \frac{CO_2 \text{ fixed}}{H_2O \text{ transpired}} \qquad (2)$$

improves considerably, it also supplies organic acid from which $CO_2$ can be regenerated during the day. This can lead to very high $CO_2$ concentrations up to a few per cent in the gas phase of the leaves behind closed stomata (see Lüttge, 1987), which provides sufficient substrate for photochemical work and thus prevents photoinhibition and photodamage. In this way CAM reveals itself as an adaptation allowing $CO_2$ acquisition under critical water relations. This has led to the cliché of CAM being generally considered as an ecophysiological mechanism for the adaptation to limited availability of water. However, although in the dilemma of starvation or desiccation, given by the fact that stomata regulate both $CO_2$ uptake and $H_2O$ loss of the leaves, reduction of $H_2O$ loss and facilitation of $CO_2$ acquisition by CAM under drought stress are just different faces of the same coin, the latter may be the primary factor. At least it explains evolution of CAM better than the former in view of the evident benefits of CAM for both fresh-water plants and drought-stressed plants (Griffiths, 1989).

Conversely, however, the dominating distribution of extant CAM plants clearly is in drought-prone habitats, especially in deserts and semideserts. Moreover, special habitats accommodate many CAM plants, e.g. the coastal sand dunes of the restinga formation in Brazil or tropical salinas. A very typical habitat of CAM plants are epiphytic sites not only in seasonally dry but also in wet tropical forests. In the canopy, epiphytes may be more or less exposed to solar irradiation, and without root contact to the soil they depend on precipitation directly or via throughfall and stemflow of the canopy. They are temporarily or permanently subject to highly restricted supply of water. That CAM provides a strong ecophysiological advantage in the epiphytic habitat is given by the sheer relation of CAM species to total species of known vascular epiphytes which is 57% while CAM plants make up only 7% of all vascular plant species (Kress 1989). Plant families with many epiphytic genera and species are the Orchidaceae, Bromeliaceae and also, to a fair extent, Cactaceae. The CAM ferns known are all epiphytic. The ferns may be a good example for Schimper's hypothesis on the evolution of epiphytism, where this was thought to be driven by the struggle for light of plants emerging from shade of the understorey towards the canopy in dense forests (Schimper, 1888). Then CAM could have evolved as an additional adaptation to cope with the water problem. However, Schimper's hypothesis has also been tested by studies of ecophysiological performance and habitat distribution within the family of Bromeliaceae, particularly based on a census of bromeliads in Trinidad (Smith, 1989). The relative abundance of CAM plants among epiphytic bromeliads is strongly correlated with the degree of exposure and the dryness of the forest sites where these epiphytes occur. Taxonomic and ecophysiological comparisons suggest that, in the bromeliads, epiphytism did not evolve from shade plants but from an ancestral stock of terrestrial $C_3$ species, which was already preadapted to open and exposed habitats imposing stress to water relations. Epiphytism and CAM must have evolved several times, i.e. polyphyletically within the family. Both must have evolved independently but nevertheless with a certain co-ordination among each other.

## THE PLASTICITY OF CAM

### Does obligatory CAM exist?

Most CAM plants have a very considerable degree of plasticity, which is given by the flexible expression of CAM phases. We may consider this starting from a medium situation with respect to water relations where all of the four CAM phases are well expressed:

If, on the one hand, water availability is sufficient, some CAM plants may extend day-time $CO_2$ acquisition considerably and take up $CO_2$ more or less continuously day and night.

If, on the other hand, water availability is increasingly restricted, activity of CAM phases is sequentially reduced as follows (Lüttge, 1987),

— phase IV is first limited and abolished entirely,

— phase II may be obliterated,

— the start of stomatal opening in phase I may be delayed for many hours to later in the night and the amplitude, i.e. the rate of $CO_2$-uptake, is much reduced,

— stomata may remain totally closed day and night.

In the later stages of this sequence of events, one aspect of CAM is becoming increasingly important, namely internal $CO_2$ recycling (Griffiths, 1989). For the night-time loop of CAM, it does not matter where the $CO_2$ is coming from, e.g. from external or internal sources. As stomata progressively close even during the night, less $CO_2$ is taken up from the atmosphere and an increasing proportion of the $CO_2$ fixed by PEPC is derived from respiration in the leaves. This $CO_2$ recycling can be determined by combining measurements of net $CO_2$ gas exchange ($J_{CO_2}$) and day–night malate oscillations ($\Delta$mal) and may be given in absolute terms ($R_{CO_2}$):

$$R_{CO_2} = \Delta\text{mal} - J_{CO_2} \quad (3)$$

or in relative terms ($r_{CO_2}$):

$$r_{CO_2} = \frac{\Delta\text{mal} - J_{CO_2}}{\Delta\text{mal}} \times 100 \quad (4)$$

where calculation of recycling is based on the fact that each $CO_2$ fixed by PEPC gives rise to one malate. When stomata are totally closed, day and night recycling, $r_{CO_2}$, is 100%. This may occur especially among succulent species having special water storage tissues like the stem succulent cacti or leaf succulent agaves and bromeliads. It is extremely important for survival of temporarily severe drought in very exposed habitats. Although, of course, recycling is not associated with carbon gain and, hence, productivity, it allows continuation of basic metabolism and – what appears most important in addition – it provides $CO_2$ from nocturnally formed malic acid during the day feeding the photosynthetic machinery to prevent photodamage (Adams & Osmond, 1988; Griffiths, 1989). Even if stomata are closed continuously day and night, some water is lost via cuticular transpiration, but this can be replaced, to some extent, to the photosynthetically active tissues from the water storage tissues.

In its extremes the plasticity, as characterized above, already may lead to the question about whether there is really something like obligatory CAM in given plant species. However, this still is within the limits of the actual biochemical scheme of CAM (p. 136). More intriguing, though, is an old experiment where the obligatory CAM plant *Kalanchoë tubiflora* was kept at 7 °C and then showed perfect $C_3$-type gas exchange with net $CO_2$ uptake during the entire light period and respiratory $CO_2$ release over the whole dark period (Kluge, 1969). Conversely, *bona fide* obligatory $C_3$-species, e.g. in the genus *Clusia*, may reveal incipient CAM patterns with some nocturnal acid accumulation under certain conditions.

## $C_3$–CAM intermediate species

A fair number of species are known which are genuinely $C_3$–CAM intermediate, i.e. they can switch between $C_3$-photosynthesis and CAM in response to environmental conditions. It appears that, as investigations develop more broadly to various taxa continuously new $C_3$–CAM intermediates are discovered. The most well-known genera and species of $C_3$–CAM intermediates (see various chapters in Winter & Smith 1996a) are:

*Mesembryanthemum crystallinum* (Aizoaceae),
*Peperomia* (Piperaceae),
*Clusia* and *Oedematopus* (Clusiaceae),
*Sedum* (Crassulaceae),
*Guzmania monostachia* (Bromeliaceae),

but even in a genus of mainly obligate CAM plants like *Kalanchoë* (Crassulaceae) there are $C_3$-CAM intermediates, of which:

*Kalanchoë blossfeldiana* cv. Tom Thumb

is studied in most detail.

Among these $C_3$–CAM intermediates, there are very different life forms like annual therophytes and perennial phanerophytes, e.g. the annual *M. crystallinum* and the trees of the genus *Clusia*, respectively. Both have been studied extensively and they show different conduct in $C_3$–CAM transitions.

### *Annuals: Mesembryanthemum crystallinum*

The annual *M. crystallinum* originates from South Africa but is now ubiquitous wherever there is a Mediterranean-type climate with winter rains and more or less severe summer drought. Its seeds germinate after sufficient rainfall, the plants grow up with $C_3$-photosynthesis and then switch to CAM as the dry season sets in and proceeds. In this way, the plants can survive much longer than competing $C_3$-annuals and bring a large number of seeds to maturity when they complete their life-cycle (see Lüttge, 1987). This $C_3$-CAM shift has been intensely investigated both at the physiological and at the molecular level (see also pp. 144–6). The shift from $C_3$-photosynthesis to CAM is triggered by salinity in the root medium. In fact, *M. crystallinum* is also a facultative halophyte (see Lüttge, 1993). However, more generally it seems to be osmotic stress or drought stress which leads to CAM expression. Simultaneously there is also an age-dependent induction of CAM. Plants kept in the laboratory perfectly well watered develop a certain degree of CAM as they age. Stress-induced CAM is reversible only to the extent that it was larger than the degree of CAM, i.e. the amplitude of day–night oscillations of malic acid levels, expressed in well-watered plants of the same age (Ratajczak, Richter & Lüttge 1994). Thus, it appears that the stress response is not a genuine induction of CAM but rather the amplification of a developmental programme. In view of the life-cycle of an annual, which may need a certain flexibility and reversibility of responses during wet season to dry season transitions but always completes its life cycle and dies in the dry season, evolvement of such a strategy appears to be quite appropriate.

### *Trees: Clusia*

For perennials like the rosettes of bromeliads (e.g. *Guzmania monostachia*) and tropical trees (e.g. *Clusia*), which may use their leaves for more than one vegetation period or rainy season – dry season cycle, an irreversible developmental programme would be problematic because much more flexibility is needed. Moreover, in the tropical environment, often stress is not so stereotyped as, for example, in deserts characterized by the one given extreme stress factor drought. Stress itself is variable in time and space in the tropics with a variety of independent or interacting stress factors. Hence, plasticity of plant responses is a highly favourable trait. Clearly, this is expressed in species of *Clusia*, where the highest flexibility of $C_3$–CAM–$C_3$ transitions has been detected so far. They are rapidly reversible within days or even hours. The major factors that determine metabolic shifts in *Clusia* are temperature, water and light (Haag-Kerwer, Franco & Lüttge, 1992):

Day–night temperature differences stimulate the expression of CAM: the cooler the nights are compared with the days, the more strongly CAM is expressed, and this is rapidly reversible.

Limited water supply also induces CAM expression. High VPD, in itself, does not appear to initiate CAM; however, it strongly amplifies CAM expression triggered by other factors, e.g. by day–night temperature rhythms.

High irradiance only stimulates CAM expression when this is associated with stressed water relations. In well-watered plants, high irradiance attenuates CAM, nocturnal $CO_2$ uptake is abolished and day-time $CO_2$ fixation is much increased.

Plasticity in *Clusia* is not restricted to transitions between modes of photosynthesis and patterns of gas exchange, and subtle metabolic modifications may also be involved. Species of *Clusia* can flexibly choose between nocturnal accumulation of malic and citric acid. This led to the question of which may be the ecophysiological benefit of day–night oscillations of citric acid in comparison to malic acid, in as much as citric acid accumulation is not associated with a net gain of carbon (see p. 136):

Organic acid molecules accumulated in the vacuole are osmotically active, and malic acid accumulation has been demonstrated to assist in water uptake by osmotic binding of the water especially towards the end of the night when malic acid levels are high and water is available from dew in tropical habitats (Lüttge, 1987). However, metabolism leading to citric acid accumulation involves less effective osmotic oscillations (Lüttge, 1996).

Malic acid oscillations have been shown to be effective in internal $CO_2$ recycling, which appeared to be particularly important for control of photodamage under more severe drought with increasing nocturnal stomatal closure (see p. 141). This function can be fulfilled equally well, or perhaps even more adequately, by citrate (p. 137, Lüttge, 1996).

It has already been mentioned (p. 136) that *Clusia* species are the strongest nocturnal acid accumulators among all CAM plants known. Under conditions where the highest amounts of accumulated titratable protons are observed, both malic acid and citric acid are contributing to this. A thermodynamic problem of vacuolar acid accumulation is that $\Delta\bar{\mu}_{H^+}$ (eqn. 1) must not exceed the free energy available from ATP hydrolysis by the V-ATPase. Since the V-ATPase has a stoichiometry of $2H^+$/ATP, the transmembrane electron potential is only slightly positive (a few tens of mV) and $pH_{cyt}$ is close to pH 7.5, according to eqn. 1, $pH_{vac}$ may not drop much below pH 3.0 during nocturnal acid accumulation, as $\Delta\bar{\mu}_{H^+}$ must remain below one-half of the free energy of ATP hydrolysis in the cytoplasm, i.e. ~55 kJ $mol^{-1}$. This was observed with various CAM species, which varied in the absolute amounts of titratable protons they could maximally accumulate. This is explained by different vacuolar buffer capacities (Lüttge *et al.*, 1981). Citrate itself is known to be an effective buffer substance. Thus, by citric acid accumulation, *Clusia* will build up vacuolar buffer capacity each night concomitantly with nocturnal acidification, and this may be the reason why it heads all records of acid accumulation. This then, of course, leads to large absolute amounts of acid available for remobilization during the day and, in this way, recycling of carbon skeletons via citrate could be particularly effective for adaptation to solar irradiance stress.

## PRODUCTIVITY OR SURVIVAL?

Generally, CAM plants are found to have a much lower productivity of net biomass than $C_3$ and $C_4$ plants. Some numbers given in the literature are 50–200 g dry weight $m^{-2}$ $day^{-1}$ for $C_3$, 400–500 g $m^{-2}$ $day^{-1}$ for $C_4$ and 1.5–1.8 g $m^{-2}$ $day^{-1}$ for CAM photosynthesis (Black, 1973). Energetically, the CAM loop with PEPC, active transport at the tonoplast and gluconeogenesis, is more costly than pure $C_3$ photosynthesis (Winter & Smith 1996*b*). This may play a certain role, although energy available from irradiance often does not appear to be the limiting factor. Moreover, rates of net $CO_2$ exchange in phase I of CAM are usually rather low, i.e. often much below 10 μmol $m^{-2}$ $s^{-1}$, while $C_3$ plants may reach rates of up to 40 μmol $m^{-2}$ $s^{-1}$ or even more. Thus, evidently under natural conditions in the field, CAM is an adaptation for survival under stress and not a character of highly productive robust competitors.

Conversely, Nobel (1996) has argued forcefully that CAM plants, in principle, can be as productive as $C_3$ plants and may even show superior performance. This was demonstrated for CAM crops like pineapple, *Aloë* and *Opuntia*. It may, however, be largely due to agricultural management allowing strong expression of phase

IV of CAM with $C_3$ photosynthetic metabolism in addition to nocturnal C acquisition.

## MOLECULAR ECOPHYSIOLOGY

As shown at the outset of this chapter, in biochemical and molecular terms, there is really nothing very much particular about CAM. This also may be one of the reasons for the high metabolic flexibility in some CAM plants like *Clusia*. On the other hand, CAM and $C_3$-CAM transitions actually do appear to be highly regulated at the molecular level. This leads to the question of molecular ecophysiology: To what extent and how are macromolecules like enzyme-polypeptides and nucleic acids involved in whole plant performance under stress in the natural habitat? Indeed, we learn most about the key elements of CAM from molecular responses and enzyme modulations in $C_3$-CAM intermediates during $C_3$-CAM transitions.

### Membranes

Spin label probes have been introduced into the membranes of isolated vacuoles of CAM plants and electron spin resonance (ESR) measurements were then performed to determine the degree of molecular order of the tonoplast membrane, or – expressed in simpler terms – membrane rigidity and fluidity, respectively.

It was found that *M. crystallinum* in the CAM state had a significantly higher degree of tonoplast membrane order than in the $C_3$ state. The temperature of CAM tonoplast vesicles needed to be elevated by 3.3 °C to reach the same membrane fluidity as those of $C_3$ tonoplast vesicles (see Lüttge *et al.*, 1995*a*). It was speculated that this may assist nocturnal accumulation of malic acid by reduced permeability of the more rigid CAM tonoplasts to malic acid leaching out of the vacuole.

This then was convincingly demonstrated in detailed comparative studies of plants of *Kalanchoë daigremontiana* grown under a lower and a higher day–night temperature regime, i.e. at 25/17 °C and 35/25 °C, respectively. There was homeoviscous adaptation of the tonoplast membranes, the tonoplast of the high-temperature plants was much more rigid, and an elevation of the temperature of membrane preparations by ~10 °C was needed to obtain the same fluidity as with tonoplasts of the low-temperature plants. This also clearly affected day-time remobilization of malic acid. Both the low temperature and the high temperature plants showed controlled gradual remobilization when kept under the temperature regime they had also experienced during growth. When high temperature plants were transferred to the low temperature regime, they remobilized their malic acid only very sluggishly, and, conversely, when low temperature plants were subjected to the high temperature regime, they lost their malic acid very rapidly and could not control vacuolar efflux (Schomburg & Kluge, 1995; Lüttge *et al.*, 1995a).

It was questioned, which was the molecular basis of these conspicuous changes of physical membrane properties? This was approached by separating membrane lipids and proteins and reforming liposomes, i.e. membrane vesicles from the lipids alone. The results of ESR measurements showed that the native membranes and the liposomes devoid of the membrane proteins for the low-temperature plants had the same state of order, i.e. the proteins did not affect the state of order. The much higher state of order of the native tonoplast membranes in the high temperature plants was completely lost, however, when the proteins were removed, and in this case the liposomes even showed a lower degree of order than those of the low-temperature plants and the native vesicles of these plants. Thus, the increased degree of membrane order was exclusively due to the proteins and not to the lipids (Schomburg & Kluge, 1995). This was also alluded to by the experiments with *M. crystallinum*, where the increased state of order in the CAM state was not associated with a change in bulk lipid composition of the tonoplast membrane (Haschke *et al.*, 1990). This does not imply that viscous adaptation of the tonoplast is due to proteins *per se*. Protein–lipid interactions and formation of membrane domains with special dynamics and phase behaviour may play a considerable role and this needs further attention.

### Key enzymes and transporter proteins

#### *Cytological enzymes*

The metabolic and molecular regulation of key enzymes of CAM has been assessed particularly by studying $C_3$–CAM transitions in $C_3$–CAM intermediates, above all

*Mesembryanthemum crystallinum* (Cushman & Bohnert, 1996). Special attention has been given to the enzymes of the most essential steps in the night-time and day-time loops of CAM, where the major focus is on PEPC as the most central element of CAM.

There are several genes for PEPC and isoforms are expressed in a tissue and mode of photosynthesis specific way. For the CAM rhythm it is essential that PEPC is feedback inhibited by malate and has a much higher affinity (i.e. ~six-fold) to $CO_2$ than RuBPC and that this is regulated on the post-translational level in a diurnal fashion. The requirement of such regulation is particularly clear if we envisage the situation during the light period in phase III of CAM when malic acid is remobilized from the vacuole and decarboxylated in the cytoplasm. With its much higher $CO_2$ affinity in competition with RuBPC, PEPC would effectively refix the $CO_2$ and resynthesize malate. The CAM loop would not work and the flow of carbon would be essentially futile. Thus, PEPC activity must be down-regulated in the light period of CAM. Another problem is malate sensitivity of PEPC. This does not hurt in the light period when PEPC activity needs to be inhibited, but it should not be too severe in the dark period when malate needs to be formed. These problems are solved by phosphorylation–dephosphorylation of PEPC mediated by a pair of enzymes, e.g. PEPC-kinase and a phosphorylase. In the phosphorylated stage ($PEPC_{phos}$) the enzyme has high $CO_2$ affinity and malate sensitivity is reduced, in the dephosphorylated stage ($PEPC_{dephos}$) malate sensitivity is higher and overall activity is lower. $PEPC_{phos}$ and $PEPC_{dephos}$ are the night and day forms of PEPC, respectively. PEPC-kinase itself shows diurnal oscillations of its activity (Carter *et al.*, 1991). During the transition from $C_3$ photosynthesis to CAM in *M. crystallinum*, PEPC is induced, and there is up-regulation at the transcriptional level which precedes the expression of CAM as given by day–night oscillations of malate levels.

### *Transporter proteins at the tonoplast*

The essential function of the tonoplast in the CAM loop is vacuolar acid accumulation by primary active transport of protons establishing $\Delta\bar{\mu}_{H^+}$ across the tonoplast membrane (eqn. 1, pp. p. 136) and secondary active transport of organic acid anions driven by $\Delta\bar{\mu}_{H^+}$. Membrane proteins which may be involved in this are two different $H^+$-pumps, i.e. the V-ATPase (p. 136) and an $H^+$-transporting membrane pyrophosphatase (V-$PP_i$ase) using the free energy of inorganic pyrophosphate ($PP_i$) hydrolysis for proton pumping, a malate transporter which presumably is a carrier, and malate channels (Lüttge *et al.*, 1995 *a, b*).

A key enzyme is the V-ATPase since the V-$PP_i$ase may only be of subordinate importance during vacuolar $H^+$-pumping in phase I of CAM (see below). The V-ATPase is a very complex multi-subunit enzyme (Lüttge & Ratajczak, 1997). There is a membrane integral proteolipid domain (called $V_0$) which is built up of several copies of a 16–20 kDa polypeptide (subunit c). Towards the cytoplasmic side there is a peripheral domain (called $V_1$) with a stalk consisting of subunits C and D, with molecular masses of 37 to 52 kDa, which attaches a head to the membrane integral proteolipid. The head is composed of subunits A and B with molecular masses of 63–72 kDa and 52–60 kDa, respectively. Depending on the materials studied, there may be additional subunits. However, the basic complement, as given here, suffices to characterize the situation in *M. crystallinum* in the $C_3$-state as a basis for the explanation of modulations during CAM induction which underline the rank of the V-ATPase as a key element in the CAM loop. The basic composition of the V-ATPase holoenzyme complex may be presented by the stoichiometry $A_3$ $B_3$ $C_1$ $D_1$ $c_6$, with a total apparent molecular mass of 584 kDa, where A is the catalytic subunit, B has regulatory functions and c subunits form the proton channel across the tonoplast membrane.

During CAM induction in *M. crystallinum*, the absolute amount of V-ATPase protein in the tonoplast increases as well as the density of V-ATPase molecules per unit of tonoplast membrane surface. The latter approximately doubles from $1.2 \times 10^3$ to $2.7 \times 10^3$ V-ATPases $\mu m^{-2}$ as deduced from determinations of activity and protein abundance and demonstrated by immunogold labelling of isolated tonoplast vesicles (Lüttge & Ratajczak, 1997).

Moreover, there are not only quantitative but also qualitative changes of the V-ATPase under salinity stress which leads to CAM expression in *M. crystallinum*. These comprise all parts of the holoenzyme (Lüttge *et al.*, 1995b; Lüttge & Ratajczak, 1997). Transcription of the DNA for subunits A, B and c varies in a tissue specific manner and is also not co-ordinated for the three subunits.

Transcription of subunit c is of particular interest in the $C_3$–CAM transition. In the $C_3$ state it shows a day–night rhythm, which is not accompanied by a rhythm of subunit c protein (Löw *et al.*, 1996). The highest transcript levels are found at the end of the day and in the early night period. They decrease during the night and increase again during the day. Thus, they seem to mirror typical diurnal water relations of a $C_3$ plant where transpiration reduces leaf water potential during the day (subunit c mRNA levels increase), which can recover during the night (subunit c mRNA levels decrease). Under stress, this is often accompanied by a so-called midday depression of photosynthesis due to stomatal closure (see p. 140). As long as there is no photodamage, the midday depression of $C_3$ photosynthesis can be readily distinguished from phase III of CAM by calculation of leaf-interior partial pressure of $CO_2$, $p^i_{CO_2}$. With limited $CO_2$ diffusion into the leaves during the midday depression, photosynthesis will deplete internal $CO_2$ reserves, and $p^i_{CO_2}$ will be low. With malate decarboxylation during phase III, $CO_2$ will be generated internally and $p^i_{CO_2}$ will be high. In this way, Winter & Gademann (1991) could demonstrate that *M. crystallinum* subjected to stress first shows repeated $C_3$ photosynthesis midday depressions before it gradually switches to CAM, and they concluded that some sort of message might build up during the recurring midday depressions. This possibly could be something like the diurnally oscillating mRNA of subunit c. Actually, the amplitude of these oscillations increases as stress proceeds and eventually an increased amount of subunit c protein is also detected as the switch to CAM is completed. Quantitative analyses of intramembrane particles seen on freeze–fracture replicas of isolated tonoplast vesicles also show that the diameter of the $V_o$-domain of the V-ATPase becomes larger during the $C_3$–CAM transition. It must be concluded that the number of copies of subunit c in the $V_o$-domain is not fixed and increases during CAM expression.

Qualitative changes in the stalk region of the V-ATPase of *M. crystallinum* are given by the appearance of the polypeptides, $D_i$ and $E_i$, during CAM induction (Ratajczak *et al.*, 1994). The head region was studied by contrast enhancement of electron microscopic images where both rotational image analyses of individual pictures and alignment of different pictures showed that it is not always a hexamer with the suggested $A_3B_3$ stoichiometry but also appears as a pentamer. The pentamer may be a rather stable (and hence detectable) state during assembly and/or breakdown in the turnover of the functional hexamer (Kramer *et al.*, 1995).

The free energy of $PP_i$ hydrolysis in the cytoplasm is only about 19 kJ $mol^{-1}$, and thus much lower than that of ATP hydrolysis with 55 kJ $mol^{-1}$ (Lüttge *et al.*, 1995a). Therefore, even with a stoichiometry of only $1H^+/PP_i$ the V-$PP_i$ase can not pump protons for very long against the increasing $\Delta\bar{\mu}_{H^+}$ at the tonoplast during phase I of CAM. At the most, it may participate only in the very beginning of phase I. In *K. blossfeldiana* cv. Tom Thumb the abundance of V-$PP_i$ase protein is increased during short day-triggered CAM enhancement. Both in this species and in *Kalanchoë daigremontiana*, the V-$PP_i$ase kinetically stimulates the V-ATPase (Lüttge *et al.*, 1995*a*). In this way it may subordinately participate in CAM. Conversely, in *M. crystallinum* the V-$PP_i$ase disappears as leaves age and switch to CAM. There, the V-$PP_i$ase has nothing to do with CAM, and rates of $H^+$-pumping by the V-ATPase are high even in the absence of kinetic stimulation by the V-$PP_i$ase.

Little is still known about the malate transporters and how they are tied in with malic acid accumulation and remobilization, respectively (Lüttge *et al.*, 1995*a*). Purification studies suggest that there is a carrier molecule in the tonoplast mediating vacuolar malate uptake (Steiger *et al.*, 1997). Gated malate channels may mediate malate efflux from the vacuole which is thermodynamically downhill and metabolically passive. However, malate may also diffuse out of the vacuole in the form of the non-dissociated malic acid, which may readily permeate the lipid phase of the membrane and whose concentration relative to malate anions builds up as the vacuole acidifies during phase I of CAM in vacuolar dissociation equilibria where $pK_1$ and $pK_2$ of malic acid are 3.18 and 4.25, respectively (Lüttge & Smith 1984).

## ENDOGENOUS CAM RHYTHMICITY: KEY CONTROL PARAMETERS AND SWITCH POINTS

CAM also functions as an endogenous, free-running, circadian rhythm (Carter *et al.*, 1996). To study this rhythm experimentally as well as by mathematical modelling and computer simulation is a good alternative to the investigation of $C_3$–CAM intermediates for learning more about key control parameters and switch points of CAM.

It was observed that net $CO_2$ exchange ($J_{CO_2}$) by the CAM plant *Kalanchoë daigremontiana* in continuous light and under otherwise constant external conditions changes from well-ordered circadian rhythmicity to arrythmic behaviour when either of the two external control parameters light intensity or temperature were raised above a certain critical threshold level. Near the threshold, very minimal changes of the control parameters are effective for the shift between rhythmic and arrythmic behaviour, and this was fully reversible when the control parameters were lowered again below their threshold levels. Theoretical analyses of extended time series of $J_{CO_2}$ under constant external conditions in continuous light have shown that the arrhythmic behaviour is not simply stochastic but reveals features of deterministic chaos (Lüttge & Beck, 1992).

A mathematical model was developed, where a very limited number of metabolite pools linked by metabolite flows and feedback loops are used and described by non-linear differential equations (Lüttge & Beck, 1992; Grams, Beck & Lüttge, 1996). It is a skeleton model of CAM with only the pools of internal $CO_2$ ($p^i_{CO_2}$), starch, glucose-6-phosphate, PEP and cytoplasmic and vacuolar malate, the control parameters light and temperature and the feedback loop of inhibition of PEPC by cytoplasmic malate. Light acts on photosynthetic starch formation and based on the strong temperature effects on the tonoplast (p. 144) temperature acts on malate fluxes between cytoplasm and vacuole in the model. A hysteresis switch based on influx and efflux of malate into, and from, the vacuole was implemented in the model. In a variant form of the model, a second hysteresis switch at the level of PEPC with activation and inactivation by phosphorylation (p. 145) was also introduced. However, in principle, the model works with, or without, the extra hysteresis switch at PEPC, which underlines the overriding importance of the tonoplast. This, of course, urges the question of which actually is the hysteresis switch in cytological reality. In this respect, the model is of extraordinary heuristic value. It not only asks this question but also allows to test the answer.

A membrane response is the best candidate. One thought considers the effects of $Ca^{2+}$ on the state of membrane order (Lüttge *et al.*, 1995*a*). The vacuoles of *K. daigremontiana* contain up to 135–160 mM total soluble $Ca^{2+}$. Calculations of free vacuolar $Ca^{2+}$ based on association and dissociation equilibria with measured levels of organic acids (malic, citric, isocitric acids, pH, $Mg^{2+}$ and $Ca^{2+}$) show that there is a day–night rhythm of free $Ca^{2+}$ in the vacuole where the highest levels are attained towards the end of the dark period and levels decline in the light period. Thus, varying amounts of free $Ca^{2+}$ are available to bind to the tonoplast membrane. Binding of $Ca^{2+}$ to the membrane is increasing the state of order. Reduced binding thus may increase fluidity and with it permeability, i.e. efflux of malic acid from the vacuole, and this may possibly satisfy the question for the molecular hysteresis switch explaining the circadian oscillation of CAM.

## REFERENCES

Adams, W. W. III & Osmond, C. B. (1988). Internal $CO_2$ supply during photosynthesis of sun and shade grown CAM plants in relation to photoinhibition. *Plant Physiology*, **86**, 117–23.

Black, C. C. (1973). Photosynthetic carbon fixation in relation to net $CO_2$ uptake. *Annual Review of Plant Physiology*, **24**, 253–86.

Carter, P. J., Nimmo, H. G., Fewson, C. A. & Wilkins, M. B. (1991). Circadian rhythms in the activity of a plant protein kinase. *EMBO Journal*, **10**, 2063–8.

Carter, P. J., Fewson, C. A., Nimmo, G. A., Nimmo, H. G. & Wilkins, M. B. (1996). Roles of circadian rhythms, light and temperature in the regulation of phospho*enol*pyruvate carboxylase in crassulacean acid metabolism. In *Crassulacean Acid Metabolism. Biochemistry, Ecophysiology and Evolution*. Ecological studies, vol. 114, ed. K. Winter & J. A. C. Smith, pp. 46–52. Berlin–Heidelberg–New York: Springer-Verlag.

Cushman, J. C. & Bohnert, H. J. (1996). Transcriptional activation of CAM genes during development and environmental stress. In *Crassulacean Acid Metabolism. Biochemistry, Ecophysiology and Evolution*. Ecological Studies, vol. 114, ed. K. Winter & J. A. C. Smith, pp. 135–58. Berlin–Heidelberg–New York: Springer-Verlag.

Grams, T. E. E., Beck, F. & Lüttge, U. (1996). Generation of rhythmic and arrhythmic behaviour of Crassulacean acid metabolism in *Kalanchoë daigremontiana* under continuous light by varying the irradiance or temperature: measurements *in vivo* and model simulations. *Planta*, **198**, 110–17.

Griffiths, H. (1989). Carbon dioxide concentrating

mechanisms and the evolution of CAM in vascular epiphytes. In *Vascular Plants as Epiphytes, Evolution and Ecophysiology*. Ecological Studies, vol. 76, ed. U. Lüttge, pp. 42–86. Berlin–Heidelberg–New York: Springer-Verlag.

Haag-Kerwer, A., Franco, A. C. & Lüttge, U. (1992). The effect of temperature and light on gas exchange and acid accumulation in the $C_3$-CAM plant *Clusia minor* L. *Journal of Experimental Botany*, **43**, 345–52.

Haschke, H. -P., Kaiser, G., Martinoia, E., Hammer, U., Teucher, T., Dorne, A. J. & Heinz, E. (1990). Lipid profiles of leaf tonoplasts from plants with different $CO_2$-fixation mechanisms. *Botanica Acta*, **103**, 32–8.

Keeley, J. E. (1996). Aquatic CAM photosynthesis. In *Crassulacean Acid Metabolism. Biochemistry, Ecophysiology and Evolution*. Ecological Studies, vol. 114, ed. K. Winter & J. A. C. Smith, pp. 281–95. Berlin–Heidelberg–New York: Springer-Verlag.

Kluge, M. (1969). Veränderliche Markierungsmuster bei $^{14}CO_2$-Fütterung von *Bryophyllum tubiflorum* zu verschiedenen Zeitpunkten der Hell/Dunkelperiode. I. Die $^{14}CO_2$-Fixierung unter Belichtung. *Planta*, **88**, 113–29.

Kramer, D., Mangold, B., Hille, A., Emig, I., Hess, A., Ratajczak, R. & Lüttge, U. (1995). The head structure of a higher plant V-type $H^+$-ATPase is not always a hexamer but also a pentamer. *Journal of Experimental Botany*, **46**, 1633–6.

Kress, W. (1989). The systematic distribution of vascular epiphytes. In *Vascular Plants as Epiphytes, Evolution and Ecophysiology*. Ecological Studies, vol. 76, ed. U. Lüttge, pp. 234–261. Berlin–Heidelberg–New York: Springer-Verlag.

Löw, R., Rockel, B., Kirsch, M., Ratajczak, R., Hörtensteiner, S., Martinoia E., Lüttge U. & Rausch, T. (1996). Early salt stress effects on the differential expression of vacuolar $H^+$-ATPase genes in roots and leaves of *Mesembryanthemum crystallinum*. *Plant Physiology* **110**, 259–65.

Lüttge, U. (1987). Carbon dioxide and water demand: crassulacean acid metabolism (CAM), a versatile ecological adaptation exemplifying the need for integration in ecophysiological work. *New Phytologist*, **106**, 593–629.

Lüttge, U. (1993). The role of crassulacean acid metabolism (CAM) in the adaptation of plants to salinity. *New Phytologist*, **125**, 59–71.

Lüttge, U. (1996). *Clusia*: Plasticity and diversity in a genus of $C_3$/CAM intermediate tropical trees. In *Crassulacean Acid Metabolism. Biochemistry, Ecophysiology and Evolution*. Ecological Studies, vol, 114, ed. K. Winter & J. A. C. Smith, pp. 296–311. Berlin–Heidelberg–New York: Springer-Verlag.

Lüttge, U. & Beck, F. (1992). Endogenous rhythms and chaos in crassulacean acid metabolism. *Planta*, **188**, 28–31.

Lüttge, U. & Ratajczak, R. (1997). The physiology, biochemistry and molecular biology of the plant vacuolar ATPase. In *The Plant Vacuole*, ed. R. Leigh & D. Sanders. New York: Academic Press, in press.

Lüttge, U. & Smith, J. A. C. (1984). Mechanism of passive malic-acid efflux from vacuoles of the CAM plant *Kalanchoë daigremontiana*. *Journal of Membrane Biology*, **81**, 149–58.

Lüttge, U., Smith, J. A. C., Marigo, G. & Osmond, C. B. (1981). Energetics of malate accumulation in the vacuoles of *Kalanchoë tubiflora* cells. *FEBS Letters*, **126**, 81–4.

Lüttge, U., Fischer-Schliebs, E., Ratajczak, R., Kramer, D., Berndt, E. & Kluge, M. (1995a). Functioning of the tonoplast in vacuolar C-storage and remobilization in crassulacean acid metabolism. *Journal of Experimental Botany*, **46**, 1377–88.

Lüttge, U., Ratajczak, R., Rausch, T. & Rockel, B. (1995b). Stress responses of tonoplast proteins: an example for molecular ecophysiology and the search for eco-enzymes. *Acta Botanica Neerlandica*, **44**, 343–62.

Nobel, P. S. (1996). High productivity of certain agronomic CAM species. In *Crassulacean Acid Metabolism. Biochemistry, Ecophysiology and Evolution*. Ecological Studies, vol. 114, ed. K. Winter & J. A. C. Smith, pp. 255–65. Berlin–Heidelberg–New York: Springer-Verlag.

Osmond, C. B. (1978). Crassulacean acid metabolism: a curiosity in context *Annual Review of Plant Physiology*, **29**, 379–414.

Ratajczak, R., Richter, J. & Lüttge, U. (1994). Adaptation of the tonoplast V-type $H^+$-ATPase of *Mesembryanthemum crystallinum* to salt stress, $C_3$–CAM transition and plant age. *Plant Cell and Environment*, **17**, 1101–12.

Schimper, A. F. W. (1888). *Botanische Mitteilungen aus den Tropen. II. Epiphytische Vegetation Amerikas*. Jena: G. Fischer.

Schomburg, M. & Kluge, M. (1995). Phenotypic adaptation to elevated temperatures of tonoplast fluidity in the CAM plant *Kalanchoë daigremontiana* is caused by membrane proteins. *Botanica Acta*, **107**, 328–32.

Smith, J. A. C. (1989). Epiphytic bromeliads. In *Vascular Plants as Epiphytes, Evolution and Ecophysiology*. Ecological Studies, vol. 76, ed. U. Lüttge, pp. 109–38. Berlin–Heidelberg–New York: Springer-Verlag.

Smith, J. A. C. & Winter, K. (1996). Taxonomic distribution of crassulacean acid metabolism. In *Crassulacean Acid*

*Metabolism. Biochemistry, Ecophysiology and Evolution.* Ecological Studies, vol. 114, ed. K. Winter & J. A. C. Smith, pp. 427–36. Berlin–Heidelberg–New York: Springer-Verlag.

Steiger, S., Ratajczak, R., Martinoia, E. & Lüttge U. (1997). The vacuolar malate transport of *Kalanchoë daigremontiana:* A 32 kDa polypeptide? *Journal of Plant Physiology*, in press.

Winter, K. & Gademann, R. (1991). Daily changes in $CO_2$ and water vapour exchange, chlorophyll fluorescence, and leaf water relations in the halophyte *Mesembryanthemum crystallinum* during the induction of Crassulacean acid metabolism in response to high NaCl-salinity. *Plant Physiology*, **95**, 768–76.

Winter, K. & Smith, J. A. C. (1996*a*). *Crassulacean Acid Metabolism. Biochemistry, Ecophysiology and Evolution.* Ecological Studies, vol. 114. Berlin–Heidelberg–New York: Springer-Verlag.

Winter, K. & Smith, J. A. C. (1996*b*). Crassulacean acid metabolism: current status and perspectives. In *Crassulacean Acid Metabolism. Biochemistry, Ecophysiology and Evolution.* Ecological Studies, vol. 114, ed. K. Winter & J. A. C. Smith, pp. 389–426. Berlin–Heidelberg–New York: Springer-Verlag.

# 11 $C_3$–$C_4$ intermediate photosynthesis

S. RAWSTHORNE AND H. BAUWE

## INTRODUCTION

As described in detail in Chapter 13, plants which use the $C_3$ photosynthetic mechanism lose $CO_2$ from their leaves in the light – the process of photorespiration – as a consequence of the metabolism of the products of the oxygenase reaction of ribulose-1,5-bisphosphate (RuBP) carboxylase/oxygenase (rubisco). The oxygenase reaction occurs because $O_2$ competes with $CO_2$ at the active site of rubisco. In $C_3$ plants, not all of the $CO_2$ produced during photorespiratory metabolism escapes from the leaf and about 50% is recaptured by the chloroplasts before it does so. Nevertheless, photorespiration represents a significant loss of carbon to the plant.

In plants which use the $C_4$ photosynthetic pathway, $CO_2$ losses due to photorespiration have been prevented by anatomical and biochemical adaptations which result in specialized functions for cells within the leaves (see Chapter 9). Only the cells which surround the vasculature (bundle sheath cells) contain rubisco, and $CO_2$ is provided to the enzyme by a biochemical cycle involving carboxylation and decarboxylation events in the mesophyll and bundle sheath cells, respectively. These modifications lead to a considerable elevation of the $CO_2$ concentration at the active site of RuBP carboxylase/oxygenase, thus strongly reducing the competition by $O_2$. These plants do not photorespire and this allows a convenient method of screening species for the $C_4$ photosynthetic mechanism by measuring their $CO_2$ compensation points ($\Gamma$). This is the $CO_2$ concentration attained after a leaf is sealed into a closed chamber and illuminated, and it represents the balance point between photosynthetic, and photorespiratory and respiratory processes. The $\Gamma$s of $C_3$ species when measured at atmospheric oxygen concentrations are between 40 and 55 $\mu l\ l^{-1}$, whereas those of $C_4$ plants are between zero and 5 $\mu l\ l^{-1}$. In the course of screening programmes (e.g. Krenzer, Moss & Crookston, 1975) some species were identified which had $\Gamma$ values intermediate between those of $C_3$ or $C_4$ species. These plants have subsequently been defined as having $C_3$–$C_4$ intermediate photosynthesis. The list of such $C_3$–$C_4$ intermediate species has grown since the initial discoveries and now includes 25 species in nine genera representing six families (Table 11.1). All of these genera include $C_3$ species and most also include $C_4$ species. The ability to make comparisons of different photo-

Table 11.1. *Phylogenetic distribution of species reported to be $C_3$–$C_4$ intermediates and the numbers of these species in the different genera.*

| Family | Genus | Number of $C_3$–$C_4$ species |
|---|---|---|
| **Monocotyledoneae** | | |
| Cyperaceae | *Eleocharis** | 1 |
| Poaceae | *Neurachne** | 1 |
| | *Panicum** | 3 |
| **Dicotyledoneae** | | |
| Aizoaceae | *Mollugo** | 2 |
| Amaranthaceae | *Alternanthera** | 2 |
| Asteraceae | *Flaveria**[a] | 9 |
| | *Parthenium* | 1 |
| Brassicaceae | *Moricandia* | 5 |
| | *Diplotaxis*[b] | 1 |

[a]The genus *Flaveria* also contains four species described as $C_4$-like.
[b]*Diplotaxis tenuifolia*: P. Apel (personal communication).
The presence of species in these genera which have been classified as $C_4$ is also indicated (*). All genera contain $C_3$ species.

synthetic mechanisms between otherwise similar plants, including the use of hybridization experiments between species which have different photosynthetic mechanisms, has been valuable in elucidating the mechanism of $C_3$–$C_4$ intermediate photosynthesis. Detailed reviews of $C_3$–$C_4$ intermediate photosynthesis have been published previously (e.g. Edwards & Ku, 1987; Rawsthorne *et al.*, 1992) and these provide a wealth of citations to the existing studies. In this chapter we will cover the well-established features of this field, and use new data to bring the subject area up to date.

## LEAF GAS EXCHANGE IN $C_3$–$C_4$ INTERMEDIATE SPECIES

Photosynthetic rates of $C_3$ and $C_3$–$C_4$ intermediate species are comparable in a range of light and atmospheric gas compositions but the responses of gas exchange parameters which provide a measure of photorespiratory activity differ widely between these two photosynthetic groups. There is a considerable range of Γs reported for $C_3$–$C_4$ intermediate species in the literature and even variation within a single species. Some of this may well be due to differences in the measurement technique used, but leaf age (Apel, Tichá & Peisker, 1978) and growing conditions can also contribute to this variation. Notwithstanding such differences, the values at high light intensity and atmospheric $O_2$ concentrations are generally between 8 and 30 μl $l^{-1}$. This clearly indicates a lower rate of apparent (i.e. measurable external to an intact leaf) photorespiration in $C_3$–$C_4$ intermediate species than in $C_3$ species.

In contrast to $C_3$ plants where Γ is essentially unaffected by light intensity, Γ is strongly light dependent in $C_3$–$C_4$ intermediate species (Brown & Morgan, 1980) (Fig. 11.1a). At photon flux densities which approach the light compensation point for photosynthesis (80–150 μmol quanta $m^{-2}$ $s^{-1}$), Γ of a $C_3$–$C_4$ intermediate species can be almost as high as that of a $C_3$ species but it declines steeply as the light intensity increases. This response has been reported for all $C_3$–$C_4$ intermediate species in which it has been studied. The light dependence of Γ shows that the mechanism which limits photorespiration in $C_3$–$C_4$ intermediate species is dependent on the rate of photosynthesis.

The Γ of $C_3$–$C_4$ intermediate species shows a biphasic response to $O_2$ concentration, rather than the linear response to $O_2$ concentration seen in $C_3$ species (Fig. 11.1b). Only limited increases occur as the $O_2$ concentration is raised to 10–15% and then a $C_3$-like, linear and steeper response is seen at higher concentrations (Apel *et al.*, 1978; Keck & Ogren, 1975). This response of Γ to $O_2$ can be explained on the basis that the mechanism(s) involved in reducing the apparent rate of photorespiration become saturated at a certain oxygen concentration and so Γ increases (von Caemmerer, 1989).

As in $C_3$ plants, net photosynthesis in $C_3$–$C_4$ species is inhibited by $O_2$. This inhibition is due in part to direct competition between $O_2$ and $CO_2$, and in part to increasing $CO_2$ release through photorespiration as oxygenation increases. The rate at 21% $O_2$ is typically 65–70% of that at 2% $O_2$ for $C_3$ species and 75–80% of that at 2% $O_2$ for a range of $C_3$–$C_4$ intermediate species (Edwards & Ku, 1987). The high concentration of $CO_2$ in the bundle sheath cells of $C_4$ species renders them almost completely insensitive to oxygen inhibition. The fact that $C_3$–$C_4$ intermediate species display oxygen inhibition approaching that of $C_3$ species reveals that the reduction in photorespiration in these species is mediated not primarily through a decrease in the oxygenase reaction of rubisco, but through a reduction in the extent to which photorespiratory $CO_2$ is released from the leaf.

Direct support for this is provided by measurements of photorespiration in leaves of $C_3$ and $C_3$–$C_4$ intermediate *Moricandia* and *Flaveria* species. Whereas about 50% of the photorespiratory $CO_2$ of a $C_3$ leaf is recaptured before it escapes from the leaf, it is estimated that between 70 and 90% is recaptured in a $C_3$–$C_4$ leaf (Bauwe *et al.*, 1987; Hunt, Smith & Woolhouse, 1987). Most of the described gas exchange data have been obtained under laboratory conditions and few data are available from field measurements. $C_3$–$C_4$ intermediate species are usually native to warm environments like Mexico and Florida (*Flaveria*) or Mediterranean countries (*Moricandia*). It has been demonstrated that *F. floridana* has distinct advantages over $C_3$ plants in its natural habitat (Monson & Jaeger, 1991). When measured in situ and at leaf temperatures of between 35 and 40 °C the Γ of *F. floridana* was estimated to be less than 20 μl $l^{-1}$ while the Γ of a $C_3$ species with similar growth form and habit was 180–200 μl $l^{-1}$ (i.e. up to four times that for a $C_3$ plant when measured at a temperature of 25 °C).

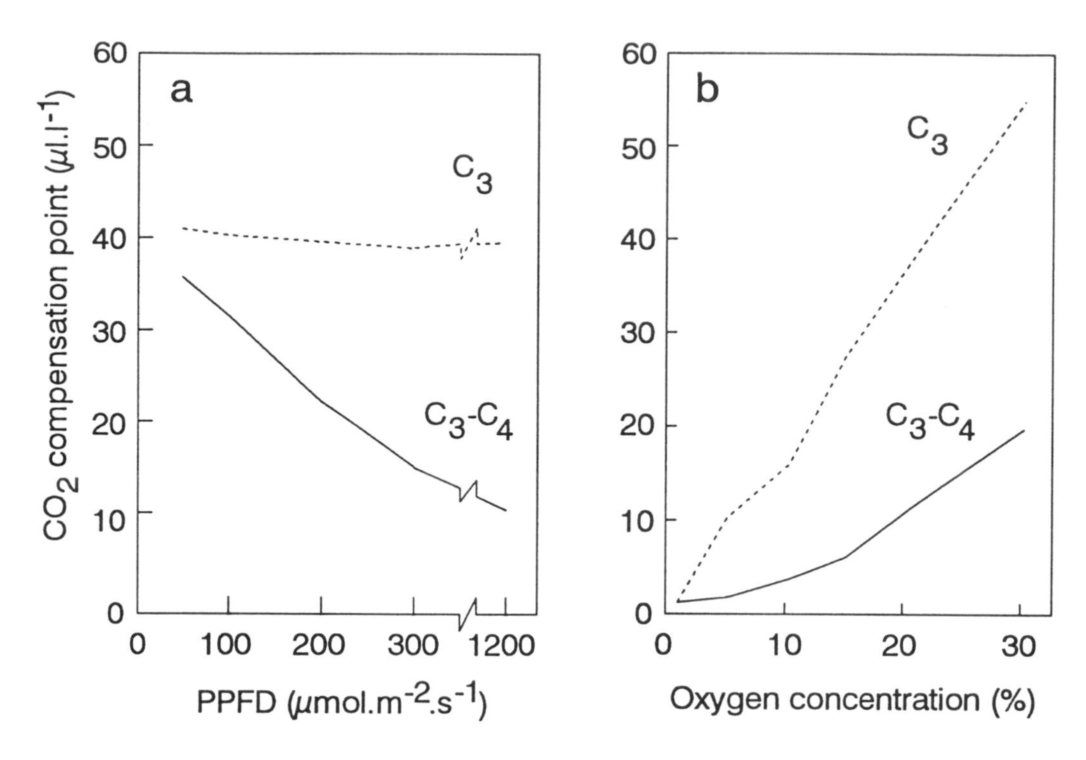


**Figure 11.1.** Typical responses of the $CO_2$ compensation point of $C_3$ (dashed lines) and $C_3$-$C_4$ intermediate (solid lines) species to (a) photosynthetic photon flux density (PPFD) and (b) $O_2$ concentration. (Data are drawn from Hunt, S. (1985) PhD thesis, University of East Anglia, UK and are for *Moricandia arvensis* ($C_3$–$C_4$) and *M. moricandioides* ($C_3$).)

Moreover, under these field conditions photosynthetic rates were up to four times greater in the $C_3$–$C_4$ intermediate species than for the $C_3$ species, giving rise to distinct improvements in water-use efficiency. These studies indicate that high recycling rates for photorespiratory $CO_2$, which lead to reduced photorespiration and a low $\Gamma$, confer advantages on photosynthetic performance, especially at high temperatures (Monson & Jaeger, 1991). In order to understand how this improved recapture of $CO_2$ occurs, the anatomical and biochemical characteristics of $C_3$–$C_4$ intermediate plants need to be understood.

## LEAF ANATOMY OF $C_3$–$C_4$ INTERMEDIATE SPECIES

Leaves of $C_3$–$C_4$ intermediate species have a distinctive anatomy (Brown & Hattersley, 1989) (Fig. 11.2). The vascular bundles are surrounded by chlorenchymatous bundle sheath cells reminiscent of the Kranz anatomy of leaves of $C_4$ plants. However, the mesophyll cells are not in a concentric ring around the bundle sheath cells as in a $C_4$ leaf, but are arranged as in leaves of $C_3$ species (Fig. 11.2) with greater interveinal distances than in the leaves of $C_4$ plants (Brown & Hattersley, 1989).

In all intermediate species, the bundle sheath cells contain large numbers of organelles. Numerous mitochondria, the peroxisomes and many of the chloroplasts are located centripetally in the bundle sheath cells (Edwards & Ku, 1987). The mitochondria are found along the cell wall adjacent to the vascular tissue and are overlain by the chloroplasts (Fig. 11.2). Quantitative studies have shown that the mitochondria and peroxisomes are four times more abundant per unit cell area than in adjacent mesophyll cells and that these mitochondria have twice the profile area of those in the mesophyll (Brown & Hattersley, 1989; Hylton *et al.*, 1988).

This leaf anatomy suggests a means by which the low

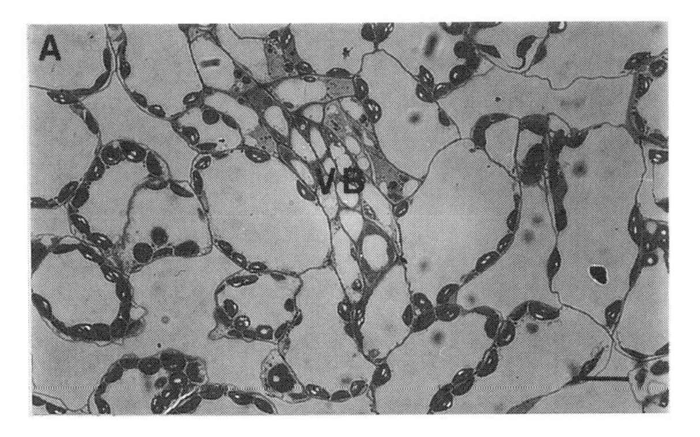


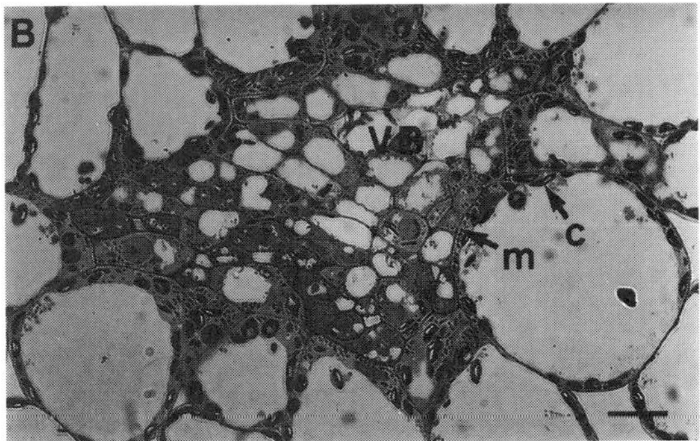


**Figure 11.2.** Transverse sections showing leaf anatomy of (A) *Brassica napus*, a $C_3$ species and (B) *Moricandia arvensis*, a $C_3$–$C_4$ intermediate species. Note the prominent appearance of the vascular bundles (VB) in the leaf of the $C_3$–$C_4$ species (numerous mitochondria (m) overlain by chloroplasts (c) can be seen in the bundle sheath cells). Scale bars are 30μm. (Photomicrographs courtesy of C. M. O'Neill.)

rates of photorespiratory $CO_2$ release of $C_3$–$C_4$ intermediate species could be achieved. Carbon dioxide is released during photorespiratory metabolism as a consequence of the activity of the mitochondrial enzyme glycine decarboxylase (GDC). In the bundle sheath cells of $C_3$–$C_4$ intermediate species, the $CO_2$ released by the mitochondria must pass out through the overlying chloroplasts in order to exit the leaf. This close association of the mitochondria and the chloroplasts should improve the extent of the $CO_2$ recapture relative to that in a typical $C_3$ species where the organelles are distributed around the cell periphery of all of the photosynthetic cells. However, given that only a small proportion of the photosynthetic cells of the $C_3$–$C_4$ leaf are bundle sheath cells, it seems extremely unlikely that enhanced recapture in these cells alone could account for the 50% improvement in recapture estimated from gas exchange studies.

## PHOTORESPIRATORY METABOLISM IN $C_3$–$C_4$ INTERMEDIATE SPECIES

There is clear evidence that $C_3$–$C_4$ intermediates in the in the genera *Alternanthera, Moricandia, Panicum* and *Parthenium* do not have a $C_4$ cycle which could account for their low rates of photorespiration. This comes from studies of the initial products of assimilation of $^{14}CO_2$ by the leaves, and of the activities and cellular locations of $C_4$ cycle enzymes (Edwards & Ku, 1987). Label from $^{14}CO_2$ is not transferred from $C_4$ compounds to Calvin cycle intermediates during photosynthesis. Activities of PEP carboxylase and the $C_4$ cycle decarboxylases are far lower than in $C_4$ leaves, and RuBP carboxylase and PEP carboxylase are both present in mesophyll and bundle sheath cells. Early work on *Mollugo verticillata* suggested that limited $C_4$ metabolism might operate in this species. However, more recent assessment of the data (Edwards & Ku, 1987) suggests that this is unlikely to account for the low $\Gamma$ of this species.

Where it has been examined, the maximum catalytic activities of the enzymes of the photorespiratory pathway are broadly comparable in leaves of $C_3$ and related $C_3$–$C_4$ intermediate species. Despite these similarities, there are clear differences in photorespiratory metabolism in the leaves of $C_3$–$C_4$ intermediate species of *Alternanthera, Flaveria, Moricandia, Mollugo, Panicum* and *Parthenium* compared to that in $C_3$ species. In leaves of $C_3$ species, all of the enzymes in the photorespiratory pathway are present in all the photosynthetic cells. However, the enzyme which releases $CO_2$ during photorespiratory metabolism, glycine decarboxylase (GDC) is not uniformly distributed throughout the photosynthetic cells in the leaves of all of the $C_3$–$C_4$ intermediate species in which it has been examined. On the basis of immunogold localization studies and measurements of enzyme activity in protoplast fractions enriched in either bundle sheath or mesophyll cells (Rawsthorne *et al.*, 1988*a*,*b*; Devi, Rajagopalen & Raghavendra, 1995), GDC is confined to the bundle sheath cells of these $C_3$–$C_4$ species. Serine hydroxymethyltransferase (SHMT) activity is also enriched in bundle sheath relative to mesophyll cells (Rawsthorne *et al.*, 1988*b*). These two enzymes together catalyse the formation of serine, ammonia and $CO_2$ from two molecules of glycine in the mitochondria (see Chapter 13).

The early work which studied the distribution of GDC in the leaf using the immunogold localization technique used antibodies raised against the P subunit of the complex. It is this subunit which catalyses the release of $CO_2$ from glycine. Subsequent studies have shown that in the leaves of $C_3$–$C_4$ intermediate species of *Panicum* and *Flaveria* the T, H, and L proteins of the GDC complex are also confined to the bundle sheath cells (Morgan, Turner & Rawsthorne, 1993). In the leaves of *Moricandia arvensis*, it is surprising that these three subunit proteins are still present in the mitochondria of the mesophyll cells (Morgan, Turner & Rawsthorne, 1993). The L protein, dihydrolipoamide dehydrogenase, is also a subunit (known as E3) of the pyruvate dehydrogenase complex (see Chapter 13) and its presence would be expected in mitochondria of all cells. However, studies have shown that the amount of the dihydrolipoamide dehydrogenase in the mitochondria which is not associated with GDC is small and the abundance of L protein in the mesophyll cell mitochondria of *M. arvensis* is not explained by this association with other multienzyme complexes. Based upon present knowledge the T and H proteins only catalyse reactions that form parts of the overall activity of GDC. In the absence of P protein glycine decarboxylation will not occur and the T, H, and L proteins appear to be redundant in the mesophyll of the latter species.

Most of the evidence on photorespiratory metabolism has been accrued from work on the $C_3$–$C_4$ intermediate species *M. arvensis*. Following on from earlier models (Monson, Edwards & Ku, 1984) and using the GDC localization data a model which described photorespiratory metabolism in the leaves of this species was proposed (Rawsthorne *et al.*, 1988*a*) (Fig. 11.3). Since mesophyll cells cannot decarboxylate glycine, the site of $CO_2$ release during photorespiration is confined to the mitochondria on the inner wall of the bundle sheath cells. These mitochondria are in close association with, and overlain by chloroplasts through which the $CO_2$ must pass to exit the leaf. This relationship between the biochemistry and anatomy of the leaf of $C_3$–$C_4$ intermediate species must enhance considerably the potential for recapture of the photorespired $CO_2$ relative to that in a $C_3$ leaf. The carbon lost to the Calvin cycle in the mesophyll as glycine must be returned to the mesophyll chloroplasts or these plants would resemble photorespiratory mutants with a lesion beyond the transamination of glyoxylate and would be unable to survive in atmospheric $O_2$ conditions. There is no direct evidence as to the nature of the compound which returns from the bundle sheath to the mesophyll: in the scheme in Figure 11.3 it is indicated that it could be serine or a product of serine metabolism in the bundle sheath cell.

Based upon enzyme assays and immunolocalization studies, the above model of photorespiratory metabolism is likely to hold true for all $C_3$–$C_4$ intermediate species and it can explain completely the high degree of light-dependent recapture and low apparent rate of photorespiration in these plants.

## PHOTORESPIRATORY NITROGEN METABOLISM

In the model of photorespiratory metabolism given in Fig. 11.3 it is proposed that glycine decarboxylation is confined to the bundle sheath cells in the leaf. Implicit in this proposal is that ammonia release from glycine during photorespiration will also be confined to the bundle sheath cells. Measurements made to date suggest that the activity of enzymes associated with ammonia recycling during photorespiration (e.g. glutamine synthetase, and aminotransferases: see Chapters 13 and 15) are not differentially distributed between mesophyll and bundle sheath cells. The measurable activities of these enzymes are, however, in excess of the ammonia fluxes which can be calculated based upon the rate of $CO_2$ release during photorespiration. The nitrogen released in the bundle sheath cells must be returned to the mesophyll cells to enable further synthesis of photorespiratory glycine through transamination of glyoxylate. At least 50% of this return could occur as serine produced by GDC/SHMT in the bundle sheath cells.

## IMPLICATIONS FOR METABOLITE TRANSPORT

The movement of glycine into the bundle sheath cells from the mesophyll may occur by diffusion in an analogous

**Figure 11.3.** A proposed pathway for photorespiration in the leaves of $C_3$–$C_4$ intermediate species (after Rawsthorne *et al.*, 1988*a*). Movement of metabolites is indicated by solid arrows while that of $CO_2$ and $O_2$ is shown by dashed arrows. The distribution of glycine decarboxylase (GDC) subunits between bundle sheath and mesophyll cells refers to *Moricandia arvensis*. In *Flaveria* and *Panicum* species all of the GDC subunits are reduced in the mesophyll cells.

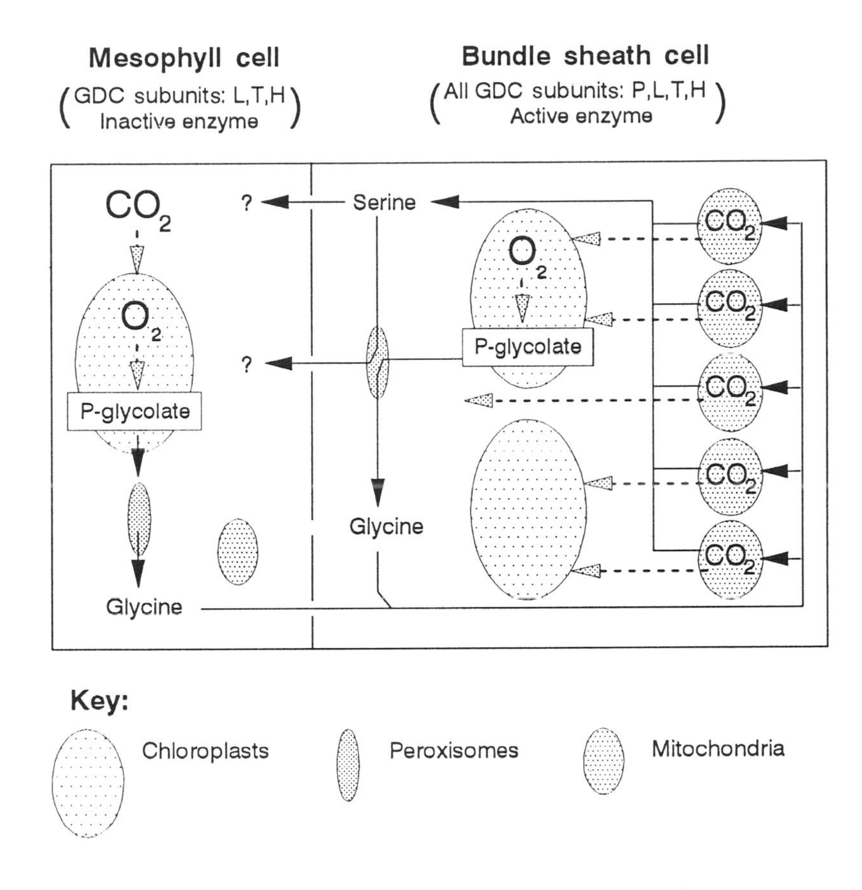


manner to the movement of metabolites in $C_4$ species (see Chapter 9). The glycine content of the leaves of *M. arvensis* ($C_3$–$C_4$) is twice that of a $C_3$ *Moricandia* and is, on a leaf area basis, up to four-fold greater than that of metabolites in $C_4$ leaves which are involved in cell to cell diffusion in the $C_4$ cycle (Leegood & von Caemmerer, 1988; Rawsthorne & Hylton, 1991). The change in glycine content of $C_3$–$C_4$ leaves during the immediate post-illumination period (Fig. 11.4), and of the labelling of glycine during pulse-chase experiments with $^{14}CO_2$, indicate that the glycine pool is a result of photorespiration and is much larger in $C_3$–$C_4$ than in $C_3$ leaves (Holaday & Chollet, 1983; Monson *et al.*, 1986; Rawsthorne & Hylton, 1991).

Whilst these data do not provide information on the size of metabolite pools in specific leaf cells, they are consistent with the ability to maintain a concentration gradient of glycine between the mesophyll and bundle sheath cells. Maintenance of a downhill concentration gradient of glycine between the mesophyll and bundle sheath cells of a $C_3$–$C_4$ species is certainly feasible given the high capacity of the bundle sheath to decarboxylate glycine. When compared on a unit chlorophyll basis, bundle sheath-enriched protoplast fractions of *M. arvensis* ($C_3$–$C_4$) decarboxylate glycine 2.5 times faster than protoplasts from mesophyll cells of pea ($C_3$) leaves.

As described for glycine above, the pool size of serine in leaves of the $C_3$–$C_4$ species *M. arvensis* is twice that in the leaves of a related $C_3$ species. This would be consistent with the differential distribution of the serine pool in the leaf of *M. arvensis* and movement of serine from bundle sheath to mesophyll cells down a diffusion gradient. Movement of serine in this way would enable the return of carbon and nitrogen, respectively, back to the Calvin and photorespiratory cycles in the mesophyll. The need for return of carbon committed to the photorespiratory pathway back to the Calvin cycle in the chloroplast is clearly demonstrated by the marked inhibition of photosynthesis in photorespiratory mutants of the $C_3$ species *Arabidopsis* when they are grown under conditions which allow photorespiration to occur. Such

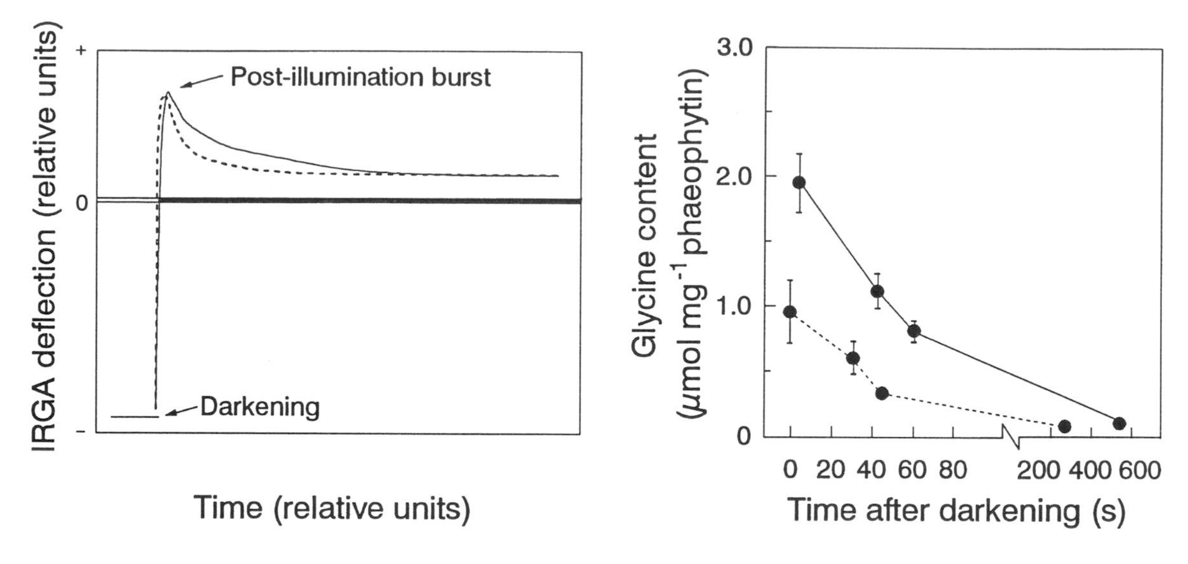


**Figure 11.4.** Changes in gas exchange and leaf glycine content during the post-illumination burst in leaves of *Moricandia arvensis* ($C_3$–$C_4$: solid lines) and *M. moricandioides* ($C_3$: dashed lines).

mutations can be overcome and wild-type rates of photosynthesis restored by feeding metabolites which are distal to the lesion in the photorespiratory pathway to leaf segments (Somerville & Somerville, 1983).

Further support for the proposed movements of metabolites between bundle sheath and mesophyll cells comes from observations of plasmodesmatal frequency between the two cell types in $C_3$, $C_3$–$C_4$ and $C_4$ species. $C_3$–$C_4$ intermediate species of *Panicum* have plasmodesmata densities at the bundle sheath/mesophyll interface which approach those of $C_4$ species and are much greater than those of the $C_3$ species studied (Brown *et al.*, 1983). In $C_4$ species the cellular anatomy is arranged so that the metabolism of the $C_4$ cycle occurs in adjacent bundle sheath and mesophyll cells (see Chapter 9). However, in $C_3$–$C_4$ species mesophyll cells may be spaced such that there are several cell distances between outlying mesophyll cells and a bundle sheath cell (e.g. see Brown & Hattersley, 1989). Diffusion distances will therefore be much longer in the intermediate than in $C_4$ species.

## $C_3$–$C_4$ INTERMEDIATE PHOTOSYNTHESIS IN THE GENUS *FLAVERIA*

All $C_3$–$C_4$ species in the genus *Flaveria* have the characteristic anatomy described above. In addition, in those intermediates in which it has been studied GDC is confined to the bundle sheath cells (Hylton *et al.*, 1988; Moore *et al.*, 1988). This suggests that the $C_3$–$C_4$ intermediate species of *Flaveria* could use the same mechanism to reduce Γ as the species in the other genera described above. However, the genus *Flaveria* contains a more-or-less continuous range of plants between those which are $C_3$ and those which are $C_4$. This range includes $C_3$–$C_4$ intermediate species which are thought to rely exclusively on the recapture of photorespired $CO_2$ to reduce Γ, since they do not show substantially greater initial incorporation of $^{14}CO_2$ into $C_4$ acids than related $C_3$ species (Ku *et al.*, 1991; Monson *et al.*, 1986). Other $C_3$–$C_4$ intermediates in the genus partition up to 50% of the $^{14}C$ label incorporated during $^{14}CO_2$ fixation directly into $C_4$ acids (Monson *et al.*, 1986) while '$C_4$-like' has been used to describe those species which incorporate between 80 and 95% of $^{14}CO_2$ into $C_4$ acids (Moore, Ku & Edwards, 1989).

In the $C_4$-like *Flaveria* species, the metabolism of the $^{14}CO_2$ incorporated, and the intercellular distribution and activities of $C_4$ cycle enzymes are consistent with operation of $C_4$ photosynthesis, but with some $CO_2$ also being assimilated directly by RuBP carboxylase (Moore *et al.*, 1989). However, despite the significant incorporation of $^{14}CO_2$ into $C_4$ acids in some $C_3$–$C_4$ *Flaveria* species, there is no convincing evidence for the transfer of this label directly into Calvin cycle intermediates as occurs for a true $C_4$ species. For example, in contrast to $C_3$ and $C_4$ species, large amounts of $^{14}C$ are found in glycine, serine and, rather unusually, fumarate in leaves of $C_3$–$C_4$ Flaverias during subsequent metabolism of the initial products of $^{14}CO_2$ incorporation (Monson *et al.*, 1986). The appearance of $^{14}C$ in fumarate in the $C_3$–$C_4$ intermediate *Flaveria* species is likely to be due to the presence of the large fumarate pool which is present in their leaves (Rawsthorne *et al.*, 1992).

The maximum catalytic activities of enzymes involved in $C_4$ photosynthesis (PEP carboxylase, NADP-malate dehydrogenase, NADP-malic enzyme and pyruvate, Pi dikinase (PPDK) are reported to be greater in a number of $C_3$–$C_4$ *Flaveria* species than in $C_3$ species or intermediates from other genera which do not have any $C_4$ fixation. However, these activities are only 10–15% of those of *Flaveria* species which have been classified as $C_4$ or $C_4$-like (Ku *et al.*, 1991). In addition, compartmentation between bundle sheath and mesophyll cells of the enzymes involved in the carboxylation and decarboxylation stages of the $C_4$ cycle is not complete or does not occur in leaves of *F. ramosissima* ($C_3$–$C_4$) or *F. brownii* ($C_4$-like) as judged from protoplast fractionation experiments. The latter species has maximum catalytic activities of PEP carboxylase, PPDK and NADP–ME which approach those in true $C_4$ or the other $C_4$-like *Flaveria* species which have more limited (5–10%) direct fixation of $CO_2$ by rubisco (Moore *et al.*, 1989). There is no marked differential distribution of rubisco or PEP carboxylase between mesophyll and bundle sheath in any $C_3$–$C_4$ *Flaveria* species examined to date (Bauwe, 1984; Reed & Chollet, 1985). Enrichment of rubisco in the bundle sheath relative to PEP carboxylase and vice versa for the mesophyll has been reported for *F. brownii*. However, without complete separation of the enzymes involved in the carboxylation and decarboxylation phases of the $C_4$ pathway, futile cycling of $CO_2$ through $C_4$ acids and hence extra energy consumption may well occur.

Whilst the genus *Flaveria* represents a unique example of photosynthetic diversity, the precise nature of photosynthetic and photorespiratory metabolism in the $C_3$–$C_4$ and $C_4$-like *Flaverias* remains to be elucidated. Nevertheless, it is apparent that the differential distribution of GDC must contribute to the observed reduction in apparent photorespiration in the $C_3$–$C_4$ species of this genus, as occurs in $C_3$–$C_4$ species of other genera.

## CARBON ISOTOPE DISCRIMINATION AND MODELLING OF $C_3$–$C_4$ INTERMEDIATE PHOTOSYNTHESIS

The application of carbon isotope discrimination has shed further light on the biochemistry of $CO_2$ assimilation in $C_3$–$C_4$ intermediate species. Based upon well-established models for $C_3$ photosynthesis, and taking into account the knowledge of photorespiratory metabolism in these plants, a model has been proposed which accounts for the patterns of carbon isotope discrimination observed *in vivo* (von Caemmerer & Hubick, 1989). Recent short-term measurements of carbon isotope discrimination occurring during photosynthesis by leaves of $C_3$–$C_4$ intermediate species are consistent with the operation of the light-dependent recapture model but they also suggest that some $C_4$-like $CO_2$ fixation occurs simultaneously in the leaves of the $C_3$–$C_4$ species *Flaveria floridana* (von Caemmerer & Hubick, 1989). However, the extent to which this short-term $C_4$-like fixation is integrated into photosynthetic $CO_2$ assimilation of $C_3$–$C_4$ *Flaveria* species remains to be determined since long-term measurements of carbon isotope discrimination (determined using leaf dry matter) of these plants give $C_3$-like values (Edwards & Ku, 1987).

## $C_3$–$C_4$ PHOTOSYNTHESIS AND THE EVOLUTION OF $C_4$ PHOTOSYNTHESIS

There has been much speculation as to the significance of $C_3$–$C_4$ intermediates in the evolutionary transition from a $C_3$ to a $C_4$ species (e.g. Monson & Moore, 1989). As discussed above, $C_3$–$C_4$ intermediate photosynthesis can give rise to improved photosynthetic performance at warm temperatures when comparisons are made with $C_3$ plants. Improved gas exchange characteristics at such temperatures could translate into improved fitness and so provide the basis for the evolution of more $C_4$-like traits (Monson & Jaeger, 1991). The similar anatomical development of bundle sheath cells and the consistent absence of GDC from mesophyll cells in all the $C_3$–$C_4$ species studied indicates that the differential distribution

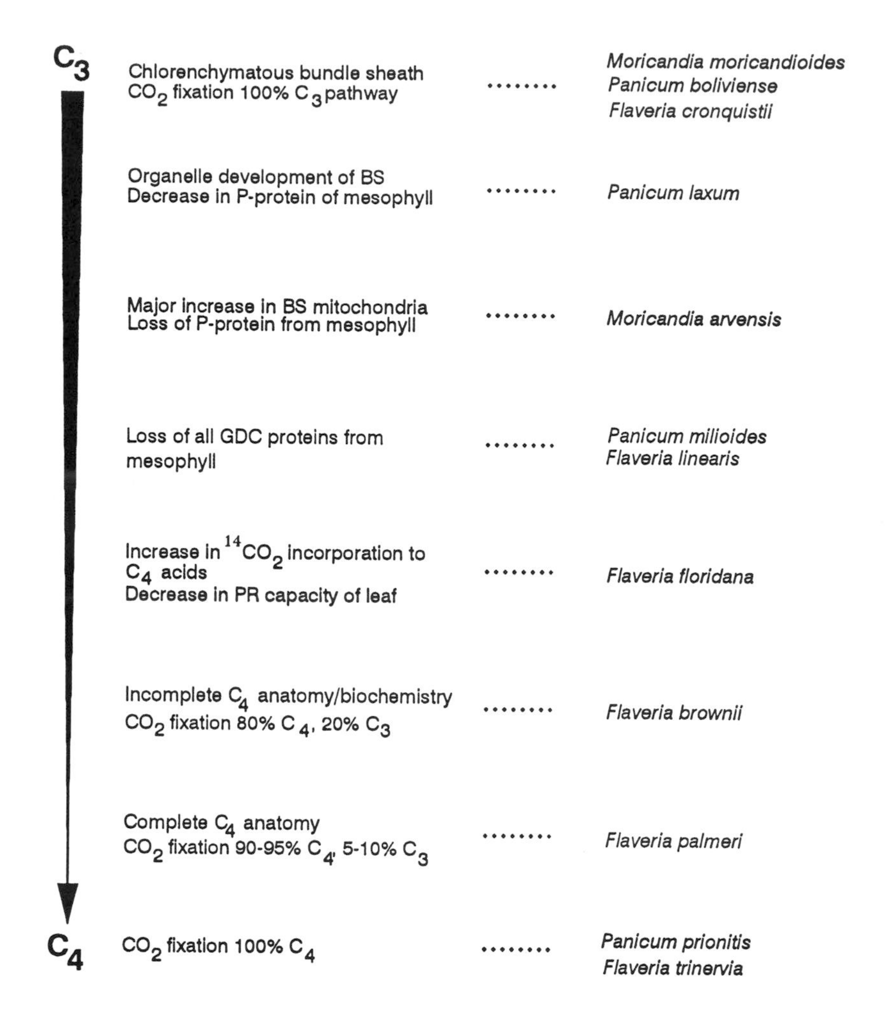


**Figure 11.5.** A proposed series of evolutionary steps between the $C_3$ and $C_4$ photosynthetic mechanisms and the position which present-day $C_3$–$C_4$ intermediate species might represent on this progression.

of GDC could have been a primary event in the evolution of $C_4$ photosynthesis. Consistent with this idea, GDC in $C_4$ species is, as in $C_3$–$C_4$ species, confined to the bundle sheath cells (Ohnishi & Kanai, 1983; Hylton *et al.*, 1988). Indeed, by drawing upon the spectrum of $C_3$–$C_4$ phenotypes which occur across the different genera a sequence of discrete steps is apparent, each representing a progressive change towards $C_4$ photosynthesis (Fig. 11.5).

Recently the use of molecular approaches has provided the first convincing evidence that $C_3$–$C_4$ intermediate plants do indeed represent intermediates in the evolution between $C_3$ and $C_4$ photosynthesis in the genus *Flaveria*. One of the GDC subunits, the moderately conserved H-protein, has now been used as molecular marker to examine the phylogenetic relationships within the genus. As yet, H-protein nucleotide sequences from 12 out of a total of 21 *Flaveria* species have been analysed (Kopriva, Chu & Bauwe, 1996). The tree shown in Figure 11.6 comprises three main clusters: (i) the $C_3$ species *F. pringlei* and *F. cronquistii*, (ii) the advanced $C_4$ species *F. australasica*, *F. trinervia*, *F. bidentis*, and *F. palmeri*, and (iii) all $C_3$–$C_4$ intermediate species including the $C_4$-like *F. brownii*. The ancient position of $C_3$ photosynthesis in the genus is clearly reflected as well as the two-fold and independent evolution of $C_4$ photosynthesis in different lineages of *Flaveria*. Although there is clearly a need for analysis of additional molecular markers, these

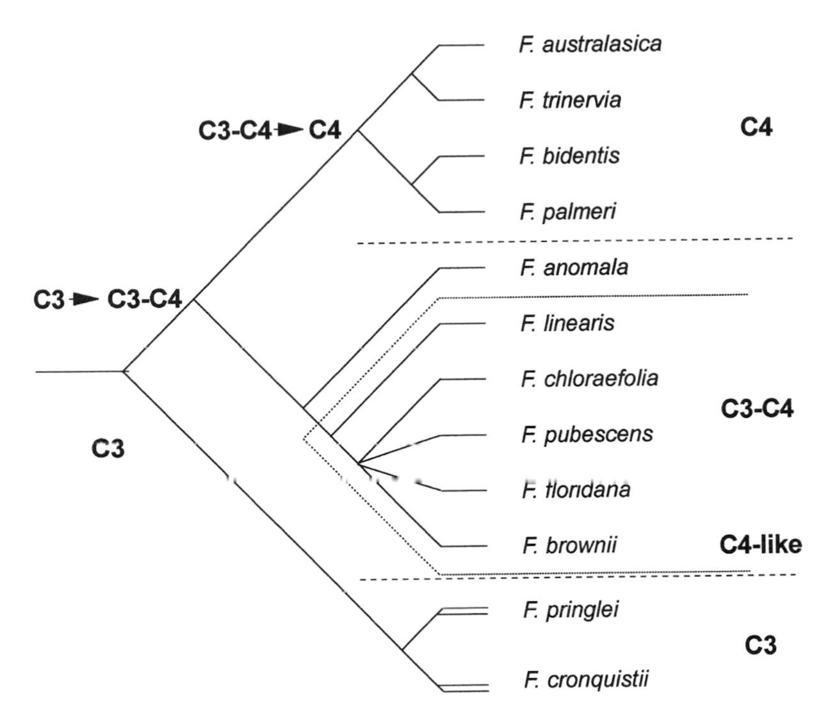


**Figure 11.6.** Reconstruction of the phylogeny of *Flaveria* by comparison of the sequences of cDNAs encoding the H protein of the glycine decarboxylase complex from 12 out of 21 species. The $C_3$ species both have multigene families. Dashed lines separate the three major groups of $C_3$, $C_3$–$C_4$/$C_4$-like, and $C_4$ species. Dotted lines indicate the separation of the genus into two major phyletic lines by flower morphology.

data provide direct evidence for a phylogenetically intermediate position of $C_3$–$C_4$ intermediate photosynthesis.

## INTRASPECIFIC AND INTRAGENERIC TRANSFER OF PHOTOSYNTHETIC TRAITS

The ability of $C_4$ plants to maintain high photosynthetic rates under conditions of limited water supply confers considerable advantages to crop production in warm environments. In contrast, periods of insufficient water supply can heavily impair the growth of $C_3$ plants. For more than two decades there has been a strong interest to introduce $C_4$ photosynthetic traits into $C_3$ crop plants, i.e. to produce plants with at least some $C_3$–$C_4$ intermediate features by genetic manipulation (for review see Brown & Bouton, 1993). Due to hybridization barriers between $C_3$ and $C_4$ crops, sexual hybrids between species of different photosynthetic types have, to date, only been obtained with non-crop plants from genera which include both $C_3$ and $C_4$ species.

The first experimental hybrids of $C_3$ and $C_4$ were between *Atriplex* species (Pearcy & Björkman, 1970), some studies with $C_3$ × $C_3$–$C_4$ hybrids have been performed both within *Moricandia* and *Panicum* and the most extensive hybridization studies have been performed within the genus *Flaveria*. In most cases, mainly due to sterility or little viability, which both indicate chromosome instability, these experiments have often been restricted to the $F_1$ generations. For most of the examined $F_1$ characteristics such as enzyme activities, photosynthesis, photorespiration, and carbon isotope discrimination, $C_3$ alleles apparently predominate. In some cases, by repeated backcrossing with the respective $C_4$ parent, photosynthetic traits of the progeny have become $C_4$-like. Of the few hybrids that have been advanced beyond the $F_1$ generation, a low correlation in the heritability of the $C_4$ photosynthetic traits was observed in the progeny. As yet, it is difficult to draw detailed conclusions from these experiments about the genetics of the $C_4$ syndrome.

The natural hybridization barriers between $C_3$ and $C_4$ species can, at least partially, be overcome by the methods of modern gene technology. There are two alternatives to the recombination of complete genomes as described above. The use of marker-assisted introgression of useful genes from one related species into another is one method. Using this approach, it is possible to select only for the parts of one genome that confer the desired trait, while retaining a recurrent parent background. Such an approach presents possibilities for the transfer of $C_3$–$C_4$ traits from *Moricandia* into the related *Brassica* crop species.

A more targeted approach is to stably integrate selected $C_4$ genes into the genome of $C_3$ plants through genetic transformation. This allows the study of effects of a defined set of gene products (enzymes, regulatory proteins, antisense mRNA) on the metabolism of the host plant. First attempts to over-express selected $C_4$ enzymes in $C_3$ plants have focused on the over-expression, under the control of the cauliflower mosaic virus 35S promoter, of the $C_4$ isoform of PEP carboxylase (Hudspeth *et al.*, 1992). The results indicated that the $C_4$ enzyme became efficiently down-regulated in the $C_3$ metabolic background. Even if this and other enzyme components of the $CO_2$ pump can be introduced with sufficient activity into a $C_3$ plant, a lack of a proper compartmentation of the respective reactions would prevent a $C_4$ cycle from operating. In contrast to $C_3$ plants, however, leaves of $C_3$–$C_4$ intermediate plants do possess a suitable anatomical structure for the stepwise introduction of genes for a $C_4$-like $CO_2$ pump mechanism. The introduction of foreign genes into $C_3$–$C_4$ plants is not yet routine. There are promising results for *M. arvensis* and a reproducible and efficient protocol for stable genetic transformation has already been established and used for promoter studies in *F. pubescens* by Bauwe and co-workers.

During the last few years, many enzyme components of the $C_4$ cycle and of the photorespiratory pathway have been cloned as cDNAs, with a number coming from $C_3$, $C_3$–$C_4$, and $C_4$ *Flaveria* species. The function and control of the 5′ promoter regions of these genes are presently being analysed using transgenic plants. Preliminary experiments using the promoter regions of *F. anomola* ($C_3$–$C_4$) GDC genes fused to the reporter gene β-glucuronidase have shown that their transcriptional activity was restricted to the bundle sheath in *F. pubescens* ($C_3$–$C_4$), or to the cell layer immediately adjacent to the vascular cells in tobacco ($C_3$). Moreover, the analysis of the gene encoding the $C_4$ isoform of phosphoenolpyruvate carboxylase from *F. trinervia* ($C_4$) has revealed that the mesophyll specificity of this promoter is retained in transgenic tobacco (Stockhaus *et al.*, 1994). Both these results show that the *trans*-acting factors required for the cell-type specificity of photosynthetic/photorespiratory genes such as PEP carboxylase and the GDC subunits are already present in $C_3$ plants. This suggests that, during the evolution of $C_3$–$C_4$, and subsequently $C_4$ photosynthesis, the cell-specific repression/induction of at least some important photosynthetic genes was achieved by modification of *cis*-elements in these genes. The 5′ promoter regions from these and other genes from $C_3$–$C_4$ intermediate and $C_4$ plants will provide useful tools for the targeting of components of the photosynthetic/photorespiratory apparatus to selected cell types in leaves of $C_3$ plants.

## FUTURE PROSPECTS

There is still much that we do not understand about $C_3$–$C_4$ intermediate photosynthesis. The details of the leaf biochemistry are still unclear with respect to the effects of confinement of glycine decarboxylation to the bundle sheath cells, including carbon and nitrogen recycling to the mesophyll. We also need to know more precisely the role of $CO_2$ fixation into $C_4$ acids in the $C_3$–$C_4$ intermediate *Flaveria* species and the extent to which this is integrated into net C assimilation. The development of reproducible and relatively efficient protocols for the genetic transformation of $C_3$–$C_4$ intermediate *Moricandia* and *Flaveria* species and $C_4$ *Flaveria* species is an important step. This will allow, by antisense repression or over-expression, an examination of the effects of selected proteins on the biochemistry and physiology of the photosynthetic apparatus, and the identification of further targets for manipulation by recombinant DNA technologies. These approaches will thus help in a better understanding of $C_3$–$C_4$ intermediate and $C_4$ photosynthesis and may open new avenues for the reduction of apparent photorespiration and the improvement of water-use efficiency in $C_3$ crop plants.

### REFERENCES

Apel, P., Tichá, I. & Peisker, M. (1978). $CO_2$ compensation concentrations in leaves of *Moricandia arvensis* (L.) DC. at different insertion levels and $O_2$ concentrations. *Biochemie und Physiologie der Pflanzen*, **172**, 547–52.

Bauwe, H. (1984). Photosynthetic enzyme activities and immunofluorescence studies on the localization of ribulose-1, 5-bisphosphate carboxylase/oxygenase in leaves of $C_3$, $C_4$, and $C_3$–$C_4$ intermediate species of *Flaveria* (Asteraeae). *Biochemie und Physiologie der Pflanzen*, **179**, 253–68.

Bauwe, H., Keerberg, O., Bassüner, R., Pärnik, T. & Bassüner, B. (1987). Reassimilation of carbon dioxide by *Flaveria* (Asteraceae) species representing different types of photosynthesis. *Planta*, **172**, 214–18.

Brown, R. H. & Bouton, J. H. (1993). Physiology and genetics of interspecific hybrids between photosynthetic types. *Annual Review of Plant Physiology and Plant Molecular Biology*, **44**, 435–56.

Brown, R. H. & Hattersley, P. W. (1989). Leaf anatomy of $C_3$–$C_4$ species as related to evolution of $C_4$ photosynthesis. *Plant Physiology*, **91**, 1543–50.

Brown, R. H. & Morgan, J. A. (1980). Photosynthesis of grass species differing in carbon dioxide fixation pathways. VI. Differential effects of temperature and light intensity on photorespiration in $C_3$, $C_4$, and intermediate species. *Plant Physiology*, **66**, 541–4.

Brown, R. H., Bouton, J. H., Rigsby, L. & Rigler, M. (1983). Photosynthesis of grass species differing in carbon dioxide fixation pathways. VIII. Ultrastructural characteristics of *Panicum* species in the *Laxa group. Plant Physiology*, **71**, 425–31.

Devi, T., Rajagopalen, A. V. & Raghavendra, A. S. (1995). Predominant localization of mitochondria enriched with glycine-decarboxylating enzymes in bundle sheath cells of *Alternanthera tenella*, a $C_3$–$C_4$ intermediate species. *Plant Cell and Environment*, **18**, 589–94.

Edwards, G. E. & Ku, M. S. B. (1987). Biochemistry of $C_3$–$C_4$ intermediates. In *The Biochemistry of Plants*, vol. 10, ed. M. D. Hatch & N. K. Boardman, pp. 275–325. London: Academic Press.

Holaday, A. S. & Chollet, R. (1983). Photosynthetic/photorespiratory carbon metabolism in the $C_3$–$C_4$ intermediate species, *Moricandia arvensis and Panicum milioides. Plant Physiology*, **73**, 740–5.

Hudspeth, R. L., Grula, J. W., Dai, Z., Edwards, G. E. & Ku M. S. B. (1992). Expression of maize phosphoenolpyruvate carboxylase in transgenic tobacco. *Plant Physiology*, **98**, 458–64.

Hunt, S., Smith, A. M. & Woolhouse, H. W. (1987). Evidence for a light-dependent system for reassimilation of photorespiratory $CO_2$, which does not include a $C_4$ cycle, in the $C_3$–$C_4$ intermediate species *Moricandia arvensis. Planta*, **171**, 227–34.

Hylton, C. M., Rawsthorne, S., Smith, A. M., Jones, D. A. & Woolhouse, H. W. (1988). Glycine decarboxylase is confined to the bundle sheath cells of leaves of $C_3$–$C_4$ intermediate species. *Planta*, **175**, 452–9.

Keck, R. W. & Ogren, W. L. (1975). Differential oxygen response of photosynthesis in soybean and *Panicum milioides. Plant Physiology*, **58**, 552–5.

Kopriva, S., Chu, C. C. & Bauwe, H. (1996). Molecular phylogeny of *Flaveria* as deduced from the analysis of nucleotide sequences encoding H-protein of the glycine cleavage system. *Plant Cell and Environment*, **19**, 1028–36.

Krenzer, E. G., Moss, D. N. & Crookston, R. K. (1975). Carbon dioxide compensation points of flowering plants. *Plant Physiology*, **56**, 194–206.

Ku, M. S. B., Wu, J., Dai, Z., Scott, R. A., Chu, C. & Edwards, G. E. (1991). Photosynthetic and photorespiratory characteristics of *Flaveria* species. *Plant Physiology*, **96**, 518–28.

Leegood, R. C. & von Caemmerer, S. (1988). The relationship between contents of photosynthetic metabolites and the rate of photosynthetic carbon assimilation in leaves of *Amaranthus edulis* L. *Planta*, **174**, 253–62.

Monson, R. K. & Moore, B. D. (1989). On the significance of $C_3$–$C_4$ intermediate photosynthesis to the evolution of $C_4$ photosynthesis. *Plant Cell and Environment*, **12** 689–99.

Monson R. K. & Jaeger, C. H. (1991). Photosynthetic characteristics of $C_3$–$C_4$ intermediate *Flaveria floridana* (Asteracae) in natural habitats: evidence of advantages to $C_3$–$C_4$ photosynthesis at high leaf temperatures. *American Journal of Botany*, **78**, 795–800.

Monson R. K., Edwards, G. E. & Ku M. S. B. (1984). $C_3$–$C_4$ intermediate photosynthesis in plants. *BioScience*, **34**, 563–74.

Monson, R. K., Moore, B. D., Ku, M. S. B. & Edwards, G. E. (1986). Co-function of $C_3$- and $C_4$-photosynthetic pathways in $C_3$, $C_4$, and $C_3$–$C_4$ intermediate *Flaveria* species. *Planta*, **168**, 493–502.

Moore, B. D., Monson, R. K., Ku, M. S. B. & Edwards, G. E. (1988). Activities of principal photosynthetic and photorespiratory enzymes in leaf mesophyll and bundle sheath protoplasts from the $C_3$–$C_4$ intermediate *Flaveria ramosissima. Plant and Cell Physiology*, **29**, 999–1006.

Moore, B. D., Ku, M. S. B. & Edwards, G. E. (1989). Expression of $C_4$-like photosynthesis in several species of *Flaveria. Plant Cell and Environment*, **12**, 541–9.

Morgan, C. L. Turner, S. R. & Rawsthorne, S. (1993). Co-ordination of the cell-specific distribution of the four subunits of glycine decarboxylase and of serine hydroxymethyl transferase, in leaves of C3–C4 intermediate species from different genera. *Planta*, **190**, 468–73.

Ohnishi, J. & Kanai, R. (1983). Differentiation of photorespiratory activity between mesophyll and bundle sheath cells of $C_4$ plants. I. Glycine oxidation by mitochondria. *Plant and Cell Physiology*, **24**, 1411–20.

Pearcy, R. & Björkman, O. (1970). Hybrids between *Atriplex* species with and without β-carboxylation photosynthesis. *Carnegie Institute Washington, Year Book*, pp. 632–40.

Peisker, M. (1985). Modelling carbon metabolism in $C_3$–$C_4$ intermediate species. 1. $CO_2$ compensation point and its $O_2$ dependence. *Photosynthetica*, **18**, 9–19.

Rawsthorne, S. & Hylton, C. M. (1991). The post-illumination $CO_2$ burst and glycine metabolism in leaves of $C_3$ and $C_3$–$C_4$ intermediate species of *Moricandia*. *Planta*, **186**, 122–6.

Rawsthorne, S., Hylton, C. M., Smith, A. M. & Woolhouse, H. W. (1988*a*). Photorespiratory metabolism and immunogold localization of photorespiratory enzymes in leaves of $C_3$ and $C_3$–$C_4$ intermediate species of *Moricandia*. *Planta*, **173**, 298–308.

Rawsthorne, S., Hylton, C. M., Smith, A. M. & Woolhouse, H. W. (1988*b*). Distribution of photorespiratory enzymes between bundle sheath and mesophyll cells in leaves of the $C_3$–$C_4$ intermediate species *Moricandia arvensis* (L.) DC. *Planta*, **176**, 527–32.

Rawsthorne, S., von Caemmerer, S., Brooks, A. & Leegood R. C. (1992). Metabolic interactions in leaves of $C_3$–$C_4$ intermediate plants. In *Plant Organelles. Compartmentation of Metabolism in Photosynthetic Cells* SEB Seminar Series, vol.50, ed. A. K. Tobin, pp. 113–39. Cambridge: Cambridge University Press.

Reed, J. E. & Chollet, R. (1985). Immunofluorescent localization of phosphoenolpyruvate carboxylase and ribulose 1,5-bisphosphate carboxylase/oxygenase proteins in leaves of $C_3$, $C_4$, and $C_3$–$C_4$ *Flaveria* species. *Planta*, **165**, 439–45.

Somerville, S. C. & Somerville, C. R. (1983). Effect of oxygen and carbon dioxide on photorespiratory flux determined from glycine accumulation in a mutant of *Arabidopsis thaliana*. *Journal of Experimental Botany*, **34**, 415–24.

Stockhaus, J., Poetsch, W., Steinmüller, K. & Westhoff, P. (1994). Evolution of $C_4$ phosphoenolpyruvate carboxylase promoter of the $C_4$ dicot *Flaveria trinervia*: an expression analysis in the $C_3$ plant tobacco. *Molecular and General Genetics*, **245**, 286–93.

Turner, S. R., Ireland, R. J., Hellens, R., Ellis, N. & Rawsthorne S. (1993). The organization and expression of the genes encoding the mitochondrial glycine decarboxylase complex and serine hydroxymethyltransferase in pea (*Pisum sativum* L.) *Molecular and General Genetics*, **236**, 402–8.

von Caemmerer, S. (1989). Biochemical models of photosynthetic $CO_2$-assimilation in leaves of $C_3$–$C_4$ intermediates and the associated carbon-isotope discrimination. I. A model based on a glycine shuttle between mesophyll and bundle sheath cells. *Planta*, **178**, 376–87.

von Caemmerer, S. & Hubick, K. T. (1989). Short-term carbon-isotope discrimination in $C_3$–$C_4$ intermediate species. *Planta*, **178**, 475–81.

# 12 Starch–sucrose metabolism and assimilate partitioning

S. C. HUBER

## INTRODUCTION

This chapter will focus on allocation of photosynthetically fixed carbon into carbohydrate end products, i.e. the utilization of triose-P generated by the reductive pentose phosphate pathway (Calvin cycle, discussed in detail in Chapter 8, this volume). Carbohydrate is the major end product of leaf photosynthesis, and in this chapter, we will restrict our attention to those species that synthesize starch (within the chloroplast) and sucrose (in the cytosol) as major products. First, we will consider the compartmentation of metabolism between the chloroplast and cytosol, and the interconnection of the two compartments by the triose-P translocator (TPT). Then we will consider starch metabolism and sucrose biosynthesis individually in more detail. Throughout the chapter, we will try to identify regulatory mechanisms involved in control of enzyme activity (allosteric control and covalent modification) and recent efforts to achieve molecular genetic manipulation of metabolism by over-expression or antisense inhibition of specific enzymes. While the chapter will primarily consider photosynthetic systems (i.e. source leaves), there will be some limited mention of starch–sucrose metabolism in sink tissues as necessary.

## DIVISION OF LABOUR BETWEEN THE CHLOROPLAST AND CYTOSOL

In the leaf of a typical $C_3$ plant, there is a single photosynthetic cell type (the mesophyll cell) which functions in photosynthetic $CO_2$ fixation. Assimilatory starch is formed within the chloroplast whereas sucrose, the other major end product of leaf photosynthesis, is synthesized exclusively in the cytosol (see Fig. 12.1). Starch plays an important role in plant metabolism as a temporary reserve form of reduced carbon, whereas sucrose plays a central role in translocation as the transport form of reduced carbon. The sucrose that is synthesized in the cytosol may have at least two fates: it can either be translocated from the leaf or temporarily stored within the leaf. In the former case, sucrose is released from the mesophyll cell to the leaf apoplast prior to active uptake into the companion cell/sieve element complex by a $H^+$:sucrose symport. The accumulation of sucrose within the phloem, often up to concentrations of 1M, is energized by the $H^+$ gradient established by an active ATPase (Fig. 12.1). In contrast, sucrose temporarily stored in the leaf can accumulate in the cytosol and vacuole. Because the transport of sucrose into the vacuole is passive and not active, no accumulation of sucrose in the vacuole is possible. None the less, the amount of sucrose accumulated in a leaf during the photoperiod can be substantial and the majority will be in the vacuole, which accounts for ~90% of the cell volume. Thus, sucrose can account for a significant fraction of the total non-structural carbohydrate available to support leaf metabolism and continued translocation during the subsequent night period. However, in some species, such as soybean (*Glycine max*), tobacco (*Nicotiana tabacum*), and *Arabidopsis thaliana*, the vacuoles of mature leaves contain high activities of soluble acid invertase (10 to 200 μmol sucrose hydrolysed $g^{-1}$ fresh wt $h^{-1}$). As a result, sucrose entering the vacuole would be rapidly hydrolysed to hexoses, which would have to be phosphorylated (via glucose and fructose kinase) in the

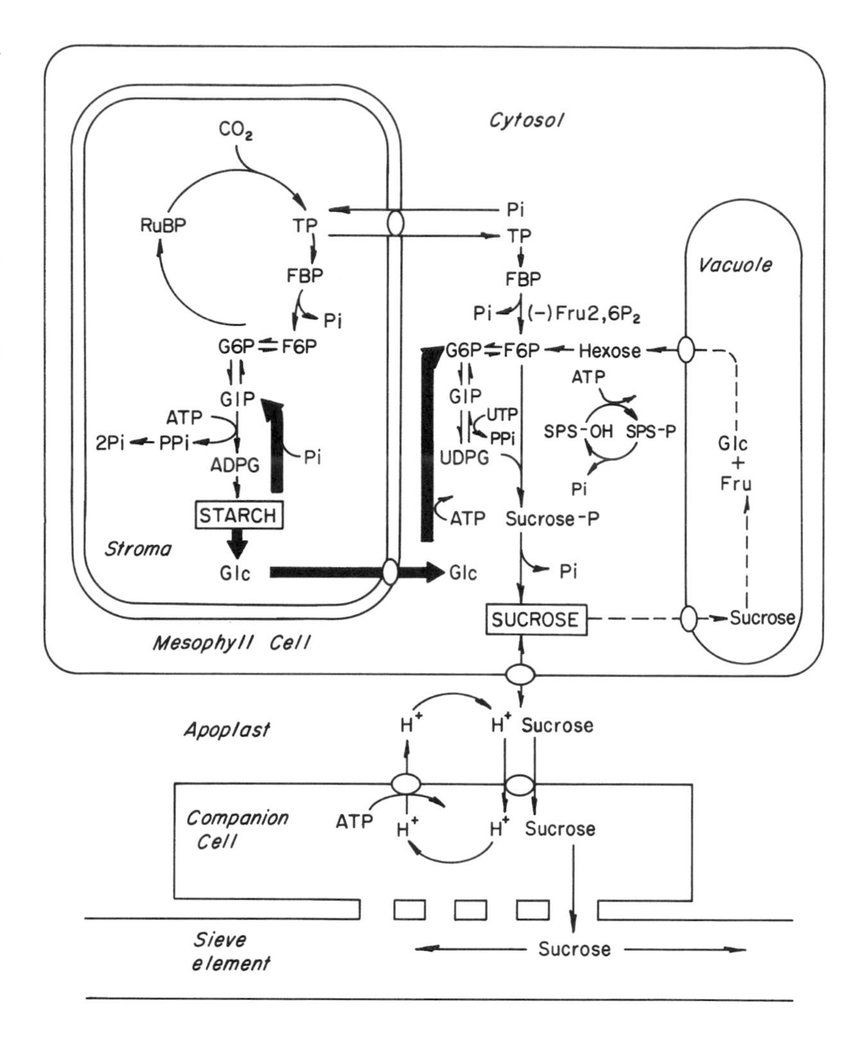


**Figure 12.1.** Simpified scheme showing pathways of starch and sucrose metabolism in mesophyll cells of a $C_3$ plant, and translocation of sucrose in the phloem. 'Sugar cycling' within the mesophyll cell is shown by the dashed lines. Alternative pathways for starch mobilization at night (phosphorolysis versus hydrolysis) are shown with solid arrows. Note that, to conserve space, metabolites have been abbreviated as follows: TP, triose-P; FBP, Fru-1,6-$P_2$, F6P, Fru-6-P; G1P, Glc-1-P; G6P, Glc-6-P; ADPG, ADPglucose; UDPG, UDPglucose.

cytosol in order to re-enter metabolism. Species with high activities of soluble acid invertase in mature leaves typically do not accumulate sucrose during the day (i.e. sucrose accumulation in the vacuole is precluded) nor do they accumulate hexose sugars, which suggests that conversion to hexose-P is efficient. These species tend to accumulate significant amounts of starch in leaves as an end product of photosynthesis. Species which accumulate substantial amounts of sucrose in leaves, such as spinach (*Spinacia oleracea*), maize (*Zea mays*) and wheat (*Triticum aestivum*), are almost entirely devoid of soluble acid invertase activity when leaves reach full expansion and become photosynthetic 'sources' and thus sucrose storage within the vacuole can occur.

## Intracellular compartmentation of metabolism

The two cellular compartments – the chloroplast stroma and the cytosol – appear to be redundant in terms of many of the metabolic intermediates and enzymes present, as each compartment contains a complete gluconeogenic pathway leading from 3-phosphoglycerate to hexose-P. However, many of the enzymes are immunologically distinct because they are the products of different genes and often have divergent properties, i.e. there are compartment-specific isozymes. Because the environments of the two compartments are quite different in certain regards, the regulation of the pathways can also be compartment specific. A good example is the chloroplast versus cytosolic fructose-1,6-bisphosphatase (FBPase; discussed also in Chapter 8, this volume). The plastid FBPase is regulated primarily by changes in sulfhydryl group redox status whereas the cytosolic enzyme is regulated primarily by the metabolite Fru-2,6-$P_2$. The chloroplast stroma undergoes large changes in [$Mg^{2+}$], pH and redox status of sulfhydryl groups during light–dark transitions and many of the Calvin cycle enzymes, including the plastidic FBPase, are regulated in response to these factors. The chloroplast stroma does not contain Fru-2,6-$P_2$ because the enzyme that synthesizes the regulatory metabolite is restricted to the cytosol and the metabolite itself cannot traverse the envelope membrane; the plastid FBPase is also not sensitive to inhibition by Fru-2,6-$P_2$, even *in vitro*. Thus, one of the major distinctions between the two compartments is the presence or absence of Fru-2,6-$P_2$.

Another difference between the two compartments is the occurrence of a significant steady-state pool of inorganic pyrophosphate (PPi). Both compartments contain pyrophosphorylase enzymes [ADPglucose pyrophosphorylase (AGPase) in the chloroplast and UDPglucose pyrophosphorylase (UGPase) in the cytosol] which generate PPi as a product in addition to the nucleotide sugar. The pyrophosphorylases are readily reversible enzymes, but the AGPase is rendered irreversible by the action of inorganic pyrophosphatase, which immediately cleaves PPi to liberate two molecules of Pi. Thus, AGPase activity, and its control by allosteric effectors, is critical to the control of carbon flux into starch. In contrast, the cytosol is devoid of inorganic pyrophosphatase activity and thus a significant pool of PPi can be found. As a result, the UGPase reaction is reversible *in vivo* and operates close to equilibrium; correspondingly, UGPase activity is not known to be regulated by allosteric effectors. Allosteric control is certainly not restricted to the chloroplast, as a number of cytosolic enzymes (e.g. sucrosephosphate synthase; SPS) are known to be controlled in this way. However, certain generalizations can be made about the control of enzymes in the two compartments. First, control of the redox status of vicinal dithiols via the ferredoxin/thioredoxin system (discussed in more detail in Chapter 8) is prevalent in the chloroplast stroma but not in the cytosol. Secondly, covalent modification of proteins by reversible phosphorylation appears to be more important in the cytosol than in the chloroplast; an important aspect of the control of carbon flux into sucrose involves the control of SPS by reversible seryl phosphorylation (see Fig. 12.1; discussed in more detail below).

Also identified in the cellular scheme presented in Fig. 12.1 is a simplified representation of two alternative pathways for starch mobilization in the dark. Starch can be broken down either by phosphorylase, which consumes Pi and generates Glc-1-P, or by hydrolytic enzymes which produce maltose and Glc. The stroma is apparently devoid of glucose kinase activity. Consequently, the Glc molecules formed during starch breakdown must be transported to the cytosol via the hexose transporter for phosphorylation to form Glc-6-P. Phosphorolytic breakdown of starch is thought to produce substrates for use within the stroma and also for export on the TPT, e.g. for respiration. In contrast, the Glc produced by hydrolytic mobilization of starch is thought to be utilized primarily for sucrose biosynthesis in the cytosol. The latter point is significant because this pathway allows sucrose synthesis to proceed at night when the dark leaves contain elevated levels of Fru-2,6-$P_2$ sufficient to completely inhibit the cytosolic FBPase.

## The triose-phosphate translocator (TPT)

The chloroplast stroma and the cytosol are delineated by the double membrane of the chloroplast envelope. The outer membrane of the envelope is relatively permeable to most compounds because of porins in the membrane which allow for passive movement of both neutral and charged molecules. In contrast, the inner membrane of the chloroplast envelope represents a permeability barrier to the free movement of many molecules, metabolic intermediates in particular. The movement of certain molecules is facilitated by specific transport proteins. For example, the TPT of the inner envelope membrane facilitates the movement of Pi, 3-phosphoglycerate and triose-P in a strict counter-exchange reaction. As shown in Fig. 12.1, a molecule of triose-P, generated in excess of that required to sustain the Calvin cycle, can be exported to the cytosol in exchange for the influx of a molecule of Pi (liberated during the conversion of four molecules of triose-P to form one molecule of sucrose). Thus, the TPT maintains a charge- and phosphate-balance on both sides of the membrane. This is significant because it renders chloroplast metabolism sensitive to cytosolic [Pi], which in turn is a function of the rate of sucrose biosynthesis. Accordingly, there is often an inverse relationship between starch and sucrose biosynthesis, i.e. a decrease in sucrose synthesis typically results in an increase in starch synthesis but not vice versa (Stitt & Quick, 1989). The importance of cytosolic [Pi] was highlighted in pioneering studies of David Walker and co-workers who used mannose to sequester cytosolic Pi (as mannose-6-P which is not metabolized further in most species). Mannose feeding to most $C_3$ plants resulted in dramatic increases in starch biosynthesis even though $CO_2$, fixation was often reduced slightly (Herold & Lewis, 1977). The effect can be readily understood based on the scheme presented in Fig. 12.1 which highlights the critical role of cytosolic [Pi] and the TPT in assimilate partitioning.

The TPT is a major protein component of the inner envelope membrane (Flügge *et al.*, 1992). The protein is nuclear encoded and is synthesized as a 40 kDa precursor protein. The native transporter is a dimer of ~29 kDa subunits that together form a single substrate binding site. Each subunit contains six helices that span the membrane; hydrophobic amino acid residues are orientated toward the outside while hydrophilic residues are orientated to the inside to form an ion-conducting channel. The 1:1 counter-exchange of metabolites involves a 'ping–pong' mechanism and there is some structural asymmetry such that affinities for transport metabolites can be different on the stromal versus cytosolic faces.

Plastids from a variety of tissues and plant species have been shown to contain a TPT; however, there are significant differences in substrate specificity which relate to the function of the plastids (Heldt, Flügge & Borchert, 1991). For example, the TPT of the $C_4$ mesophyll chloroplast is specialized to transport phospho*enol*-pyruvate, which is required for operation of the $C_4$ pathway. Similarly, the TPT of the pea root plastid is distinct from the chloroplast TPT in that Glc-6-P is readily transported.

### *Molecular genetic manipulation of the TPT*

The gene encoding of the TPT translocator has been cloned from several species, which makes possible the manipulation of the transporter using biotechnological approaches. Antisense inhibition has been used to achieve up to a 60% reduction in TPT activity in transgenic potato plants (Riesmeier *et al.*, 1993). As a result, metabolite levels and normal assimilate partitioning patterns were altered. The whole leaf level of 3-phosphoglycerate (largely in the chloroplast) was increased whereas Glc-6-P (largely in the cytosol) was decreased, consistent with a greater retention of carbon within the plastid during photosynthesis. Accordingly, leaf starch accumulation was increased substantially relative to wild-type plants. Interestingly, plant growth was relatively unaffected because the increased starch formation resulted in enhanced sucrose biosynthesis and export at night. Sucrose formation at night was apparently not restricted by the reduction in TPT because carbon was being exported to the cytosol in the form of Glc (from starch breakdown), which involves the hexose transporter and not the TPT. This work highlights the role of the hexose transporter in carbohydrate interconversion in the dark, and the phenotypic plasticity of plants to adapt to a change in metabolism without adversely affecting growth.

### *The hexose transporter*

Another example of the role of the hexose transporter comes from recent studies of the starch over-producing 'TC26' mutant of *Arabidopsis thaliana*. Leaves of the mutant contain significantly more leaf starch than the wild-type (Caspar *et al.*, 1991); starch synthesis is not increased but rather mobilization in the dark is somehow restricted. The basis for this interesting mutant has recently been elucidated by Trethewey & ap Rees (1994). In a series of experiments, they established that chloroplasts of the TC26 mutant contain a functional TPT but lack a functional hexose transporter; as a result, starch cannot be completely mobilized in the dark. The envelope membranes from the TC26 mutant also lacked a 40 kDa protein, which was postulated to be the subunit of the hexose transporter. This is another elegant example which highlights the essential role of the hexose transporter in moving products of starch breakdown to the cytosol for sucrose synthesis at night.

## STARCH BIOSYNTHESIS

The insoluble starch granules found in leaves (assimilatory starch) and sink tissues (storage starch) consist of amylose (unbranched) and amylopectin (branched) molecules (for review, see Martin & Smith, 1995). Considerable evidence suggests that starch biosynthesis in leaves and sink tissues such as potato tubers is controlled primarily at the level of the AGPase. Briefly stated, early studies from Preiss and co-workers established that the activity of AGPase is allosterically regulated by the ratio of 3-phosphoglycerate (activator) to $P_i$ (inhibitor) (Preiss *et al.*, 1995). During formation of ADPglucose, the $PP_i$ formed is rapidly removed by inorganic pyrophosphatase rendering the enzyme irreversible *in vivo* (see Fig. 12.1). Starch synthases then utilize the ADPglucose and transfer the glucosyl moiety to the nonreducing end of an $\alpha$(1–4)-linked glucan chain. Starch branching enzymes can cut an $\alpha$(1–4)-linked glucan chain and form an $\alpha$(1–6) linkage with the cut chain and the 6-carbon of another Glc residue in the glucan, thereby forming a 'branch'. The starch branching enzymes show specificity for the length of the glucan chain that is used as substrate, and thereby some order or characteristic structural features emerge for the granule.

Direct evidence that the activity of plastid AGPase controls the rate of starch accumulation has come from recent attempts to use molecular genetic manipulation. Stark *et al.* (1992) transformed potato plants with a mutant *E. coli* gene (designated *GlgC16*) that encodes a regulatory variant of AGPase. The mutant enzyme is much less sensitive to allosteric effectors and is thus always in the active state. Tubers derived from the transgenic potato plants expressing the *GlgC16* gene accumulated significantly more starch than wild-type tubers, indicating that AGPase catalyses a rate-limiting step in starch biosynthesis in sink tissues such as the potato tuber.

## SUCROSE BIOSYNTHESIS

In the cytosolic sucrose synthesis pathway, triose-P molecules exported from the chloroplast on the TPT, are converted to the level of hexose-P and then to sucrose. The Pi molecules released are recycled back to the chloroplast to continue the release of triose-P molecules (see Fig. 12.1). Control of carbon flux through the pathway is shared among several steps; in particular, the co-ordinated regulation of the cytosolic FBPase and SPS catalysed reactions is recognized to be quite important. Co-ordinate control of these two enzymes provides a mechanism to regulate sucrose synthesis both in terms of 'feedforward' and 'feedback' control. Feedforward control relates to the situation where photosynthetic rates are rising and the pool of triose-P available for utilization is increasing, e.g. with increasing irradiance early in the photoperiod in a natural environment. Increased flux of carbon into sucrose is accommodated by allosteric control of cytosolic FBPase and covalent modification of SPS, as described below. In contrast to feedforward control, feedback control involves restriction of carbon flux into sucrose when production exceeds demand of the plant for assimilates. Under these conditions, the export of triose-P from the chloroplast is ultimately controlled by the demand for sucrose. As end products of leaf photosynthesis (including, but not limited to, sucrose) accumulate in leaves under feedback conditions, the flux of carbon into sucrose is reduced, and that into starch is increased. Once again, allosteric control of cytosolic FBPase and covalent modification of SPS underlie the response.

### Regulation of cytosolic FBPase

One of the major factors controlling cytosolic FBPase activity is the concentration of the allosteric inhibitor Fru-2,6-$P_2$. This regulatory metabolite occurs in the cytosol of plant cells at extremely low concentrations (1 to 10 $\mu$M), and as already mentioned, is not present in the chloroplast. In the plant cell cytoplasm, there are at least two enzymes that are regulated by Fru-2,6-$P_2$ (see Fig. 12.2A): the pyrophosphate-linked phosphofructokinase (PPi-PFK), which catalyses a readily reversible reaction that is activated by Fru-2,6-$P_2$, and the cytosolic FBPase, which catalyses an essentially irreversible reaction and is inhibited by Fru-2,6-$P_2$. The ATP-PFK, which is the classical enzyme involved in synthesis of Fru-1,6-$P_2$, is unaffected by the regulatory metabolite. The relative roles of the ATP-PFK and $PP_i$-PFK in plant metabolism are not entirely clear; however, the cytosolic FBPase is clearly involved in photosynthetic sucrose formation and it appears that Fru-2,6-$P_2$ plays an

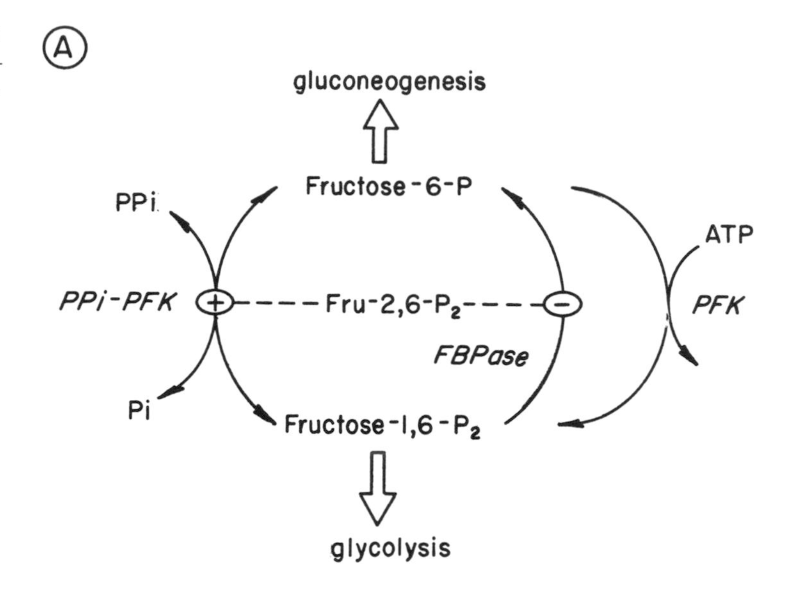


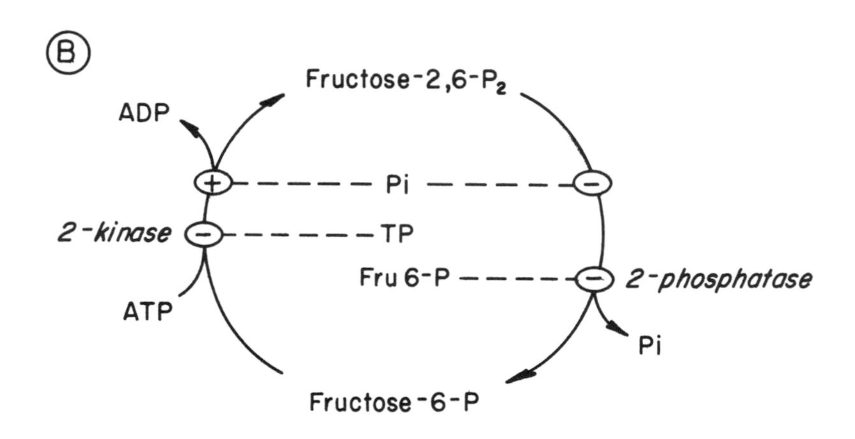


**Figure 12.2.** The Fru-2,6-$P_2$ system in plants. (A) Cytosolic enzymes regulated by Fru-2,6-$P_2$. (B) Metabolite regulation of Fru-2,6-$P_2$ synthesis (via the 2-kinase) and degradation (via the 2-phosphatase).

important role by virtue of the regulation of the FBPase.

Fru-2,6-$P_2$ is an extremely potent inhibitor of the cytosolic FBPase. *In vitro*, it is active at the nM concentration range and inhibits the enzyme by decreasing the affinity for the substrate Fru-1,6-$P_2$ and by inducing a sigmoidal substrate saturation profile. In addition, the regulatory metabolite increases the sensitivity of the enzyme to inhibition by 5′-AMP and $P_i$. Consequently, cytosolic FBPase activity will be a function of the concentrations of the regulatory metabolite as well as several metabolites, including the substrate molecule.

### *Regulation of Fru-2,6-$P_2$ level*

The regulatory metabolite is synthesized from Fru-6-P by the requisite 2-kinase and degraded by a specific 2-phosphatase to generate a cycle of synthesis and degradation (see Fig. 12.2(B)). The activities of both the 2-kinase and 2-phosphatase are regulated by metabolites (for review, see Stitt, Huber & Kerr, 1987). As shown, the 2-kinase is activated by $P_i$ and inhibited by triose-P (and 3-phosphoglycerate), whereas the 2-phosphatase is inhibited by both of its products, $P_i$ and Fru-6-P. In addition, the 2-kinase can be considered to be stimulated by Fru-6-P, because the affinity of the enzyme for its substrate is rather low.

In feedforward control, when photosynthetic rates are rising, the leaf [Fru-2,6-$P_2$] is decreasing, largely reflecting the regulation of the 2-kinase by decreasing [$P_i$] and increasing [triose-P]. During feedback control, the leaf [Fru-2,6-$P_2$] is increasing as a result of activation of the 2-kinase and inhibition of the 2-phosphatase; both would occur in response to increased cytosolic [Fru-6-P]. The Fru-6-P is thought to increase as a result of inactivation of SPS (discussed below), thereby resulting in at least a transient increase in the concentration of its substrates. Whether the 2-kinase and 2-phosphatase are also regulated by covalent modification has not yet been established; however, the metabolite effects mentioned above can provide an adequate model to explain many aspects of the regulation of leaf Fru-2,6-$P_2$ level.

### *Molecular genetic manipulation of Fru-2,6-$P_2$*

As one approach to study the role of Fru-2,6-$P_2$ during photosynthesis, Kruger and co-workers produced transgenic tobacco plants expressing a modified mammalian gene encoding the 2-kinase. The gene was modified to produce a 2-kinase protein that lacked 2-phosphatase activity (in liver, the 2-kinase and 2-phosphatase activities reside on a single bifunctional protein) and was also not subject to inactivation by protein phosphorylation. Transgenic plants had increased 2-kinase activity and as a result, increased levels of Fru-2,6-$P_2$. Importantly, the flux of carbon into sucrose (measured early in the photoperiod) was restricted providing direct evidence for a role of Fru-2,6-$P_2$ in regulating sucrose synthesis (Scott *et al.*, 1995).

## Regulation of SPS

There are at least three levels at which SPS is regulated in leaves (for reviews see Huber & Huber, 1992, 1996). First, there is control of the steady-state level of SPS enzyme *protein*, which suggests control of SPS gene expression and/or protein turnover (e.g. during leaf development, and in response to changes in irradiance and N-nutrition). Secondly, SPS enzyme *activity* can be modulated by the allosteric activator, Glc-6-P, and inhibitor, $P_i$. Thirdly, enzyme *activity* is regulated by covalent modification, e.g. reversible protein phosphorylation (Huber, Huber & McMichael, 1994). Phosphorylation of SPS reduces enzyme activity by decreasing affinities for the substrate Fru-6-P and the allosteric activator Glc-6-P, and increasing the affinity for the allosteric inhibitor, $P_i$. Thus, regulation of SPS by protein phosphorylation can be 'over-riden' both *in vivo* and *in vitro* by increasing the concentrations of substrates and activator and decreasing the concentration of inhibitor.

### *Control of SPS by protein phosphorylation*

Covalent modification of SPS by reversible seryl phosphorylation probably plays a major role in the regulation of this enzyme in both feedforward and feedback situations. SPS has been shown to be phosphorylated *in vivo* on multiple seryl residues, and it has been recently demonstrated that the phosphorylation of a specific site (Ser 158 of spinach SPS) plays the major role in regulating the activation state of the enzyme. In darkened leaves, the regulatory site is phosphorylated, and SPS is in the less active state. After a dark – light transition, photosynthesis is induced and SPS is activated by dephosphorylation of the regulatory site, which is catalysed by a type 2A protein phosphatase. The increased activity of SPS is one of the major factors responsible for increased capacity for sucrose biosynthesis (feedforward control). In leaves of some species (including spinach), as the photoperiod progresses, the flux of carbon into sucrose slows and there is a commensurate increase in starch accumulation. These 'diurnal changes' are thought to reflect progressive feedback regulation after extended periods of illumination. The reduction in sucrose biosynthesis is paralleled by reduced SPS activity that appears to occur as a result of phosphorylation of the regulatory site. The inactivation of SPS in the light can be accelerated if export from the leaf is reduced, e.g. by phloem girdling or leaf excision.

It is not entirely clear whether it is the accumulation of sucrose in the leaf that causes inactivation of SPS. Apparently, some 'signal metabolite' either activates SPS-kinase and/or inhibits SPS-protein phosphatase but it is likely that sucrose *itself* is not the signal metabolite. However, there is some evidence to suggest that amino acids might play such a role. Amino acids are synthesized in leaves from concurrent nitrate assimilation and are normally exported from the leaf along with sucrose. Thus, a restriction in export will result in elevated levels

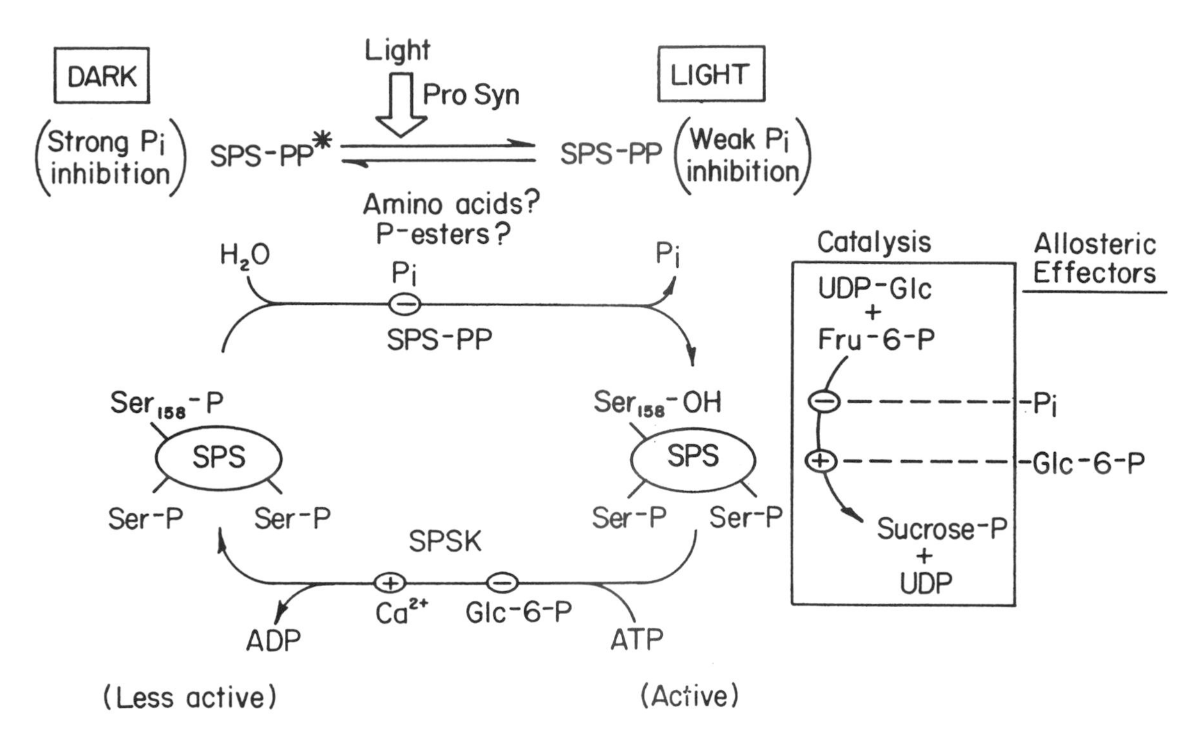


**Figure 12.3.** Schematic representation of the regulation of spinach leaf SPS by reversible seryl phosphorylation in relation to light/dark signals. Effects of metabolites on the interconverting enzymes (SPSk and SPS-PP) and on the catalytic reaction of SPS are shown. Apparent light activation of SPS-PP, including an alteration of regulatory properties, is also shown. (Adapted from Huber & Huber (1996) with permission.)

of sucrose as well as many amino acids, and it is the latter group of compounds that could signal to the cell the status of source–sink interactions. The role of amino acids, if any, remains to be conclusively established.

There appear to be elements of both 'fine' and 'coarse' control involved in the regulation of the phosphorylation status of SPS. Fine control refers to metabolic regulation of the requisite protein kinase(s) and protein phosphatase(s). In spinach leaves, SPS-kinase is inhibited by Glc-6-P and SPS-protein phosphatase is inhibited by Pi, and possibly some P-esters and amino acids (see Fig. 12.3). The regulation by $P_i$ and Glc-6-P may be particularly important for at least two reasons. First, it is likely that cytosolic [$P_i$] will decrease and cytosolic [Glc-6-P] will increase during a dark – light transition; these changes would promote dephosphorylation of SPS as phosphatase activity would increase, while kinase activity would be reduced. Secondly, the postulated changes in metabolite concentrations would have a direct stimulatory effect on SPS catalytic activity, because Glc-6-P is an allosteric activator and $P_i$ is an allosteric inhibitor (see Fig. 12.3). Thus, the metabolite effects would reinforce one another. Another potentially very important effector of the system is $Ca^{2+}$, because SPS-kinase from several species is $Ca^{2+}$-dependent and there are suggestions that cytosolic [$Ca^{2+}$] may change with light – dark transitions.

Coarse control of SPS phosphorylation status refers to an apparent light-activation of SPS-protein phosphatase itself that involves some increase in activity as well as (and perhaps more importantly) a decrease in sensitivity to $P_i$ inhibition. Thus, small changes in cytosolic [$P_i$] will have a much larger effect on SPS activity *in vivo* because of the manifold steps affected. The light-activation of the phosphatase is dependent upon cytoplasmic protein synthesis, because the effect can be blocked by cycloheximide. It is not known which specific protein component is involved and whether the dependence on protein synthesis is direct or indirect.

### *Molecular genetic manipulation of SPS*

The gene encoding SPS was first cloned from maize, and recently has been cloned from several additional species. Thus, deduced sequences are now available for SPS from several species, which will facilitate identification of important regions of the molecule (Salvucci, van de Loo & Klein, 1995). In addition, availability of cDNA clones allows the potential for SPS activity to be altered by genetic manipulation in transgenic plants (Stitt & Sonnewald, 1995). Over-expression of the enzyme has now been achieved, and at least in some cases, the result is increased partitioning of fixed carbon into sucrose. In other cases, SPS protein can be over-expressed but an increase in activity is not realized apparently because of down-regulation by protein phosphorylation. Conversely, reduction of SPS expression can be achieved by antisense inhibition and has clearly been shown to reduce sucrose synthesis. Collectively, these results confirm the important role of SPS in sucrose biosynthesis but also highlight the fact that SPS *per se* is not the only determinant of the rate of carbon flow into sucrose.

## ASSIMILATE PARTITIONING IN LEAVES

In species that synthesize and accumulate starch and sucrose as the principal end products of photosynthesis, the partitioning of photosynthate between starch and sucrose (e.g. as measured with $^{14}CO_2$) generally correlates rather closely with the pattern of carbohydrate accumulation in leaves over the course of the photoperiod (Lunn & Hatch, 1995). It is generally recognized that assimilate partitioning in leaves varies among species; some tend to partition a significant amount of fixed-C into starch (e.g. soybean, tobacco, potato) whereas others partition much more C into sucrose (e.g. small grains). Moreover, within a given species or genotype, assimilate partitioning can be influenced by a variety of factors including time of day, environmental factors (e.g. temperature and photoperiod), and mineral nutrition.

It is not at all clear how differences in assimilate partitioning, or different carbohydrate storage strategies, in leaves are of any advantage. However, it has been postulated that the concentration of non-structural carbohydrates (and/or nitrate) may be involved in regulating whole plant allocation between the shoot and root. For example, leaf [sucrose] has been correlated negatively with shoot growth and correlated positively with root growth (Schulze *et al.*, 1994). Some of the effects on plant growth may be related to the timing of assimilate export from the leaves, since greater partitioning of C into starch will dictate greater export at night compared to the day (provided that the starch reserve is mobilized). This could be important because sucrose exported from leaves at night (derived from starch hydrolysis) may be preferentially used for growth of shoots as opposed to roots. However, recent studies with transgenic plants are highlighting the notion that plant growth can be extremely adaptive with respect to changes in the timing of assimilate export. Clearly, the production and metabolism of carbohydrates in leaves as well as sink tissues are inter-related in the growth process.

## FUTURE OUTLOOK

In future studies of plant carbohydrate metabolism, the impact of molecular biology and genetic manipulation will continue to increase in importance as an experimental approach, along with biochemistry and physiology. There are at least three areas in which we may expect progress. First, assimilate partitioning and growth may be related in certain ways. Molecular genetic approaches may make it possible to identify the causal factors underlying such correlations. A second area concerns the 'sugar cycling' postulated to occur in leaves of some species (see Fig. 12.1) and also certain sink tissues such as the potato tuber. It will be important to determine to what extent such cycling actually occurs *in vivo* and whether it plays any physiological role. Again, molecular genetic approaches might be useful to elucidate the relation between sugar cycling and possible alterations to the efficiency of metabolism. A third area concerns the impact of sink metabolism on source leaf assimilate partitioning. Enhancing the growth and yield of crop plants may require manipulation of carbohydrate metabolism in both sources *and* sinks. Advances in our understanding of potential control points will almost certainly be closely linked to efforts to manipulate them using modern genetic approaches.

REFERENCES

Caspar, T. C., Lin, T -P., Kakefuda, G., Benbow, L., Preiss, J. & Somerville, C. (1991). Mutants of *Arabidopsis* with altered regulation of starch degradation. *Plant Physiology*, **95**, 1181–8.

Flügge, U -I., Weber, A., Fischer, K., Loddenkötter, B. & Wallmeier, H. (1992). Structure and function of the chloroplast triose phosphate-translocator. In *Research in Photosynthesis*, vol. III, ed. N. Murata, pp. 667–74. The Netherlands: Kluwer Academic Publishers.

Heldt, H. W., Flugge, U-I. & Borchert, S. (1991). Diversity of specificity and function of phosphate translocators in various plastids. *Plant Physiology*, **95**, 341–43.

Herold, A. & Lewis, D. H. (1977). Mannose and green plants: occurrence, physiology and metabolism, and use as a tool to study the role of orthophosphate. *New Phytologist*, **79**, 1–40.

Huber, S. C. & Huber, J. L. (1992). Role of sucrose-phosphate synthase in sucrose metabolism in leaves. *Plant Physiology*, **99**, 1275–8.

Huber, S. C. & Huber, J. L. (1996). Sucrose-phosphate synthase. *Annual Review of Plant Physiology and Plant Molecular Biology*, **47**, 431–44.

Huber, S. C., Huber, J. L. & McMichael, R. W., Jr. (1994). Control of plant enzyme activity by reversible protein phosphorylation. *International Review of Cytology*, **149**, 47–98.

Lunn, J. E. & Hatch, M. D. (1995). Primary partitioning and storage of photosynthate in sucrose and starch in leaves of $C_4$ plants. *Planta*, **197**, 385–391.

Martin, C. & Smith, A. M. (1995). Starch biosynthesis. *Plant Cell*, 7, 971–85.

Preiss, J., Ballicora, M. A., Laughlin, M. J., Fu, Y., Okita, T. W., Barry, G. F., Guan, H. & Sivak, M. N. (1995). Studies on the starch biosynthetic enzymes for manipulation of starch content and quality. In *Carbon Partitioning and Source–Sink Interactions in Plants*, vol. 13, ed. M. A. Madore & W. J. Lucas, pp. 91–101. Rockville, MD: American Society of Plant Physiologists.

Riesmeier, J. W., Flugge, U-I., Schulz, B., Heineke, D., Heldt, H-W., Willmitzer, L. & Frommer, W. B. (1993). Antisense repression of the chloroplast triose phosphate translocator affects carbon partitioning in transgenic potato plants. *Proceedings of the National Academy of Sciences, USA*, **90**, 6160–4.

Salvucci, M. E., van de Loo, F. J. & Klein, R. R. (1995). The structure of sucrose phosphate synthase. In *Sucrose Metabolism, Biochemistry, Physiology and Molecular Biology*, ed. H. G. Pontis, G. L. Salerno & E. Echeverria. Rockville, MD: American Society of Plant Physiologists.

Schulze, W., Schulze, E-D., Stadler, J., Heilmeier, H., Stitt, M. & Mooney, H. A. (1994). Growth and reproduction of *Arabidopsis thaliana* in relation to storage of starch and nitrate in the wild-type and in starch-deficient and nitrate-uptake-deficient mutants. *Plant, Cell and Environment*, **17**, 795–809.

Scott, P., Lange, A. J., Pilkis, S. J. & Kruger, N. J. (1995). Carbon metabolism in leaves of transgenic tobacco (*Nicotiana tabacum* L.) containing elevated fructose 2,6-bisphosphate levels. *The Plant Journal*, 7, 461–9.

Stark, D. M., Timmerman, K. P., Barry, G. F., Preiss, J. & Kishore, G. M. (1992). Regulation of the amount of starch in plant tissues by ADP glucose pyrophosphorylase. *Science*, **258**, 287–92.

Stitt, M., Huber, S. C. & Kerr, P. S. (1987). Control of photosynthetic sucrose formation. In *The Biochemistry of Plants*, vol. 8, ed. M. D. Hatch & N. K. Boardman, pp. 327–409. New York: Academic Press.

Stitt, M. & Quick, W. P. (1989). Photosynthetic carbon partitioning: its regulation and possibilities for manipulation. *Physiologia Plantarum*, **77**, 633–41.

Stitt, M. & Sonnewald, U. (1995). Regulation of metabolism in transgenic plants. *Annual Review of Plant Physiology and Plant Molecular Biology*, **46**, 341–68.

Trethewey, R. N. & ap Rees, T. (1994). A mutant of *Arabidopsis thaliana* lacking the ability to transport glucose across the chloroplast envelope. *Biochemical Journal*, **301**, 449–54.

# 13 Photorespiration and the $C_2$ cycle

D. J. OLIVER

## INTRODUCTION: EARLY OBSERVATIONS AND EXPLANATIONS

Photosynthesis is the process by which green plants use solar energy to produce carbohydrates from $CO_2$. Some of the carbohydrates are expected to be oxidized through dark respiration to yield $CO_2$ and energy, while much of the carbon is converted into biomass. The first data suggesting that this scheme was incomplete came with the observation that respiratory rates immediately after the end of illumination were substantially higher than the rates exhibited a few minutes later by the same plant. This high rate of respiration, often called the post-illumination burst, suggested that there was a very rapid rate of respiration occurring in the light. When the lights were extinguished, photosynthesis stopped immediately while this unique form of respiration, present only in the light, was carried over into the first moments of darkness. This unusual respiration, $O_2$ uptake and $CO_2$ release, that occurs only in the light has come to be known as photorespiration (Fig. 13.1).

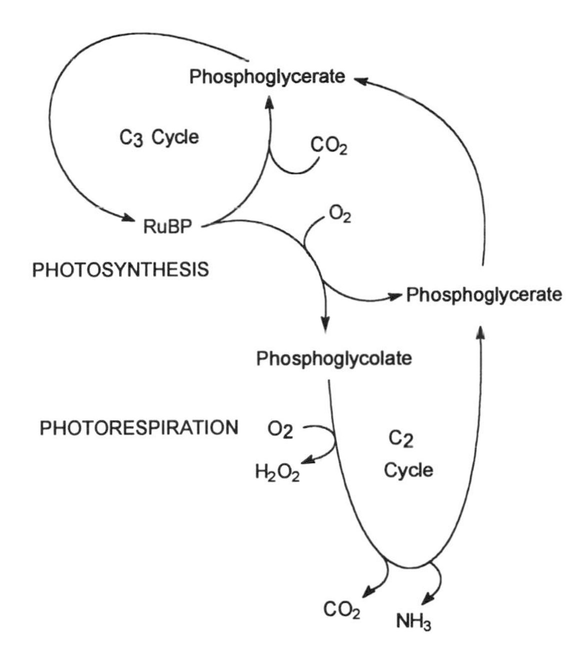


**Figure 13.1** The interactions of the $C_2$ and the $C_3$ cycles. The photorespiratory $C_2$ cycle is an adjunct of the photosynthetic $C_3$ cycle that has the function of returning 75% of the carbon that is directed into 2-phosphoglycolate back to the $C_3$ cycle.

Photorespiration is inherently difficult to measure, although its rate is substantially higher than that of dark respiration. Net photosynthesis, measured as either $O_2$ release or $CO_2$ uptake, is the difference between gross photosynthesis and the sum of the rates of two respiratory processes: photorespiration and dark respiration.

$$\text{Net photosynthesis} = \text{gross photosynthesis} - (\text{photorespiration} + \text{dark respiration})$$

Since gross photosynthesis is larger than photorespiration, net photosynthesis is positive and tends to hide the rate of photorespiration.

In addition to measuring the post-illumination burst, several other methods have been developed to estimate photorespiration. Plants have been labelled with $^{14}CO_2$ for a long enough period to bring their soluble carbohydrates to a steady-state specific radioactivity. The plants are then placed in a stream of $CO_2$-free air and the rates of $^{14}CO_2$ released by respiration measured in the light and the dark. The rates of $^{14}CO_2$ release are five to seven times greater in the light than in the dark. These experiments confirmed the conclusions drawn earlier that a higher rate of respiration in the light was responsible for the post-illumination burst. A semi-quantitative measure of the relative amount of photorespiration is the $CO_2$ compensation point. Plants illuminated in a closed environment will deplete the $CO_2$ down to a steady-state level where the rate of photosynthesis is equal to the rate of respiration (primarily photorespiration). This $CO_2$ compensation point is about 60 ppm for $C_3$ plants, about 5 ppm for $C_4$ plants, and 10 to 30 ppm $CO_2$ for $C_3/C_4$ intermediates. The $CO_2$ compensation point is inversely related to the rate of photorespiration exhibited by the plant.

A more sophisticated technique involves measuring the steady-state level of net photosynthesis with an infrared gas analyser. A brief pulse of $^{14}CO_2$ is then added and the tissue rapidly killed. If the $^{14}CO_2$ pulse is short enough, the $^{14}C$ will not have equilibrated with the photosynthate pool and very little will have been respired. The rate of $^{14}CO_2$ fixation therefore, will approximate gross photosynthesis. Photorespiration can then be estimated as the difference between the rate of net photosynthesis (measured by gas exchange) and gross photosynthesis (measured by $^{14}CO_2$ fixation).

All measurements of photorespiration will, by their very nature, underestimate the actual rate of the reaction. This under-estimation occurs because it is very difficult to measure the portion of respiratory $CO_2$ that is refixed by photosynthesis before it can escape the leaf and be measured. Given these limitations, in $C_3$ plants under standard atmospheric conditions (400 ppm $CO_2$ and 21% $O_2$) the rate of photorespiration is estimated at one-fourth to one-third of the photosynthetic rate. Under conditions where the intracellular concentration of $CO_2$ drops or the concentration of $O_2$ rises, as would happen when stomates close in the middle of the day, photorespiration rates can approach those of photosynthesis.

## THE BIOCHEMISTRY OF PHOTORESPIRATION

### The oxygenase activity of rubisco – directing carbon to photorespiration

Carbon is committed to photorespiration by the oxygenase reaction of rubisco. In the oxygenation reaction $O_2$ reacts with the C-2 carbon of the enediol form of RuBP to form the transient intermediate proposed in Fig. 13.2. After rearrangement and the breaking of the bond between carbons 2 and 3, the final products of the reaction are 2-phosphoglycolate and 3-phosphoglycerate. The enzyme produces a mono-oxygenase reaction where one atom of the oxygen molecule is incorporated into the carboxyl carbon of 2-phosphoglycolate and the second atom appears in water.

The fixation of $O_2$ and $CO_2$ by rubisco are mutually competitive. After the activation of rubisco by carbamylation of an active site lysine residue and the binding of $Mg^{+2}$, the activated enzyme binds the first substrate of the reaction, RuBP. Neither of the gaseous substrates, $O_2$ and $CO_2$, form a Michaelis complex with the activated enzyme, but directly attack carbon 2 of the RuBP molecule. If the first gas to arrive with sufficient energy to react is $CO_2$, the carboxylation reaction occurs. If $O_2$ arrives first, the oxygenation reaction results. Kinetically, $O_2$ is a competitive inhibitor of the carboxylation reaction and $CO_2$ is a competitive inhibitor of the oxygenase. Those metabolic mechanisms that control the rate of rubisco activity (e.g. carboxyarabinitol 1-phosphate and rubisco activase) do not alter the ratio of carboxylase to oxygenase, but do control the two reactions in parallel.

The ratio of carboxylase to oxygenase activity is defined by the specificity factor ($\tau$).

$$\tau = V_c K_o / V_o K_c$$

where, $V_c = V_{max}$ carboxylase, $K_o = K_m$ oxygenase, $V_o = V_{max}$ oxygenase, $K_c = K_m$ carboxylase
Remember that the $K_m$ is inversely proportional to the affinity of the enzyme for the substrate, either $CO_2$ or $O_2$ (the higher the $K_m$ the lower the affinity). Typical values for $K_c$ are 10 to 20 μM and for $K_o$ are 400 to 600 μM. Under standard conditions of water in equilibrium with air, the concentration of $CO_2$ is about 12 μM and the $O_2$ concentration is approximately 250 μM. The actual ratio of velocity for the carboxylase to that for the oxygenase reaction is defined by the specificity factor for the particular rubisco and the concentration of $O_2$ and $CO_2$.

$$V_c/V_o = \tau([CO_2]/[O_2])$$

Rubisco molecules from different organisms have different specificity factors. $C_3$ plants have evolved rubisco

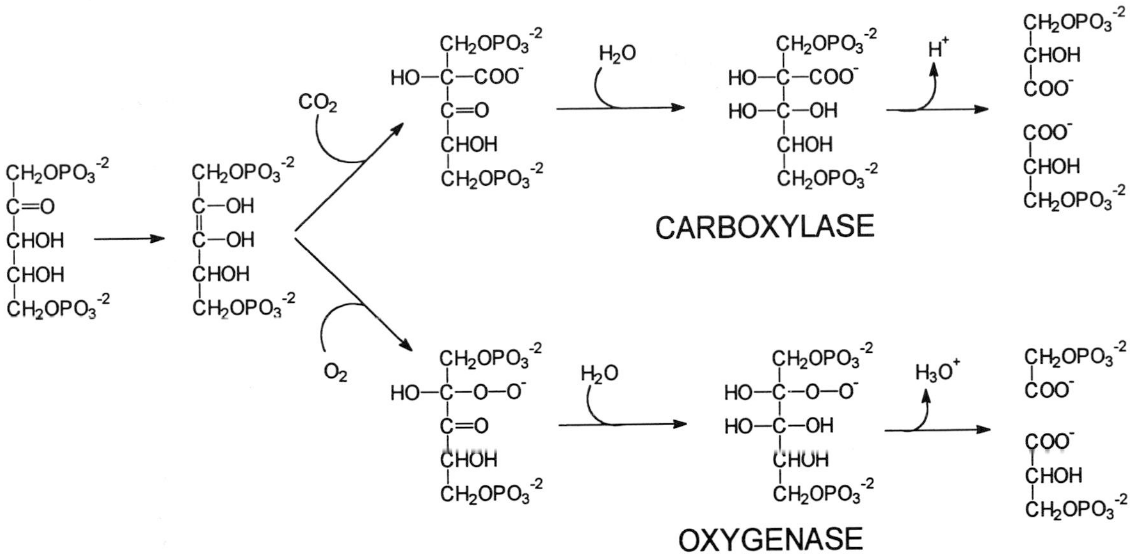

**Figure 13.2.** The carboxylase and oxygenase reactions of rubisco. $O_2$ and $CO_2$ compete to react with the enediol form of RuBP on the enzyme. Use of a common substrate results in $CO_2$ being a competitive inhibitor of the oxygenase activity and $O_2$ acting as a competitive inhibitor for the carboxylase.

that discriminates most strongly against $O_2$ ($\tau = 80$). The $V_c/V_o$ for $C_3$ plants, therefore, is about 4 under physiological conditions. $C_4$ plants have evolved mechanisms for concentrating $CO_2$ at the site of rubisco and therefore, possess an enzyme that has not experienced as much selection to exclude $O_2$ ($\tau = 50$). Anaerobic bacteria grow in the absence of $O_2$ and possess a form of rubisco that discriminates little against $O_2$ ($\tau = 10$). Rubisco from these organisms has a $V_c/V_o$ value of about 0.5 in air-equilibrated water. Substantial effort has been directed towards using biochemical and genetic techniques to generate a rubisco that discriminates more strongly against $O_2$ (increased $\tau$) than the values exhibited by the $C_3$ enzyme. To date, this has largely been unsuccessful and most changes have decreased the $\tau$ value.

This competition that is based on substrate access to the enzyme is also exhibited with whole plants. High $O_2$ concentrations inhibit photosynthesis and stimulate photorespiration. Conversely, high $CO_2$ levels increase photosynthesis and decrease photorespiration. Net rates of $CO_2$ fixation in plants are decreased by about 30% when the atmospheric $O_2$ concentration is increased from 5% to 21%. This inhibition was first reported by the German biochemist, Otto Warburg, and is referred to as the Warburg Effect.

Some plants have evolved mechanisms to repress photorespiration. In addition to the changes in $\tau$ noted above, photorespiration in plants is inhibited by concentrating $CO_2$ at the active site of rubisco such that the carboxylase reaction out-competes the oxygenase reaction. $C_4$ plants use phospho*enol*pyruvate carboxylase as the primary carboxylase. This enzyme uses $HCO_3^-$ as the substrate and is not inhibited by $O_2$. Rubisco is limited to the bundle sheath cells where decarboxylation of the four carbon acids causes elevated $CO_2$ levels, thus decreasing oxygenation. Algae and cyanobacteria contain one or more pumps for dissolved inorganic carbon (DIC) that also prevent photorespiration by concentrating $CO_2$ in the same compartment as rubisco. These DIC pumps are induced by light and low $CO_2$ concentrations. Organisms grown at 5% $CO_2$ lack the DIC pumps and exhibit photorespiration when first transferred to air levels of $CO_2$. As exposure to low $CO_2$ concentrations continues, the pump is induced and photorespiration drops below

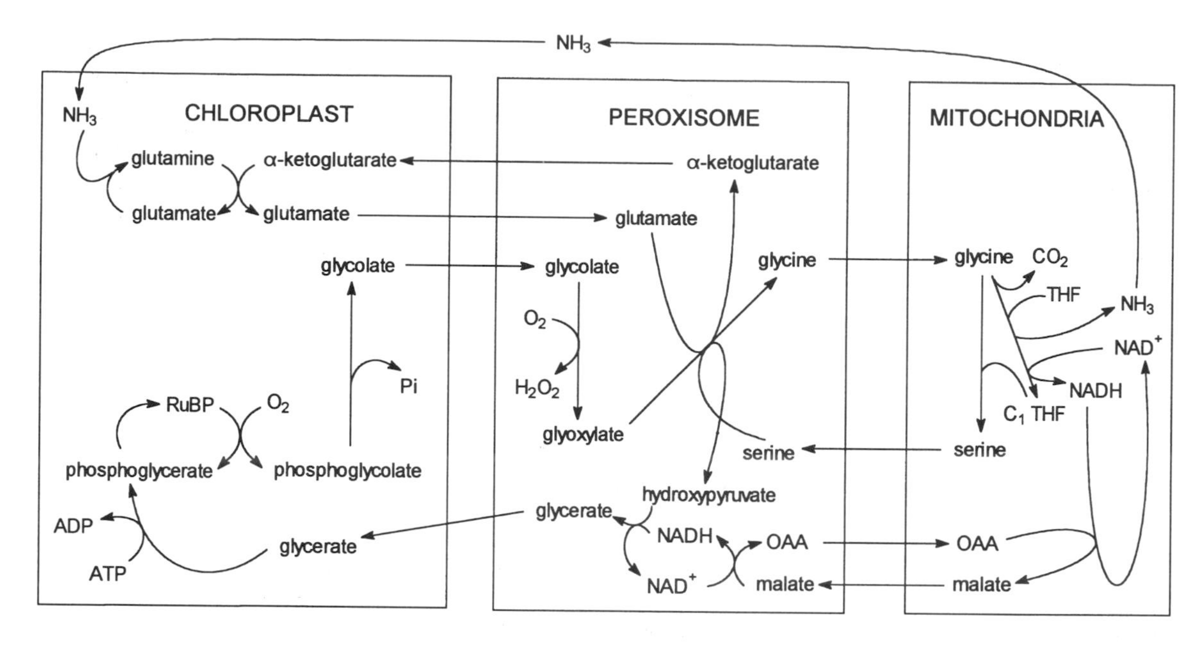


**Figure 13.3.** The $C_2$ cycle showing the distribution of reactions within the three organelles.

detectable levels. Plants that avoid photorespiration have done so more by concentrating $CO_2$ at the site of rubisco than by eliminating the oxygenase reaction.

## The $C_2$ cycle – metabolism of 2-phosphoglycolate and photorespiratory $CO_2$ release

In $C_3$ plants grown at air levels of $CO_2$ and $O_2$ the rate of 2-phosphoglycolate synthesis is substantial. The amount of carbon converted to 2-phosphoglycolate is almost equal to the amount of carbon fixed by photosynthesis. It is essential therefore, that in order to maintain rates of net $CO_2$ fixation that will allow the plant to survive, most of the carbon in 2-phosphoglycolate must be returned to the $C_3$ cycle. The photorespiratory $C_2$ cycle (sometimes called the photosynthetic carbon oxidation or PCO cycle) returns 75% of the carbon in 2-phosphoglycolate back to the photosynthetic $C_3$ cycle, thus allowing its continued function.

The $C_2$ cycle involves enzymes located in three different organelles, the chloroplast, the peroxisome, and the mitochondria (Fig. 13.3). The production of 2-phosphoglycolate by rubisco occurs in the chloroplast. A specific phosphatase, 2-phosphoglycolate phosphatase, converts 2-phosphoglycolate to glycolate. The glycolate is transported out of the chloroplast and enters the peroxisome. Within this organelle, the glycolate is oxidized to glyoxylate by the enzyme glycolate oxidase. This flavoprotein (the co-factor is flavin mononucleotide (FMN)) uses molecular oxygen as the terminal electron acceptor reducing it to hydrogen peroxide. This is the major site of hydrogen peroxide production within leaves in the light, and peroxisomal catalase is needed to prevent accumulation of $H_2O_2$ to toxic levels. Glyoxylate can be oxidized by $H_2O_2$ to produce $CO_2$ and formate. This has been shown to occur when glycine metabolism is blocked by inhibitors or mutations. Under these conditions, all of the available nitrogen in the system accumulates in glycine. As a result, glyoxylate accumulates and decarboxylates by direct reaction with $H_2O_2$. It is generally accepted that this reaction does not occur at substantial rates under normal physiological conditions.

The α-keto acid, glyoxylate, is transaminated to the α-amino acid, glycine. This reaction is catalysed by two transaminases, serine : glyoxylate transaminase and glutamate : glyoxylate transaminase. Serine and glutamate each contribute about 50% of the amino groups

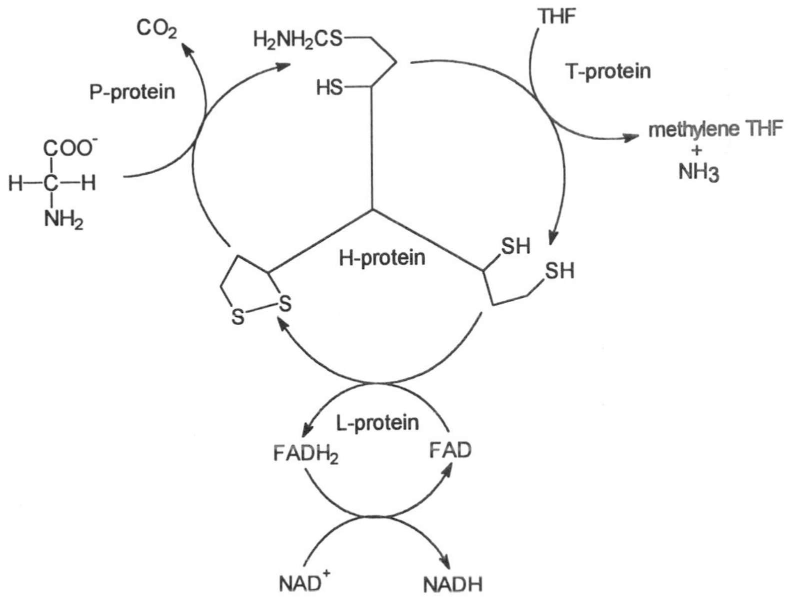

**Figure 13.4.** The reaction mechanism of the glycine decarboxylase multienzyme complex.

donated. Both transaminases, however, are sufficiently promiscuous that several other amino acids can also act as amino donors. Glycine then leaves the peroxisome and is imported into the mitochondria.

Within the mitochondria two molecules of glycine are converted to one each of serine, $NH_3$, and $CO_2$. This reaction is catalysed by the sequential reaction of two enzymes, the glycine decarboxylase multienzyme complex, and serine hydroxymethyltransferase. Mitochondria in the leaves of $C_3$ plants have largely been converted to photosynthetic organelles. Nearly half of the soluble protein of leaf mitochondria are involved in photorespiratory glycine metabolism.

Glycine decarboxylase is an extremely large enzyme complex composed of multiple copies of four different component proteins. Glycine decarboxylase catalyses the oxidative decarboxylation of glycine where the $\alpha$-carboxyl group is released as $CO_2$, the $\alpha$-amino group is released as $NH_3$, and the methylene carbon is transferred to tetrahydrofolate to make $N^5$, $N^{10}$-methylene tetrahydrofolate. The electrons removed from glycine are given to $NAD^+$ to reduce it to NADH. The reaction mechanism for glycine decarboxylase is shown in Fig. 13.4. The reaction begins with the pyridoxal 5-phosphate co-factor of the P-protein of the glycine decarboxylase complex forming a Schiff base with the first molecule of glycine. The oxidation of glycine occurs as electrons are passed to the lipoamide co-factor of the H-protein of the complex. The glycine is decarboxylated (this is the site of photorespiratory $CO_2$ release) and the remaining carbon and nitrogen of glycine are passed to the lipoamide. The lipoamide carries the reaction intermediates from the reactive site of the P-protein to the active site of the T-protein. Here the remaining carbon from glycine is transferred from the lipoamide on H-protein to tetrahydrofolate to make $N^5$, $N^{10}$-methylene tetrahydrofolate. The amino group of glycine is lost as ammonia. The lipoamide now leaves the T-protein and moves to the active site of the L-protein where it is oxidized and thereby able to resume its catalytic role. Within the L-protein (dihydrolipoamide dehydrogenase) the electrons are passed through flavin adenine dinucleotide (FAD) and eventually used to reduce $NAD^+$ to NADH + $H^+$.

The product of the glycine decarboxylase reaction, $N^5$, $N^{10}$-methylene tetrahydrofolate, donates its one carbon fragment to a second molecule of glycine to form serine. This reaction is catalysed by serine hydroxymethyltransferase. Working together, the mitochondrial enzymes glycine decarboxylase and serine hydroxymethyltransferase convert two carbon intermediates that have relatively limited metabolic flexibility into three carbon metabolites that can re-enter central metabolism in the cell.

Serine now leaves the mitochondria and enters the peroxisome. Here, it acts as an amino donor for the transamination of glyoxylate. The product of this reaction is hydroxypyruvate. Hydroxypyruvate is reduced with NADH to form glycerate by the enzyme $NAD^+$-hydroxypyruvate reductase. The glycerate now leaves the peroxisome and enters the chloroplast. The three-carbon organic acid, glycerate, is phosphorylated by glycerate kinase and the 3-phosphoglycerate re-enters the $C_3$ carbon reduction cycle.

For every four carbons that leave the chloroplast in the form of two molecules of glycolate, three carbon atoms re-enter photosynthesis as 3-phosphoglycerate and the fourth is lost as $CO_2$. NADH is produced by the oxidation of glycine within the mitochondria at rates that are substantially higher than the predicted rates of mitochondrial electron transport and oxidative phosphorylation. Much of the reducing equivalents generated by photorespiratory glycine oxidation are shuttled out of the mitochondria and consumed by the reduction of hydroxypyruvate in the peroxisome.

## Photorespiratory nitrogen metabolism

The rate of photorespiratory $NH_3$ release in mitochondria is approximately ten times greater than the rate of primary nitrogen fixation. Unless the plant nitrogen supply is to be rapidly depleted, this ammonia must be recycled with near perfect efficiency. Photorespiratory $NH_3$ is refixed by the chloroplastic glutamine synthetase/glutamate synthase system. Glutamine synthetase fixes the ammonia by reacting it with glutamate to form glutamine in an ATP-dependent reaction. The glutamine acts as an amino donor for the transamination of $\alpha$-ketoglutarate to glutamate in a reaction catalysed by glutamate synthase (alternatively referred to as ferredoxin-dependent glutamate: $\alpha$-ketoglutarate amino transferase or GOGAT). The $\alpha$-ketoglutarate is produced when glutamate acts as an amino donor during the transamination of glyoxylate in the peroxisome. Glutamate synthase uses reduced ferredoxin from the photosynthetic electron transport chain. Because $NH_3$ is produced in the mitochondria and refixed in the chloroplast there is a diffusion path that must be transversed during the reaction. This provides the opportunity for loss of volatile ammonia gas and some loss has been documented during this reaction.

## Substrate flux between organelles

The photorespiratory reactions require rapid rates of substrate transport between three different organelles. This rapid rate of substrate flux is supported by a group of transport proteins. These include two chloroplastic glycolate transporters, a glycolate/glycerate exchanger and a glycolate/$OH^-$ exchanger. There are also two dicarboxylate exchangers in the chloroplast membrane, a glutamate/malate and a $\alpha$-ketoglutarate/malate exchanger, that together exchange cytosolic $\alpha$-ketoglutarate for chloroplastic glutamate. The mitochondrial membrane appears to contain a glycine/serine and a glycine/OH– exchanger, although they have proven difficult to study. Glycolate/glycerate exchange in the chloroplast and glycine/serine exchange within the mitochondria both show 2 : 1 stoichiometries (2 glycolate : 1 glycerate and 2 glycine : 1 serine) thus necessitating dual transporters. Mitochondria also contain a very active oxaloacetate/malate exchanger in the mitochondrial membrane. The oxaloacetate/malate exchanger combined with $NAD^+$-dependent malate dehydrogenase in the mitochondrial matrix and the cytosol support a redox shuttle that moves NADH equivalents across the mitochondrial membrane. Interestingly, a similar transport system occurs in the chloroplast, thus allowing reducing equivalent exchange between these three compartments. Substrate entry and egress from the peroxisome probably does not require a specific transporter but the specificities of the reactions are maintained by porins in the single peroxisomal membrane and substrate channeling between enzyme active sites.

## Energetics of photorespiration

Every $CO_2$ fixed by the $C_3$ cycle uses three molecules of ATP and two molecules of NADPH (reducing equivalents). Since one reducing equivalent can make 3 ATP molecules when oxidized in the mitochondria, the total energy bill for $CO_2$ fixation is equal to 9 ATP (Table 13.1). The fixation of $O_2$ produces 1.5 PGA, $\frac{1}{2}$ $CO_2$, and $\frac{1}{2}$ $NH_3$ (Fig. 13.3). Conversion of the 1.5 PGA back to RuBP takes 2.25 ATP and 1.5 NADPH (equivalent to 4.5 ATP). The $C_2$ cycle itself requires 0.5

Table 13.1. *Energetics of photorespiration*

| Reaction | $C_3$ cycle 1 $CO_2$ | $C_2$ cycle 1 $O_2$ |
|---|---|---|
| $RuBP + CO_2 \rightarrow 2$ PGA | | |
| 2 PGA + 2 ATP → 2 1,3-bisPGA + 2 ADP | −2 ATP | −1.5 ATP |
| 2 1,3-bisPGA + 2 NADPH → 2 glyceraldehyde 3-P + 2 $NADP^+$ | −6 ATP | −4.5 ATP |
| 2 glyceraldehyde 3-P → Ribose 5-P + 1C | | |
| Ribose 5-P + ATP → RuBP + ADP | −1 ATP | −0.75 ATP |
| $RuBP + O_2 \rightarrow$ PGA + P-glycolate | | |
| P-glycolate → glycine | | |
| glycine + $\frac{1}{2}$ $NAD^+$ → $\frac{1}{2}$ serine + $\frac{1}{2}$ $CO_2$ + $\frac{1}{2}$ $NH_3$ + $\frac{1}{2}$ NADH | | +1.5 ATP |
| $\frac{1}{2}$ serine → $\frac{1}{2}$ hydroxypyruvate | | |
| $\frac{1}{2}$ hydroxypyruvate + $\frac{1}{2}$ NADH → $\frac{1}{2}$ glycerate + $\frac{1}{2}$ $NAD^+$ | | −1.5 ATP |
| $\frac{1}{2}$ glycerate + $\frac{1}{2}$ ATP → $\frac{1}{2}$ PGA + $\frac{1}{2}$ ADP | | −0.5 ATP |
| $\frac{1}{2}$ $NH_3$ + $\frac{1}{2}$ glutamate + $\frac{1}{2}$ ATP → $\frac{1}{2}$ glutamine + $\frac{1}{2}$ ADP + $\frac{1}{2}$ Pi | | −0.5 ATP |
| $\frac{1}{2}$ glutamine + $\frac{1}{2}$ α-ketoglutarate + $\frac{1}{2}$ $Fd_{(red)}$ → glutamate + $\frac{1}{2}$ $Fd_{(ox)}$ | | −1.5 ATP |
| refixation of $\frac{1}{2}$ $CO_2$ | | −4.5 ATP |
| Total | −9 ATP | −13.75 ATP |

ATP per $O_2$ fixed (the $\frac{1}{2}$ NADH generated from glycine oxidation formally balances the $\frac{1}{2}$ NADH used for hydroxypyruvate reduction). The rest of the energy, 0.5 ATP and 0.5 reducing equivalents (as reduced ferredoxin), is needed to recycle the $\frac{1}{2}$ $NH_3$ released. The fixation of one $O_2$ also results in the release of 0.5 $CO_2$. An additional 4.5 ATP are needed to refix this 0.5 $CO_2$. The total amount of energy needed for fixing one molecule of $O_2$ is 13.75 ATP. The fixation of $O_2$ requires half again more energy than the fixation of $CO_2$. In addition, the energy spent on fixing $O_2$ does not result in any positive gain to the plant where the fixation of $CO_2$ results in the contribution of one atom of reduced carbon to the total plant biomass.

## THE CONSEQUENCES OF PHOTORESPIRATION

Photorespiration is an oxidative pathway that results in the destruction of newly synthesized carbohydrates. Every time a molecule of RuBP serves as the substrate for the oxygenase activity of rubisco, one 2-phosphoglycolate is produced. For each 2-phosphoglycolate produced, 0.5 $CO_2$ is released. The oxygenase reaction together with the $C_2$ pathway result in the loss of net carbohydrate accumulation in two ways. First, the destruction of newly fixed carbohydrate by photorespiration results in the direct loss of a portion of the $CO_2$ fixed by photosynthesis. This $CO_2$ entered the $C_2$ cycle when 2-phosphoglycolate was formed from RuBP and was lost from the cycle during the glycine decarboxylase reaction. Secondly, the competition between $O_2$ and $CO_2$ for binding to rubisco means that less $CO_2$ is fixed (decreased gross photosynthesis) because during some reaction cycles $O_2$ is fixed. In these two ways, photorespiration decreases the rate of net $CO_2$ fixation. This means that there is less surplus carbohydrate available after the respiratory needs of the plant have been satisfied to be directed to biomass accumulation. Under conditions where plant growth is not limited by environmental factors (water, nutrients, pests, etc), decreased photosynthesis caused by photorespiration can result in decreased crop yields. $C_4$ crop plants that have elevated photosynthetic rates due to decreased photorespiratory $CO_2$ loss, lessened competition by $O_2$ at the site of rubisco, and diminished limitation that low $CO_2$ levels places on the rate of rubisco turnover, have higher yields on the average than $C_3$ crop plants.

## INHIBITORS AND MUTATIONS

The conclusion that photorespiration decreases net photosynthesis and under some circumstances limits crop yields has led to interest in inhibitors and mutants in the $C_2$ cycle. A number of inhibitors have been identified that block the $C_2$ cycle. These include the glycolate oxidase inhibitors, α-hydroxy-2-pyridinemethane sulfonate and 2-hydroxy-3-butynoate, and the glycine decarboxylase inhibitors, aminooxyacetate, aminoacetonitrile, and isonicotinic acid hydrazide. Long-term exposure of plants to these chemicals causes an inhibition of both photorespiration and photosynthesis and eventually is lethal. It was not possible to resolve whether the inhibition of photosynthesis resulted from the blocking of the $C_2$ cycle, thus preventing the flow of carbon back into photosynthesis, or from a lack of specificity by the inhibitors. This problem was resolved more directly by screening for plants that were mutant in enzymes of the $C_2$ cycle. The mutants were selected for their ability to grow normally under conditions where photorespiration was inhibited (1500 ppm $CO_2$) but to turn chlorotic under conditions where photorespiration was active (350 ppm $CO_2$). This ground-breaking research identified some of the first well-characterized biochemical mutants in *Arabidopsis*. Biochemical characterization of these plants, as well as additional mutant in barley, tobacco, and peas, have identified mutations in 2-phosphoglycolate phosphatase, catalase, serine : glyoxylate aminotransferase, glycine decarboxylase, serine hydroxymethyltransferase, hydroxypyruvate reductase, glutamine synthetase, glutamate synthase, and the dicarboxylate transporter from the chloroplast envelope. Interestingly, glycolate oxidase and glutamate : glyoxylate aminotransferase mutants have not been selected. In addition to proving the reactions of the $C_2$ cycle and establishing the usefulness of *Arabidopsis* as a genetic system in plants, these mutants have been very important in establishing the interactions between the $C_2$ and the $C_3$ cycle.

The phenotype of these plants is that they have normal growth rates and photosynthetic rates with high levels of $CO_2$. When the $CO_2$ concentration drops and photorespiration becomes important, the photosynthetic rate drops precipitously. These results show that the $C_2$ cycle is essential for returning the carbon diverted into 2-phosphoglycolate by the oxygenase reaction of rubisco back into the $C_3$ cycle. When the flow of carbon through the $C_2$ cycle is disrupted, carbon (and possibly nitrogen and phosphate) accumulates in the intermediates of the pathway, the concentration of $C_3$ cycle intermediates drops, and photosynthesis is inhibited. The observations with inhibitors and mutations in the $C_2$ cycle conclusively showed that it would not be possible to inhibit photorespiration and divert the carbon lost back to net photosynthesis and yield by blocking reaction within the cycle. If such progress is to be made, it will require that the enzyme that dedicates carbon into the photorespiratory pathway, rubisco, be altered such that the ratio of oxygenase to carboxylase is decreased under physiological conditions.

The second important observation from these mutants is that photorespiration and the reactions of the $C_2$ cycle, are not essential for normal growth and development of plants. Given the caveat that the mutants selected may not be eliminating 100% of the affected enzyme, it is obvious that none of these enzymes is essential for non-photorespiratory reactions. This is surprising, particularly in the case of glycine decarboxylase. Mutations in this enzyme in humans results in the lethal genetic disease, hyperglycinaemia.

## FUNCTION OF PHOTORESPIRATION

If photorespiration is not essential, and this does seem to be true, given that plants with mutations in the $C_2$ cycle grow and develop normally as long as the deflection of carbon into the pathway is restricted, then why does it exist? Over the years, two major arguments have surfaced. Photorespiration might function to prevent photodestruction of chloroplasts under conditions where intracellular $CO_2$ levels are low but light levels are high (closed stomates during the middle of the day). Under these conditions, photorespiration keeps the $CO_2$ level within the leaf at the $CO_2$ compensation point and allows for continued turnover of the photosynthetic apparatus. Restricting photorespiration under these conditions in the laboratory by lowering the concentration of $O_2$ does result in increased oxidative stress.

An alternative view that has been widely received was originally postulated by George Lorimer. He suggested that the oxygenase activity of rubisco is an inevitable consequence of the chemistry of the carboxylase reaction. Since the enzyme does not bind either gaseous substrate, but rather they react directly with an activated form of RuBP, there are no enzymatic mechanisms available to prevent $O_2$ from having access to the substrate. Support for this idea comes from several sources. The rubisco from photosynthetic obligate anaerobic bacteria like

*Rhodospirillum rubrum*, still has oxygenase activity. This oxygenase activity can not have a physiological function because exposure to oxygen kills the bacteria. On an evolutionary scale, rubisco predates the presence of high concentrations of $O_2$ in the atmosphere. Following the development of oxidative conditions, although rubisco has evolved to have an increased specificity factor, the oxygenase activity has never been eliminated. Meanwhile, other mechanisms, $C_4$ photosynthesis in higher plants and the dissolved inorganic carbon pump of algae, have evolved independently several times to repress photorespiration. Both observations support the idea that the oxygenase activity is a consequence of the carboxylase reaction mechanism and does not have another physiological function in plants.

## CONTROL OF THE $C_2$ CYCLE ENZYMES

### Metabolic control

If we accept the idea that the oxygenase activity of Rubisco is an inevitable consequence of the carboxylase reaction, and that the $C_2$ cycle exists as a means of recycling most of the carbon in 2-phosphoglycolate back into $C_3$ cycle intermediates, we would not expect to see much metabolic regulation in the pathway. In fact, little control occurs within the pathway. Some of the intermediates of the $C_2$ cycle, however, might work to regulate overall photosynthetic rates. 2-phosphoglycolate inhibits triose-phosphate isomerase; glyoxylate can inhibit a number of enzymes including rubisco; glycerate inhibits sedoheptulose bisphosphatase and fructose bisphosphatase. The consequence of all three potential inhibitors is that they will decrease the rate of photosynthesis and photorespiration. If any of these $C_2$ intermediates accumulate, they will not alter the ratio of carbon commitment to the $C_3$ versus the $C_2$ cycle, but will decrease carbon flow in parallel through both pathways. As a consequence, an inability of the $C_2$ cycle to accommodate the amount of carbon directed into this pathway will result in a partial inhibition of photosynthesis and thus prevent accumulation of the $C_2$ intermediates above acceptable levels.

Glycine decarboxylase is inhibited by the two products of mitochondrial glycine oxidation, serine and NADH. Serine ($K_i$ = 4 mM) is a competitive inhibitor of glycine ($K_m$ = 6 mM) binding to the P-protein of the complex, and NADH ($K_i$ = 15 μM) competes with $NAD^+$ ($K_m$ = 75 μM) for access to the L-protein. However, since estimates of the rate of glycine metabolism based on *in vitro* enzyme activities and rates with isolated mitochondria suggest that there is not much excess capacity above what is needed for the rates of photorespiratory $CO_2$ released, it seems unlikely that this regulation is important in intact leaves. Rates of glycine oxidation in isolated mitochondria are controlled by the rate of electron transport and oxidative phosphorylation which are largely limited by ADP availability. Since much of the NADH produced by glycine decarboxylase is exported from the mitochondria, regulation via coupling is probably not very important.

### Gene control

Photorespiration is a subset of the reactions of photosynthesis. It is not, therefore, surprising that the expression of the proteins of the $C_2$ cycle is controlled in much the same way as the expression of the proteins for the $C_3$ cycle. The mRNA for glycolate oxidase, serine : glyoxylate aminotransferase, hydroxypyruvate reductase, three of the four component proteins for glycine decarboxylase, and serine hydroxymethyltransferase are all found at low levels in etiolated tissues and the level of each increases during greening. In those cases where it has been examined, control is at the transcriptional level. The primary photoreceptor is probably phytochrome. In addition to the phytochrome-based response, a secondary signal is needed to indicate that the leaves contain mature chloroplasts before full transcriptional activation occurs. Thus, the development of the enzymes of the $C_2$ cycle is co-ordinated with the expression of photosynthetic competence in $C_3$ plants.

#### FURTHER READING

Calvin, D. T. (1990). Photorespiration and $CO_2$-concentrating mechanisms. In *Plant Physiology, Biochemistry and Molecular Biology*, ed. D. T. Dennis & D. H. Turpin, pp. 253–73. Essex: Longman Scientific & Technical.

Hartman, F. C. & Harpel, M. R. (1994). Structure, function, regulation, and assembly of D-ribulose-1,5-bisphosphate carboxylase/oxygenase. *Annual Review of Biochemistry*, **63**, 197–234.

Husic, D. W., Husic H. D. & Tolbert, N. E. (1987). The

oxidative photosynthetic carbon cycle. *Critical Reviews in Plant Science*, **5**, 45–100.

Leegood, R. C., Lea, P. J., Adcock M. D. & Hausler, R. E. (1995). The regulation and control of photorespiration. *Journal of Experimental Botany*, **46**, 1397–414.

Lorimer, G. H. & Andrews, T. J. (1981). The $C_2$ chemo- and photorespiratory carbon oxidation cycle. In *The Biochemistry of Plants*, vol.8, ed. M. D. Hatch & N. K. Boardman, pp. 329–74. New York: Academic Press.

Oliver, D. J. (1994). The glycine decarboxylase complex from plant mitochondria. *Annual Review of Plant Physiology and Plant Molecular Biology*, **45**, 323–37.

# 14 Assimilation of non-carbohydrate compounds

G. SCHULTZ

## INTRODUCTION

Chloroplasts synthesize several non-carbohydrate compounds (e.g. fatty acids and $C_3$-derived amino acids) from phosphoenolpyruvate (PEP), pyruvate and acetyl-CoA ($C_3$- and $C_2$-compounds). This chapter describes how a developing chloroplast during a short time of ontogeny forms non-carbohydrate compounds from PEP, pyruvate and acetyl-CoA supplied directly by photosynthetic carbon fixation. At this metabolically autonomous stage, i.e. about one day after the assembly of thylakoid membranes, chloroplasts (with about one-fifth of their final volume) exhibit low activity in photosynthetic carbon fixation. After development, within one to two days, into mature chloroplasts, these organelles show maximum activity in photosynthetic carbon fixation. At this stage, the synthesis of $C_3$- and $C_2$-compounds requires import of substrates from the extraplastidic space. Comparative enzyme studies on plastids give further evidence for the presence of a 3-phosphoglycerate→acetyl-CoA ($C_3$→$C_2$) pathway, which transitorily functions at low rates during development of non-green and green plastids.

## DEVELOPING CHLOROPLASTS

### Photosynthetic assimilation of $^{14}CO_2$ into non-carbohydrate compounds

$^{14}C$-Incorporation from photosynthetically fixed $CO_2$ into non-carbohydrate compounds has been a matter of debate during the last two decades. In most cases, chloroplasts isolated from fully expanded leaves are unable (or able only at negligible extent) to form fatty acids, PEP- and pyruvate-derived amino acids and other compounds from $CO_2$ in the light, and are dependent on an external application of substrates/intermediates. For instance, an effective fatty acid formation is found when acetate has been given. This is explained by the used plant material which represents differentiated tissues containing mature chloroplasts. These chloroplasts are not active in the glycolytic reaction sequence and, thus, are unable to form acetyl-CoA from 3-phosphoglycerate (3-PGA).

On the other hand, developing chloroplasts exhibit relatively low activities of $CO_2$ fixation but are able to form fatty acids, amino acids and other compounds derived from photosynthetically fixed $CO_2$ (Heintze *et al.*, 1990). This suggests the presence of a transitorily active pathway, by which acetyl-CoA is formed from 3-PGA via 2-PGA, PEP and pyruvate at a sufficient rate. Such a pathway – described in the next paragraph – is named 3-PGA→acetyl-CoA ($C_3$→$C_2$) pathway since it provides intermediates at low rates sufficient for a synthesis of compounds essential in plant metabolism but not in energy metabolism through glycolysis.

The graminean leaf serves as an ideal model to demonstrate all stages – from developing to mature chloroplast. Sections cut near the base of the first leaf of barley seedlings (about one-tenth of the leaf length; containing chloroplasts of the post-meristematic zone) effectively incorporate $CO_2$ in the light into fatty acids, amino acids, isoprenoids, etc. (Hoppe *et al.*, 1993). The efficiency of $CO_2$-incorporation declines asymptotically in leaf sections towards the apical region, i.e. the leaf tip. Figure 14.1(*a*) demonstrates the situation for chloroplast isoprenoid synthesis.

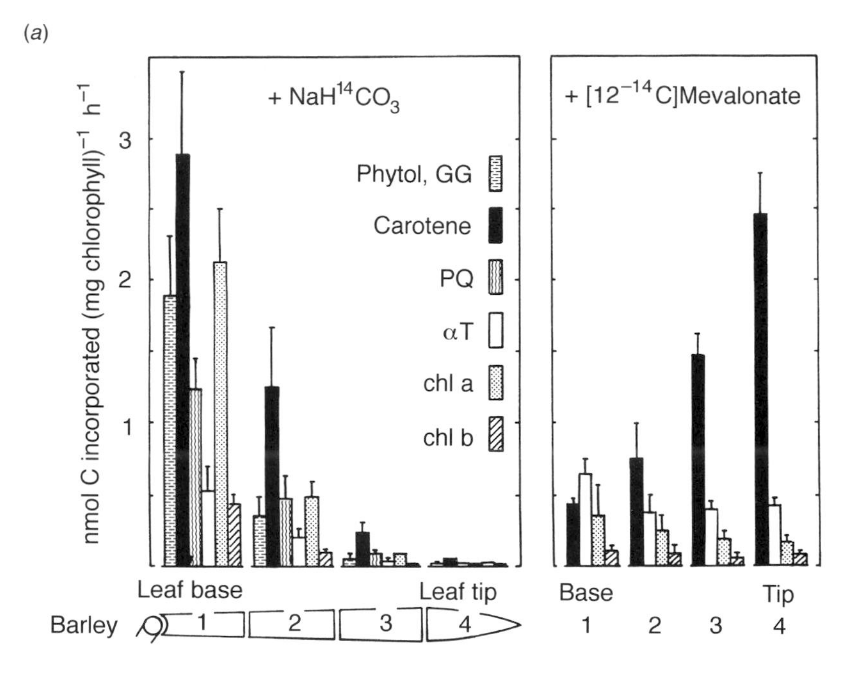


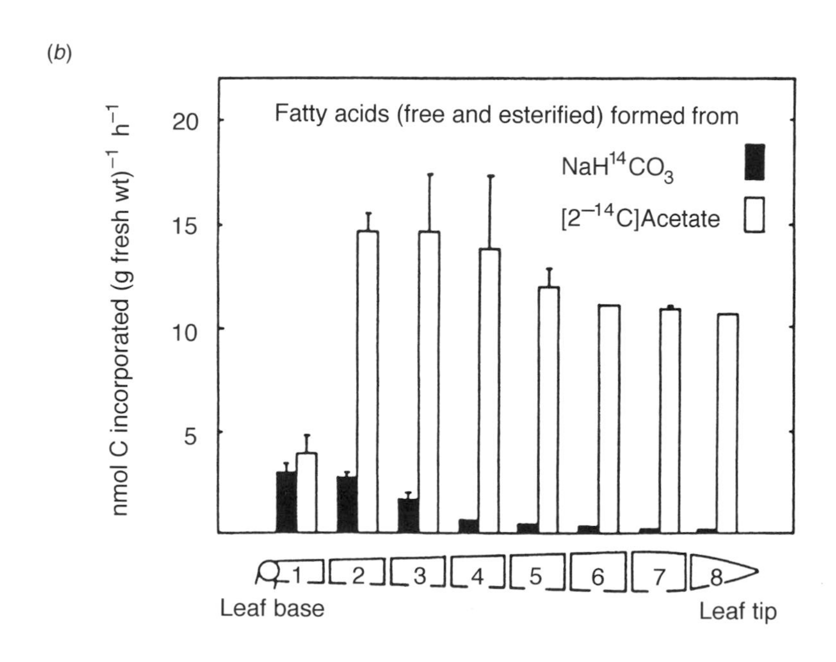


**Figure 14.1.** Synthesis of (*a*) plastidic isoprenoid compounds from NaH$^{14}$CO$_3$ and [2-$^{14}$C] mevalonate, respectively, and (*b*) total (free and esterified) fatty acids from NaH$^{14}$CO$_3$ and [2-$^{14}$] acetate, respectively, by different sections of the first foliage leaf of barley seedlings. Basically, the same results were obtained with chloroplasts isolated from different sections of the leaf (data not shown). Reproduced from G. Schultz *et al.* (1991). In *Active Oxygen/Oxidative Stress an Plant Metabolism*, pp. 156–170, ed. E. J. Pell & K. L. Steffen, by permission of the American Society of Plant Physiologists.

The decline in isoprenoid synthesis during leaf maturation suggests a decrease in activity for the formation of acetyl-CoA (from 3-PGA) or/and subsequently isopentenyl diphosphate (IPP), and implies a change to an import of respective intermediates across the chloroplast envelope membranes. Evidence for this assumption is shown in Fig. 14.1(*a*). Mevalonate, an intermediate of IPP synthesis (permeable across cell membranes but impermeable through chloroplast envelope membranes; Goodwin, 1965) was applied to leaf sections by vacuum infiltration. Using this technique, IPP is formed at the endoplasmic reticulum (ER) and transferred across chloroplast envelope membranes by a carrier-mediated transport (Soler *et al.*, 1993). The Figure shows that apical leaf sections with mature chloroplasts use mevalonate as effectively as basal sections with developing chloroplasts, thus confirming the above-mentioned prediction. The metabolic autonomy of developing chloroplasts changes to a division of labour in mature chloroplasts during leaf differentiation (see scheme in Fig. 14.2).

Turnover of pyruvate by the chlp pyruvate dehydrogenase complex (PDC) in developing chloroplasts (chlp) is sufficient to support isoprenoid synthesis (Fig. 14.1(a)) but fatty acid synthesis, having rates about one order of magnitude higher, works only at the earliest stage of developing chloroplasts (Fig. 14.1(*b*)). In comparison to fatty acid synthesis, isoprenoid synthesis has a much higher affinity to substrates of the chlp $C_3 \rightarrow C_2$ pathway (obviously to acetyl-CoA; Schulze-Siebert & Schultz, 1987).

Experiments using intact chloroplasts isolated from basal, middle and apical leaf sections of barley seedling confirmed the results obtained from whole leaf sections (Hoppe *et al.*, 1993). Solely developing chloroplasts were able to form fatty acids, isoprenoids and PEP- and pyruvate-derived amino acids from photosynthetically fixed $CO_2$.

## Formation of acetyl-CoA from 3-phosphoglycerate: ($C_3 \rightarrow C_2$) pathway

The presence of plastidic $C_3 \rightarrow C_2$ pathway enzymes which differ from the ones of the cytosolic pathway has been established by studies from Dennis's laboratory on leukoplasts isolated from endosperm of developing castor bean seeds (Dennis *et al.*, 1985; Dennis, 1989). Plastidic isoenzymes of phosphoglycerate mutase, enolase, pyruvate kinase and PDC have been isolated and characterized. Some other features point to a genetical identity of leukoplasts and chloroplasts implying differential expression systems for the development of the diverse functioning plastids. The small subunit of rubisco is imported and processed by leukoplasts. Steady-state levels of mRNAs transcribed from various plant genes, e.g. the large subunit of rubisco and cytochrome f from leukoplasts and chloroplasts do not differ remarkably. Yet, the expression of these genes varies, dependent on plastid development stage. Generally, the activities of the glycolytic enzymes, i.e. also of the enzymes of the $C_3 \rightarrow C_2$ pathway decrease in the late stage of seed growth suggesting a decline in gene expression.

Since 'plastids constitute a family of organelles in plants that have differentiated from identical proplastids' (Dennis, 1989), it appears conceivable that not only leukoplasts but also chloroplasts primarily possess the enzymes for conversion of hexoses into acetyl-CoA via 3-PGA. A conclusive evidence that enzymes of a chlp $C_3 \rightarrow C_2$ pathway are present at the stage of developing chloroplasts has been given by Hoppe *et al.* (1993). Chloroplast isoenzymes of phosphoglycerate mutase, enolase, pyruvate kinase and PDC have been analysed in the stromal (soluble protein) phase obtained from intact chloroplasts of barley seedlings by osmotical lysis. Since the corresponding cytosolic enzymes are present at much higher activities, chloroplasts used have to be highly purified by Percoll gradient centrifugation to remove the respective cytosolic enzymes. Contamination by PEP carboxylase serving as a marker enzyme for cytosol was $3.5 \pm 1\%$ (based on chlorophyll content of the leaf homogenate).

The enzymes of chlp $C_3 \rightarrow C_2$ pathway are characterized by some special features: (i) Activity of chlp $C_3 \rightarrow C_2$ pathway enzymes (determined on dry weight basis of leaf) is low and amounts to about one-tenth of the activity of the corresponding cytosolic enzymes. (ii) During development of the first leaf, the activity of $C_3 \rightarrow C_2$ pathway enzymes maintains an almost constant level or decreases slightly whereas the activity of some Calvin cycle enzymes tested, i.e. NADP-glyceraldehyde 3-phosphate dehydrogenase, phosphoglycerate kinase (Shah & Bradbeer, 1991), rubisco etc. increases about half an order of magnitude. (iii) chlp Phosphoglycerate

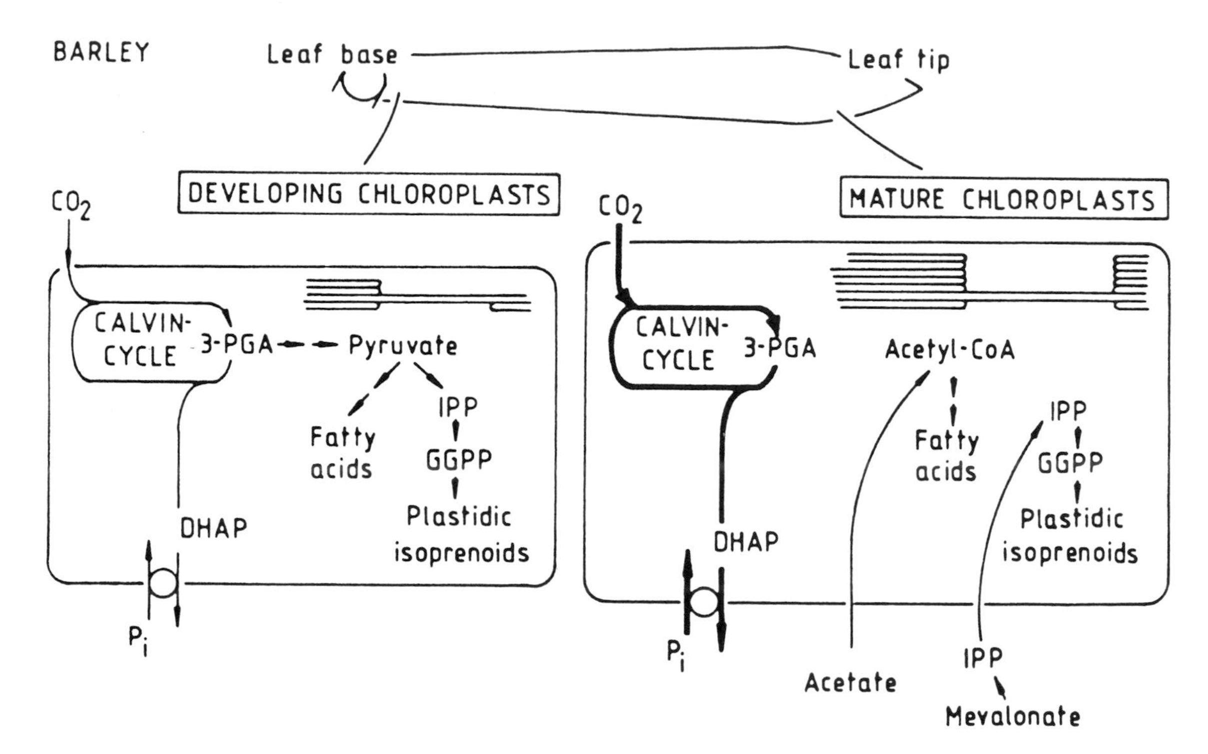


**Figure 14.2.** Scheme to illustrate the change from the autonomous $C_3 \rightarrow C_2$ metabolism in developing chloroplasts to the division-of-labour situation in mature chloroplasts. Based on a scheme from G. Schultz *et al.* (1991). In *Active Oxygen/Oxidative Stress and Plant Metabolism*, pp. 156–170, ed. E. J. Pell & K. L. Steffen, by permission of the American Society of Plant Physiologists.

mutase, with an apparent $K_m$ of 1.6 to 1.8 mM for 3-PGA, shows only a low affinity to substrate. This means the enzyme may act as a key enzyme into the $C_3 \rightarrow C_2$ pathway only functioning at high concentration of 3-PGA. Unlike the cytosolic enzyme it has a pH optimum in the alkaline, at pH 8.4, and is independent of 2,3-bis-PGA as generally known for all plant mutases. The enzyme was partially purified from barley and spinach; its localization in chloroplasts was evidenced by latency experiments. (iv) The chlp PDC might be the second regulating step in the $C_3 \rightarrow C_2$ pathway. Compared with the mitochondrial enzyme, it has a pH optimum in the alkaline, at pH 8.2, and is only slightly inhibited by ATP and AMP. Activities of chlp PDC differ species specifically: those ones found in barley and pea are about one order of magnitude higher than those found in spinach (Preiss *at al.*, 1993).

Reactions of other chloroplast enzymes may also contribute $C_3$ intermediates especially pyruvate to the $C_3 \rightarrow C_2$ pathway. Andrews and Lorimer (1987) described a side-reaction of rubisco converting 3-PGA to pyruvate by $P_i$ elimination. Expression of enzymes of $C_4$-photosynthesis occurs at a late stage of chloroplast development (Sheen & Bogorad, 1987), though pyruvate formed by the malic enzyme reaction in bundle sheath chloroplasts may serve only as substrate for the synthesis of pyruvate-dependent amino acids. The activity of chlp PDC in maize is too low to form effectively acetyl-CoA. There appears only one exception for a co-ordinated malic enzyme / PDC reaction in acetyl-CoA formation of plastids: leukoplasts from castor beans synthesize fatty acids from malate and pyruvate at higher rates than with added acetate. Apparently, malate import, malic enzyme and chlp PDC are more effective than acetate import and acetyl-CoA synthetase (Smith *et al.*, 1991).

A stringent concept for the decline in $C_3 \rightarrow C_2$-pathway-enzyme activities and/or gene expression during chloroplast development is a problem to be resolved in the future. A possible role of specific proteolytic effects on enzymes has been reported (Plaxton, 1991, Miernyk & Dennis, 1992). From the biophysical view, a suppression of substrate supply for $C_3 \rightarrow C_2$ pathway enzymes is to be explained by a specific and more effective by unspecific (adsorptive) binding of substrates to Calvin cycle and energy-converting enzymes. These enzymes are massively formed during maturation of chloroplasts. The effect is a decrease in free substrate concentrations, reducing their availability to $C_3 \rightarrow C_2$ pathway enzymes (Hoppe *et al.*, 1993).

## PHOSPHOENOLPYRUVATE-, PYRUVATE- AND ACETYL-COA-DEPENDENT SYNTHESES IN CHLOROPLASTS

### Fatty acid formation

The backbones of fatty acids are formed in chloroplasts. Chapter 4 specifically deals with fatty acids and their syntheses. For comparison, data of total (free and esterified) fatty acid synthesis and isoprenoid synthesis are given on p. 185 of this chapter.

### Formation of chloroplast isoprenoids (including prenylquinones)

Smoothly hypotonical treated spinach chloroplasts incorporate labeled acetate and to a lower extent pyruvate into plastidic isoprenoids (Heintze *et al.*, 1994). This indicates the presence of a plastidic mevalonate (MVA) pathway for the formation of IPP. Although this pathway is constitutively expressed at the ER, an equivalent pathway appears to be active in developing chloroplasts preferably for the synthesis of chloroplast isoprenoids like carotene and the prenyl moieties for the synthesis of plastoquinone-9 (PQ), tocopherols (vitamin E), phylloquinone (vitamin $K_1$) etc.

Since enzyme expression appears to be restricted to early stages of chloroplast development, only some enzymes of chloroplast MVA pathway have been identified. A chlp 3-hydroxy-3-methylglutaryl-CoA (HMG-CoA) reductase has been partially purified from buds of pea seedlings. The enzyme is localized at the chlp envelope membranes (Russell, 1985). It differs from the enzyme localized at the ER by a higher substrate affinity (app. $K_m$ of 0.77 μM for HMG-CoA) and a pH optimum in the alkaline (pH 7.9). The enzyme from pea is activated by 2 min red light irradiation, and activation is reversed by far red indicating a control by phytochrome. Phosphorylation by a putative chlp-stromal kinase leads to inactivation. Dephosphorylation by phosphatase reverses the effect and leads to enzyme activation. Indications that chlp MVA kinases differ from the corresponding kinases at the ER have been obtained by Arebalo & Mitchell (1984) by comparing the fraction pattern of enzymes from whole leaf cell extract and purified chloroplasts from *Nepeta cataria*. The fact that chlp envelope membranes are impermeable for MVA has been applied to discriminate the activities of chlp kinases against those ones of the ER, which contaminate the preparations of developing barley chloroplasts (M. Preiss & G. Schultz, unpublished results). The remarkable increase of activities after hypotonical treatment of intact chloroplasts is a further indication for the presence of MVA kinases in this organelle.

Only poor indications exist for the introductory enzyme system, an acetoacetyl-CoA lyase (synthase) and HMG-CoA synthase as these proteins are unstable and/or expressed only at early stages of development. Similar assumptions can be made for MVA 5-diphosphate decarboxylase as final enzyme of the chlp MVA pathway. At the ER in radish, a single monomeric enzyme, an acetoacetyl-CoA lyase/HMG-CoA synthase with a $M_r$ of 56 kDa converts acetyl-CoA to HMG-CoA in a concerted reaction (Weber & Bach, 1994). The purified enzyme system is activated by a redox couple like $Fe^{2+}$/quinone, which apparently mimics the *in vivo* co-factors. Apparently, a radical (e.g. $OH^{\cdot}$) plays a role in the proton abstraction from the acetyl-CoA molecule, which reacts as nucleophil in the first partial reaction of HMG-CoA synthesis. The energetically unfavourable Claisen condensation leading to the formation of acetoacetyl-CoA, which is apparently not liberated from the enzyme system, could be thereby facilitated.

Very recently, a non-mevalonate pathway for the IPP synthesis, earlier established in some bacteria by Rohmer *et al.* (1993), has also been evidenced in green algae and higher plant plastids (Schwender *et al.*, 1996; Lichtenthaler *et al.*, 1997). In this pathway, pyruvate and glyceraldehyde 3-phosphate form an intermediate, presumably

1-deoxyxylulose 5-phosphate, in a reaction analogous to acetohydroxyacid synthesis. 1-Deoxyxylulose 5-phosphate is converted to IPP in a reaction sequence, which is currently under investigation. The above mentioned mevalonate pathway in plastids may be an additional pathway in early developmental stages.

The formation of prenyl diphosphates occurs as an electrophilic reaction of an allylic carbocation (and concomitant abstraction of the diphosphate anion) with IPP (Poulter & Rilling, 1978). In chloroplasts, the enzymes involved in these reactions, IPP isomerase and GGPP synthase (more or less aggregated with phytoene synthase of carotene formation) are integrated or associated with the inner envelope membrane (Camara *et al.*, 1989; Fig. 14.3). The nuclear-encoded enzymes are synthesized as preproteins of higher molecular weight and are processed post-translationally at the membranes. The enzymes exhibit extreme high substrate affinities (in the micromolar range). In the first reaction at the IPP isomerase, IPP is isomerized to form dimethylallyl diphosphate (DMAPP, an allylic diphosphate); $Mg^{2+}$ or $Mn^{2+}$ are co-factors in the reaction. In the following prenylation reaction at the GGPP synthase, GGPP, a diterpene, is the only product, which leaves the active site of the enzyme. The counterpart to plastidic GGPP formation is the formation of farnesyl diphosphate in the cytosol, which is the central intermediate in the synthesis of sterols and numerous sesquiterpenes.

GGPP is the central intermediate in the carotenoid formation in plastids (Fig. 14.3; for details see Chapter 4) and the synthesis of abscisic acid derived from it (see Bartley & Scolnik, 1995). Putatively, the first steps of gibberellin formation from GGPP, the synthesis of copalyl-PP and *ent*-kaurene, occurs in chloroplasts (Sun & Kamya, 1994).

For the formation of phytyl diphosphate (PPP), GGPP is hydrogenated in three discrete steps at the inner chlp envelope membrane. NADPH is assumed to function as electron donor (Soll *et al.*, 1983). As a main reaction, PPP forms chlorophyll *a* from chlorophyllide *a* by an esterification reaction (for details, see Chapter 4). A parallel reaction for chlorophyll *a* formation is the hydrogenation of chlorophyllide *a*-GG-ester. Additionally, PPP is the prenylating moiety in the formation of tocopherols and phylloquinone (Fig. 14.3).

For the synthesis of PQ and tocopherols, the prenylquinones in chloroplasts, homogentisate is the immediate aromatic intermediate (see Pennock & Threlfall, 1983; Fig. 14.3). The prenyl moiety (as carbocation) is inserted into the phenolic nucleus in a strict regioselective substitution reaction with the concomitant loss of $CO_2$ yielding a 2-methyl-1,4-benzoquinol prenylated at position C-6. The involved prenyltransferases have high substrate specificities. The substrate for the synthesis of PQ is solanesyl- (nonaprenyl-) diphosphate, for tocopherol it is PPP and for tocotrienols in latex it is GGPP.

Only the quinol (although of low stability) and not the quinone of the prenylated compound is favoured in the following methylation reaction with S-adenosylmethionine (SAM) as an electrophilic substitution reaction (Soll, 1987). PQ is formed from 2-methyl-6-solanesyl-(nonaprenyl-) 1,4-benzoquinol by C-methylation in the *ortho*-position. In the tocopherol synthesis, an analogous reaction occurs with 2-methyl-6-phytyl-1,4-benzoquinol to form 2,3-dimethyl-5-phytyl-1,4-benzoquinol (Step 1). The quinol is then cyclicized to form the corresponding tocol, γ-tocopherol ( Step 2). γ-Tocopherol is finally C-methylated to yield α-tocopherol (Step 3). Trimethylphytyl-1,4-benzoquinol is ineffective for the formation of α-tocopherol. This reaction sequence is preferred in plants like spinach and lettuce chloroplasts. Other reaction sequences, shown in Fig. 14.3, may act in some plants, e.g. the sequence via δ-tocopherol probably in *Hedera helix*, which contains δ-, γ- and α-tocopherol, and another one via β-tocopherol probably in gramineae with higher content of β- and α-tocopherol.

The prenylation step in phylloquinone synthesis (see Inouye & Leistner, 1988) occurs also at the chlp envelope membrane. 1,4-Dihydroxy-2-naphthoate is prenylated with PPP to form 2-phytyl-1,4-naphthoquinol, the desmethyl-vitamin K (Fig. 14.3).

## Biosynthesis of essential amino acids

Synthesis of essential amino acids in the cell (except for the last step of methionine synthesis) and some others occurs in chloroplasts and other plastid types. Synthesis of aspartate- and glutamate-derived amino acids in chloroplasts has been studied earlier by Miflin's group (Miflin & Lea, 1977; Bryan, 1990; Lea, 1993; scheme in

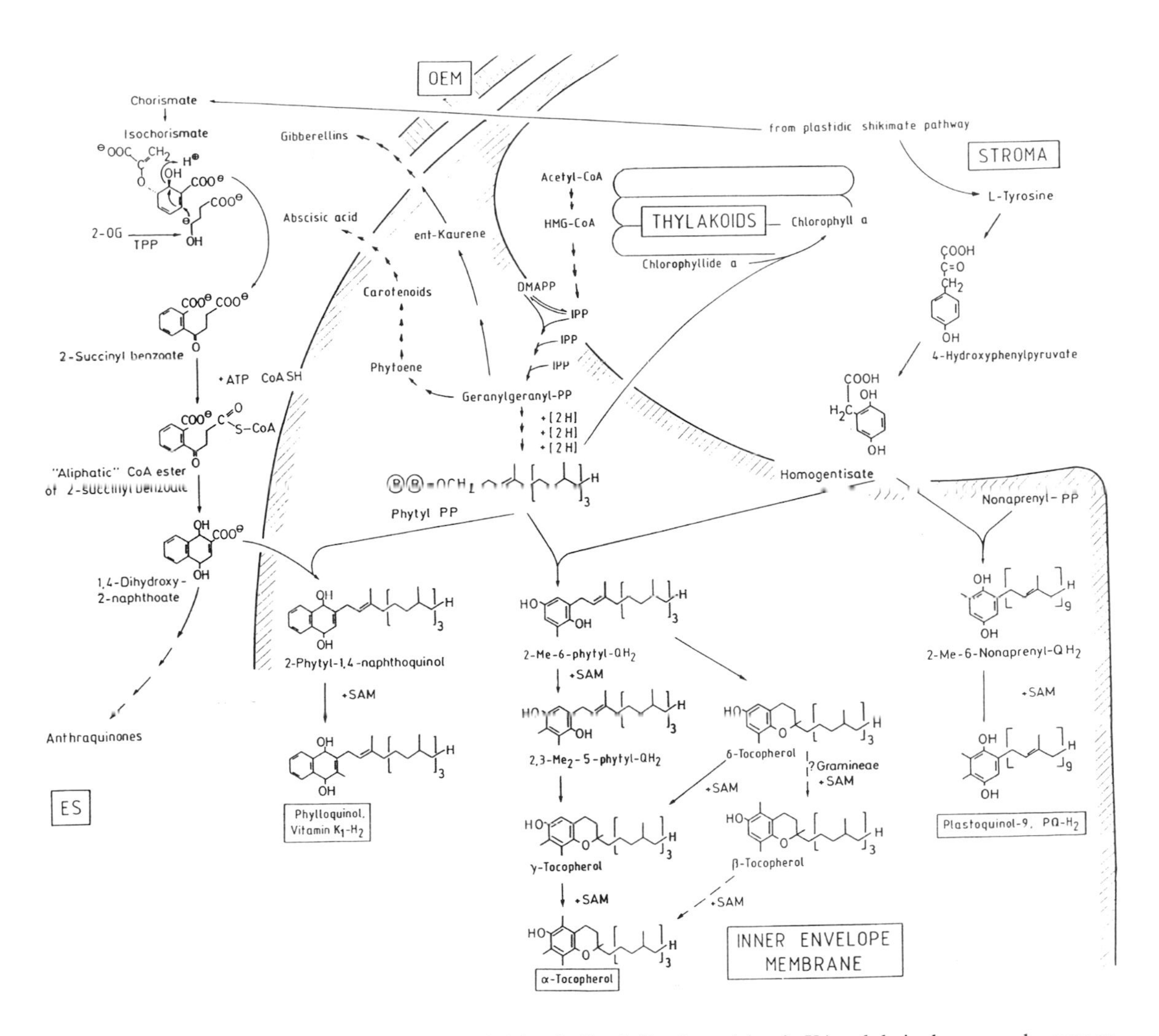


**Figure 14.3.** Synthesis of plastoquinone-9, tocopherols (vitamin E), phylloquinone (vitamin $K_1$) and derived compounds: compartmental view of reaction sequences. The methylation reaction in phylloquinol synthesis was studied with thylakoid fractions. The compartmentation of 1,4-dihydroxy-2-naphthoate synthesis is not yet clear and is assumed to occur in the extraplastidic site. ES, extraplastidic site (cytosol, ER, mitochondria etc); OEM, outer envelope membrane (for clarity, only indicated in the upper part of the Figure).

Fig. 14.4). Aspartate and glutamate are transported across the inner chloroplast envelope membrane by a dicarboxylate translocator shuttle, which uses also malate, 2-oxoglutarate and glutamine (Lehner & Heldt, 1978). For oxaloacetate transport across chloroplast membranes, a highly specific translocator exists having an extreme high substrate affinity (app. $K_m$ value 50 μM; Hatch *et al.*, 1984).

### *Aromatic amino acids*

The chloroplast shikimate pathway is the only site of aromatic amino acid synthesis in higher plants (Bickel &

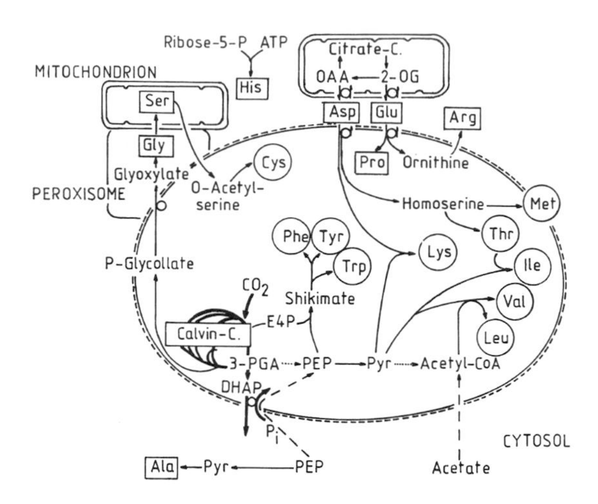


**Figure 14.4.** Scheme to illustrate the compartmentation of amino acid synthesis in a higher plant cell. O, essential amino acids; □, non-essential amino acids; . . . ., reactions occurring only in developing chloroplasts; ——, reactions occurring in mature chloroplasts.

Schultz, 1979; Fig. 14.5). No indications exist for the presence of cytosolic isoenzymes in higher plants. Exceptions are the DAHP synthase (Co), which reacts predominately with glycolaldehyde, chorismate mutase CM II, an enzyme not regulated by tryptophan, isochorismate synthase (for vitamin $K_1$ synthesis; Poulsen *et al.*, 1993; Fig. 14.3) and quinate oxidoreductase, which is not directly involved in the main pathway.

Most of the higher plant enzymes of the pre- and post-chorismate trunk of the pathway were purified or at least enriched (see Bentley, 1990; Haslam, 1993; Herrmann 1995). Of the tryptophan branch in plants, only the anthranilate synthase was isolated (Poulsen *et al.*, 1993). Complementation of an *E. coli* or a yeast mutant with an *Arabidopsis* cDNA library led to identification of the cDNAs encoding the genes of additional enzymes of the tryptophan synthesis (see Radwanski & Last, 1995).

All genes of enzymes involved in the chloroplast shikimate pathway are localized in the nucleus. The translation products possess a transit peptide sequence for the import into plastids (see Herrmann, 1995). An alternative splicing during transcription procedure of chorismate synthase gene LeCS2 of tomato has been observed for the first time by Görlach *et al.* (1995*a*). The homology of DNA sequences of higher plant and bacterial genes is higher than that for fungi (see Herrmann, 1995). In analogy to bacteria and in contrast to fungi, the enzymes of pre- and post-chorismate trunk of the higher plant pathway are present as single enzymes except that of 3-dehydroquinase/shikimate oxidoreductase. All these findings allowed the suggestion that higher plants possess a shikimate pathway originating from their prokaryotic endosymbionts. A question remains whether an eukaryotic pathway as established in fungi was originally present in higher plants and has been eliminated during the evolution.

There are two variations of post-chorismate pathway. In higher plants an aminotransferase forms arogenate from prephenate (Jung, Zamir & Jensen, 1986), whereas in some bacteria prephenate is converted to form 4-hydroxyphenylpyruvate and phenylpyruvate by an aromatization reaction. For the synthesis of homogentisate in PQ and tocopherol formation (p. 188) 4-hydroxyphenylpyruvate is formed from tyrosine by an chlp L-amino acid oxidase.

In higher plant chloroplasts, the post-chorismate pathway is regulated at the step of chorismate-mutase CM I reaction by less than 10 μM tryptophan. The arogenate dehydrogenase is feedback-inhibited by tyrosine and the dehydratase by phenylalanine. A very sensitive feedback inhibition by tryptophan is acting on the anthranilate synthase. In contrast to bacteria, a feedback regulation of the DAHP synthase by intermediates and endproducts of the pathway is lacking. However, the whole pathway seems to be responsive towards stress factors (Görlach *et al.*, 1995*b*), and in particular the DAHP synthase is transcriptionally regulated by wounding, elicitor treatment and pathogenic attack (Keith *et al.*, 1991).

Aromatic amino acids like branched chain amino acids, both known as relatively lipophilic acids, are transferred by unknown mechanisms across chloroplast membranes. Aromatic amino acids are the substrate not only for protein synthesis but also for a massive synthesis of numerous secondary compounds occurring in the cytosol and apoplastic space (lignins, etc.). A transport of aromatic amino acids into vacuoles driven by a tonoplast $H^+$-ATPase and a $H^+$-pyrophosphatase has been shown by Homeyer *et al.* (1989). A lipophilic L-amino acid carrier at the tonoplast recognizes specifically aromatic amino acids as well as L-leucine and L-isoleucine but not valine.

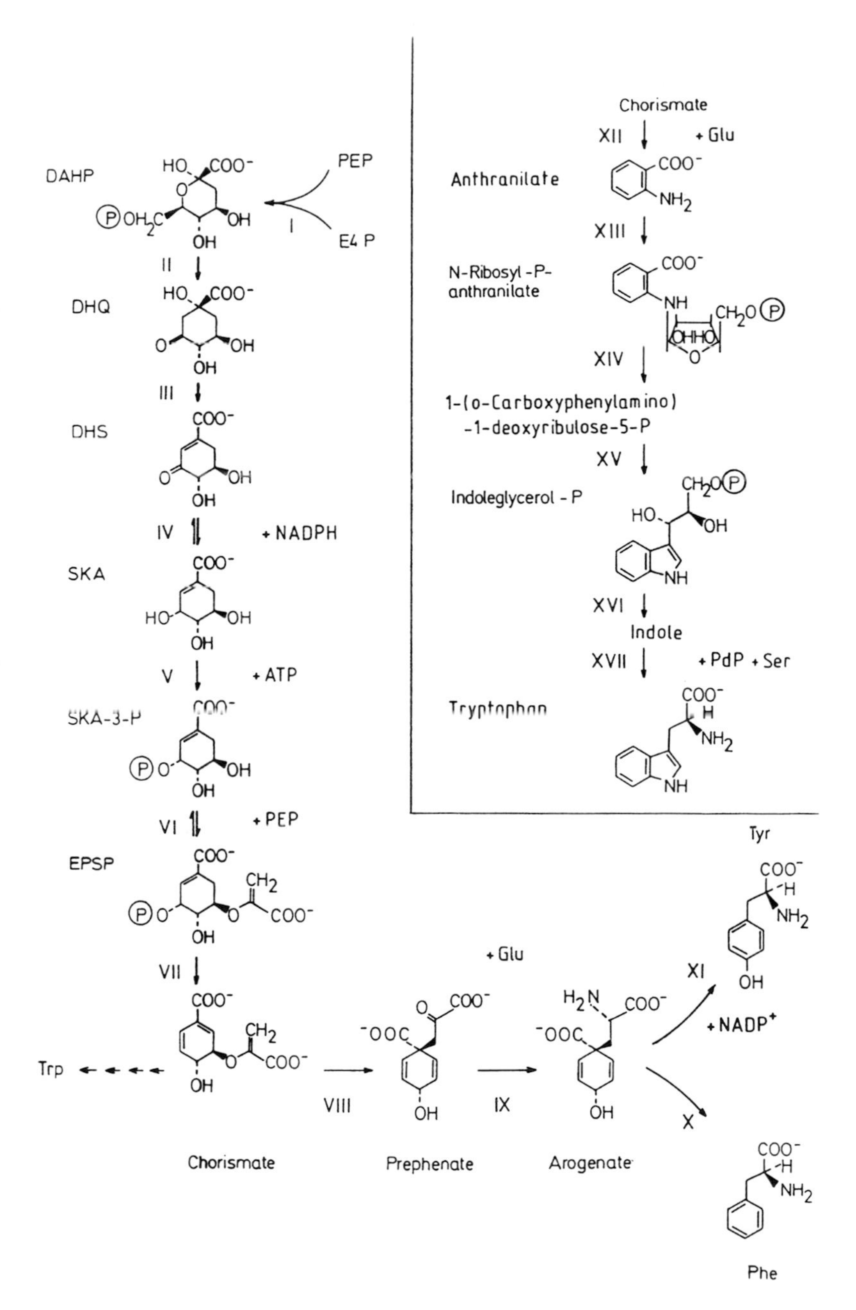


**Figure 14.5.** Reaction sequence of chlp shikimate pathway in aromatic amino acid synthesis. DAHP, 3-deoxy-D-*arabino*-heptulosonate 7-phosphate; DHQ, 3-dehydroquinate; DHS, 3-dehydroshikimate; EPSP, 5-enolpyruvyl-shikimate 3-phosphate; SKA, shikimate; SKA-3-P, shikimate-3-phosphate. I, DAHP synthase; II, 3-dehydrochinate synthase ($NAD^+$, $Co^{2+}$); III, 3-dehydroquinate dehydratase (3-dehydroquinase); IV, shikimate dehydrogenase (NADPH); the latter enzymes form a bifunctional protein III/IV in higher plants; V, shikimate kinase; VI, 5-enolpyruvyl-shikimate 3-phosphate synthase; VII, chorismate synthase; VIII, chorismate mutase; IX, prephenate aminotransferase; X, L-arogenate dehydratase; XI, L-arogenate dehydrogenase ($NADP^+$), XII, anthranilate synthase; XIII, anthranilate :5-phosphoribosyl-1-diphosphate phosphoribosyl transferase; XIV, phosphoribosyl-anthranilate isomerase; XV, indol-3-glycerolphosphate synthase; XVI, tryptophan synthase α; XVII, tryptophan synthase β.

### *Branched chain amino acids (BCAA)*

BCAA synthesis appears to be localized solely in plastids (Schulze-Siebert *et al.*, 1984) of all plant tissues. Highest enzyme activities were found in early stages of the organ development. All genes isolated so far (i.e. those of acetohydroxyacid (acetolactate) synthase, ketolacid (acetohydroxyacid) reductoisomerase and 3-isopropylmalate dehydrogenase; Fig. 14.6) contain a putative chloroplast-specific transit peptide sequence. In higher plants, the BCAA synthesis is feedback regulated at several

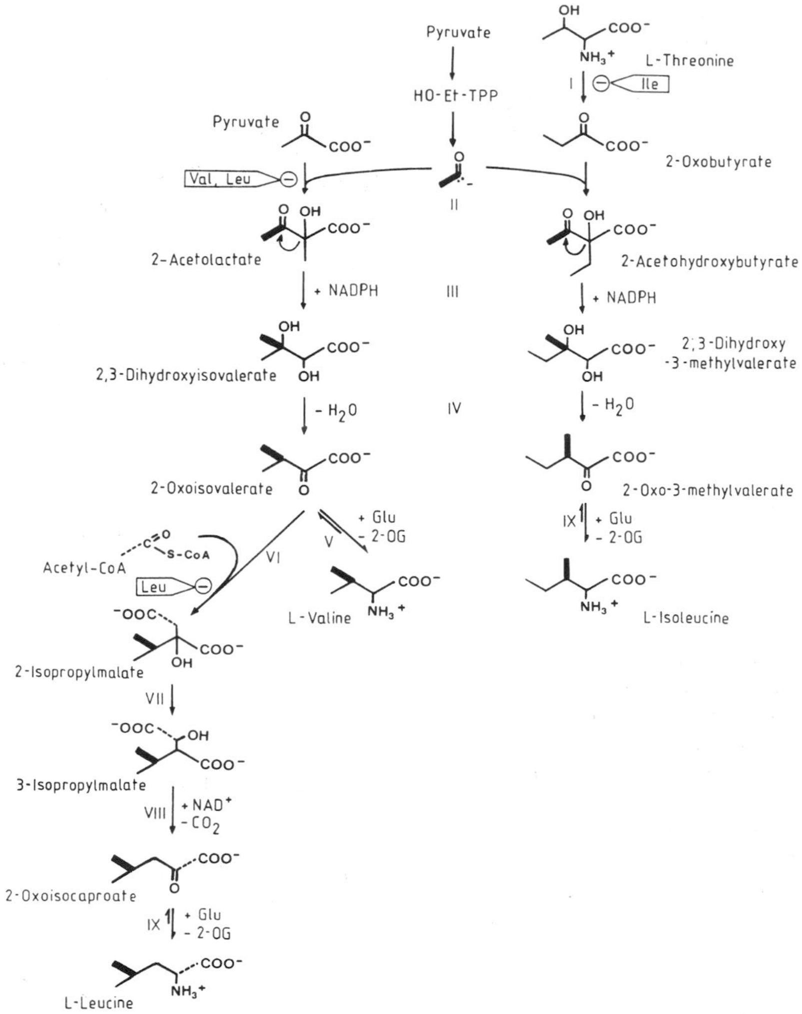

**Figure 14.6.** Reaction sequence of branched chain amino acid synthesis. I, threonine deaminase (dehydratase; PLP); II, acetohydroxyacid (acetolactate) synthase; III, ketolacid reductoisomerase (NADPH); IV, dihydroxyacid dehydratase; V, valine aminotransferase (PLP); VI, 2-isopropylmalate synthase; VII, isopropylmalate isomerase (dehydratase); VIII, 3-isopropylmalate dehydrogenase ($NAD^+$); IX, leucine/isoleucine aminotransferase (PLP); HO-Et-TPP, hydroxyethylthiaminediphosphate; (–), feedback inhibition. Enzymes II to IV or isoenzymes are involved in valine and isoleucine formation.

steps: the threonine deaminase reaction is regulated by isoleucine, and the acetohydroxyacid synthase reaction by valine and leucine (see Singh & Shaner, 1995), and the 2-isopropylmalate synthase reaction by leucine (Hagelstein & Schultz, 1993).

The branching reaction in the synthesis of the three amino acids are carried out in parallel pathways by three enzymes, acetohydroxyacid synthase, ketolacid reductoisomerase and dihydroxyacid dehydratase. Isoforms of acetohydroxyacid synthase differing in specificity (Delfourne *et al.*, 1994), and tetrameric and monomeric forms of the enzyme have been characterized suggesting a more complex regulation of the pathways.

In the acetohydroxyacid-synthase reaction, pyruvate and thiamine diphospate (TPP) form an intermediate that undergoes decarboxylation to produce a stabilized anion of hydroxyethyl-TPP. The anion acts as a nucleophil on the 2-oxogroup of a second molecule of pyruvate or 2-oxobutyrate (Fig. 14.6). The enzyme is dependent on FAD to stabilize the active oligomeric form. The same dependence on FAD is shown for the bacterial pyruvate oxidase. The genes of both enzymes exhibit strong amino acid sequence similarities, which suggests that genes of both enzymes originate from common ancestors.

Acetohydroxyacid synthase is the target site of two potent herbicides, sulfonylurea and imidazolinone herbicides, being different in chemical structure. From studies on herbicide resistance, it can be concluded that binding sites of the two herbicides are distinct but close to one another. Herbicides and feedback inhibitors bind to separate sites on the enzyme.

Ketolacid reductoisomerase was highly purified and crystallized from spinach (Dumas, Job & Douce, 1994). Only one gene per haploid genome was found. Consistent with the NAD(P)H requirement of the enzyme for catalysis, the deduced amino acid sequence of the enzyme contains a fingerprint region of an NAD(P)H binding site that is known for several NAD(P)H-dependent oxidoreductases.

Dihydroxyacid dehydratase was purified to homogeneity from spinach. The enzyme has a distinct brown color because it contains a [2Fe–2S] cluster being involved in dehydration reaction. A similar reaction mechanism occurs in the isopropylmalate-isomerase reaction of the leucine synthesis branch. This enzyme catalyses a dehydration/rehydration reaction analogous to that of aconitase. Aconitase has an unusual [4Fe–4S] cluster that participates in catalysis. EPR spectroscopy of isopropylmalate isomerase from yeast evidenced a similar structure in this enzyme (Emptage, 1990). The $NAD^+$-dependent 3-isopropylmalate dehydrogenase is the target of another herbicide, 0-isobutenyl oxalylhydroxamate (Wittenbach *et al.*, 1994).

In the final step of BCAA synthesis, two aminotransferases, which differ in specificity, are involved (Hagelstein *et al.*, 1997). Both enzymes were purified from spinach chloroplasts. One enzyme, a valine aminotransferase, forms solely valine from 2-oxoisovalerate. The other enzyme, a leucine/isoleucine aminotransferase, forms leucine and isoleucine from the corresponding branched chain 2-oxoacids but is ineffective in valine synthesis.

### *Lysine*

In lysine synthesis, 4-aspartyl semialdehyde (the intermediate formed from aspartyl phosphate) condenses with pyruvate to form dihydrodipicolinate (DHP). The involved enzyme, DHP synthase, is extremely sensitive to feedback inhibition by lysine, more sensitive than the lysine-sensitive isoenzyme of aspartate kinase (see Galili, 1995). However, there appears to exist some limitations in substrate supply. Whereas aspartate is imported by the chlp dicarboxylate translocator (see p. 139), pyruvate is formed in chloroplasts of $C_3$ plants only at limited rates: in developing chloroplasts by the $C_3 \rightarrow C_2$ pathway, and in mature chloroplasts from imported PEP (see Fig. 14.4). Since pyruvate concentration in chloroplasts is low (about 100 μM; Schulze-Siebert *et al.*, 1984) and the $K_m$ value of DHP synthase for pyruvate is remarkably high (about 1.7 mM), higher than that for aspartyl semialdehyde (about 0.4 mM; Dereppe *et al.*, 1992), a rate limitation of lysine synthesis by pyruvate appears likely.

## CONCLUDING REMARKS

This short review must remain incomplete and should present only some outlines of a feature under-represented in considerations of chloroplast metabolism. Figures 14.2 and 14.4 illustrate the situation in a $C_3$ chloroplast. Some supplementary data are added. The basic idea of this chapter is to show that chloroplasts (or plastids in general) in early developmental stages behave in many directions as metabolically autonomous organelles in respect to the supply of a series of plastidic syntheses with $C_3 \rightarrow C_2$ pathway intermediates. It appears that chloroplast ontogeny, particularly at this stage, reflects some of its phylogeny. In the mature stage with maximum photosynthetic activity, the syntheses of $C_3$- and $C_2$-derived compounds is sustained but the supply of these syntheses by $C_3 \rightarrow C_2$ pathway is replaced by an import of intermediates from the extraplastidic space, and chloroplasts undergo a division of labour with other cellular compartments. chlp Pyruvate kinase is the only enzyme, which remains active in mature chloroplasts to form pyruvate for chlp syntheses. Phosphoenolpyruvate needed for that is imported by the triosephospate/phosphate translocator at the chlp envelope membranes (Fischer *et al.*, 1994; See Fig. 14.4).

ACKNOWLEDGEMENTS

This chapter is dedicated to Professor Dr h. c. Helmut Holzer, Freiburg, on the occasion of his 75th birthday.

Author's work was supported by the Deutsche Forschungsgemeinschaft, Bonn, and Fonds der Chemischen Industrie, Frankfurt/Main (literature supply). I wish to express sincere thanks to all collaborators in my laboratory, who have made contributions through the years in studying the chloroplast syntheses. I thank Dr Karin M. Jäger for constructive criticism of this paper.

## REFERENCES

Andrews, T. J. & Lorimer, G. H. (1987). Rubisco: structure, mechanism and prospect for improvement. In *The Biochemistry of Plants*, vol. 10, ed. M. D. Hatch & N. K. Boardman, pp. 131–218. New York: Academic Press.

Arebalo, R. E. & Mitchell E. D., Jr (1984). Cellular distribution of 3-hydroxy-3-methylglutaryl coenzyme A reductase and mevalonate kinase in leaves of *Nepeta cataria*. *Phytochemistry*, **23**, 13–18.

Barak, Z., Chapman, D. M. & Schloss, J. V. (eds). (1990). *Biosynthesis of Branched Chain Amino Acids*. Weinheim: VCH Verlagsgesellschaft.

Bartley, G. E. & Scolnik, P. A. (1995). Plant carotenoids: pigments for photoprotection, visual attraction, and human health. *The Plant Cell*, **7**, 1027–38.

Bentley, R. (1990). The shikimate pathway – a metabolic tree with many branches. *Critical Reviews in Biochemistry and Molecular Biology*, **25**, 307–84.

Bickel, H. & Schultz, G. (1979). Shikimate pathway regulation in suspensions of intact spinach chloroplasts. *Phytochemistry*, **18**, 498–9.

Bryan, J. K. (1990). Advances in the biochemistry of amino acid biosynthesis. In *The Biochemistry of Plants*, vol. 16, ed. B. J. Miflin & P. J. Lea, pp. 161–96. New York: Academic Press.

Camara, B., Bousquet, J., Cheniclet, C., Carde, J -P., Kunitz, M., Evrard, J -L. & Weil, J -H. (1989). Enzymology of isoprenoid biosynthesis of plastid and nuclear genes during chromoplast differentiation in pepper fruits (*Capsicum annuum*). In *Physiology, Biochemistry, and Genetics of Nongreen Plastids*, ed. C. D. Boyer, J. C. Shannon & R. C. Hardison, pp. 141–56. Rockville, MD: American Society of Plant Physiologists.

Delfourne, E., Bastide, J., Bandon, R., Rachon, A. & Genix, P. (1994). Specificity of acetohydroxyacid synthase: formation of products and inhibition by herbicides. *Plant Physiology and Biochemistry*, **32**, 473–7.

Dennis, D. T. (1989). Fatty acid biosynthesis. In *Physiology, Biochemistry, and Genetics of Nongreen Plastids*, ed. C. D. Boyer, J. C. Shannon & R. C. Hardison, pp. 120–9. Rockville, MD: American Society of Plant Physiologists.

Dennis, D. T., Hekman, W. E., Thomson, A., Ireland, R. J., Botha, F. C. & Kruger, N. J. (1985). Compartmentation of glycolytic enzymes in plants. In *Regulation of Carbon Partitioning in Photosynthetic Tissues*, ed. R. L. Heat & J. Preiss, pp. 127–46. Rockville, MD: American Society of Plant Physiologists.

Dereppe, C., Bold, G., Ghisalba, O., Ebert, E. & Schür, H-P. (1992). Purification and characterization of dihydrodipicolinate synthase from pea. *Plant Physiology*, **98**, 813–21.

Dumas, R., Job, D. & Douce, R. (1994). Crystallization and preliminary crystallographic data for acetohydroxy acid isomeroreductase from *Spinacia oleracea*. *Journal of Molecular Biology*, **242**, 578–81.

Emptage, M. H. (1990). Yeast isopropylmalate isomerase as an iron sulphur protein. In *Biosynthesis of Branched Chain Amino Acids*, ed. Z. Barak, D. M. Chapman & J. V. Schloss, pp. 315–28. Weinheim: VCH Verlagsgesellschaft.

Fischer, K., Arbinger, B., Kammerer, B., Busch, C., Brink, S., Wallmeier, H., Sauer, N., Eckerskorn, C. & Flügge, U. I. (1994). Cloning and *in vivo* expression of functional trisose phospate / phosphate translocators from $C_3$- and $C_4$-plants: evidence for the putative participation of specific amino acid residues in the recognition of phosphoenolpyruvate *The Plant Journal*, **5**, 215–26.

Galili, G. (1995). Regulation of lysine and threonine synthesis. *The Plant Cell*, **7**, 899–906.

Goodwin, T. W. (1965). Regulation of terpenoid synthesis in higher plants. In *Biosynthetic Mechanisms in Higher Plants*, ed. J. B. Pridham & T. Swain pp. 57–71. London: Academic Press.

Görlach, J., Raesecke, H. -R., Abel, G., Wehrli, R., Amrhein, N. & Schmid, J. (1995*a*). Organ-specific differences in the ratio of alternatively spliced chorismate synthase (LeCS2) transcripts in tomato. *The Plant Journal*, **8**, 451–6.

Görlach, J., Raesecke, H., Rentsch, D., Regenass, M., Roy, P., Zala, M., Keel, C., Boller, T., Amrhein, N. & Schmid, J. (1995*b*). Temporarily distinct accumulation of transcripts encoding enzymes of the prechorismate pathway in elicitor-treated, cultured tomato cells. *Proceedings of the National Academy of Sciences, USA*, **92**, 3166–70.

Hagelstein, P. & Schultz, G. (1993). Leucine synthesis in spinach chloroplasts. Partial characterization of 2-isopropylmalate synthase. *Biological Chemistry Hoppe-Seyler*, **374**, 1105–8.

Hagelstein, P., Sieve, B., Klein, M., Jans, H. & Schultz, G. (1997). Leucine synthesis in chloroplasts: leucine/ isoleucine aminotransferase and valine aminotransferase are different enzymes in spinach chloroplasts. *Journal of Plant Physiology*, **150**, 23–30.

Haslam, E. (1993). *Shikimic Acid. Metabolism and Metabolites.* Chichester: John Wiley.

Hatch, M. D., Dröscher, L., Flügge, U. I. & Heldt, H. W. (1984). A specific translocator for oxaloacetate transport in chloroplasts. *FEBS Letters*, **178**, 15–17.

Heintze, A., Görlach, J., Leuschner, C., Hoppe, P., Hagelstein, P., Schulze-Siebert, D. & Schultz, G. (1990). Plastidic isoprenoid synthesis during chloroplast development. Change from metabolic autonomy to division-of-labor stage. *Plant Physiology*, **93**, 1121–7.

Heintze, A., Riedel, A., Aydogdu, S. & Schultz, G. (1994). Formation of chloroplast isoprenoids from pyruvate and acetate by chloroplasts from young spinach plants. Evidence for a mevalonate pathway in immature chloroplasts. *Plant Physiology and Biochemistry*, **32**, 791–7.

Herrmann, K. M. (1995). The shikimate pathway: early steps in the biosynthesis of aromatic compounds. *The Plant Cell*, **7**, 907–19.

Homeyer, U., Litek, K., Huchzermeyer, B. & Schultz, G. (1989). Uptake of phenyl alanine into barley vacuoles is driven by the tonoplast adenosinetriphosphatase and pyrophosphatase. Evidence for a hydrophobic L-amino acid carrier system. *Plant Physiology*, **89**, 1388–93.

Hoppe, P., Heintze, A., Riedel, A., Creuzer, C. & Schultz, G. (1993). The plastidic 3-phosphoglycerate→acetyl-CoA pathway in barley leaves and its involvement in the synthesis of amino acids, plastidic isoprenoids and fatty acids during chloroplast development. *Planta*, **190**, 253–62.

Inouye, H. & Leistner, E. (1988). Biochemistry of quinones. In *The Chemistry of Quinonoid Compounds*, vol II, ed. S. Patal & Z. Rappoport, pp. 1294–349. Chichester: John Wiley.

Jung, E., Zamir, L. O. & Jensen, R. A. (1986). Chloroplasts of higher plants synthesize L-phenylalanine via L-arogenate. *Proceedings of the National Academy of Sciences, USA*, **83**, 7231–5.

Keith, B., Dong, X., Ausubel, F. M. & Fink, G. R. (1991). Differential induction of 3-deoxy-D-arabinitol-heptulosonate 7-phosphate synthase genes in *Arabidopsis thaliana* by wounding and pathogenic attack. *Proceedings of the National Academy of Sciences, USA*, **88**, 8821–5.

Lea, P. J. (1993). Nitrogen metabolism. In *Plant Biochemistry and Molecular Biology*, ed. P. J. Lea & R. C. Leegood, pp. 155–80. Chichester: John Wiley.

Lehner, K. & Heldt, H. W. (1978). Dicarboxylate transport across the inner membrane of chloroplast envelope. *Biochimica et Biophysica Acta*, **501**, 531–44.

Lichtenthaler, H. K., Schwender, J., Disch, A. & Rohmer, M. (1997). Biosynthesis of isoprenoids in higher plant chloroplasts proceeds via a mevalonate independent pathway. *FEBS Letters*, **400**, 271–4.

Miernyk, J. A. & Dennis, D. T. (1992). A developmental analysis of the enolase isoenzymes from *Ricinus communis*. *Plant Physiology*, **99**, 748–50.

Miflin, B. J. & Lea, P. J. (1977). Amino acid metabolism. *Annual Review of Plant Physiology*, **28**, 299–329.

Pennock, J. F. & Threlfall, D. R. (1983). Biosynthesis of ubiquinone and related compounds. In *Biosynthesis of Isoprenoid Compounds*, ed. J. W. Porter & S. L. Spurgeon, pp. 191–303. Chichester, John Wiley.

Plaxton, W. C. (1991). Leucoplast pyruvate kinase from developing castor oil seeds. Characterization of the enzyme's degradation by cysteine endopeptidase. *Plant Physiology*. **97**, 1334–8.

Poulsen, C., Bongaerts, R. J. M. & Verpoorte, R. (1993). Purification and characterization of anthranilate synthase from *Catharanthus roseus*. *European Journal of Biochemistry*, **212**, 131–40.

Poulter, C. D. & Rilling, H. C. (1978). The prenyl transfer reaction, enzymatic and mechanistic studies of the 1–4 coupling reaction in the terpene biosynthetic pathway. *Accounts of Chemical Research*, **11**, 307–13.

Preiss, M., Rosidi, B., Hoppe, P. & Schultz, G. (1993). Competition of $CO_2$ and acetate as substrates for fatty acid synthesis in immature chloroplast of barley seedlings. *Journal of Plant Physiology*, **142**, 525–30.

Radwanski, E. R. & Last, R. L. (1995). Tryptophan biosynthesis and metabolism: biochemical and molecular genetics. *The Plant Cell*, **7**, 921–34.

Rohmer, M., Knani, M., Simonin, P., Sutter, B. & Sahm, H. (1993). Isoprenoid synthesis in bacteria: a novel pathway for the early steps leading to isopentenyl diphosphate. *The Biochemical Journal*, **295**, 517–24.

Russell, D. W. (1985). 3-Hydroxy-3-methylglutaryl-CoA reductase from pea seedlings. Plastid HMG-CoA reductase: assay, isolation and properties. *Methods in Enzymology*, **110**, 36–40.

Schulze-Siebert, D., Heineke, D., Scharf. H. & Schultz, G. (1984). Pyruvate-derived amino acids in spinach chloroplasts. Synthesis and regulation during photosynthetic carbon metabolism. *Plant Physiology*, **76**, 465–71.

Schulze-Siebert, D. & Schultz, G. (1987). α-Carotene synthesis in isolated spinach chloroplasts. Its tight linkage to photosynthetic carbon metabolism. *Plant Physiology*, **84**, 1233–7.

Schwender, J., Seemann, M., Lichtenthaler, H. K. & Rohmer, M. (1996). Biosynthesis of isoprenoids (carotenoids, sterols, prenyl side-chains of chlorophylls and plastoquinone) via a novel pyruvate/glyceraldehyde-3-phosphate non-mevalonate pathway in the green alga *Scenedesmus obliquus*. *The Biochemical Journal*, **316**, 73–80.

Shah, N. & Bradbeer, J. W. (1991). The development of the chloroplast and cytosolic isoenzymes of phosphoglycerate kinase during barley leaf ontogenesis. *Planta*, **185**, 401–8.

Sheen, J. Y. & Bogorad, L. (1987). Differential expression of $C_4$ pathway genes in mesophyll and bundle sheath cells of greening maize leaves. *Journal of Biological Chemistry*, **262**, 11726–30.

Singh, B. K. & Shaner, D. L. (1995). Biosynthesis of branched chain amino acids: from test tube to field. *The Plant Cell*, **7**, 935–44.

Smith, R. G., Gauthier, D. A., Dennis, D. T. & Turpin, D. H. (1991), Malate- and pyruvate-dependent fatty acid synthesis in leucoplasts from developing castor endosperm. *Plant Physiology* **98**, 1233–8.

Soler, E., Clastre, M., Bantignies, B., Marigo, G. & Ambid, C. (1993). Uptake of isopentenyl diphosphate by plastids isolated from *Vitis vinifera* L. cell suspensions. *Planta*, **191**, 324–9.

Soll, J. (1987). α-Tocopherol and plastoquinone synthesis in chloroplast membranes. *Methods in Enzymology*, **148**, 383–92.

Soll, J., Schultz, G., Rüdiger, W. & Benz, J. (1983). Hydrogenation of geranylgeraniol. Two pathways exist in spinach chloroplasts. *Plant Physiology*, **71**, 849–54.

Sun, T. P. & Kamya, Y. (1994). The *Arabidopsis* GA1 locus encodes the cyclase ent-kaurene synthase A of gibberellin biosynthesis. *The Plant Cell*, **6**, 1509–18.

Tobin, A. K., ed. (1992) *Plant Organelles – Compartmentation of Metabolism in Photosynthetic Cells* (Society for Experimental Biology Seminar Series 50). Cambridge: Cambridge University Press.

Weber, T. & Bach, T. J. (1994). Conversion of acetyl-coenzyme A into 3-hydroxy-3-methylglutaryl-coenzyme A in radish seedlings. Evidence for a single monomeric protein catalyzing a FeII/quinone stimulated double condensation reaction. *Biochimica et Biophysica Acta*, **1211**, 85–96.

Wittenbach, V. A., Teaney, P. W., Hanna, W., Rayner, D. & Schloss, J. (1994). Herbicidal activity of an isopropylmalate dehydrogenase inhibitor. *Plant Physiology*, **106**, 321–8.

# 15 Interaction with respiration and nitrogen metabolism

K. PADMASREE AND A. S. RAGHAVENDRA

## INTRODUCTION

Photosynthesis is traditionally considered to be an autonomous process, since carbon can be assimilated even by a system of isolated chloroplasts. However, in recent years, it has become evident that photosynthetic carbon assimilation is strongly dependent on mitochondrial activity and that carbon partitioning is regulated significantly by nitrogen metabolism. High photosynthetic efficiency helps to achieve maximal plant growth and plant productivity. Similarly, high rates of respiration are characteristic of rapidly growing tissues. Both photosynthesis and respiration therefore, form essential components of plant growth (Amthor, 1989, 1994).

Photosynthesis results in $O_2$ evolution and the generation of ATP and NADPH, which are then used for the reduction of $CO_2$ (or other compounds like $NO_2^-$ or $SO_4^{2-}$). On the other hand, respiration accomplishes oxidation of carbon compounds and evolution of $CO_2$. NADH produced in these reactions is utilized for ATP production and oxygen consumption. Thus, ATP is generated in both processes, while pyridine nucleotides are reduced during photosynthesis, but are oxidized in respiration. The biochemical nature of photosynthetic and respiratory reactions implies that these two processes are complementary to each other.

The requirements of ATP and reducing power (NADH or NADPH) for the cells are met not only by photosynthetic reactions in chloroplasts but also by oxidative metabolism of mitochondria. Besides mitochondria and cytosol, the reduced equivalents are used up during photorespiration as well as nitrogen metabolism. As a result, in a plant cell, the processes of photosynthesis, respiration, nitrogen metabolism and photorespiration become dependent on each other, underscoring the concept of organelle autonomy (Turpin & Weger, 1990; Azcón-Bieto, 1992; Raghavendra, Padmasree & Saradadevi, 1994; Gardeström & Lernmark, 1995; Krömer, 1995).

## LONG- AND SHORT-TERM INTERACTIONS

The rate of dark respiration in leaves is higher after a few hours of illumination than that during the steady-state in prolonged darkness. Such an increase in dark respiration is often proportional to the period of illumination and depends on temperature. The high rates of respiration decline slowly to reach the lower level of a steady state, characteristic of the dark period. The increase in respiratory rate is believed to be primarily due to the accumulation of carbohydrates. This effect of photosynthetic activity on respiration should be treated as a long-term effect since the interaction becomes pronounced only after hour(s) of illumination. A long-term (spanning hours or days) illumination increases the carbohydrate content and the respiratory capacity of the tissue. This occurs in several tissues, e.g. in roots with increased carbohydrate flux from the shoots or in cells growing in culture. Thus, plant respiration responds readily to long-term changes in substrate availability (via photosynthesis).

Net $O_2$ evolution during the steady state of photosynthesis is a result of the combined effect of (a) photosynthetic $O_2$ evolution, (b) $O_2$ uptake by the Mehler reaction, (c) respiratory $O_2$ uptake, and (d) $O_2$ consumption by oxygenase activity of rubisco and glycolate oxidase, while net $CO_2$ evolution is a result of (a) $CO_2$ release from tricarboxylic acid cycle (TCA cycle), (b) glycine decarboxylation, (c) photosynthetic $CO_2$ fixation, and (d) phospho*enol*pyruvate carboxylase (PEPC) activity. If respiration is defined as a process of $CO_2$ release and/or $O_2$ consumption, several such metabolic activities occur in leaves/green cells besides the dark respiration and interact rapidly with photosynthesis. These can be treated as short-term interactions and include photorespiration, chlororespiration, Mehler reaction and the mitochondrial respiration.

## Photorespiration

Photorespiration involves combined cycling of carbon and nitrogen between chloroplasts, mitochondria and cytosol (Husic, Husic & Tolbert, 1987; Canvin, 1990; see also Chapter 13 for detailed description). A part of the carbon from photorespiratory glycolate is supplied back to the chloroplast in the form of glycerate. Ammonia released during the conversion of glycine to serine in mitochondria is reassimilated in the cytosol by glutamine synthetase (GS) and further metabolised in the chloroplast by glutamine: 2-oxoglutarate aminotransferase (GOGAT). If reassimilation of ammonia is blocked by methionine sulfoxime, which inhibits GS, then ammonia accumulates in tissues and photosynthesis stops. Nitrogen turnover during photorespiration is much greater than the net nitrogen (e.g. $NO_3^-$) reduction in the chloroplast.

The photorespiratory pathway requires NADH for hydroxypyruvate reduction and draws it in the form of malate from mitochondria and chloroplasts. Such consumption of reductant may be important when photosynthesis is severely restricted, e.g. water-stressed plants with closed stomata, when the electron transport chain is fully reduced and chlorophyll absorbs excess energy forming excited states and damaging products (e.g. superoxide, $H_2O_2$). Photorespiration, by draining the energy burden, may function together with the carotenoids in quenching excited state of chlorophyll and protecting PSII, which is sensitive to photoinhibition.

## Chloroplast respiration

Although the reactions of dark and photorespiration occur mostly outside the chloroplast, a modified system of respiratory pathway with an $O_2$ dependent $CO_2$ efflux occurs in chloroplasts of higher plants and *Chlamydomonas*. This phenomenon, which is termed as chloroplast respiration or chlororespiration, includes (i) reactions which oxidize glucose and (ii) electron transport pathway similar to respiratory chain, fed with electrons from reduced plastoquinone. These pathways result in $CO_2$ efflux and $O_2$ uptake and may generate limited quantity of reduced equivalents to be either used within the chloroplast during darkness or exported to cytosol (Garab *et al.*, 1989; Gibbs *et al.*, 1989).

## Mehler reaction

It involves light-dependent PSI driven $O_2$ uptake in chloroplasts. Under conditions when the rate of $CO_2$ fixation is limiting, that is, when the NADPH/NADP ratio is high, a significant amount of the reductant of photosystem I is oxidized by oxygen resulting in the formation of $H_2O_2$ (Carrier, Chagvardieff & Tapie, 1989). In the presence of catalase, $H_2O_2$ undergoes rapid dismutation to water and $O_2$. Thus uptake, as well as evolution, of $O_2$ occurs in the chloroplasts and both these processes require light. Under active photosynthesis, $O_2$ evolution exceeds $O_2$ consumption, but, if catalase is removed by washing, or inhibited by azide, the Mehler reaction leads to a net $O_2$ uptake.

Among these three types of respiration, only photorespiratory rates exceed those of dark respiration. Photorespiration is three to eight times greater than typical dark respiration. On the other hand, the rates of chlororespiration and Mehler reaction are about 20 and 30%, respectively, of dark respiration. Respiration of leaves, measured in darkness, is 5–10% of net photosynthesis in plants of brightly lit environments, but from those of shade it may be of a much larger proportion.

# BENEFICIAL EFFECTS OF PHOTOSYNTHESIS ON DARK RESPIRATION

The status of dark respiration in illuminated green cells/leaves has been a topic of interest. There is a considerable disagreement in the earlier literature (until the last decade) on the effect of light on respiration, since dark respiration was reported to be either hardly affected or even stimulated, while in other reports there was an inhibition of up to 100% (Graham, 1980; Graham & Chap-

Table 15.1. *LEDR in different plant tissues*

| Material | $O_2$ uptake: Before illumination (μmol mg $Chl^{-1}$ $h^{-1}$) | $O_2$ uptake: After illumination (μmol mg $Chl^{-1}$ $h^{-1}$) | Stimulation (fold) | Reference |
|---|---|---|---|---|
| Pea mesophyll protoplasts | 7.9 | 22.8 | 2.9 | Reddy *et al.*, 1991 |
| Barley protoplasts | 5.4 | 6.8 | 1.3 | Gardeström *et al.*, 1992 |
| Barley protoplasts | 3.6 | 29 | 8 | Hill & Bryce, 1992 |
| Spinach leaf discs | 1.2 | 8.4 | 7 | Stokes *et al.*, 1990 |
| *Dianthus caryophyllus* cells | 21.6 | 26.4 | 1.2 | Avelange *et al.*, 1991 |

man, 1979). Such a large variation is due to several factors: the component of dark respiration being monitored, the experimental technique being used, and finally the type of plant tissue being studied. For example, it is difficult to assess the operation of photosynthesis or respiration in leaves based on net uptake/evolution of either $O_2$ or $CO_2$, since the measurements are compromised by inter- and intracellular recycling of the gases. Another problem is the technical difficulty of monitoring precisely the different types of oxidative reactions besides respiration (e.g. photorespiration, Mehler reaction, pseudocyclic electron transport). All these processes result in $O_2$ uptake and operate concurrently in light. Nevertheless, a promising solution appears to be the technique of mass spectrometry which distinguishes between uptake and efflux of $CO_2$ or $O_2$, which occurs simultaneously (Carrier, Chagvardieff & Tapie, 1989; Avelange *et al.*, 1991).

## Light enhanced dark respiration

The beneficial or stimulatory effect of photosynthesis on mitochondrial respiration is not restricted only to the light period but also can be seen in the successive dark period. There is a marked upsurge in respiratory oxygen uptake after even short periods of illumination (5 to 15 minutes). This phenomenon termed as 'light-enhanced dark respiration' (LEDR) is documented in leaves (spinach) and mesophyll protoplasts (pea and barley). Although there is a large variation, the extent of LEDR in mesophyll protoplasts can be quite high, up to eight-fold of the rate of dark respiration (Table 15.1).

The respiratory rate is not altered if the protoplasts or leaf discs are kept in continuous darkness for similar periods. The extent of LEDR progressively increases with longer duration of preillumination. The high respiratory rate following illumination persists for only a few minutes, before subsiding subsequently to reach a lower rate similar to that in the dark control. The marked stimulation of respiratory activity within a few minutes of illumination suggests that the interaction between respiration and light is quite rapid and involves early photosynthetic products. The stimulation of LEDR by the presence of bicarbonate, or inhibition by classic photosynthetic inhibitors (e.g. DCMU or DL-glyceraldehyde) during pre-illumination suggests that the upsurge of respiratory $O_2$ uptake was dependent on the products of photosynthetic carbon assimilation/electron transport (Stokes *et al.*, 1990; Reddy, Vani & Raghavendra, 1991). The high rate of LEDR in the presence of even saturating $CO_2$ and its insensitivity to aminoacetonitrile, an inhibitor of mitochondrial glycine metabolism, demonstrate that LEDR is distinct from the photorespiratory post-illumination burst of $CO_2$ (Gardeström, Zhou & Malmberg, 1992).

During LEDR of barley mesophyll protoplasts, the levels of sucrose, glucose and fructose did not change significantly while malate accumulated during illumination and was metabolized rapidly in subsequent darkness (Hill & Bryce, 1992). Thus, LEDR appears to be primarily due to malate oxidation. The enzyme pyruvate dehydrogenase complex (PDC), is known to be inactivated in the light. On transfer to darkness both NAD–malic enzyme (ME) and PDC are reactivated and photosynthetically generated malate is oxidized via both malate dehydrogenase (MDH) and NAD–ME. Pyruvate produced by the latter is oxidized by PDC and the resulting acetyl-CoA combines with oxaloacetate (OAA) to enhance TCA cycle operation. This rapid malate metabolism leads to the respiratory burst observed as LEDR. A transient increase in the level of pyruvate on transfer to darkness was observed in also leaves of *Elodea* and spinach. Malate formation during photosynthesis in $C_3$ plants may represent a reserve of reducing power to mediate the transfer from photosynthetically generated reducing equivalents to the cytosol and later to mitochondria.

When photoautotrophic cells of carnation (*Dianthus caryophyllus*) or *Euphorbia characias* were darkened after a few minutes of illumination, the rate of respiratory $O_2$ uptake in darkness (soon after switching off the light) was enhanced (Avelange *et al.*, 1991), once again confirming the phenomenon of LEDR.

## Modified/partial TCA cycle in light

Dark respiration is comprised of three components: (i) glycolytic reactions, (ii) decarboxylation of carbon compounds to produce $CO_2$ and reduced nucleotides (NADH and FADH); and (iii) oxidation of NADH/FADH leading to $O_2$ consumption and ATP production. A strong indication that the processes of $CO_2$ efflux and $O_2$ uptake are not as tightly coupled in light as in darkness came from observations on carnation cells. Some reactions of TCA cycle activity and substrate decarboxylation, in particular, are inhibited in the light, although to varying extents. In mung bean leaves, a comparison in light and dark of the effects of two inhibitors of the TCA cycle, malonate and fluoroacetate, on carbon traffic in the cycle led to the conclusion that in transition from dark to light there was an initial inhibition of the TCA cycle and a subsequent recovery.

Mass spectrometric studies using $^{13/12}CO_2$ and $^{18/16}O_2$ also demonstrate that respiratory $CO_2$ efflux is inhibited upon illumination, while $O_2$ uptake is either unaffected or even stimulated. In photoautotrophic cells of carnation, in presence of high light and saturating levels of $CO_2$, rate of $CO_2$ influx represented 75% of the rate of $O_2$ evolution. After a dark-to-light transition, the rate of $CO_2$ efflux was inhibited whereas the $O_2$ uptake was not affected. Thus, TCA cycle activity (decarboxylation of TCA cycle compounds) is inhibited but oxidative electron transport is unaffected by illumination. Such inhibition of $CO_2$ efflux indicating a significant decrease in TCA cycle activity, has been observed also in different systems: cells of *Euphorbia characias*, *Commelina communis*, algal cells of *Selenastrum minutum* and *Chlamydomonas reinhardtii*.

The adenylate ratio (relative level of ATP to ADP) is an important regulatory factor of mitochondrial respiration. An increase in the ratio of ATP/ADP could be the reason for the decrease in mitochondrial TCA cycle, as ATP levels in both the chloroplast and cytosol increase during illumination. However, the adenylate levels do not appear to be crucial since the cytosolic ATP/ADP ratios are not much affected by light/dark transitions (compared with the NAD(P)H to NAD(P) ratio, for example) and they are usually not high enough to inhibit mitochondrial oxidative metabolism. The CN-resistant alternative pathway which operates in plant mitochondria helps to maintain oxidative electron transport, independent of ATP formation.

The main reason for the decrease in TCA cycle activity appears to be the marked inactivation of mitochondrial PDC on illumination, which catalyses the primary event of oxidative decarboxylation of pyruvate to acetyl CoA. The inactivation of PDC is reversed, when the tissue is returned to the darkness. This reversible inactivation by light was dependent on photosynthetic activity as indicated by its sensitivity to DCMU (an inhibitor of PSII of photosynthesis) and absence of the phenomenon in etiolated seedlings. Not only photosynthesis, but also photorespiratory metabolism, can stimulate PDC inactivation. Conditions that reduce the photorespiration (i.e. increasing the $CO_2$ concentration or reducing the $O_2$ concentration) inhibited the initial light-dependent drop in mitochondrial PDC activity.

The biochemical reason for this regulatory mechanism is covalent modification of PDC by phosphorylation (inactivation) in light and dephosphorylation (reactivation) in darkness (Luethy *et al.*, 1996; Randall *et al.*, 1996). Illumination increases the level of ATP in the cytosol or mitochondria and this enhances protein phosphorylation (by a PDC-protein kinase), which inactivates the PDC enzyme. Reactivation of phosphorylated PDC is catalysed by PDC-protein phosphatase.

The reversible phosphorylation of PDC is fine tuned by the inhibition of pyruvate or NADH and stimulation by $NH_4^+$ of PDC-protein kinase. The inactivation of

PDC can be reversed by pyruvate or NADH. If the pH of the incubation media is 7 or less, ME is able to generate sufficient pyruvate to inhibit the kinase and maintain a high level of PDC activity. Further, NADH produced by glycine decarboxylase supports mitochondrial ATP production. This ATP production in combination with the $NH_4^+$-stimulation of the PDC kinase would result in very rapid phosphorylation and inactivation of PDC.

A major requirement of carbon skeletons is still met by TCA cycle, even in light. Although a few reactions of TCA cycle are inhibited or suppressed, a partial or modified TCA cycle operates in cells during illumination. Mitochondria take up readily OAA from cytosol, because of their highly active OAA-translocator. A major supply of OAA appears to be through the activity of cytosolic PEPC. Within mitochondria, OAA is converted to malate. A part of malate is decarboxylated to form pyruvate and then to acetyl coA. Although at a subdued rate, acetyl coA is condensed with OAA to form citrate. As a result, mitochondria can effectively keep up cycling of reduced nucleotides and export citrate to cytosol (Hanning & Heldt, 1993). Citrate is then converted to 2-oxoglutarate (2-OG) which becomes the major source of carbon for amino acid biosynthesis in chloroplasts (Fig. 15.1).

## ESSENTIALITY OF MITOCHONDRIAL METABOLISM FOR PHOTOSYNTHESIS

The contribution of mitochondrial oxidative phosphorylation to the ATP demands of a photosynthesizing cell is not in doubt anymore. Recent reports establish that mitochondrial activity is essential for the optimal performance of photosynthesis in green cells (Amthor, 1994; Raghavendra *et al.*, 1994; Gardeström & Lernmark, 1995; Krömer, 1995).

Oligomycin, an inhibitor of mitochondrial oxidative phosphorylation (but not photophosphorylation) suppressed photosynthesis by 30–40% in illuminated barley and pea mesophyll protoplasts at a concentration, as low as 0.05 μg $ml^{-1}$. Similarly, oligomycin when fed through transpiration stream of barley leaves, inhibits photosynthesis by up to 60%. The inhibition of photosynthesis by oligomycin, however, is observed only with intact protoplasts. When protoplasts, whose photosynthesis is inhibited by oligomycin are disrupted by forcing through a 5 μm nylon net, a procedure leaving chloroplasts and mitochondria intact, photosynthesis recovers back to the rate found in those without oligomycin. These results indicate that the strong inhibition of photosynthesis observed with oligomycin was due not to an effect on chloroplast photosynthesis as such, but to interference with reactions outside the chloroplast in the cytosol.

In plant cells, most of the photosynthate is converted to sucrose, an ATP consuming process located in the cytosol. The addition of oligomycin caused a drastic decrease in the cellular ATP/ADP ratio, which is mainly due to a decrease in the extrachloroplastic compartment. A significant part of the ATP requirement for cytoplasmic sucrose synthesis appears to be met by mitochondrial oxidative phosphorylation. The sharp decrease in photosynthesis of intact protoplasts upon addition of oligomycin may be therefore, explained partly by a decrease in sucrose synthesis caused by metabolic regulation of fructose-1,6-bisphosphatase and sucrose-6-phosphate synthase (SPS) in response to the decreased availability of cytoplasmic ATP. The increase in cellular contents of glucose-6-phosphate, triose phosphates (triose-P), precursors of sucrose synthesis on addition of oligomycin further confirm this hypothesis. Thus mitochondrial oxidative phosphorylation appears to be an efficient way for providing the cytosol with ATP.

A marked increase in the cellular dihydroxyacetone phosphate/phosphoglyceric acid (DHAP/PGA) ratios is observed in presence of oligomycin. The increase in the DHAP/PGA ratio on addition of oligomycin in leaves and protoplasts reflects an increased reduction of the stromal NADP-system. Under steady state conditions of photosynthesis, redox equivalents are exported from the chloroplasts by malate–OAA shuttle or DHAP-PGA shuttle to the cytosol where they are oxidised by the external NADPH dehydrogenase system of mitochondrial electron transport chain. The addition of oligomycin may reduce the capacity of mitochondria to oxidize surplus redox equivalents from the chloroplasts and thus cause an increase in the reductive state of the stromal NADPH/NADP system and of the chloroplast electron transport carriers.

Further experiments with oligomycin and barley leaf protoplasts revealed that the importance of mitochondrial metabolism to photosynthesis may depend on light intensity (Krömer, 1995; Krömer, Malmberg & Garde-

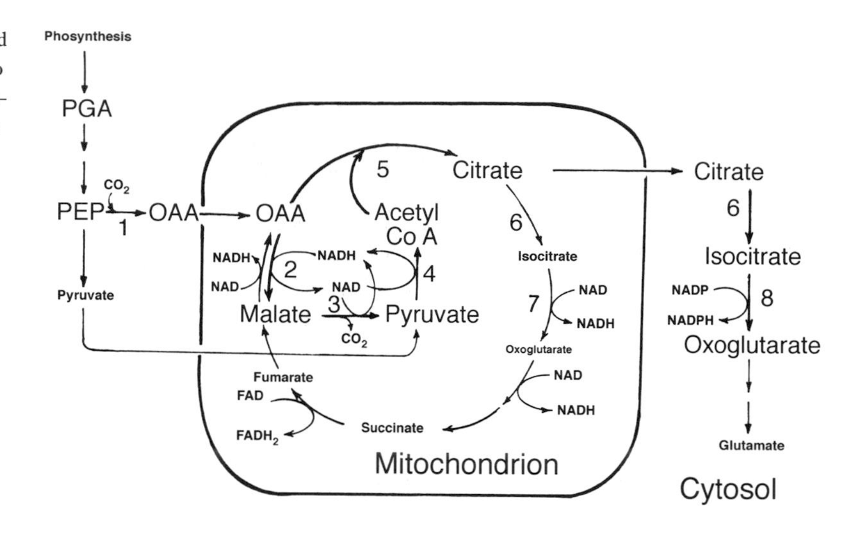


**Figure 15.1.** The partial or modified TCA cyle (thick arrows) expected to operate in mitochondria of photosynthetic cells during illumination. The main input into mitochondria is OAA, derived from photosynthetic products through glycolytic reactions. A part of OAA is then converted to malate, which on decarboxylation yields pyruvate. Acetyl CoA, derived from pyruvate combines with OAA to form citrate, which is exported to cytosol and converted to 2-OG, to be used for nitrogen metabolism. The other reactions of TCA cycle, which are subdued or restricted, are indicated by thin arrows. The numbers indicate the important enzymes. 1, PEPC; 2, NAD-MDH; 3, NAD-ME; 4, PDC; 5, citrate synthase; 6, *cis*-aconitase; 7, NAD-IDH; and 8, NADP-IDH (Adapted from Hanning & Heldt, 1993.)

ström, 1993). Mitochondrial ATP synthesis was active under limiting as well as saturating light intensities and under both photorespiratory and non-photorespiratory conditions. A major function of the mitochondrion in a photosynthesizing cell, particularly under low light intensities, seems to be the supply of ATP for cytosolic carbon metabolism, i.e. sucrose synthesis. In high light, mitochondria take on the additional role of oxidising the excess reducing equivalents generated by photosynthesis, preventing over-reduction of chloroplastic redox carriers and thus maintaining high rates of photosynthesis.

Studies with a starchless mutant of *Nicotiana sylvestris* suggest that a respiratory supply of ATP could affect assimilate partitioning into sucrose and thereby, modulate photosynthesis. The mutant NS 458 contains a defective plastid phosphoglucomutase and accumulates only trace amounts of starch. Treating mutant leaf protoplasts and young leaves with oligomycin-reduced photosynthesis by as much as 25% and 40%, respectively. The wild-type failed to show inhibition by oligomycin, i.e. its effect was masked when starch and sucrose synthesis could interact. Maximal $CO_2$ assimilation in the mutant thus appears to be fine-tuned by mitochondrial metabolism such that any interruption processes would generate oscillations in photosynthesis.

## MUTUALLY BENEFICIAL INTERACTION BETWEEN PHOTOSYNTHESIS AND RESPIRATION

The interaction between photosynthesis and respiration within the leaf tissue is striking but a matter of intense debate. It has been difficult to establish the extent and mechanism of interaction between photosynthesis and respiration, based on gas exchange characteristics of leaves. However, recent experiments using the system of mesophyll protoplasts or the use of mass spectrometry with cell suspensions, illustrate nicely the marked interaction between photosynthesis and respiration.

The processes of photosynthetic $O_2$ evolution and respiratory $O_2$ uptake in mesophyll protoplasts benefited each other significantly, even during short cycles of light and darkness (Vani, Reddy & Raghavendra, 1990). The rate of dark respiratory $O_2$ uptake was enhanced markedly after each cycle of illumination. The extent of respiration was unaffected if the protoplasts were kept in darkness. The rate of photosynthetic $O_2$ evolution by these protoplasts decreased under continuous illumination. On the other hand, if illumination was interrupted by short periods of darkness, the photosynthetic rate was stimulated. The stimulation in photosynthesis was much higher (nearly 50%), if compared to the rate of protoplasts under continuous illumination.

The interdependence of photosynthesis and respiration was even more striking in the presence of metabolic inhibitors. For example, DCMU or glyceraldehyde, the inhibitors of photosynthesis, restricted the respiratory $O_2$ uptake in the dark. Similarly respiratory inhibitors, like antimycin A, sodium azide or oligomycin reduced markedly the rates of photosynthetic $O_2$ evolution as well. The concentration of oligomycin used in these experiments does not affect the chloroplast function. Thus, the inhibitors of photosynthesis decreased the extent of LEDR while the respiratory inhibitors suppressed also photosynthesis. The statistical significance of interaction between photosynthesis and respiration only in the presence of $CO_2$ (but not in the absence) suggests that carbon assimilation is a prerequisite.

## Turnover of redox equivalents between chloroplasts, mitochondria, peroxisomes and cytosol

The turnover of redox equivalents within a photosynthetic plant cell extends between at least four compartments: chloroplasts, mitochondria, cytosol and peroxisomes (Fig. 15.2). Chloroplasts are the major sources of reducing power as their photochemical activity is very high. The reductants, in excess of the requirements of the Calvin cycle, are exported out of chloroplasts through the shuttling of either OAA–malate (by dicarboxylate translocator) or PGA–DHAP ($P_i$ translocator).

Glycine and malate, both of which are formed during active photosynthesis, form the substrates for leaf mitochondrial oxidation *in vivo*. However, the main substrate for mitochondrial respiration in the light is probably glycine, which is produced at high rates during photorespiration. At least 25% of the NADH formed during oxidation of these metabolites is expected to be used for extramitochondrial requirements, particularly hydroxypyruvate reduction in peroxisomes and $NO_3^-$ reduction in cytosol. Studies with spinach leaf mitochondria postulated that export of reducing equivalents from mitochondria proceeds by a malate–aspartate shuttle. In pea leaf mitochondria, no activity of a malate–aspartate shuttle is detected, but a high activity of a malate–OAA shuttle is found. Alternatively, cytosolic nitrate reductase (NR) and peroxisomal hydroxypyruvate reductase can be served via a chloroplastic malate–OAA shuttle with reducing equivalents generated from photosynthetic electron transport (Heupel & Heldt, 1992). Thus, mitochondrial metabolism becomes a link between photosynthesis, photorespiration and nitrogen assimilation, in recycling pyridine nucleotides (Fig. 15.2).

### FURTHER BENEFITS OF MITOCHONDRIAL RESPIRATION

The major advantage of mitochondrial metabolism is to provide an outlet for excess reduced equivalents in chloroplasts. This is essential because the capacity of photochemical electron transport usually exceeds that of biochemical carbon fixation. The beneficial effect of mitochondrial respiration on photosynthesis is further reflected under conditions which either over-stimulate photochemical activity (e.g. excess light) or limit carbon fixation (e.g. low temperature, deficiency of Calvin cycle enzymes).

## Protection against photoinhibition

The photosynthetic rates of protoplasts decline when exposed to supra-optimal light intensity of 2500 $\mu E\ m^{-2}\ s^{-1}$, demonstrating the photoinhibition of photosynthesis. Not only photosynthesis but also dark respiration gets reduced under such photoinhibitory conditions. Mitochondrial respiration forms an additional defence mechanism to protect illuminated leaf cells against photoinhibition (Saradadevi & Raghavendra, 1992). This suggestion is based on three observations: restriction of respiration by test compounds, decrease in respiratory rates due to photoinhibitory light and marked promotion of photoinhibition even at very low concentrations of

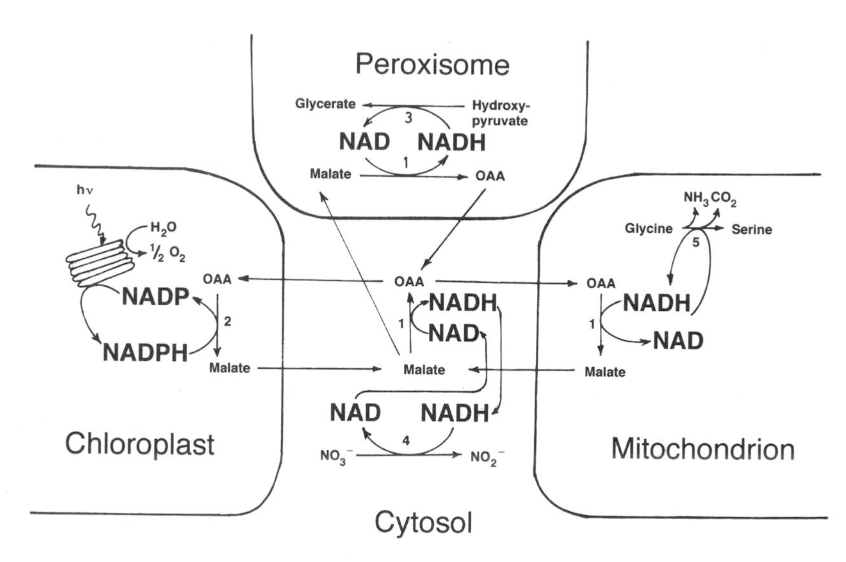


**Figure 15.2.** Exchange of redox equivalents (indicated by bold letters) between four compartments within a photosynthetic cell. Reduced equivalents generated from photosynthesis in chloroplasts or glycine oxidation in mitochondria are exported to peroxisomes and cytosol where they are used to reduce either hydroxypyruvate or $NO_3^-$, respectively. The relevant and key enzymes are indicated by numbers. 1, NAD-MDH; 2, NADP-MDH; 3, hydroxypyruvate reductase; 4, NR; and 5, glycine decarboxylase complex. See also Fig. 15.3.

classic inhibitors of mitochondrial metabolism, such as oligomycin or antimycin A.

There are three possible factors that could facilitate respiration to protect the mesophyll protoplasts against photoinhibition. Respiratory metabolism could either (a) elevate the level of intracellular $CO_2$, particularly at $CO_2$ limiting conditions, (b) provide extra energy towards the turnover of PSII-associated D1 protein within the chloroplast, by meeting cytosolic demands for ATP or (c) help to maintain an optimal redox state in the chloroplasts or cytosol of the cells. Even a marginal interference of photosynthetic metabolism by respiratory inhibitors made protoplasts highly susceptible to photoinhibitory light. Thus, mitochondrial oxidative metabolism appears to be a protective valve to prevent the over-reduction of the photosynthetic electron transport chain in chloroplasts.

A slightly different situation is observed in the cyanobacterium *Anacystis nidulans*, where dark respiration modulates the process of photoinhibition of photosynthesis and its reactivation (Shyam, Raghavendra & Sane, 1993). Photoinhibition is a result of the interference with the degradation and repair of the D1 protein which requires energy. Any limitation in ATP supply would accelerate photoinhibition. Decreased ATP production due to restricted respiration is proposed as the primary cause for stimulation of photoinhibition in *Anacystis*.

## Photosynthesis at low temperature

Mitochondrial respiration plays an important role in optimizing photosynthesis during cold hardening in winter rye (see also Chapter 18, this volume).

Cold hardening of winter rye increases dark respiration by about 75% and the capacity of photosynthetic $O_2$ evolution by about two-fold, compared to those in non-hardened plants. Oligomycin inhibited photosynthesis in non-hardened and cold-hardened leaves by 14 and 25%, respectively, and decreased photochemical quenching of chlorophyll *a* fluorescence to a greater degree in cold-hardened than that in non-hardened leaves. These results indicate an increase both in the rate of respiration in the light and the importance of respiration to photosynthesis following cold hardening (Hurry *et al.*, 1995). Metabolite analysis indicated that oligomycin inhibited photosynthesis by limiting regeneration of ribulose-1, 5-bisphosphate and this was particularly severe in cold-hardened leaves. In winter rye, mitochondria function in the light (at least at higher photon flux density) by using the reducing equivalents generated by non-cyclic photosynthetic electron transport in spite of increasing the availability of cytosolic ATP. The co-operative action of chloroplastic and mitochondrial electron transport may become more important after prolonged growth at low, cold-hardening temperatures. Although the functional importance of this co-operative activity is unclear, it may be to balance electron flow into reducing equivalents with ATP production as a result of non-cyclic electron transport.

## Interaction in specialized cells

The interaction between respiration and photosynthesis is quite pronounced in cells which are deficient in Rubisco/Calvin cycle activity, such as stomatal guard cells and mutants of *Chlamydomonas* (Raghavendra *et al.*, 1994).

Guard cells have high rates of respiratory activity but contain very low levels of rubisco and consequently limited carbon metabolism through Calvin cycle. Despite the limited $CO_2$ fixation in guard cells, the reduced equivalents produced by their chloroplasts are exported to the cytosol through OAA–malate or PGA–DHAP shuttles. The reduced pyridine nucleotides (NADH) formed in the cytosol from the oxidation of malate and/or DHAP may act as the respiratory substrates for mitochondrial ATP production, needed for $K^+$ uptake. A very strong interaction between respiration and photosynthesis has been shown in guard cell protoplasts of *Vicia faba* and *Brassica napus* at varying $O_2$ concentrations. A strong cooperation between chloroplasts and mitochondria appears to be essential for the maintenance of guard-cell bioenergetic processes.

A similar situation appears to operate in two mutants of *Chlamydomonas reinhardtii*, one devoid of rubisco and the other lacking functional chloroplast ATP synthase. The *C. reinhardtii* mutant FUD50 lacks β-subunit of chloroplast ATP synthase and cannot produce ATP during photophosphorylation. A modified strain of this mutant FUD50su can grow under photoautotrophic conditions, although it still showed no synthesis of the β-subunit of coupling factor. Photosynthesis in FUD50su mutant was extremely sensitive to inhibitor antimycin A, a specific inhibitor of mitochondrial electron transport. Photosynthesis in the FUD50su strain is achieved through an unusual interaction between mitochondria and chloroplast. Export of reduced compounds, made in light, from the chloroplast to the mitochondria elicits ATP formation in the latter, and ATP is subsequently imported to the chloroplast.

## BIOCHEMICAL BASIS FOR INTERACTION: METABOLITE EXCHANGE BETWEEN CHLOROPLASTS, CYTOSOL AND MITOCHONDRIA

The biochemical basis of the mutually beneficial interaction between the process of photosynthesis and dark respiration is the rapid exchange of metabolites between chloroplasts, cytosol and mitochondria (Fig. 15.3). The transport of PGA/DHAP and OAA/malate between chloroplasts and the cytosol is a well-known phenomenon (Heineke *et al.*, 1991). Not only a rapid transport of ATP, ADP and Pi, but also metabolite shuttles of malate and OAA occur between mitochondria and the cytosol (Douce, 1985; Douce & Neuburger, 1990). Such metabolite transport systems can achieve a favourable balance of adenine and pyridine nucleotides in these compartments.

A photosynthetic cell has two different systems to produce ATP: photophosphorylation and oxidative phosphorylation. For a long time it was believed that photophosphorylation would drain all the ADP away from the mitochondria and restrict oxidative phosphorylation. At least three factors suggest that this may not occur. First, the non-phosphorylating path in leaves would allow electron transport to continue even at low levels of available ADP. Secondly, the levels of ADP in the mitochondria are not very low in the light. And thirdly, glycine is preferred substrate for mitochondria in light.

An attractive means of possible regulation is the mitochondrial oxidation of photosynthetically reduced pyridine nucleotides. Illuminated chloroplasts are expected to have excess NADPH or related metabolites since their electron transport activity exceeds the capacity of carbon fixation. The excess reducing equivalents are transported from the chloroplasts (in the form of DHAP and malate) to the cytosol to generate NAD(P)H. Mitochondria are capable of oxidising external NADH or NADPH under certain conditions. The oxidation could also be indirect through the shuttles of related metabolites formed during photosynthesis.

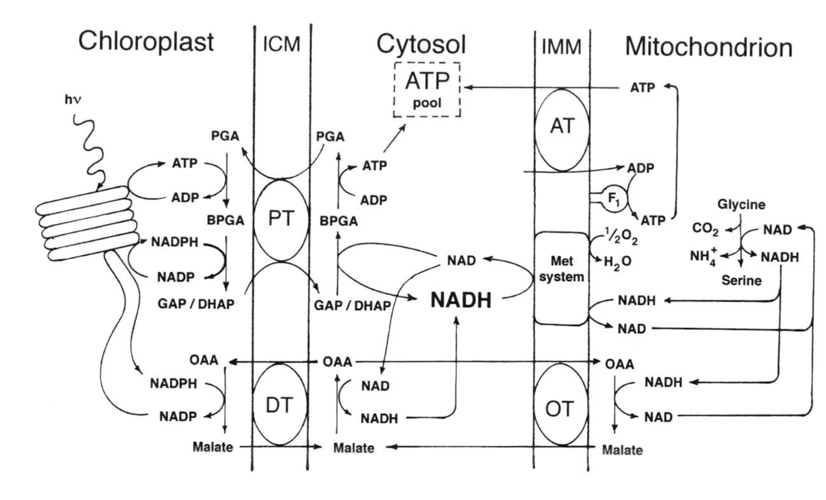


**Figure 15.3** Rapid movement of metabolites between chloroplasts and mitochondria via the cytosol form biochemical basis for the interaction between photosynthesis and respiration (see also Fig. 15.2). Chloroplasts export ATP and NADPH to the cytosol via the phosphate translocator (PT) and/or dicarboxylate translocator (DT) located in the inner chloroplast envelope membrane (ICM). Mitochondria may either export reduced pyridine nucleotides equivalents via the OAA translocator (OT) or use cytosolic NADH directly by the external NADH dehydrogenase and components of the mitochondrial electron transport system (MET system) located in the inner mitochondrial membrance (IMM). The direct export of ATP by the mitochondrial adenylate translocator (AT) contributes to the intracellular ATP pool, along with the supply from the chloroplasts. GAP, glyceraldehyde-3-phosphate; BPGA, glycerate-1,3-bisphosphate (Adapted from Raghavendra *et al.*, 1994.)

The difference in redox potentials of the stroma (NADPH/NADP) and cytosol (NADH/NAD) is known to be quite large, particularly in light. Redox equivalents can be transferred from the chloroplast stroma to the cytosol by two different metabolites: the triose-P-PGA mediated by the phosphate translocator and the malate–OAA shuttle facilitated by the dicarboxylate translocator. The operation of a malate–OAA shuttle by mitochondria facilitates further the exchange of reducing equivalents between mitochondria and the cytosol or peroxisomes (Fig. 15.3). The triose-P-PGA shuttle is controlled by $P_i$ availability for counter-exchange by the phosphate translocator, chloroplastic PGA reduction and cytosolic triose-P oxidation. The malate–OAA shuttle is regulated by stromal NADP–MDH and the [NADPH]/[NADP], and also by the translocating step across the inner chloroplast envelope membrane.

On the basis of the metabolite movements described above, the photosynthetic and respiratory activity in chloroplasts and mitochondria, respectively, appears to be modulated by one or both of the following factors: (a) the redox state due to the relative levels of NAD(P) or NAD(P)H, and (b) inter-organelle movement of metabolites such as PGA, DHAP, malate and OAA. Adenine nucleotides (ATP, ADP, AMP) and/or cytosolic pH also could regulate indirectly mitochondrial and/or chloroplastic reactions.

## INTERACTION OF PHOTOSYNTHESIS AND NITROGEN METABOLISM

Photosynthesis involves assimilation and metabolism of not only carbon but also nitrogen. The complex interactions between the photosynthetic assimilation of

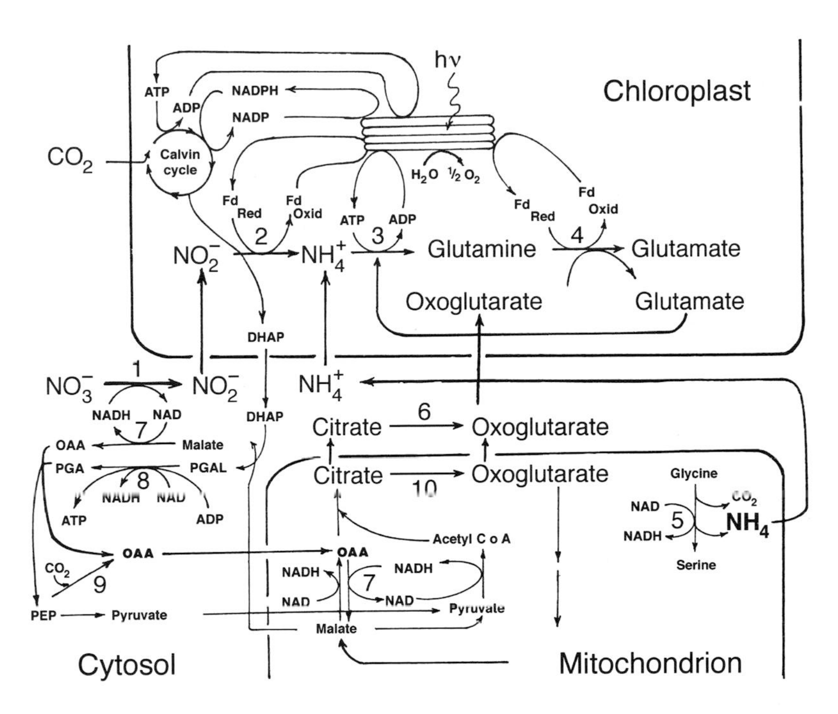


**Figure 15.4.** Photosynthetic nitrogen metabolism in chloroplasts supported by exchange of metabolites from mitochondria and cytosol (key components indicated by bold letters). The formation and assimilation of NH4$^+$ into Glu is supported by reduced ferredoxin and ATP from photosynthetic electron transport chain. Mitochondria supply 2-OG (the major carbon source), besides ammonia (released during glycine decarboxylation, details in Fig. 15.1). The reduction of $NO_3^-$ occurs in cytosol, drawing reductant from either chloroplasts or mitochondria (further details in Fig. 15.2). Carbon skeletons may come from photosynthetically generated PGA through glycolysis (in the form of pyruvate) and/or PEPC (OAA). Key enzymes are indicated by numbers. 1, NR; 2, nitrite reductase; 3, GS; 4, GOGAT; 5, Glycine decarboxylase; 6, aconitase and NADP-IDH; 7, NAD-MDH; 8, NAD-phosphoglyceraldehyde dehydrogenase; 9, PEPC; 10, aconitase and NAD-IDH.

$NO_3^-$ and that of $CO_2$ in photosynthetic cells and tissues is exerted at different levels. The reducing power generated by photosynthetic electron transport is used for reduction of $NO_3^-$ to $NO_2^-$ and 2-OG to glutamate (Glu) in chloroplasts. Mitochondria and cytosol collaborate in this process by recycling the carbon skeletons (Fig. 15.4). These processes benefit each other, with a major result being the prevention of over-reduction of chloroplasts whose photochemical activities are usually in excess of their capacity of carbon assimilation (Turpin & Weger, 1990; Lara, 1992; Weger *et al.*, 1992; Lea, 1993).

## Nitrogen and ammonia assimilation in chloroplasts

$NO_3^-$ assimilation is a genuine photosynthetic process. A significant part of nitrogen metabolism occurs within the chloroplasts and is dependent on ATP and reducing power, generated by photosynthetic electron transport chain. Among these chloroplastic reactions are: (i) reduction of $NO_2^-$ to ammonia in a light-dependent manner, (ii) rapid reassimilation of ammonia, through GS and GOGAT (iii) conversion of 2-OG to Glu using reduced ferredoxin and subsequent transamination of pyruvate and OAA to alanine and aspartate, respectively (iv) light-dependent reduction and assimilation of $SO_4^{2-}$ into cysteine. Nitrite is one of the end electron acceptors, like $CO_2$, during photochemical electron transport. About 20% of photosynthetic electron transport capacity is used for $NO_3^-/NO_2^-$ reduction.

Besides, the chloroplasts can make several of the essential amino acids, most of them in light-dependent reactions (see Chapter 14, this volume). All these processes require either ATP or reduced power (in the form of reduced ferredoxin or NADPH or reduced membrane-thiols).

## Regulatory role of $CO_2$ fixation on nitrate assimilation

In higher plants, $NO_3^-$ cannot be assimilated unless $CO_2$ fixation is operative or a carbohydrate source is available. In cyanobacterial cells, such as *Anacystis*, $NO_3^-$ transport is inhibited by $NH_4^+$ addition which suggests that $NO_3^-$ transport could be subject to feedback control by $NO_3^-$ assimilation. In plant leaves, a different strategy seems to operate. In spinach leaves, limitation on $CO_2$ fixation by either stomatal closure under mild stress or restriction of $CO_2$ supply, results in decreased $NO_3^-$ reduction with practically no effect on net $NO_3^-$ uptake into the leaf. Although it is possible that $CO_2$ stress restricts the $NO_3^-$ efflux from the vacuole, it has been shown that, upon $CO_2$ removal, the extractable NR activity of the leaves decreases rapidly and recovers when $CO_2$ is added back, suggesting that in spinach leaves NR activity is sensitive to $CO_2$ levels.

## Effects of nitrogen assimilation on $CO_2$ fixation

The operativity of the GS–GOGAT pathway and transamination reactions under conditions of nitrogen assimilation would accelerate the carbon flow through the TCA cycle and the glycolytic pathways at the expense of carbon flow to starch and sucrose. This diversion of carbon flow has no negative effects on the operativity of the Calvin cycle and might even increase $CO_2$ fixation under light and $CO_2$ saturating conditions. Enhanced operation of the glycolytic pathway can effectively contribute to inorganic phosphate recycling, an intrinsic factor limiting $CO_2$ fixation under these conditions. It is also possible that, when the rates of $NO_3^-$ or $NH_4^+$ assimilation are higher than normal, as occurs after nitrogen starvation, competition for carbon skeletons may result in severe depression of $CO_2$ fixation.

The requirement of reductant for $NO_3^-$ assimilation and $CO_2$ fixation leads to a competition between these two processes in photosynthesizing tissues. However, this does not happen, as the capacity of photosynthetic electron transport is far in excess of that required for carbon fixation.

## ROLE OF MITOCHONDRIA IN NITROGEN ASSIMILATION

In leaves, Glu and glutamine (Gln) are major products of $NO_3^-$ assimilation via the GS-GOGAT pathway. Mitochondria provide the carbon skeletons in the form of 2-OG. For this pyruvate and/or OAA (formed from either PGA, photosynthetic product or PEPC, both in the cytosol) act as precursors for the formation of citrate in the mitochondria (Hanning & Heldt 1993; Krömer *et al.*, 1996*a*,*b*). Then citrate is exported to the cytosol and converted to 2-OG via cytosolic aconitase and NADP-isocitrate dehydrogenase (IDH) (Figs. 15.1 and 15.4).

When the green alga *Selenastrum minutum* is transferred from the N-limited growth conditions to excess $NH_4^+$, there is a dramatic increase in TCA cycle carbon flow and $CO_2$ release (in both light and darkness). This increase in TCA cycle activity implies that the subsequent consumption of NADH and $FADH_2$ via the mitochondrial electron transport chain, results in enhanced rates of $O_2$ consumption. In *C. reinhardtii*, studies with respiratory inhibitors and $^{18}O_2$ discrimination experiments indicated that respiratory electron flow was mediated entirely via the cytochrome pathway in both the light and dark, despite a large capacity for the alternative pathway. The alternative pathway serves to maintain TCA cycle carbon flow in support of biosynthesis when mitochondrial electron transport chain activity is restricted by ADP supply. This may be especially important in the light, where photophosphorylation may meet cellular ATP demands. These different mechanisms of oxidation of TCA cycle reductant are reflected by high sensitivity to salicylhydroxamic acid of $NH_4^+$ assimilation by cytochrome oxidase-deficient cells as compared to wild type.

Unlike $NH_4^+$, assimilation of $NO_3^-$ or $NO_2^-$ was relatively insensitive to anaerobiosis. These results indicated that operation of the mitochondrial chain was not required to maintain TCA cycle activity during $NO_3^-$ and $NO_2^-$ assimilation, suggesting as alternative sink for TCA cycle-generated reductant. Apparently, TCA cycle is capable of supplying both carbon skeletons for amino acid synthesis and reducing power for $NO_3^-$ and $NO_2^-$

reduction not only in the dark but also during photosynthesis. Evaluation of changes in gross $O_2$ consumption during $NO_3^-$ and $NO_2^-$ assimilation suggest that reductant from TCA cycle was exported to the chloroplast during photosynthesis and used to support $NO_3^-$ and $NO_2^-$ reduction. Further these results imply that mitochondrial respiration can supply reducing power to the chloroplast during photosynthetic conditions where the production of photogenerated reductant is inadequate to meet metabolic demands. The possibility of such transfer of reducing power from the mitochondrion to the chloroplast for use in $NO_2^-$ reduction during photosynthesis highlights the complexity of the interactions between photosynthesis, respiration and nitrogen assimilation.

## INTEGRATION OF CARBON PARTITIONING DURING PHOTOSYNTHESIS AND NITROGEN METABOLISM

The formation of both carbohydrate and amino acids require energy and carbon which are provided by photosynthetic process. The direction of carbon flow towards either sucrose or amino acids is strikingly regulated within the leaves so as to sustain both $NO_3^-$ assimilation and carbohydrate formation. When the leaves of $C_3$ or $C_4$ plants are fed with nitrogen ($KNO_3$, $KNO_2$ or $NH_4Cl$), the extent of $CO_2$ fixation or photochemical $O_2$ evolution are either unaffected or slightly increased. However, the sucrose formation is reduced, while the amino acid formation is stimulated on addition of nitrogen source. In summary, nitrogen availability causes a marked shift in the partitioning of carbon towards amino acid biosynthesis (Champigny, 1995).

Such interesting integration of photosynthetic carbon and nitrogen metabolism in higher plants is brought out by the modulation of three key enzymes: PEPC, SPS and NR. PEPC catalyses the anapleurotic pathway of carbon fixation into OAA, leading to the formation of other keto acids and finally the corresponding amino acids. Sucrose-P synthesis controls the carbon flux into sucrose and modulates the photosynthetic rate. NR is the first enzyme in the process of $NO_3^-$ assimilation and regulates the flow of nitrogen into amino acids. All these three enzymes are located in the cytosol of mesophyll cells.

Post-translational modification of PEPC and SPS contributes to the modulation and synchronization of $N_2$ assimilation, C-fixation and partitioning of carbon. The phosphorylated form of PEPC is active, while SPS is active in dephosphorylated state. On exposure to $NO_3^-$ (a nitrogen source) the phosphorylation status of these two enzymes is increased due to enhanced activity of protein kinase. As a result, the flow of carbon towards amino acids (through PEPC) is enhanced while sucrose formation (via SPS) is restricted. Although NR is also subject to phosphorylation/dephosphorylation cascade, this seems to be not so critical. On the other hand, there is a marked increase in the synthesis of NR-protein directing nitrogen towards amino acid formation. Thus, protein phosphorylation appears to be the primary event allowing the short-term adaptation of leaf carbon metabolism in response to changes in nitrogen supply. It is possible, that Gln, the primary product of $NO_3^-$ assimilation is the metabolite effector for short-term modulation of PEPC and SPS.

## CONCLUDING REMARKS

A rapid and mutually beneficial interaction between chloroplast photosynthesis and mitochondrial respiration occurs in plant cells. The 'coarse' (long-term) control of the interaction between photosynthesis and respiration appears to be through the levels of soluble sugars, while the 'fine' (short-term) control is exerted by intracellular redox state or adenine/pyridine nucleotide levels. There is a marked turnover of redox equivalents between chloroplasts, mitochondria, cytosol and peroxisomes, particularly under photorespiration/$CO_2$ limiting conditions. The dependence of Gln and Glu biosynthesis on chloroplasts (for ATP and reducing power) and mitochondria (for 2-OG) makes the triangular interaction between photosynthesis, nitrogen assimilation and respiration extremely interesting. However, the available information on such an interaction in higher plants is much less than that in algal cells.

Respiration can prevent the over-reduction of the photosynthetic electron transport chain in illuminated chloroplasts by providing the outlet for excess reducing equivalents to the cytosol or mitochondria. As a result, oxidative electron transport and phosphorylation in mitochondria not only benefit photosynthesis but also protect isolated leaf protoplasts against photoinhibition. It should be of great interest to examine the consequence of such marked interaction between respiration and photosynthesis under other situations, such as $CO_2$ enrichment or sunflecks or shaded environment.

ACKNOWLEDGEMENTS

Work in our laboratory on photosynthesis and respiration in mesophyll and guard cell protoplasts was supported by grants from Career Award of University Grants Commission, New Delhi, Council of Scientific and Industrial Research, New Delhi and Department of Atomic Energy, Bombay. KP is a recipient of SRF from University Grants Commission. We wish to thank Professors J. Amthor, D. D. Randall, H. G. Weger, S. Krömer and P. Gardeström for kindly providing us with reprints of their work.

REFERENCES

Amthor, J. S. (1989). *Respiration and Crop Productivity*. Berlin: Springer-Verlag.

Amthor, J. S. (1994). Higher plant respiration and its relationship to photosynthesis. In *Ecophysiology of Photosynthesis, Ecological Studies*, vol. 100, ed. E. D. Schulze & M. M. Caldwell, pp. 71–101. Berlin: Springer-Verlag.

Avelange, M-H., Thiéry, J. M., Sarrey, F., Gans, P. & Rébeillé, F. (1991). Mass-spectrometric determination of $O_2$ and $CO_2$ gas exchange in illuminated higher-plant cells. Evidence for light-inhibition of substrate decarboxylations. *Planta*, **183**, 150–7.

Azcón-Bieto, J. (1992). Relationship between photosynthesis and respiration in the dark in plants. In *Trends in Photosynthesis Research*, ed. J. Barber, M. G. Guerrero & H. Medrano, pp. 241–53. Andover, UK: Intercept.

Canvin, D. T. (1990). Photorespiration and $CO_2$-concentrating mechanisms. In *Plant Physiology, Biochemistry and Molecular Biology*, ed. D. T. Dennis & D. H. Turpin, pp. 253–73. Harlow, UK: Longman Scientific & Technical.

Carrier, P., Chagvardieff, P. & Tapie, P. (1989). Comparison of the oxygen exchange between photosynthetic cell suspensions and detached leaves of *Euphorbia characias* L. *Plant Physiology*, **91**, 1075–9.

Champigny, M-L. (1995). Integration of photosynthetic carbon and nitrogen metabolism in higher plants. *Photosynthesis Research*, **46**, 117–27.

Douce, R. (1985). *Mitochondria in Higher Plants*. Orlando: Academic Press.

Douce, R. & Neuburger, M. (1990). Metabolite exchange between the mitochondria and the cytosol. In *Plant Physiology, Biochemistry and Molecular Biology*, ed. D. T. Dennis & D. H. Turpin, pp. 173–90. Harlow, UK: Longman Scientific & Technical.

Garab, G., Lajko, F., Mustardy, L. & Marton, L. (1989). Respiratory control over photosynthetic electron transport in chloroplasts of higher-plant cells: evidence for chlororespiration. *Planta*, **179**, 349–58.

Gardeström, P. & Lernmark, U. (1995). The contribution of mitochondria to energetic metabolism in photosynthetic cells. *Journal of Bioenergetics and Biomembranes*, **27**, 415–21.

Gardeström, P., Zhou, G. & Malmberg, G. (1992). Respiration in barley protoplasts before and after illumination. In *Molecular, Biochemical and Physiological Aspects of Plant Respiration*, ed. H. Lambers & L. H. W. van der Plas, pp. 261–5. The Hague: SPB Academic Publishing.

Gibbs, M., Willeford, K., Ahluwalia, K. J. K., Gombos, Z. & Jun, S-S. (1989). Chloroplast respiration. In *Perspectives in Biochemical and Genetic Regulation of Photosynthesis*, ed. I. Zelitch, pp. 339–53. New York: Alan R. Liss.

Graham, D. (1980). Effects of light on dark respiration. In *Biochemistry of Plants. A Comprehensive Treatise*, vol. 2, ed. D. D. Davies, pp. 525–79, New York: Academic Press.

Graham, D. & Chapman, M. D. (1979). Interactions between photosynthesis and respiration in higher plants. In *Encyclopaedia of Plant Physiology, New series*, vol. 6, ed. M. Gibbs & E. Latzko, pp. 150–62. Berlin: Springer-Verlag.

Hanning, I. & Heldt, H. W. (1993). On the function of mitochondrial metabolism during photosynthesis in spinach leaves (*Spinacia oleracea* L.). Partitioning between respiration and export of redox equivalents and precursors for nitrate assimilation products. *Plant Physiology*, **103**, 1147–54.

Heineke, D., Riens, B., Grosee, H., Hoferichter, P., Peter U., Flugge, U. I. & Heldt, H. W. (1991). Redox transfer across the inner chloroplast envelope membrane. *Plant Physiology*, **95**, 1131–7.

Heupel, R. & Heldt, H. W. (1992). Redox transfer between mitochondria and peroxisomes. In *Molecular, Biochemical and Physiological Aspects of Plant Respiration*, ed. H. Lambers & L. H. W. van der Plas, pp. 243–7. The Hague: SPB Academic Publishing.

Hill, S. A. & Bryce, J. H. (1992). Malate inhibition and light-enhanced dark respiration in barley protoplasts. In *Molecular, Biochemical and Physiological Aspects of Plant*

*Respiration*, ed. H. Lambers & L. H. W. van der Plas, pp. 221–30. The Hague: SPB Academic Publishing.

Hurry, V., Tobiaeson, M., Krömer, S., Gardeström, P. & Oquist, G. (1995). Mitochondria contribute to increased photosynthetic capacity of leaves of winter rye (*Secale cereale* L.) following cold-hardening. *Plant, Cell and Environment*, **18**, 69–76.

Husic, D. W., Husic, H. D. & Tolbert, N. E. (1987). The oxidative photosynthetic carbon cycle or $C_2$ cycle. *Critical Reviews in Plant Science*, **5**, 45–100.

Krömer, S. (1995). Respiration during photosynthesis. *Annual Review of Plant Physiology and Plant Molecular Biology*, **46**, 45–70.

Krömer, S., Malmberg, G. & Gardeström, P. (1993). Mitochondrial contribution to photosynthetic metabolism. *Plant Physiology*, **102**, 947–55.

Krömer, S., Gardeström, P. & Samuelsson, G. (1996*a*). Regulation of the supply of cytosolic oxaloacetate for mitochondrial metabolism via phospho*enol*pyruvate carboxylase in barley leaf protoplasts. I. The effect of covalent modification on PEPC activity, pH response, and kinetic properties. *Biochimica et Biophysica Acta*, **1289**, 343–50.

Krömer, S., Gardestöm, P. & Samuelsson, G. (1996*b*). Regulation of the supply of oxaloacetate for mitochondrial metabolism via phospho*enol*pyruvate carboxylase in barley leaf protoplasts. II. Effects of metabolites on PEPC activity at different activation states of the protein. *Biochimica et Biophysica Acta*, **1289**, 351–61.

Lara, C. (1992). Photosynthetic nitrate assimilation. Interactions with $CO_2$ fixation. In *Trends in Photosynthesis Research*, ed. J. Barber, M. G. Guerrero & H. Medrano, pp. 195–208. Andover, UK: Intercept.

Lawlor, D. W. (1993). Metabolism of photosynthetic products. In *Photosynthesis: Molecular, Physiological and Environmental Process*, 2nd edn, pp. 155–76. Essex, England: Longman Scientific & Technical.

Lea, P. J. (1993). Nitrogen metabolism. In *Plant Biochemistry and Molecular Biology*, ed. P. J. Lea & R. C. Leegood, pp. 155–80. New York: John Wiley.

Luethy, M. H., Miernyk, J. A., David, N. R. & Randall, D. D. (1996). Plant pyruvate dehydrogenase complexes. In *Alpha-keto Acid Dehydrogenase Complexes*, ed. M. S. Patel, T. E. Roche & R. A. Haris, pp. 71–92. Basel/Switzerland: Birkhäuser Verlag.

Raghavendra, A. S., Padmasree, K. & Saradadevi, K. (1994). Interdependence of photosynthesis and respiration in plant cells: interactions between chloroplasts and mitochondria. *Plant Science*, **97**, 1–14.

Randall, D. D., Miernyk, J. A., David, N. R., Gemel, G. & Luethy, M. H. (1996). Regulation of leaf mitochondrial pyruvate dehydrogenase complex activity by reversible phosphorylation. In *Protein Phosphorylation in Plants*, ed. P. R. Shewry, N. G. Halford & Hooley, pp. 87–103. Oxford: Clarendon Press.

Reddy, M. M., Vani, T. & Raghavendra, A. S. (1991). Light enhanced dark respiration in mesophyll protoplasts from leaves of pea. *Plant Physiology*, **96**, 1338–71.

Saradadevi, K. & Raghavendra, A. S. (1992). Dark respiration protects photosynthesis against photoinhibition in mesophyll protoplasts of pea (*Pisum sativum*). *Plant Physiology*, **99**, 1232–7.

Shyam, R., Raghavendra, A. S. & Sane, P. V. (1993). Role of dark respiration in photoinhibition of photosynthesis and its reactivation in the cyanobacterium *Anacystis nidulans*. *Physiologia Plantarum*, **88**, 446–52.

Stokes, D., Walker, D. A., Grof, C. P. L. & Seaton, G. G. R. (1990). Light enhanced dark respiration. In *Perspectives in Biochemical and Genetic Regulation of Photosynthesis*, ed. I. Zelitch, pp. 319–38. New York: Alan R. Liss.

Turpin, D. H. & Weger, H. G. (1990). Interactions between photosynthesis, respiration and nitrogen metabolism. In *Plant Physiology, Biochemistry and Molecular Biology*, ed. D. T. Dennis & D. H. Turpin, pp. 422–33. Harlow, UK: Longman Scientific & Technical.

Vani, T., Reddy, M. M. & Raghavendra, A. S. (1990). Beneficial interaction between photosynthesis and respiration in mesophyll protoplasts of pea during short light–dark cycles. *Physiology Plantarum*, **80**, 467–71.

Weger, H. G., Vanlerberghe G. C., Guy, R. D. & Turpin, D. H. (1992). Respiratory carbon flow to nitrogen assimilation. In *Molecular, Biochemical and Physiological Aspects of Plant Respiration*, ed. H. Lambers & L. H. W. van der Plas, pp. 149–65. The Hague: SPB Academic Publishing.

PART III

# Agronomy and environmental factors

# 16 Leaf/canopy photosynthesis and crop productivity

R. ISHII

## INTRODUCTION – COMPONENTS OF CROP YIELD

The economic yield (EY) of crop plants can be expressed as the function of biological yield (BY) and harvest index (HI) as follows:

$$EY = BY \times HI$$

A large increase of EY which has been achieved in this century, is attributed to increase in BY, HI, or both depending on the crop species. The HI is the ratio of photoassimilates partitioned to the harvesting organs to total dry matter yield, as defined by Donald in 1962. HI shows a remarkable increase not only in the breeding progress for high-yielding varieties but also in the domestication process from the wild-type to the cultivated one in various crop species. Further, improvement of HI has been realized not only in grain crops but also in root crops and other crop species. For BY, on the other hand, increase has not necessarily been realized through breeding progress for high yielding varieties. In wheat, for example, the semi-dwarf gene was introduced, and the accumulation of carbohydrate in the vegetative organs was reduced. Therefore, positive correlation cannot be observed between EY and BY in wheat. In rice plants, which were also introduced with the semi-dwarf gene, however, there were many reports of positive correlation between EY and BY (Ashraf, Akbar & Salim, 1993). In addition, in maize or sorghum, in which EY was improved by the development of F1 hybrid varieties, increase of EY seemed to be attributed to the increase of total dry matter yield. Thus, the relation between EY and BY is different with the crop species, and hence the relation is not so simple.

Even if not correlated with EY, BY is an essential and necessary condition to the high yielding of the crop plants. Since BY is the net amount of dry matter accumulation at the harvest time, BY is determined as the integrated crop growth rate (CGR) through the growing season. The CGR can be determined theoretically as the difference of the integrals of canopy photosynthesis (CPS) and canopy respiration (CRS) through the cultivated season, as follows:

$$CGR = 30/44\ (CPS - CRS)$$

where 30/44 is the conversion coefficient from $CO_2$ to carbohydrate formulated as $CH_2O$. BY, therefore, can be theoretically obtained from the measurement of CPS and CRS. The CPS has shown a good agreement with the crop yield in many crop species (Christy & Porter 1982; Shibles, Secor & Ford, 1987; Pereira, 1993). We begin this chapter with an examination of the determination of canopy photosynthesis (CPS).

## CANOPY PHOTOSYNTHESIS AND ITS COMPONENTS

The canopy photosynthesis (CPS) is composed of three parameters: leaf area index (LAI), light intercepting efficiency (LIE) and photosynthetic rate per unit leaf area (leaf photosynthesis, LPS).

### Leaf area index

#### *Optimum leaf area index for crop growth rate*

High crop growth rate is associated with high LAI in crop plants, because the leaves are the photosynthetic organs,

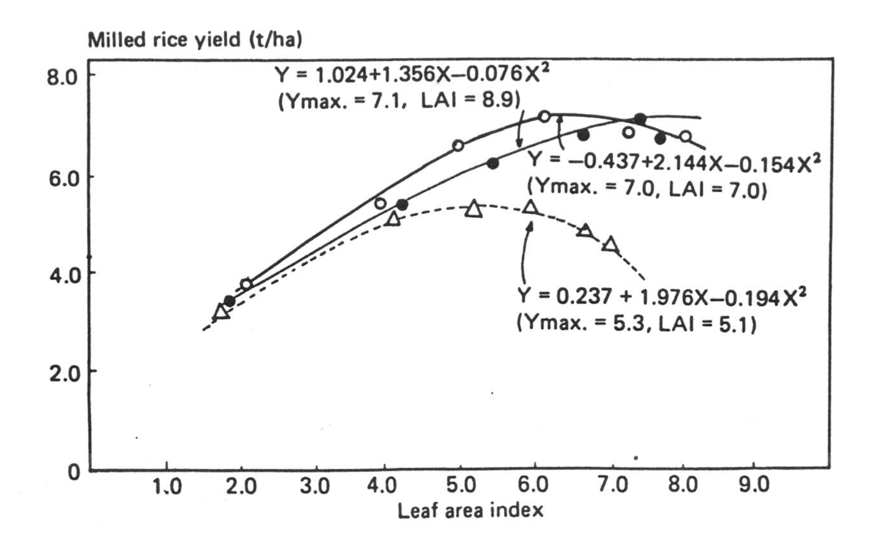


**Figure 16.1.** Relationship between yield and leaf area index in the improved rice varieties. Open triangle, Jinheung; Closed circle, Tongil; Open circle, Miryan 23. (Adapted from Crop Experimental Station of RDA, 1985.)

and hence high LAI means greater quantity of photosynthetic organs. However, there is an optimum LAI where crop growth rate reaches the maximum level. Crop growth rate increases with increment of LAI, attains the maximum level, and then decreases with increase of LAI. The mechanism of the existence of optimum LAI to crop growth rate is considered as follows. Canopy photosynthesis (CPS) increases proportionally with LAI in the range of low LAI, but the increasing rate of CPS becomes gradually slow, attaining saturation due to the mutual shading of the leaves. Canopy respiration (CRS), on the other hand, increases proportionally with the increment of LAI. Crop growth rate, the difference of CPS and CRS therefore, presents an optimum LAI around where CPS is saturated against LAI. If the mutual leaf shading occurs at higher LAI, the optimum LAI for crop growth rate will shift to a higher value. Furthermore, if optimum LAI is shifted to a higher value, the level of CGR attained at the optimum LAI will also increase, because the more photosynthetic organs that exist, the higher the CGR that will be attained. Judging from the close association of CGR with BY, eventually EY, there will also be an optimum LAI to EY.

Figure 16.1 shows the relationship between LAI and the yield of rice variety bred in Korea. Jinheung, a traditional Korean rice variety, showed an optimum LAI of 5.1 and high grain yield of 5.3 t/ha at the optimum LAI. On the other hand, Tongil, and Milyang, which were high yielding indica–japonica hybrid varieties, showed the optimum LAI as 8.9, and 7.0, respectively, and also such high grain yield as 7.1, and 7.0 t/ha, respectively, at each optimum LAI. In the improved high yielding varieties, the optimum LAI would be high, and they could achieve high level of grain yield under the high optimum LAI. To get high optimum LAI and consequently high yield, reduction of mutual shading of the leaves is essential. Therefore, the light-intercepting efficiency or canopy structure would be important as a determinant of CPS, and eventually of crop productivity.

Planting density and basal dressing of nitrogen fertilizer can control LAI. There should be optimum planting density and optimum amount of nitrogen fertilizer to get optimum LAI in each variety of each crop species. To get high crop yield, the crop canopy should attain the optimum LAI as early as possible in the growing season, and keep it up for as long as possible. Crop canopy with too low LAI or over-luxurious leaves can not attain high productivity.

### *Area of other photosynthetic organs*

There are photosynthetic organs other than leaves in some crop species. In barley, which has long awns in the spike, photosynthesis by awns is contributing greatly to CPS, compared with that in wheat (Rasmusson & Gengenbach 1983). This is because the total surface area of the barley awns are almost equal to that of a terminal leaf. Also in rapeseed plants, pods play an important role to supply photoassimilates to the seeds in the pod-filling stage. The glumes of rice spikelets also photosynthesize, but the extent of contribution to CPS or grain yield is still obscure. When we consider LAI in relation to CPS or crop yield, not only leaf area but also the surface area of other photosynthetic organs should be taken into consideration.

## Light intercepting efficiency (LIE)

### *Canopy structure*

Light extinction coefficient can be used to quantify the extent of light penetration into the plant canopy of various species. The light extinction within the canopy follows Lambert–Beer's Law.

$$\ln I/Io = -kL$$

where, $Io$ is the incident light intensity on the canopy, $I$ is the light intensity at a certain level in the canopy where the cumulative leaf area index from the top of canopy is $L$. The value of $k$ is the light extinction coefficient which is specific to the respective plant canopy. The decrease in the extinction coefficient represents an increase in light penetration into the canopy. Plants with the smaller values of $k$ were monocotyledonous plant canopies with long, thin and upright leaves, such as *Phragmites communis* or *Miscanthus sinensis*, while the plant species with larger $k$ were dicotyledonous ones with round leaves, such as *Chenopodium album* or *Helianthus tuberosus*. These observations gave a theoretical background to the breeding for high yielding varieties, and to the agronomic research (especially of rice), where a cultivation method for high yield was developed.

Besides the theoretical studies, practical improvement of light intercepting efficiency was achieved during breeding for high-yielding varieties in many crop species. The LIE is strongly associated with morphological leaf traits and leaf distribution in the canopy.

Leaf erectness is the most distinct determinant of LIE. This effect is particularly remarkable in rice. Grain yield of rice tends to increase with decrease of plant height, and short plants usually have erect leaves. The CPS can be modified in the rice canopy with mechanically drooped leaf, by putting a small weight on the top of leaf blades, and comparing with the untreated erect leaf canopy. High CPS in the untreated canopy was found to be due to leaf erectness. In rice, the highest CPS was found in the new cultivar group, indicating that the flag leaves were getting more erect with varietal improvement from old to new cultivars. Thus, breeding for high yielding in rice had been achieved in part through the improvement of leaf inclination.

Leaf erectness is often linked with other traits of the plant. In maize, the plants with liguleless genes had erect leaves, and hence light penetration into the canopy was greater in the liguleless plant canopy than in the plant canopy with ligule leaves, leading to increase of the yield. This positive effect of upright leaf on the grain yield seems to be great under high leaf area index population not only in maize but also in rice. High LAI is usual in the conditions of heavy nitrogen application and high planting density, which are typical cultivating conditions of modern cultivars in many crop species.

In contrast to rice and maize, barley shows a quite different tendency in the relationship between leaf erectness and grain yield. Wu of Jiling Agricultural University of China compared the grain yield between the variety groups with the horizontal, and the erect leaf type, and found that the average grain yield of horizontal, and erect leaf type variety group was 363, and 345 $g/m^2$, respectively (personal communication), showing practically no difference in yield between different leaf-type varieties. In soybean too, Wells *et al.* (1982) reported that light interception efficiency had no influence on canopy photosynthesis. Therefore, LIE is a determinant factor to canopy photosynthesis of maize and rice, but it is not so important in other crop species.

Trenbath and Angus (1975) reported in the review of the relationship between the leaf inclination and crop productivity that the effect of leaf erectness was particularly large in rice and maize, while it seemed to be not so important in other cereal crops (Table 16.1). Furthermore, the effect of leaf erectness is large in the canopies with higher LAI than 4 at least. Light-intercepting efficiency is important in high LAI conditions which can be achieved by high planting density. In the case of rice, the introduction of semi-dwarf gene led to high planting density, and erect leaf-type plants were selected in such conditions. In the case of wheat, the plant canopy reaches the maximum LAI in spring when solar elevation is still low, especially in the higher latitude regions. Trenbath and Angus thought this might be the reason for the

Table 16.1 *Results of field trials comparing the grain yield of cultivars with different leaf inclination.*

| Crop | % Advantage of erectophile type | Method of estimation of leaf inclination |
|---|---|---|
| Wheat ($C_3$) | −10 | Visual |
| Wheat ($C_3$) | +19 | Visual |
| Barley ($C_3$) | +10 | Light[a] |
| Barley ($C_3$) | −5 | Light |
| Rice ($C_3$) | +68 | Visual |
| Rice ($C_3$) | +49 | Light |
| Rice ($C_3$) | +34 | Visual |
| Maize ($C_4$) | +41 | Visual |
| Maize ($C_4$) | +7 | Measured |
| Maize ($C_4$) | +22 | Visual |
| Maize ($C_4$) | +8 | Visual |
| Maize ($C_4$) | −18[b] +38[c] | Visual |
| Maize ($C_4$) | −13 | Visual |
| Maize ($C_4$) | +4 | Measured |

[a]'Light' indicates the use of an indirect method involving light interception. Wheat and barley were grown during spring or summer as stated, but other crops were grown during the temperate summer or in the tropics.
[b]At low density.
[c]At high density.
(Modified from Trenbath & Angus, 1975.)

absence of any link between high yield and leaf erectness in wheat and some other winter crops.

### *Panicle position*

The LIE is influenced also by shading of the panicles. In maize, grain yield is increased when the tassels are removed, because the tassels shade the canopy, resulting in reduction of CPS. Also in rice, high-positioned panicles reduce CPS in the same way as the tassels of maize. So, low position of the panicles is preferable for maintaining high CPS in rice.

### *Leaf orientation adjustment*

High light intensity, such as full sunlight in the summer season, is harmful to plant leaves, especially under the water-stressed conditions. The crop leaves are often damaged by high intensity of the sunlight as termed by photoinhibition (refer to Chapter 20). To avoid photoinhibition, some crop species, especially leguminous crop plants, change their leaf orientation (Hirata *et al.*, 1983*a*). The leaves on the top of the canopy, orientate themselves parallel to the incident sunlight. By such movement, the top leaves can avoid receiving too much light, and simultaneously, the sunlight can penetrate into the crop canopy through the top layer. So, leaf orientation adjustment is also a factor of light-intercepting efficiency. Moreover, chloroplasts will be arranged in the cell in parallel to the transmitted light to make the light pass through a leaf, which might be one of the avoiding mechanisms of photoinhibition (Hirata *et al.*, 1983*b*).

### *Leaf thickness and chlorophyll content*

Light-intercepting efficiency is determined not only by the plant canopy but also by the traits of single leaf. An adaptative change of leaf traits can be observed in leaf thickness and chlorophyll content in a leaf. If the plants are grown in weak light conditions, leaves are large and thin. Theoretically the chlorophyll content per unit leaf area should be low in thin leaves. But, in fact, chlorophyll content is high in a weak light condition. This may be an adaptive strategy for the efficient trapping of photons for the performance of photosynthesis under the regime of low light intensity. In addition, the ratio of chlorophyll *a* to *b* is high in the leaves of shade plants, which is also considered as an adaptation to drive the photochemical process efficiently.

## Photosynthetic rate per unit leaf area

The apparent rate of photosynthetic $CO_2$ uptake per unit leaf area is usually called leaf photosynthesis (LPS). When the rice plant canopy was shaded at different growth stages, the effect of the shading on rice yields was high at reproductive and grain filling stage, but was low at the vegetative stage. Thus LPS, is an important determinant of grain yield of rice through CPS, particularly in the grain-filling period. When rice plants were subjected to $CO_2$ enrichment for just 10 days, which was considered not to affect any component of CPS other than LPS, the grain yield increased by increase of LPS even in such a short duration. Thus, activated leaf photosynthesis can raise the grain yield of rice. Further, top-dressing of nitrogen at the grain-filling period will increase the grain yield by slowing the leaf senescence and maintaining high LPS.

It is obvious that LPS is an important factor for

determining the grain yield through CPS. Genetic improvement of LPS can therefore be an important breeding target for high yielding varieties. Many reports are made on the positive relationship between LPS and yield in various crop species. However, it is not so simple to relate LPS to plant productivity or crop yield. In the later section of this chapter, interspecific or intraspecific differences of LPS are discussed in relation to crop yield.

## CANOPY RESPIRATION

Crop growth rate is the function of CPS in the day and canopy respiration (CRS) in the night. It can then be predicted that low night temperature will increase crop growth rate. If we grow crop in the day/night temperature regimes as 25 °C/15 °C, 25 °C/25 °C and 25 °C/35 °C, the plants in 25 °C/15 °C regime will give the best crop growth rate, as predicted by $CO_2$ budget. But, actually the plant in 25 °C/35 °C is the best in the short period experiment with rice (unpublished data). This suggests that CRS activated by high temperature has a positive effect on crop growth. During the active growth stages, CRS is occupied by growth respiration, in the later growth stage such as after heading, and the greatest part of the respiration will be maintenance respiration, since growth is stopped after heading. Therefore, in rice, respiration should not be reduced before heading, while that after heading can be reduced to increase crop growth rate. Grain yield of rice is positively correlated with the integrated daily temperature before heading, but negatively correlated with that after heading.

## RELATIONSHIP BETWEEN LEAF PHOTOSYNTHESIS AND CROP GROWTH AND YIELD

### Leaf photosynthesis of $C_3$ and $C_4$ species

The development and use of the infra-red $CO_2$ gas analyzer in the time from late 1950s to early 1960s was a great help in determining LPS in various crop species with different potential productivity. During this research, species such as maize and sugarcane were found to have remarkably high LPS compared with common crop species, such as rice, wheat and soybean. Along with the progress in our knowledge of interspecific differences of LPS, physiological and biochemical studies on $CO_2$ fixation mechanism revealed the occurrence of $C_4$ photosynthesis. Figure 16.2 demonstrates the positive correlation between leaf photosynthesis and crop growth rate in various crop species. The $C_4$ plants have high LPS, and hence high crop growth rate, and $C_3$ plants have comparatively low LPS, and hence low crop growth rate.

The possibility of introducing the high LPS ability of $C_4$ plant into $C_3$ plant evoked the scientific and practical interests of both crop scientists and plant physiologists. $C_3$ or $C_4$ or $C_4$-like species or $C_3$–$C_4$ intermediates are reported in the genera of *Panicum*, *Atriplex*, *Flaveria*, and *Moricandia*. Among the several trials for hybridization between $C_3$ and $C_4$ species, a few were successful (Brown & Bouton, 1993), but most of the hybrids were $C_3$-like. Although these investigations are of extreme scientific interest, the introduction of high LPS of $C_4$ plants into $C_3$ crops appears to be unrealistic at present.

Ueno *et al.* (1988) and Uchino *et al.* (1995) recently discovered that *Eleocharis vivipara* and *E. baldwinii* showed $C_3$ and $C_4$-type characteristics not only in $CO_2$ fixation pattern but also in leaf anatomy, in the plant grown under submerged, and terrestrial condition, respectively. When the plants grown in the terrestrial conditions were transferred to the submerged condition, they changed both $CO_2$ fixation pattern and leaf anatomy from being $C_4$-like to $C_3$-like. This finding shows the possibility still exists for the modulation of gene expression of $C_3$ or $C_4$ photosynthesis.

The reason for low LPS in $C_3$ plants could be attributed to photorespiratory $CO_2$ loss. If photorespiration is suppressed, apparent LPS should increase. Environmental or chemical control of photorespiration was therefore attempted to increase apparent LPS of $C_3$ crop plants, such as rice and soybean. When the $O_2$ concentration was reduced to as low as 2%, LPS of $C_3$ plants like rice or soybean increased to the level of $C_4$ plants. Although vegetative growth was increased tremendously, grain yield was not increased in grain crops. This could be because reduced $O_2$ concentration directly hampered the fertilization and produced many aborted grains.

As the chemical regulation, several inhibitors of photorespiratory glycolic acid metabolism were discovered (Cho *et al.*, 1983). Among them, aminoacetonitrile (AAN) was found to be a specific inhibitor of

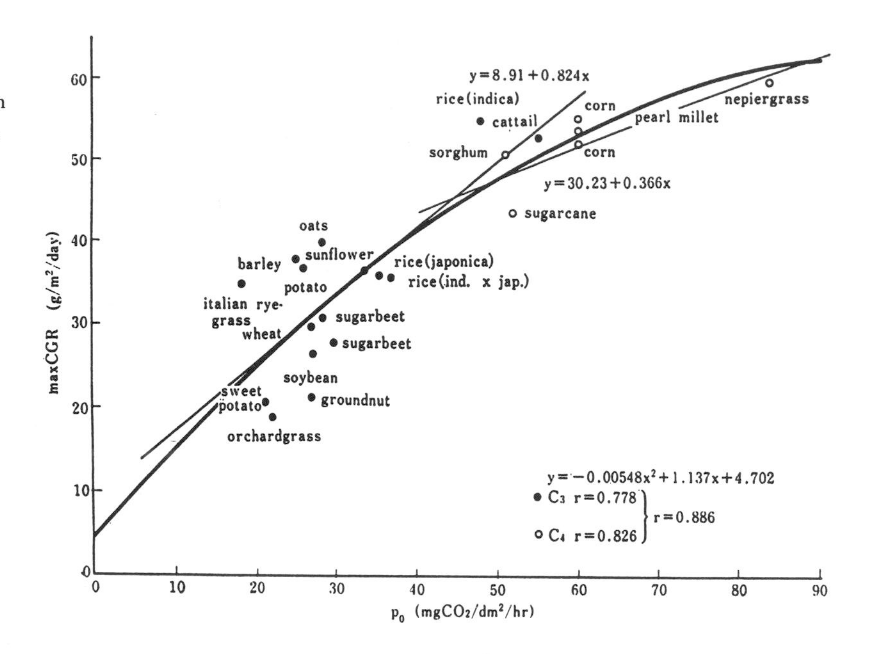


**Figure 16.2.** Relationship between leaf photosynthesis (Po) and maximum crop growth rate (maxCGR) in various crop species. (Adapted from Murata, 1981.)

photorespiratory glycolate pathway without any effect on other carbon metabolism, such as TCA respiration, photosynthetic $CO_2$ fixation (Cho *et al.*, 1985). From this, it would be natural to consider that apparent LPS would be enhanced by application of AAN. But, when tried, the apparent rate of LPS was not increased by treatment with AAN. It is possible that intermediates, which accumulated in the presence of AAN, were harmful to $CO_2$ fixation, leading to decrease of apparent LPS. Therefore, increase of LPS by chemical inhibition of photorespiration is quite difficult.

## Evolution of leaf photosynthesis in crop plants

During the evolution of wheat, the grain yield improvement through the domestication process is suggested to have been achieved mainly by the increase of leaf area, rather than that of LPS. A similar trend was observed also in rice. Thus the LPS had not necessarily been improved through domestication process of rice (Cook & Evans, 1983).

The following paragraphs describe how the photosynthetic characteristics have evolved differently among the wild and cultivated species. At the seedling stage, LPS is high in *O. sativa*, the Asian cultivated species, compared with the Asian and African wild species, or with the African cultivated species, *O. glaberrima*. Between the subspecies in *O. sativa*, upland indica rice and javanica rice have a relatively low LPS. Cook and Evans (1983) suggested that the interspecific difference of LPS in *Oryza* species could be attributed to the difference in nitrogen content per unit leaf area, and to the specific leaf weight (SLW), which was an index of the leaf thickness. This means that high LPS can be realized in the thick leaves which contain more nitrogen per unit leaf area. During domestication of rice, leaf became thick, resulting in high LPS in Asian cultivated rice.

The strategy of the plant during its evolution or domestication process, would vary among different crop species. Such variations are due to the environmental conditions, such as soil moisture, solar radiation and temperature in the regions where the crops evolved or domesticated. For example, the activities and kinetic characteristics of ribulose 1,5 bisphosphatecarboxylase (rubisco) among *Oryza* species, imply that, during evolution of *Oryza*, the improvement of biochemical reaction of photosynthesis takes place toward the direction of the increase in $CO_2$ fixing reaction catalysed by RuBP carboxylase (Makino *et al.*, 1987).

## Varietal difference of LPS in relation to crop growth and yield

LPS is improved in the process of breeding for high yielding varieties in various crop species, and that improvement of LPS greatly contributed to increase of the yield. Further, LPS can continue to be one of the most important breeding criteria for high yielding varieties also in the future.

Good correlation was found between LPS and yield in a few cases (Zelitch 1982; Van der Werf, 1996). However, in many cases, LPS was not necessarily correlated with crop growth or yield (Gifford, 1987; Shibles *et al.*, 1987; Poorter & Remkes, 1990). The reason for this would be that there are too many limiting factors to the crop yield, and that LPS has been masked by stronger limiting factors of the yield, such as sink capacity.

In our studies on flag leaves of rice among old, middle and new variety groups, significant positive correlation was obtained between LPS of a flag leaf and grain yield only at the grain filling stage and in addition, in nitrogen top-dressed conditions (Ishii, 1988). We could not find a positive correlation between flag leaf photosynthesis and grain yield in the heading stage. Moreover, flag leaf photosynthesis in the grain filling stage showed a significantly high correlation coefficient ($r = 0.54^{**}$) against the release year of variety. The maintenance of high LPS in the grain-filling stage therefore, can be the breeding target for high yielding varieties also in the future.

One of the problems here is that LPS in the senescing leaves is not stable and is easily influenced by the environmental conditions (Sasaki & Ishii, 1992). Therefore, efforts should be concentrated on the development of varieties with stability of LPS against the climatic and environmental variations, as well as the varieties with high potential LPS and slow progress of senescence.

Similar to LPS, the leaf morphology is one of the important leaf characteristics. Specific leaf area (SLA) was smaller in the nitrogen tolerant varieties than in the intolerant ones. This means that the tolerant varieties have thicker leaves, while the intolerant varieties have the thinner ones. The thick leaves naturally lead to the narrow and erect leaves, a common trait included for high yielding rice varieties in Japan. This trend can be observed also in the indica varieties, suggesting a common direction in rice breeding.

The morphological characteristics of a leaf are associated with the physiological ones. In wheat and rice, LPS shows a positive correlation with specific leaf weight, an indicator of leaf thickness. This is due to the fact that thick leaves have a high nitrogen per unit leaf area, leading to high LPS. In other words, photosynthetic enzymes are diluted in thin and large leaves, and hence LPS shows a positive correlation with leaf thickness and a negative correlation with leaf size. The nitrogen content can be replaced by soluble protein content which usually corresponds to enzymic protein. Furthermore, about 50% of the soluble protein in the leaf is rubisco. Therefore, the varietal difference of LPS would be attributed to RuBP carboxylase content.

The LPS is said to be limited by two resistances, the stomatal (*R*s) and mesophyll (*R*m) resistance. The latter resistance involves those in the $CO_2$ diffusion process from the stomatal cavity to the chloroplast and in the $CO_2$ fixation process at the chloroplasts. Sasaki, Samejima & Ishi (1996) attempted to separate the *R*m into two components, one is the $CO_2$ diffusion or $CO_2$ transportation resistance (*R*r), and the other is the $CO_2$ fixation or carboxylation resistance (*R*c). They estimated the relative magnitude of each resistance to the total one (*R*t) in 31 Japanese rice varieties, and found that *R*s, *R*r, and *R*c occupied 20–37%, 4–17%, and 55–67%, against *R*t respectively. This suggests that the most important determinant in the varietal difference of LPS is carboxylation resistance. The carboxylation resistance is considered to be regulated by RuBP carboxylase activity. Actually, high correlation was observed between LPS and RuBP carboxylase in rice varieties.

## Heterosis in leaf photosynthesis

F1 heterosis is frequently used in crop species, to increase the crop yield. Also in rice, heterosis is observed for LCP. Murayama, Miyazato and Nose (1987) found in the crossing trial of many rice varieties, strong hybrid vigour as high as 57% against the mid-parent, and 51% against the higher parent. On the other hand, Xuan and Ishii failed to find the heterosis in potential photosynthesis in the flag leaves of F1 hybrid, in the experiment using male sterile, maintenance, and restoration lines of Chinese rice (unpublished data).

But they found an evident heterosis in LPS in the senescing period of the flag leaves, and consequently the integrated LPS of the flag leaves through the senescing period was larger in F1 hybrids than in the parents. From this, they inferred that the high yielding of F1 hybrids would be attributed to the large integrated LPS of the flag leaves during the ripening period.

## PHYSIOLOGICAL FACTORS OF LPS DETERMINATION

### Nitrogen responsiveness

Nitrogen responsiveness is an important criterion for high yielding varieties in crop plants, although the breeding for this criterion is presumably made unconsciously in many crops. New cultivars of rice have a high response of LPS to heavy nitrogen application. Among nitrogen responsive new varieties of rice, the marked response of LPS, but not that of respiration, resulted in the increased ratio of photosynthesis to respiration. The japonica–indica hybrid varieties recently developed in Korea, which were known for their high responsiveness to nitrogen in grain yield, also showed a large responsiveness to nitrogen in LPS (Cho & Murata, 1980). These data suggest that high responsiveness to nitrogen fertilizer was introduced in the breeding process for the high yielding varieties, and was accompanied by the improvement of LPS.

### Senescence

The LPS reaches a maximum rate, a couple of days after leaf emergence, and thereafter decreases gradually with the progress of leaf senescence. Since LPS in senescing process shows a significant positive correlation with the yield, it is important to delay the senescence in the leaves. During senescence, degradation of functional proteins, such as rubisco, occurs in leaves. This proteolysis is suppressed by cytokinin which is produced in the roots and transported to the leaves. However, if the root activity is decreased due to disease such as root rot, or senescence, the supply of cytokinin from the root to the leaves is restricted and hence, senescence progresses fast. This mechanism is supported by the experiments where application of artificial cytokinin, benzyl adenine enabled delayed leaf senescence. Therefore, maintenance of root activity is important to LPS particularly in the grain filling period, a determinant stage for grain yield.

### Sink–source relationship

Sink demand for photosynthetic assimilates can regulate LPS (Lauer & Shibles, 1987). In sweet potato plant, deep application of potassium fertilizer in the soil (absorbed by the main tap root) enlarges the sink capacity of the tuber root and increases LPS. This is because yield of sweet potato can be increased by potassium application. In soybean, thinning the plant stand increased the pod set and growth of the remaining plants, resulting in increment of LPS from controls. These examples show that increase of sink capacity will induce activation of LPS. This shows how important the secure or enlargement of sink capacity is for the increase of crop yield.

Change of LPS due to the reduction of sink capacity is obscure. The LPS did not decrease by depodding treatment which would reduce the sink demand. Moreover, progress of senescence was delayed, and LPS was maintained for a long time. This could be due to the decrease in translocation of degraded leaf protein on reduction of sink organs.

## LEAF PHOTOSYNTHESIS AND ENVIRONMENTS

This section considers the relationship between environmental factors and LPS in relation to crop productivity and yield in the following two aspects; one is LPS and crop yield under suboptimal environment conditions, and the other is also LPS and crop yield in changing global environments.

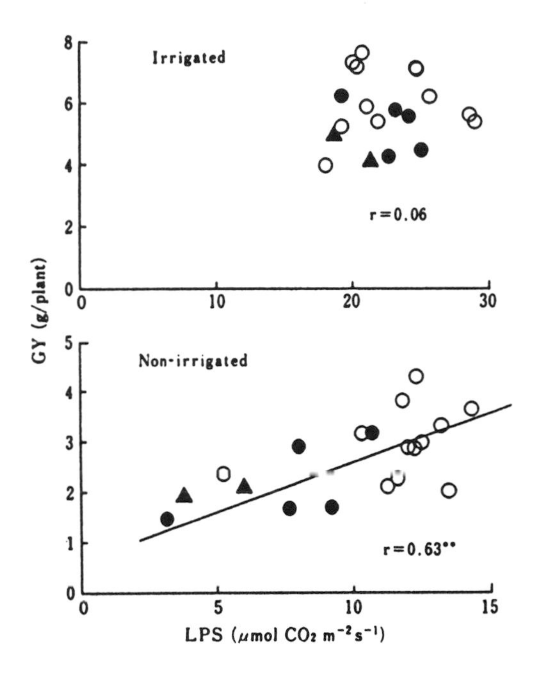


**Figure 16.3.** Relationship between leaf photosynthesis (LPS) and grain yield (GY) under the irrigated and non-irrigated conditions (Adapted from Wada *et al.*, 1994.)

## LPS under suboptimal environment conditions and its relation with crop productivity

In many crop species, a close association of crop yield with potential LPS, measured in favourable environmental conditions with young leaves, is difficult to find (Frederick & Hesketh, 1993). However, LPS under unfavourable conditions often shows a good correlation with crop yield. Figure 16.3 shows the relationship between LPS and grain yield of wheat under irrigated and non-irrigated condition (Wada *et al.*, 1994). Under irrigated condition, the leaves can demonstrate their potential LPS, and it is not correlated with grain yield. However, LPS limited by water relations in the leaves showed a significantly positive correlation with grain yield. Thus, LPS in suboptimal conditions is closely associated with crop yield. This is because LPS in such conditions becomes a limiting factor to crop yield, while in favourable conditions, components other than LPS limit the yield, masking the effect of LPS. Therefore, LPS in suboptimal conditions is more important as the breeding target for stable yields of crop species rather than potential LPS.

## LPS and crop productivity in changing global environments

Among the environmental factors which have been changing fast in the recent years is the marked rise in atmospheric $CO_2$, which is due to the burning of fossil fuels. Such $CO_2$ rise, will produce the following physiological change (Acock 1990; Wittwer, 1994).

(i) Net increase of photosynthetic carbon gain due to competitive inhibition of photorespiration and acceleration of photosynthetic $CO_2$ uptake.
(ii) Rise of plant tissue temperature by suppressed transpiration due to increased stomatal diffusion resistance.
(iii) Maintenance of water potential in the leaf tissue also due to increased stomatal diffusion resistance.

These three factors will lead to an increase in growth rate, eventually crop yield. So, as far as restricted to the specific effects of $CO_2$ on the crop yield, we can be optimistic, ignoring various side-effects of $CO_2$. Particularly, there is greater benefit of $CO_2$ rise to $C_3$ crop plants compared with $C_4$ crop plants, because $C_3$ species can respond to $CO_2$ concentration up to about 1000 l/l, while LPS of $C_4$ species will saturate around the present $CO_2$ concentration. In addition to this, rise of $CO_2$ can decrease transpiration by reducing stomatal aperture. High LPS with low transpiration rate would provide $C_3$ crop species with high water use efficiency, which would provide favourable traits to water-deficient conditions.

## CONCLUDING REMARKS

The remarkable increase in food production is attributed to development of high yielding varieties in various crop species, often called 'Green Revolution'. However, the rate of increase in food production is now getting slow, and production is reaching a plateau. The advances in the areas of crop physiology and photosynthesis, gave a scientific background to the development of new varieties or new cultivation methods in the first Green Revolution. The following research topics should be urgently addressed to usher in the second Green Revolution.

(i) Establishment of the concept for the new plant type of high yielding varieties. Increase of canopy photosynthesis would be the fundamental step for the increase of crop yield.
(ii) Increase of potential LPS is essential for high yield, although not final. Introduction of high LPS genes of C4 photosynthesis into C3 crop species could be one of the possibilities.
(iii) Crop production under suboptimal conditions should get more importance. Photosynthetic criteria for environmental stress tolerance should be known.
(iv) Enlargement of sink capacity (both morphological and physiological) for photoassimilate would be important target for increase of crop yield.
(v) Additional criteria for breeding to adapt to the changing global environmental conditions should be established.

## REFERENCES

Acock, B. (1990). Effects of carbon dioxide on photosynthesis, plant growth and other processes. In *Impact of Carbon Dioxide, Trace Gases, and Climate Change on Global Agriculture*, ed. G. H. Heichel *et al.*, pp. 45–60. Madison: American Society of Agronomy, Crop Science Society of America and Soil Science Society of America, Inc.

Ashraf, M., Akbar, M. & Salim M. (1993). Genetic improvement in physiological traits of rice yield. In *Genetic Improvement of Field Crops*. ed. G. A. Slafer, pp. 413–55. New York: Marcel Dekker, Inc.

Brown, R. H. & Bouton, J. H. (1993). Physiology and genetics of interspecific hybrids between photosynthetic types. *Annual Review of Plant Physiology and Plant Molecular Biology*, **44**, 435–56.

Cho, D. S. & Murata, Y. (1980). Studies on the photosynthesis and dry matter production of rice plants. I. Varietal differences in photosynthetic activity induced by nitrogen top-dressing. *Japanese Journal of Crop Science*, **49**, 88–94.

Cho, C., Sugimoto, Y., Kim, J-M., Usuda, H., Ishii, R., Hyeon S-B. & Suzuki, A. (1983). Search for photorespiration inhibitors; glycine and serine derivatives. *Agricultural and Biological Chemistry*, **47**, 2685–7.

Cho. C., Kim, J-M., Ishii, R., Hyeon, S-B. & Suzuki, A. (1985). Effect of aminoacetonitrile on the $CO_2$ exchange rate in rice leaves. *Agricultural and Biological Chemistry*, **49**, 2847–50.

Christy, A. L. & Porter C. A. (1982). Canopy photosynthesis and yield in soybean. In Photosynthesis, vol. 2, ed. Govindjee, pp. 499–511. New York: Academic Press.

Cook, M. G. & Evans, L. T. (1983). Some physiological aspects of the domestication and improvement of rice (*Oryza* spp.). *Field Crops Research*, **6**, 219–38.

Crop Experiment Station of RDA (1985). *Rice Varietal Improvement in Korea*, Suweon: Crop Experiment Station of Rural Development Authority of Korea.

Frederick, J. R. & Hesketh, J. D. (1993). Genetic improvement in soybean: physiological Attributes. In *Genetic Improvement of Field Crops*, ed. G. A. Slafer, pp. 237–86. New York: Marcel Dekker, Inc.

Gifford, R. M. (1987). Barriers to increasing crop productivity by genetic improvement in photosynthesis. In *Progress in Photosythesis Research*, vol.4, ed. J. Biggins, pp. 377–84. Dordrecht: Martinus Nijhoff Publishers.

Hirata, M., Ishii, R., Kumura, A. & Murata, Y. (1983*a*). Photoinhibition of photosynthesis in soybean leaves. 2. Leaf orientation-adjusting movement as a possible avoiding mechanism of photoinhibition. *Japanese Journal of Crop Science*, **52**, 319–22.

Hirata, M., Ishii, R., Kumura, A. & Murata, Y. (1983*b*). Photoinhibition of photosynthesis in soybean leaves. 3. Leaf transmittance change in response to incident light intensity. *Japanese Journal of Crop Science*, **52**, 430–4.

Ishii, R. (1988). Varietal differences of photosynthesis and grain yield in rice (*Oryza sativa* L.). *Korean Journal of Crop Science*, **33**, 315–21.

Ishii, R. (1995). Roles of photosynthesis and respiration in the yield-determining process. In *Science of the Rice Plant* vol. 2, *Physiology*, ed. T. Matsuo *et al.*, pp. 691–6. Tokyo: Food and Agriculture Policy Research Center.

Lauer, M. J. & Shibles, R. (1987). Soybean leaf photosynthetic response to changing sink demand. *Crop Science*, **27**, 1197–201.

Makino, A., Mae, T. & Ohira, K. (1987). Variations in the contents and kinetic properties of ribulose-1,5-bisphosphate carboxylases among rice species. *Plant Cell Physiology*, **28**, 799–804.

Murata, Y. (1981). Dependence of potential productivity and efficiency for solar energy utilization on leaf photosynthetic capacity in crop species. *Japanese Journal of Crop Science*, **50**, 223–32.

Murayama, S., Miyazato, K. & Nose, A. (1987). Studies on matter production of F1 hybrid in rice. I. Heterosis in the single leaf photosynthetic rate, *Japanese Journal of Crop Science*, **56**, 198–203.

Pereira, J. S. (1993). Gas exchange and growth. In *Ecophysiology of Photosynthesis*, ed. E-D. Schulze & M. M. Caldwell, pp. 147–84. Berlin: Springer-Verlag.

Poorter, H. & Remkes, C. (1990). Leaf area ratio and net assimilation rate of 24 wild species differing in relative growth rate. *Oecologia*, **83**, 553–9.

Rasmusson, D. C. & Gengenbach, B. G. (1983). Breeding for physiological traits. In *Crop Breeding*, ed. D. R. Wood *et al.* pp. 231–54. Madison: American Society of Agronomy, Crop Science Society of America and Soil Science Society of America, Inc.

Rawson, H. M., Hindmarsh, J. H., Fischer, R. A. & Stockman, Y. M. (1983). Changes in leaf photosynthesis with plant ontogeny and relationships with yield per year in wheat cultivars and 120 progeny. *Australian Journal of Plant Physiology*, **10**, 503–14.

Sasaki, H. & Ishii, R. (1992). Cultivar differences in leaf photosynthesis of rice bred in Japan. *Photosynthesis Research*, **32**, 139–46.

Sasaki, H., Samejima, M. & Ishii, R. (1996). Analysis by $\delta^{13}C$ measurement on mechanism of cultivar difference in leaf photosynthesis of rice (*Oryza sativa* L.). *Plant and Cell Physiology*, **37**, 1161–6.

Shibles, R., Secor, J. & Ford, D. M. (1987). Carbon assimilation and metabolism. In *Soybeans: Improvement, Production, and Uses*, ed. J. R. Wilcox, pp. 535–88. Madison: American Society of Agronomy, Crop Science Society of America and Soil Science Society of America, Inc.

Slafer, G. A., Satorre, E. & Andrade, F. H. (1994). Increases in grain yield in bread wheat from breeding and associated physiological changes. In *Genetic Improvement of Field Crops*, ed. G. A. Slafer, pp. 1–68. New York: Marcel Dekker, Inc.

Stitt, M. & Schulze, D. (1994). Does rubisco control the rate of photosynthesis and plant growth? An exercise in molecular ecophysiology. *Plant, Cell and Environment*, **17**, 465–87.

Trenbath, B. R. & Angus, J. F. (1975). Leaf inclination and crop production. *Field Crop Abstracts*, **28**, 231–44.

Uchino, A., Samejima, M., Ishii, R. & Ueno, O. (1995). Photosynthetic carbon metabolism in an amphibious sedge, *Eleocharis baldwinii* (Torr.) Chapman: modified expression of C4 characteristics under submerged aquatic conditions. *Plant and Cell Physiology*, **36**, 229–38.

Ueno, O., Samcjima, M., Muto, S. & Miyachi, S. (1988). Photosynthetic characteristics of an amphibious plant, *Eleocharis vivipara*: expression of C4 and C3 modes in contrasting environments. *Proceedings of National Academy of Sciences, USA*, **85**, 6733–7.

Van der Werf, A. (1996). Growth analysis and photoassimilate Partitioning. In *Plants and Crops*, ed. E. Zamski & A. A. Schaffer, pp. 1–20. New York: Marcel Dekker, Inc.

Wada, M., Carvalho, L. J. C. B., Rodrigues, G. C. & Ishii, R. (1994). Cultivar differences in leaf photosynthesis and grain yield of wheat under soil water deficit conditions. *Japanese Journal of Crop Science*, **63**, 339–44.

Wells, R., Schulze, L. L., Ashley, D. A., Boerma, H. R. & Brown, R. H. (1982). Cultivar differences in canopy apparent photosynthesis and their relationship to seed yield in soybeans. *Crop Science*, **22**, 886–90.

Wittwer, S. H. (1994). Impact of the greenhouse effect on plant growth and crop productivity. In *Mechanisms of Plant Growth and Improved Productivity*, ed. A. S. Basra, pp. 199–228. New York: Marcel Dekker, Inc.

Zelitch, I. (1982). The close relationship between net photosynthesis and crop yield. *Bioscience*, **32**, 796–802.

# 17 Water and salt stress

G. A. BERKOWITZ

## INTRODUCTION: DEFINING THE PROBLEM

Understanding the biophysical, biochemical, and physiological basis for impairment of photosynthesis in plants which experience internal water deficits has been a controversial and vexing problem for plant physiologists over the last several decades. Even more challenging is the problem of how to engineer a new generation of crop plants which display relatively greater photosynthetic capacity (leading to yield enhancement) in the face of the typical plant water deficits encountered during growth. A better understanding of this particular stress should begin with a clear characterization of the nature of the problem encountered by plants.

Internal plant water deficits develop under a variety of environmental conditions. The development of plant water deficits occurs whenever transpirational water loss (by leaves) exceeds water uptake (by roots). This situation typically develops when the soil $\Psi w$ in the rhizosphere declines to suboptimal levels and/or the saturation water vapour pressure deficit between the leaf and surrounding air is high. Low soil $\Psi w$ in the rhizosphere can result from (a) a reduction in the amount of soil water (i.e. depletion of water in the soil profile from prior uptake by roots), (b) a build-up of salts in the soil solution (causing a reduction in the $\Psi s$ component of soil $\Psi w$), or (c) a decrease in soil hydraulic conductivity (thereby restricting water movement to the site of uptake at the root). The leaf:air saturation vapour pressure deficit is the driving force for transpirational water loss from the leaf. Environmental factors which can lead to high transpiration rates and therefore induce plant water deficits include (a) high ambient air temperature and/or low atmospheric water vapour pressure (relative humidity), (b) high radiant heat load on the leaf (resulting in a build-up of leaf temperature over the ambient air temperature), and (c) wind, which creates turbulence in the canopy microclimate boundary layer, reducing the localized water vapour pressure outside the leaf.

Internal water deficits develop in crop plants as the growing plant depletes the soil profile of available water. In this case, leaf $\Psi w$ declines and stays low until the soil profile is recharged with irrigation or precipitation. This type of plant water deficit is considered a long-term stress, as the leaf $\Psi w$ in the plant stays low until soil water becomes available again. It is important to realize that crop plants routinely also experience transitory, short-term water deficits as they grow. Short-term water deficits can occur daily, as soil water absorption lags behind transpiration of plants with open stomata. Growing plants, then, typically experience a diurnal fluctuation in leaf $\Psi w$. Maximal (i.e. least negative) leaf $\Psi w$ occurs pre-dawn, transpiration commences as stomata open upon illumination. The air temperature increase, relative humidity decrease, and increasing radiant heat load on the leaf which typically occur as the day progresses all increase the saturation vapour pressure deficit. The resulting increase in transpiration rate can deplete the plant of stored water (in the stem, tree trunk, etc.) and exceed the rate of water uptake. This can lead to a decline in afternoon leaf $\Psi w$ by as much as 1.5 to 2 MPa from pre-dawn values. During the night when stomata are closed, the leaf : soil $\Psi w$ gradient drives water absorption, and leaf $\Psi w$ increases again to pre-dawn levels (unless soil water is limiting).

Thus, plant water status can fluctuate substantially over the course of a growing season, and also diurnally. Water stress can be defined as a reduction in plant $\Psi w$

and/or water content to a point at which physiological function and cell metabolism (for the purposes of this text, we can focus specifically on processes related to photosynthesis) are impaired. This definition of plant water stress can serve as a beginning to an elucidation of how photosynthesis is adversely affected under water deficits, and what endogenous mechanisms in the plant interact with sensitivity of photosynthesis to this stress. This definition of water stress is rather limited in scope. In fact, water deficit (i.e. leaf $\Psi w$ decline due to environmental factors which result in transpirational water loss exceeding adsorption) effects on plant growth and development are dynamic. The severity, duration, and timing (with regard to ontological development) of leaf $\Psi w$ decline, in addition to the stress pre-history and nutritional status of the plant, all modulate the deleterious effects the imposed stress has on plant performance.

## IMPACT OF WATER DEFICITS ON PHOTOSYNTHESIS: STOMATAL CLOSURE

Photosynthetic inhibition is one of the primary and most obvious repercussions of water (and salt) stress effects on plant metabolism. Observations made many decades ago led to the conclusion that stomatal closure occurs in water stressed plants, limiting gas (i.e. internal water vapour and external $CO_2$) exchange between the leaf and atmosphere. (The mechanistic basis for water and salt stress effects on stomatal conductance will be delineated later in this chapter.) Restriction of gas exchange upon stomatal closure was envisioned to be the major 'lesion' in the photosynthetic process induced by water stress because this response would limit $CO_2$ supply to the site of carboxylation in the chloroplast.

Even though decreased stomatal conductance in response to water deficits would inhibit photosynthetic capacity, it was long considered to be a stress adaptive response. Specifically, stomatal closure reduces transpirational water loss, and can positively impact the potential for survival of plants which must reach reproductive maturity on a limited supply of water. From an evolutionary perspective, a temporary (and reversible) restriction of photosynthetic capacity can be viewed as an acceptable consequence of an adaptation which could lead to the production of seed instead of premature death of the plant. However, adaptations which confer an evolutionary advantage are not necessarily beneficial to crop plants which are grown for an economic yield.

Depending in part on the harvested portion of the plant (i.e. vegetative structures vs. fruit or seed) and on the economic use of the plant (for example, turfgrass which is mowed regularly), the photosynthetic inhibition which occurs as a consequence of water stress-induced stomatal closure generally has a negative impact on crop growth and yield. However, it should be noted that (partial) stomatal closure typically increases the water-use efficiency, or transpiration ratio of crop plants. Both transpiration ratio (g $H_2O$ transpired per g $CO_2$ fixed on an instantaneous basis) and water-use efficiency (increase in plant dry weight per unit water used, on a long-term basis) refer to the 'cost' of carbon gain in terms of water usage. Ambient stomatal aperture of most crop plants grown under well-watered conditions is near-maximal, and facilitates a $C_i$ ($\approx$300 µl/l) which supports photosynthetic rates which are RuBP regeneration limited. A unit decrease from this $C_i$ will not inhibit photosynthesis to as great an extent as would occur at lower $C_i$ (<200 µl/l), when photosynthetic rates are RuBP saturated and therefore affected to a greater extent by changes in substrate (i.e. $CO_2$) availability. As stomatal conductance decreases from ambient levels (e.g. due to water stress), the effect on photosynthesis is curvilinear. The relationship between stomatal conductance and transpiration, on the other hand, is linear. Therefore, initial decreases from ambient stomatal conductance have a relatively greater effect on transpiration than on photosynthesis, leading to increased water-use efficiency as water stress induces stomatal closure. The identification of a causal relationship between stomatal closure and increased water-use efficiency has not, however, led to the development (through breeding programs) of crop plants which display relatively greater yield maintenance under water stress conditions. This is likely because stomatal closure invariably leads to photosynthetic inhibition, despite the enhancement in efficiency of water-use.

## NON-STOMATAL VS. STOMATAL EFFECTS OF WATER STRESS ON PHOTOSYNTHESIS

Work undertaken about a decade ago by G. Farquhar, T. Sharkey and colleagues led to the development of a mathematical model delineating the various plant parameters which contribute to the limitation of photosynthesis. This modelling work allowed for the estimation of $C_i$ from measurement of photosynthetic rates and stomatal conductance. Application of this model to water-stressed plants led to the intriguing finding that, in most crop plants, internal water deficits which resulted in stomatal closure and photosynthetic inhibition are associated with an increase in $C_i$. This finding led to the conclusion that stomatal closure was not the only (or, in some cases, even the primary) lesion in the photosynthetic process of plants which experience internal water deficits. The build-up of ambient $C_i$ suggested that use of intracellular $CO_2$ (i.e. by the enzymes involved in photosynthetic carbon assimilation and reduction) was inhibited to a greater extent than supply of $CO_2$ to the leaf interior.

Numerous studies with isolated chloroplasts, protoplasts, and cells exposed to low Ψw *in vitro*, along with studies of chloroplasts and crude enzyme extracts prepared from water-stressed plants had suggested that cell water deficits affect the chloroplast machinery involved in photosynthesis. However, $C_i$ increases in water stressed plants offered the best evidence that *in situ*, cell dehydration effects on chloroplast biochemistry are an important limitation to photosynthesis of plants experiencing internal water deficits. Many researchers working in this field were alarmed, therefore, when the validity of $C_i$ estimations in water stressed plants was challenged. In the later 1980s and early 1990s, it was noted that when exogenous ABA (endogenous ABA is known to be involved in water stress-induced stomatal closure) was supplied to leaves, or leaves were allowed to dehydrate rapidly, $CO_2$ uptake across the surface of a leaf was not uniform. It was determined that this non-uniform $CO_2$ uptake pattern was due to heterogeneous stomatal closure across the leaf surface. This phenomenon of 'patchy' stomatal closure could lead to artefacts in the development of an A:$C_i$ curve for water stressed leaves. However, more recent studies have demonstrated that, when water deficits develop slowly in the plant, patchy stomatal closure does not occur. Currently, $C_i$ estimates developed from gas exchange measurements taken on water-stressed plants are cautiously accepted as evidence that internal water deficits substantially affect chloroplast capacity for photosynthesis. Researchers have also demonstrated that supplying leaves of water stressed plants with air containing very high (e.g. 10 000 μl/l) [$CO_2$] does not fully overcome the stress-induced inhibition of photosynthesis. The high [$CO_2$] should overcome any stomatal limitation to photosynthesis (despite the known effect of high $CO_2$ causing increased stomatal closure). These results offer additional *in situ* evidence that photosynthetic inhibition under water stress is caused, at least in part, by lesions in chloroplast biochemistry.

It should be noted that non-stomatal effects of water stress on the photosynthetic process have, in the past, been collectively referred to as increases in mesophyll resistance. All of the (plant) factors other than stomatal resistance which contribute to rate-limitation of the photosynthetic process were pooled together and referred to as 'mesophyll resistance'. This term has been taken in older literature to include the physical resistances in the diffusional pathway for $CO_2$ to move from the air space in the substomatal cavity to mesophyll cells, and intracellular $CO_2$ and/or bicarbonate movement through the cytosol and across the chloroplast envelope. It is currently envisioned that the major component of total 'mesophyll resistance' is the photochemical and biochemical processes of the chloroplast involved in $CO_2$ fixation.

## WHAT IS THE TRANSDUCING MECHANISM?

How is a change in the physical state of cellular water 'sensed' by the enzymes and membrane systems responsible for carrying out metabolic functions such as photosynthesis? Our understanding of how plant water deficits lead to stomatal closure is relatively clear.

Due to hardened interior cell walls of guard cells, turgor pressure in these cells is anisotropic and thus leads to the outward bending of the guard cell walls farthest away from the stomatal pore. This unequal distribution of pressure pulls the guard cell away from the central axis of the stomatal pore, leading to stomatal opening. Thus, any physiological factor which facilitates turgor loss in guard cells leads to a restriction in $CO_2$ supply to the leaf interior and resultant photosynthetic inhibition. Changes in the bulk leaf water status are directly transmitted to the guard cell through the continuum of liquid water running along the leaf mesophyll walls. In the absence of acclimation mechanisms (as discussed later), declining leaf Ψw leads to turgor loss in the bulk leaf, and in guard cells. It should be noted that in some plant species, guard cell Ψp is only loosely related to the bulk leaf water status. This is thought to be due to species differences in the degree of hydraulic connection between the leaf mesophyll and the guard cells in the epidermis. In most cases, however, Ψp in the bulk leaf mesophyll and guard cells are intimately related.

Several other mechanisms (in addition to changes in bulk leaf water status which are linked in a physical sense to guard cell Ψp) also act to transduce changes in plant water status into guard cell turgor loss. The plant hormone ABA is known to induce stomatal closure. ABA-induced stomatal closure is a well-known response to imposed plant water deficit in a large number of crop species. Although recent work has led to a fairly clear understanding of the molecular basis for ABA-induced stomatal closure, there are still some aspects of ABA physiology that remain to be resolved.

ABA arriving at the guard cell in the transpiration stream initiates a signal transduction system which involves an increase in guard cell cytosolic [$Ca^{2+}$]. It is currently believed that ABA binds to a receptor on the exterior of the guard cell membrane, and also accumulates in the guard cell cytosol. Recent studies have demonstrated that either ABA bound to the plasmalemma, or cytosolic ABA is sufficient to cause cytosolic $Ca^{2+}$ to increase. Movement of $Ca^{2+}$ to the cytosol is known to occur from a number of different pools; it is therefore possible that plasmalemma-bound ABA and cytosolic ABA both (independently) initiate processes which allow for movement of $Ca^{2+}$ to the guard cell cytosol. $Ca^{2+}$ accumulation in the cytosol, which is a general signal transduction mechanism acting as an intermediate step in the regulation of a number of cell processes, has been shown to result in stomatal closure. Current evidence suggests that ABA-induced increase in guard cell $Ca^{2+}$ results from both influx of extracellular $Ca^{2+}$ through plasmalemma ion channels, and movement of $Ca^{2+}$ to the cytosol from the vacuole through tonoplast $Ca^{2+}$ channels. The mechanism linking ABA to the opening of $Ca^{2+}$ channels is not resolved. ABA-induced increase in cytosolic $Ca^{2+}$ is known to then lead to the opening of guard cell plasmalemma anion channels, which allow for $Cl^-$ and malate efflux from the cytosol. This anion efflux depolarizes the membrane potential across the plasmalemma (inside negative). This depolarization then activates outwardly rectified plasmalemma $K^+$ channels, leading to $K^+$ efflux from the guard cell. The loss of $K^+$, and the anions $Cl^-$ and malate through ion channels which open in response to ABA is the biophysical response to this signal transduction cascade which directly leads to guard cell turgor loss and stomatal closure. ABA also causes inwardly rectified $K^+$ channels to close as $K^+$ anions are lost from the guard cell cytosol. During the cascade of events induced by ABA which leads to stomatal closure, tonoplast ion channels which facilitate $K^+$ and anion release from the vacuole also open, allowing for this major pool of osmotic solutes to eventually exit the guard cell. ABA is also known to induce solute ($K^+$, $Cl^-$, and malate) efflux from the guard cell cytosol via a $Ca^{2+}$-independent system; less is known about this signal transduction pathway.

To this point, examination of the association of ABA with stomatal closure has focused on the signal transduction pathway linking ABA which arrives at the guard cell plasmalemma to turgor loss in guard cells. ABA action as a signal which generates a plant response (i.e. stomatal closure, reduced transpiration, and concomitant photosynthetic inhibition) to water deficits also involves ABA movement within the plant to the site of action (at the guard cell plasmalemma). There are several aspects of ABA movement in water-stressed plants which have not been fully resolved, and are the focus of current research.

Leaf ABA levels are known to increase in response to plant water deficits. However, initial stomatal closure occurs prior to the synthesis of new ABA. Some experimental evidence, along with theoretical analyses, suggests that water stress (along with a number of other environmental stresses including salinity) can cause a redistribution in pools of sequestered ABA within the leaf. This redistribution is thought to result in the release of ABA from mesophyll cells and an increase in ABA arriving at the guard cell through the transpiration stream. Some investigators have suggested that stress-induced redistribution of ABA in leaves occurs due to pH changes in cell compartments. ABA is a weak acid, and is thought to move across membranes as the protonated, uncharged moiety. As a weak acid, then, ABA is thought to accumulate (in the unprotonated form) in the chloroplast stroma (the most alkaline compartment of the cell). Theoretical analysis has led some to postulate that under water or saline stress impairment of leaf cell plasmalemma proton pump activity leads to an acidification of the chloroplast stroma (and the cytosol to a lesser extent). This model predicts that the acidification of these cell compartments is a transducing mechanism leading to the release of sequestered ABA and, thus, stomatal closure.

The release of sequestered ABA from mesophyll pools upon development of leaf water deficits typically occurs concomitantly with turgor loss as leaf Ψw declines. The same situation would develop in plants subjected to saline stress. It is unclear therefore, to what extent ABA contributes to initial stomatal closure in plants subjected to field conditions (low soil Ψw; induced by soil water deficits or salt build-up) which cause leaf Ψw and Ψp decline. The demonstration that ABA alone can cause stomatal closure does not necessarily mean that the increase in ABA in the transpiration stream is causal to initial stomatal closure which occurs in water-stressed leaves.

Some investigators in this field believe that ABA action on guard cells is more physiologically relevant with regard to the 'after-effects' of water stress. It is known that after relief from a period of water stress, stomata can remain closed despite the return of leaf Ψw and Ψp to pre-stress levels. Whether or not ABA is involved in initial water stress-induced stomatal closure, it most certainly could be involved in the maintenance of low stomatal conductance during a period after a stress has been relieved.

Review of plant physiological mechanisms involved in transducing a signal of water deficit into altered plant function should mention the emerging concept of root signalling. Some investigators believe that root signalling (to leaves) upon exposure to soil water deficits is a more significant modulator of plant response to stress than redistribution of ABA sequestered in leaf cells. The experimental basis underlying the concept of root signalling involves split-root system work. When a root system is separated, with part of the roots kept in well-watered soil and part subjected to low soil Ψw, the plant can be exposed to soil water deficits without any decline in leaf Ψw. Stomatal closure has been documented to occur under these conditions, leading to the concept that the growing root tip is a sensitive and important water/salt stress sensor. Most of the evidence from these split-root experiments indicates that ABA movement from the root (exposed to low soil Ψw) to the leaf is the signal which results in the adaptive response of stomatal closure when plants are exposed to low soil Ψw. However, under most field conditions, low soil Ψw does lead directly to leaf Ψw decline. Therefore, it is not clear how much root signalling mediates the initial plant response to exposure to soil water deficits.

Water and salt stress effects on photosynthesis clearly involve both decreases in stomatal conductance and impairment of chloroplast biochemistry. As discussed, ABA movement to the guard cell and leaf Ψw decline both are likely involved in transducing a stress signal into the plant response of stomatal closure. The mechanisms involved in transducing a water/salt stress signal to the chloroplast machinery involved in $CO_2$ assimilation are not yet clearly understood.

One of the more important concepts which has recently been developed in the field of water stress effects on photosynthesis regards what is not involved as a transducing mechanism. It appears that leaf Ψw, and stress-induced leaf Ψw decline have no physiological relevance to the functioning of organelles such as the chloroplast, and the enzymes and membrane systems which are involved in their metabolic activity. Leaf Ψw is an important indicator of plant water status specifically with regard to water movement; the plant : soil Ψw gradient is the driving force for water uptake. However, the Ψw of a cell likely has no direct impact on metabolism. Rather, it is the water relations parameters associated with Ψw decline which likely impair chloroplast function. It is the relative water content (RWC) changes, or more specifically the reduction in protoplast and/or chloroplast volume which occurs concomitantly with Ψw decline during the development of plant water deficits which can be directly linked with impaired biochemical function. The experimental evidence supporting this contention is as follows. Depending on stress pre-history (i.e. extent of acclimation), a given leaf Ψw can be associated with widely varying degrees of chloroplast-mediated photosynthetic inhibition. Conversely, a specific extent of cell water removal (i.e. reduction in protoplast and/or chloroplast volume) is intimately linked to a particular degree of nonstomatal-mediated photosynthetic inhibition. This linkage between metabolic inhibition, and volume change is demonstrated by the results shown in

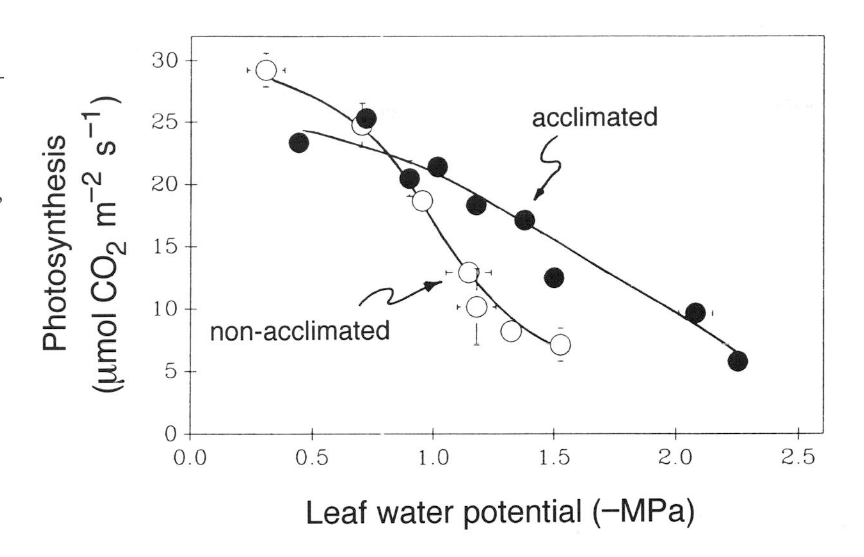


**Figure 17.1.** Infra-red gas analyser measurement of *in situ* net photosynthesis in water-stressed spinach (*Spinacea oleracea*) plants plotted as a function of declining leaf Ψw during the stress period. These data, and those shown in Figs. 17.2–17.4 are adapted from Santakumari and Berkowitz, 1991. Results are shown for control (non-acclimated) plants which had been kept well watered, and for acclimated plants which had been subjected to a prior stress period (leaf Ψw decline to −1.5 MPa over ~10 days) and then re-watered. Results shown in Figs. 17.2–17.4 are from the same set of plants.

Fig. 17.1–17.3. Control (nonacclimated) and stress-preconditioned (acclimated) spinach plants demonstrate different degrees of photosynthetic sensitivity to stress imposition, when metabolic impairment (photosynthetic inhibition) is plotted as a function of declining leaf Ψw (Fig. 17.1). Control and acclimated plants also show differences in the extent of *in situ* chloroplast volume reduction during the stress period (Fig. 17.2). However, in both control and acclimated plants, there appears to be a similar, fundamental, and apparently causal relationship between the extent of chloroplast volume reduction as Ψw declines, and water stress inhibition of photosynthesis (Fig. 17.3).

Results of other studies support the assertion of a causal relationship between cell volume changes and water/salt stress impairment of photosynthetic metabolism. The relationship between leaf cell Ψw and volume can vary amongst genotypes of a particular crop plant species. Plants which demonstrate variability in the volume relationship also have very different degrees of photosynthetic inhibition at low leaf Ψw. However, when photosynthetic inhibition during an imposed water

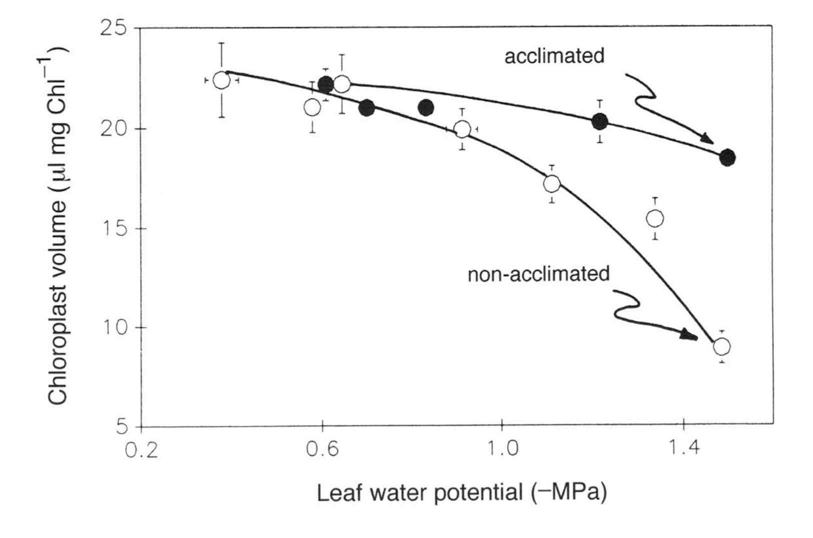


**Figure 17.2.** Estimation of *in situ* chloroplast volume in non-acclimated and acclimated plants as leaf Ψw declined during a period of water stress.

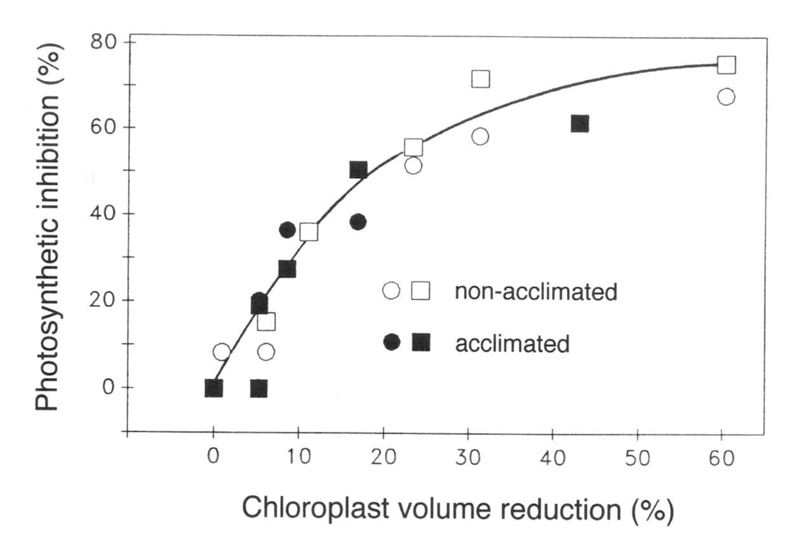


**Figure 17.3.** The relationship between the extent of photosynthetic inhibition and chloroplast volume reduction which occurred as leaf Ψw declined during a period of water stress.

deficit is plotted as a function of volume reduction, the relationship is similar in the different genotypes. Plant growth conditions (e.g. $K^+$ fertilization levels) can alter ambient leaf cell Ψs, resulting in an altered Ψw : volume relationship. These management practices can result in an altered Ψw : photosynthetic inhibition relationship during imposed water deficits; the volume: photosynthetic inhibition relationship is much less dynamic. Thus, there appears to be a causal relationship between the extent of volume reduction (induced by cell and/or organelle dehydration) and chloroplast photosynthetic capacity. Although most studies of this nature focused on the protoplast volume : Ψw relationship in plants subjected to water stress, it is most likely the chloroplast volume changes which 'transduce' developing plant water deficits into an impairment of photosynthetic metabolism.

## STRESS-INDUCED LESIONS IN CHLOROPLAST FUNCTION

A fundamental issue related to photosynthesis of crop plants (which, as mentioned, routinely experience water deficits) still unresolved is the identification of the enzymatic step(s) which is most severely affected by the level of cell dehydration typically associated with water and salt stress. Most of the published studies focusing on this issue do not provide definitive evidence, but are rather inferential in nature. For example, there is a plethora of published studies involving the assay of photosynthetic enzymes in crude extracts of leaf tissue from water-stressed plants. Implications that can be drawn from this work are limited because the *in vitro* enzyme assay is undertaken under standard conditions (e.g. saturating substrate and optimized cofactor concentrations) which have no relation whatsoever to the chloroplast stromal milieu conditions that will limit the steady-state activity of the enzyme *in situ*. Another problem with these analyses is that they do not necessarily distinguish between a primary site of stress inhibition, and secondary ramifications of stress effects on the photosynthetic process. For example, numerous studies have demonstrated depressed ribulose bisphosphate carboxylase (rubisco) activity in extracts of water-stressed leaves. However, rubisco represents a substantial (≈25%) portion of leaf protein. A primary lesion in some other partial reaction of photosynthesis could limit carbon supply to the cell, which could depress protein synthesis on a long-term basis. Stress inhibition of protein synthesis could then be causal to depressed rubisco activity. A second approach which has been used to probe the photosynthetic step(s) which represents the major water stress lesion involves subjecting isolated protoplasts or chloroplasts to hyperosmotic Ψs *in vitro* (thus exposing photosynthetic machinery to instantaneous dehydration). Implications that can be made from these studies are also limited because the nature of the stress imposition does not match the situation *in vivo*, when the plant is subjected to more slowly developed stress. Of course, there are a number of other factors related to *in situ* imposition of plant water deficits which are not replicated in these *in vitro* studies. For example, *in situ* water/salt stress which leads to stomatal closure could negatively impact chloroplast metabolism due to concomitant heat stress. This linking of water/salt stress to high temperature damage of photosynthetic machinery would not be maintained with *in vitro* studies. In addition, water/salt stress inhibition of the photosynthetic process can lead to photoinhibitory conditions in intact plants subjected to the high light intensities typically encountered in the field. Under *in vivo* water/salt stress conditions which cause impairment of chloroplast photosynthetic capacity (i.e. capability to use light energy), over-reduced photosystems can generate oxygen radicals which then cause further damage to the photosynthetic apparatus. The association of water/salt stress with photoinhibition would not necessarily be maintained with *in vitro* studies. Probing the chloroplast photosynthetic machinery *in vivo*, in attached leaves of water-stressed plants, is also problematic because not all necessary studies (such as steady-state photosynthetic metabolite analysis) can be undertaken in whole plants.

Due to the forementioned technical challenges, the emerging picture of water/salt stress effects on chloroplast photosynthetic capacity is not fully resolved. None the less, we can make some inferences from the available data. Photosynthetic electron transport and coupling factor activity (i.e. capacity for ATP synthesis) has been probed in studies of chlorophyll fluorescence. These studies have been undertaken on leaf tissue removed from water-stressed plants, and also *in situ*. It can be concluded unequivocally from this body of work that coupling factor activity (i.e. ATP synthesis) and the photochemical apparatus are not substantially damaged under plant water deficits until very low leaf $\Psi w$ is reached.

Due to the centrality of the role rubisco plays in the photosynthetic process, perturbation in the function of this enzyme under water deficits has received much attention. Results of much of this work are inconclusive. However, studies recently undertaken with rubisco antisense plants provide definitive evidence indicating that rubisco activity is not a major lesion in the photosynthetic process under water stress. Sensitivity of photosynthesis to water deficits was not altered in plants which (due to antisense expression) contained only 30% of the level of functional rubisco found in wild-type plants. If rubisco activity was a major site of inhibition, reducing its expression level should have increased the extent of water stress inhibition of photosynthesis.

If rubisco activity and photochemical production of ATP are not major sites of stress inhibition, then some conclusions can be formulated about the major lesion induced in the photosynthetic process by plant water deficits. One can speculate that either the regeneration of RuBP (by the photosynthetic carbon reduction cycle enzymes), or feedback inhibition (e.g. inhibited starch and/or sucrose synthesis/utilization) must be rate-limiting the capacity of chloroplasts to assimilate $CO_2$ in plants subjected to water deficits. Although some have speculated in the pertinent literature that feedback inhibition does rate-limit chloroplast metabolism under water stress, there is little experimental evidence to support this assertion.

One of the most powerful means of elucidating which step of an integrated, multi-enzyme pathway is most impaired by a given stress imposition is to monitor *in situ* perturbations in the levels of metabolites which are intermediates in the pathway when the plant is subjected to the stress. Carbon flow through a multi-enzyme pathway will be associated with steady-state levels of the intermediate metabolites. If stress imposition is associated with the increase in the steady-state level of a metabolite, then the enzyme which uses the metabolite as a substrate can be identified as the most sensitive site of stress inhibition. Conversely, the steady-state level of the metabolite which is the product of the sensitive enzyme should decrease. A full analysis of this sort for the photosynthetic carbon reduction cycle pathway would require the monitoring of compartmentalized metabolites within chloroplasts which are isolated from leaf tissue using non-aqueous fractionation techniques. The reason for this is as follows. Many of the photosynthetic cycle intermediates (3-phosphoglycerate, the triose-phosphates, hexose mono- and bisphosphates, and pentose monophosphates) are also intermediates in cytosol-localized metabolic pathways (e.g. glycolysis, gluconeogenesis, pentose phosphate shunt). Therefore,

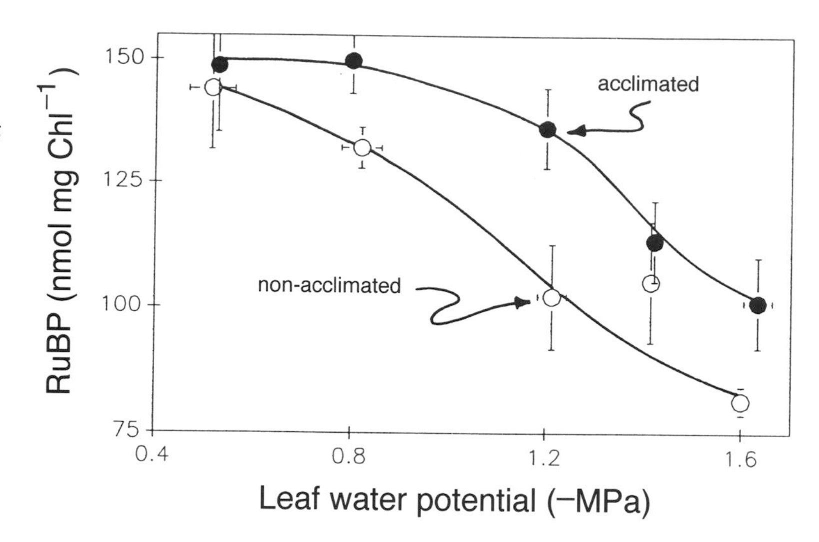


**Figure 17.4.** Steady-state *in situ* RuBP levels in non-acclimated and acclimated plants presented as a function of declining leaf Ψw during a period of water stress.

metabolite analysis of whole-leaf extracts will be confounded by the mixing of cytosolic and chloroplast stromal pools of these intermediates. Non-aqueous fractionation techniques allow for purification of exclusively stromal metabolites from leaf extracts. These sorts of studies could provide convincing evidence about water/salt stress-induced lesions in individual steps of the photosynthetic carbon reduction cycle.

Thorough studies of chloroplast stromal metabolites in water stressed plants have not yet been undertaken. However, some work germane to this issue has been published. Measurement of RuBP levels in water-stressed plants can provide an initial evaluation of site-specific lesions. RuBP is only found in the chloroplast stroma; non-aqueous fractionation techniques are not required for the analysis of stromal concentrations of this metabolite. A build-up of RuBP concomitant with photosynthetic inhibition and increased internal leaf [$CO_2$] would suggest that impairment of rubisco function rate limits photosynthesis in the water stressed plant. Conversely, a decline of steady-state RuBP levels under similar conditions would indicate that the major stress-induced lesion in chloroplast metabolism would be involved in RuBP regeneration (from the initial photosynthetic product, 3-phosphoglycerate). Data from several different laboratories indicate that RuBP levels decline when plants are subjected to water deficits. An example of such findings is shown in Fig. 17.4. The differential extent of photosynthetic inhibition demonstrated by spinach plants subjected to water stress (Fig. 17.1) is associated with different degrees of reduction in the steady-state level of RuBP (Fig. 17.4). These results support the contention that rubisco activity is not much affected by cell dehydration, and that RuBP regeneration rate limits photosynthetic capacity of chloroplasts in water stressed plants.

Other than RuBP decline, there is little evidence supporting the contention that one of the enzymes in the regenerative phase of the photosynthetic carbon reduction cycle is the major lesion induced in chloroplast biochemistry under cell dehydration *in situ*. In addition, not much definitive evidence has been generated to identify which of the photosynthetic enzymes is impaired (leading to the decline in steady-state RuBP levels), or how dehydration (volume change) adversely affects the particular enzyme. *In vitro* metabolite analyses from several laboratories with isolated chloroplastes subjected to instantaneous dehydration have identified FBPase as rate-limiting carbon flow to RuBP in chloroplasts subjected to this stress. It is intriguing to re-examine these findings in light of the newly developed models of stress-induced ABA remobilization (discussed above). As mentioned previously, some scientists believe that acidification of the chloroplast stroma occurs under both water and saline stress, and that this pH change leads to a release of ABA sequestered in the chloroplast. FBPase is a highly regulated enzyme which has a high pH optimum; matching the pH which is established in the chloroplast stroma during illumination. Even minor (~0.1 pH unit) decreases in stromal pH would severely reduce FBPase activity. Maximum extractable FBPase activity is often not much greater than photosynthetic rates; FBPase can rate-limit photosynthesis by limiting the rate at which RuBP is regenerated. Thus, FBPase is a candidate for future study as one possible site at which cell dehydration negatively impacts chloroplast function. Precise quantitative measurements of chloroplast stromal pH are not currently possible (in the intact leaf). Thus, the assertion that organelle volume reduction (i.e. dehydration) which occurs during water stress leads to an increased proton concentration in the stroma, restricting FBPase activity and hence, photosynthesis is highly speculative. However, leading scientists in this field such as W. Kaiser have suggested that the extent of dehydration which occurs concomitant with typical levels of water/salt stress would have a concentrating effect on chloroplast ions to an extent which could impair enzyme function. Perhaps, stress-induced volume reduction also concentrates the protons in the stroma to an extent beyond the buffering capacity in this compartment, leading to a decline in ambient pH. A second prospective mechanism linking chloroplast volume reduction to stromal acidification is $K^+/H^+$ exchange across the chloroplast envelope. $K^+$ movement across the limiting membrane of the chloroplast (i.e. the inner envelope) is known to be (indirectly) linked to $H^+$ counterflux. The chloroplast stromal [$K^+$], at ~200 mM, is higher than cytosolic [$K^+$]. Perhaps the concentrating effect of volume reduction shifts the stroma : cytosol $K^+$ gradient away from its equilibrium potential, resulting in $K^+$ efflux, which then indirectly results in $H^+$ influx. Of course, there is little experimental evidence at the whole plant level which addresses these questions and this model therefore, remains highly speculative.

## ACCLIMATION TO WATER/SALT STRESS

When it comes to environmental stresses, plants cannot run, nor can they hide. They have, as a consequence, developed numerous strategies which allow for some degree of stress acclimation. A wide number of crop plants demonstrate substantial capability for acclimation of photosynthesis to low $\Psi$w. In addition, substantial genotypic variation (within a species) has been found in water/salt stress acclimation capability. This acclimation capability has been demonstrated to influence stress effects on stomatal responses, and nonstomatal mediated photosynthetic factors. Thus, we can conclude that (through either genetic engineering, or conventional breeding) incorporating maximal acclimation capacity into cultivars of crop plants which are grown in drought-prone areas could improve plant photosynthetic performance under stress.

In our analysis of photosynthetic acclimation to water/salt stress, a distinction should first be made between stress avoidance, stress tolerance, and stress resistance. Cultivars of crop plants which demonstrate relatively high yield maintenance under drought often do so because they demonstrate stress avoidance. In this case, the plant has typically been bred to mature faster, or shift the period of reproductive growth such that it occurs prior to the peak drought period. In this manner, the plant avoids exposure to environmental conditions which can lead to leaf $\Psi$w decline during a developmental period critical to the attainment of maximum yield.

Stress tolerance is another acclimation strategy which is based on the avoidance of leaf $\Psi$w decline. In this case, environmental conditions are encountered by the plant which would typically result in leaf $\Psi$w decline. However, acclimation to stress is facilitated by altered morphology, anatomical features, and/or physiology such that leaf $\Psi$w decline is prevented. Stress tolerance strategies of acclimation either facilitate enhanced water uptake, or reduced transpirational water loss; in either case, leaf $\Psi$w remains high. Stress tolerance can be afforded by shifts in root : shoot ratios (leading to the development of deeper root systems which allow for extraction of water from deeper zones of the soil profile), leaf rolling, increased leaf pubescence, leaf abscission, or altered leaf morphology (all of which would reduce transpirational water loss).

Acclimation to water/salt stress, with particular reference to photosynthetic capacity, can also be facilitated by physiological mechanisms which impart a degree of stress resistance to plants. Stress resistance refers to the ability to acclimate to water stress conditions such that low leaf $\Psi$w has less of a deleterious effect on photosynthesis. There are many examples of crop plants which acclimate to water/salt stress by stress resistance strategies, both with regard to reducing low $\Psi$w effects on leaf conductance, and chloroplast photosynthetic capacity.

A very common acclimation response to the development of low leaf Ψw during periods of water stress is a shift in the relationship between stomatal resistance and Ψw. Many plants acclimate to water/salt stress by maintaining greater leaf conductance at a given low Ψw. This, of course, leads to the maintenance of relatively greater photosynthesis during subsequent episodes of water stress. Greater leaf conductance is facilitated by physiological changes in leaf cells which allow for the maintenance of greater Ψp at low Ψw. The shift in the Ψp : Ψw relationship during subsequent periods of low Ψw typically occurs due to the acclimation mechanism of osmotic adjustment. Osmotic adjustment occurs when there is a net accumulation and/or synthesis of solutes in cells. As cells dehydrate during an imposed water stress, the loss of water will have a concentrating effect on the cell solutes within the cell. Ψs decline will follow the Boyle–Mariotte relationship; the inverse of the Ψs will demonstrate a linear change with volume reduction. Osmotic adjustment is evidenced by Ψs decline to an extent greater than that predicted by the Boyle–Mariotte relationship. Osmotic adjustment can be facilitated by either the *de novo* synthesis of organic solutes such as glycine betaine, proline, or sucrose, or the accumulation (from the soil solution) of inorganic ions such $K^+$, $Cl^-$, or $Na^+$. In the case of salt stress, the low soil Ψw induced by the build-up of $Na^+$ in the soil solution is often balanced by accumulation of $K^+$, $Cl^-$, or even $Na^+$ in cells. Often, the accumulated ions are compartmentalized within the vacuoles of cells undergoing osmotic adjustment, with concomitant generation of organic solutes in the cytosol and metabolically active organelles such as the chloroplast. It is thought that these organic solutes (in contrast to inorganic ions) are less harmful at high concentrations to enzymes.

The acclimation response of turgor (and therefore, leaf conductance) maintenance at low Ψw can also occur in response to other alterations in cell water relations parameters. As water is lost from fully turgid cells, the rate at which turgor is lost is influenced by the elasticity of the cell walls. Cells with highly elastic cell walls are able to maintain positive Ψp to a relatively lower Ψw as cells dehydrate. It has been demonstrated that some plants, upon exposure to an episode of water stress involving leaf Ψw decline, alter the elasticity properties of the leaf mesophyll cell walls, allowing for a greater degree of turgor maintenance at low Ψw during subsequent stress imposition. Another stress acclimation response which results in turgor maintenance at low Ψw involves alterations in the relative degree of apoplastic or 'bound', and symplastic water. Some plants which are exposed of water stress generate new (leaf mesophyll) cells which are smaller, and have thicker cell walls than the cells in leaves of non-stressed plants. The water which is bound tightly to the microfibrils of the cell wall is apoplastic, and, therefore, not osmotically active. This apoplastic water can range from 5 to 35% of the total cell water. When a plant is subjected to water stress, water evaporates from the film of water covering the cell walls. Localized Ψw gradients drive movement of symplastic water from within the cell to the site of evaporation. Thus, leaf Ψw decline during stress is associated with loss of symplastic water; the water bound tightly to the cell wall remains. Smaller cells with thicker cell walls in stress-adapted leaves have a relatively greater proportion of total cell water in the apoplastic fraction. Thus, a given amount of dehydration (loss of total cell water) will have a greater concentrating effect on the solutes within the cell (i.e. in the symplastic water fraction). Thus, Ψs will decline more rapidly in these cells as water is removed. This more rapid Ψs decline will, therefore, allow for a relatively greater Ψp to be maintained at a given degree of cell dehydration. Although increased cell wall elasticity and apoplastic water fraction can theoretically impart a degree of stress resistance to plants (in terms of allowing for greater leaf conductance at low Ψw), they are not very common acclimation mechanisms. Osmotic adjustment is much more readily encountered in crop plants.

A second area of stress acclimation involves alterations in cell water relations parameters and/or physiology which allow for the maintenance of relatively greater chloroplast photosynthetic activity at low Ψw. As discussed previously, impairment of chloroplast function during the development of internal plant water deficits is linked more fundamentally with a given degree of symplastic volume change than a given degree of Ψw decline. Therefore, factors which shift the symplast volume : Ψw relationship can lead to the maintenance of greater chloroplast photosynthetic activity at low Ψw. In addition to the effect osmotic adjustment has on Ψp maintenance at low Ψw, this stress acclimation mechanism also can greatly alter the volume: Ψw relationship in leaf cells. Thus, osmotic adjustment can afford a degree of stress resistance to plants exposed to water/salt stress in two complementary fashions; allowing for relatively greater $CO_2$ supply to the chloroplast, and chloroplast capacity to assimilate $CO_2$ at low Ψw.

## CONCLUDING REMARKS

The information provided in this review of water and salt stress effects on photosynthesis provides a new context within which drought resistance strategies of crop plants can be re-evaluated. The development of low leaf Ψw in plants subjected to water or salt stress should not be envisioned as necessarily deleterious to photosynthetic capacity. Cultural management, breeding, and genetic engineering strategies which facilitate enhanced osmotic adjustment capability can be thought of as imparting a degree of stress resistance to plants which experience internal water deficit conditions. Both stomatal limitation of $CO_2$ supply to the leaf interior, and chloroplast metabolic activity limit photosynthesis of water/salt stressed plants. Both of these stress lesions can be positively impacted by osmotic adjustment.

## FURTHER READING


Davies, W. J. & Zhang, J. (1991). Root signals and the regulation of growth and development of plants in drying soil. *Annual Review of Plant Physiology and Plant Molecular Biology*, **42**, 55–76.

Jones, H. G. & Corlett, J. E. (1992). Current topics in drought physiology. *Journal of Agricultural Science*, **119**, 291–96.

Kaiser, W. M. (1987). Effects of water deficit on photosynthetic capacity. *Physiologia Plantarum*, **71**, 142–9.

Niu, X., Bressan, R. A., Hasegawa, P. M. & Pardo, J. M. (1995). Ion homeostasis in NaCl stress environments. *Plant Physiology*, **109**, 735–42.

Santakumari, M. & Berkowitz, G. A. (1991). Chloroplast volume: cell water potential relationships and acclimation of photosynthesis to leaf water deficits. *Photosynthesis Research*, **28**, 9–20.

Sinclair, T. R. & Ludlow, M. M. (1985). Who taught plants thermodynamics? The unfulfilled potential of plant water potential. *Australian Journal of Plant Physiology*, **12**, 213–17.

Slovik, S. & Hartung, W. (1992). Compartmental distribution and redistribution of abscisic acid in intact leaves. *Planta*, **187**, 37–47.

Ward, J. M., Pei, Z. M. & Schroeder, J. I. (1995). Roles of ion channels in initiation of signal transduction in higher plants. *Plant Cell*, **7**, 833–44.

# 18 Photosynthesis at low growth temperatures

V. HURRY, N. HUNER, E. SELSTAM, P. GARDESTRÖM AND G. ÖQUIST

## INTRODUCTION

Plants from high latitudes and high altitudes are faced with short growing seasons and the need to grow at low temperatures to extend the growing season for as long as possible. Thus, the capacity for active photosynthesis during prolonged exposure to low growth temperatures is essential in determining their successful site occupation and subsequent productivity. All these plants, however, are exposed to a range of temperatures, and survival and active growth at low temperatures requires specific acclimation. Short-term exposure of plants to low temperature usually causes inhibition of net photosynthesis. This is due to the accumulation of soluble carbohydrates and to reduced orthophosphate cycling from the cytosol back to the chloroplast. In contrast, long-term acclimation allowing new growth and development at low temperatures results not only in the accumulation of soluble carbohydrates but also in an increase in the capacity for photosynthesis. Besides growing at low temperatures, these plants must also acclimatize to survive freezing winter temperatures. We propose that restoration of the balance of photosynthesis and carbon metabolism is a mechanism facilitating the accumulation of sugars with possible cryoprotective functions, and ones that are essential substrates for basal metabolism during winter. Thus, the restoration of photosynthesis and carbon metabolism and the development of freezing tolerance represents a continuum in the low temperature acclimation response.

In this chapter, we focus on the long-term acclimative responses of photosynthesis that result from growth and developments at low temperatures. We summarize alterations to the structure, function and composition of the photosynthetic apparatus, and the acclimative responses of cellular sugar metabolism. We also briefly discuss the role of photosynthetic acclimation to low growth temperatures in the development of resistance to photoinhibition. We suggest that the different responses of cold-tolerant perennial evergreens and herbaceous annuals represent different developmental strategies to cold acclimation. For a more extensive review of photoinhibition, see Chapter 20 of this volume. For more extensive reviews of the effects of cold hardening on photosynthesis and photoinhibition, see recent reviews by (Huner *et al.*, 1993; Ball, 1994; Krause, 1994) and the references cited therein.

## ACCLIMATION OF THE PHOTOSYNTHETIC APPARATUS TO LOW GROWTH TEMPERATURES

### Thylakoid composition and organization

Exposure of conifers, herbaceous plants and mosses to low growth temperatures typically leads to a decrease in the ratio of the bulk thylakoid lipids, monogalactosyl-diacylglycerol (MGDG) to digalactosyl-diacylglycerol (DGDG), without causing major changes to the content of sulphoquinovosyl-diacylglycerol (SQDG) or phosphatidyl-diacylglycerol (PG) (see Öquist & Martin, 1986; Selstam & Widell Wigge, 1993). In contrast to the other membrane lipids that are bilayer forming, MGDG forms a reversed hexagonal phase. The more MGDG that is present in a mixture of MGDG and DGDG, the more non-lamellar phase is formed. Thus, a change in the ratio of MGDG to DGDG could have significant effects on thylakoid stability. A temperature-induced adjustment in the ratio of MGDG to DGDG may reflect alterations to maintain the functional integrity of thylakoids in response to the accumulation of solutes and an increase in cellular osmoticum. The decrease in the ratio of MGDG to DGDG is also correlated with an increase in the ratio of lipid to chlorophyll (E. Selstam unpublished observations). This suggests that the more lipid per protein there is in the membrane, the less MGDG is needed to maintain the correct balance between lamellar and hexagonal lipids. However, the structural and functional significance of changes to the ratio of MGDG to DGDG at low temperature is not yet fully understood.

Many organisms show an increase in the level of fatty acid unsaturation following growth at low temperature. Lipid unsaturation increases generally by 5 to 15% in the major thylakoid lipids of cold-tolerant dicotyledonous annuals. The importance of a general increase in the level of unsaturation of the thylakoid lipids was shown in the *fad6* mutant of *Arabidopsis*, which contains reduced levels of both 18 : 3 and 16 : 3 fatty acids and concomitantly higher levels of the 18 : 1 and 16 : 1 precursors. This mutant showed chilling injury when shifted to low temperatures but was still able to complete its life cycle at 6 °C (see Browse & Somerville, 1994). The results of experiments with transgenic *Nicotiana tabacum* and *Arabidopsis thaliana* suggest that the content of high melting point molecular species of PG is a major factor determining the chilling sensitivity of plants (see Somerville, 1995). Cold-tolerant varieties of wheat and rye, however, do not show significant changes in the level of 18 : 3 in PG upon growth at low temperature. Further, Browse and co-workers have shown that the *fab1* mutant of *Arabidopsis*, which has increased levels of 16 : 0 at the expense of 18 : 3 in PG, was no more sensitive to chilling than wild-type over a range of chilling treatments. Long-term growth of the *fab1* mutant at 2 °C resulted in slower growth rates compared with wild-type, but this temperature did not induce necrotic lesions indicative of chilling injury (Wu & Browse, 1995).

Recent data from tobacco transformed to produce lower levels of unsaturated fatty acids in PG, showed that these plants were not more susceptible than wild-type to low-temperature-induced photoinhibition. However, they did show impaired recovery from photoinhibition (Moon *et al.*, 1995). The authors suggest that the lower levels of thylakoid fatty acid unsaturation may have

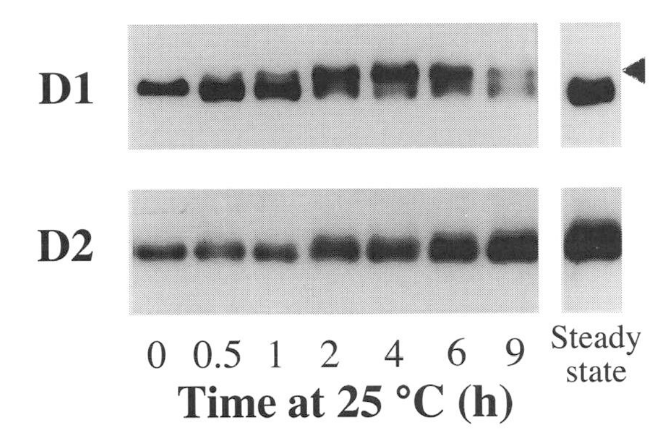


**Figure 18.1.** Accumulation of PSII core polypeptides D1 and D2 in the cyanobacterium *Synechococcus* sp. PCC 7942 following a shift from 37 to 25 °C. Proteins were detected by immunoblotting after SDS-PAGE. Note the accumulation of an unprocessed form of the D1 protein (*arrow*), and its eventual disappearance once the cells had fully acclimated to the lower growth temperature. (From Campbell *et al.*, 1995.)

reduced the rate at which the D1 protein could be processed and inserted into the PSII core complex during repair. Earlier experiments with chilling sensitive maize have shown that a precursor of CP29, a chlorophyll *a*/*b* binding protein component of LHCII, accumulates in thylakoids during chilling in the light (Hayden, Covello & Baker, 1988). This may represent another example of interrupted protein processing at low temperature due to thylakoid fatty acid composition in a chilling sensitive plant. Further, Campbell and co-workers (Campbell *et al.*, 1995) have reported recently that the cyanobacterium *Synechococcus* suffers a transient interruption in D1 protein processing when exposed to a sudden drop in temperature (Fig. 18.1). Upon prolonged low temperature incubation, processing of the D1 protein resumes in parallel with an increase in membrane lipid unsaturation (Porankiewicz, unpublished observations).

Thus, the biological significance of changes in thylakoid lipid unsaturation in response to low temperature is not clear. Mutations in chloroplast fatty acid biosynthesis, at least in *Arabidopsis*, appear to result in only a limited loss in the function of existing membranes following a shift to low temperature. The principal effect may be to membrane biogenesis, and chloroplast ultrastructure during development at low temperature (Somerville, 1995). While these effects may compromise long-term plant performance, they appear to play only a minor direct role in chilling sensitivity (Browse & Somerville, 1994). The long-term detrimental effects of lower levels of fatty acid unsaturation at low temperature may also include interference with protein processing or insertion into the thylakoid membrane. This is a potentially important aspect of cold acclimation that has generally been overlooked in studies of low temperature-induced photoinhibition.

Prolonged exposure of Scots pine and winter rye to low growth temperatures results in the development of normal chloroplasts differentiated into grana and stromal thylakoids. Similarly, exposure of cold-tolerant plants such as pine, spinach and winter rye to low growth temperatures does not result in major changes in the pigment or stromal and thylakoid membrane polypeptide composition (Öquist & Martin, 1986; Huner *et al.*, 1993). However, ultrastructural examination of rye chloroplasts in leaf cross-sections shows a reduction in the number of thylakoids *per granum* in cold-grown chloroplasts. Similarly, the thylakoid ultrastructure of Scots pine shows de-stacking of grana following exposure to low temperatures (Öquist & Martin, 1986). Thus, although exposure of cold-tolerant plants to low growth temperatures does not result in major changes in the quantitative composition of proteins and pigments, qualitative changes occur which we relate to modified organization. For example, during autumn and winter, Scots pine appears to reorganize the light-harvesting complexes *in vivo* into large aggregates. These aggregates are specifically enriched in 22, 7, 8, and 10 kDa polypeptides associated with the LHCII proteins, chlorophylls and xanthophylls of the antennae complexes (Fig. 18.2). The formation of these aggregates may contribute to the de-stacking of grana observed in over-wintering conifers.

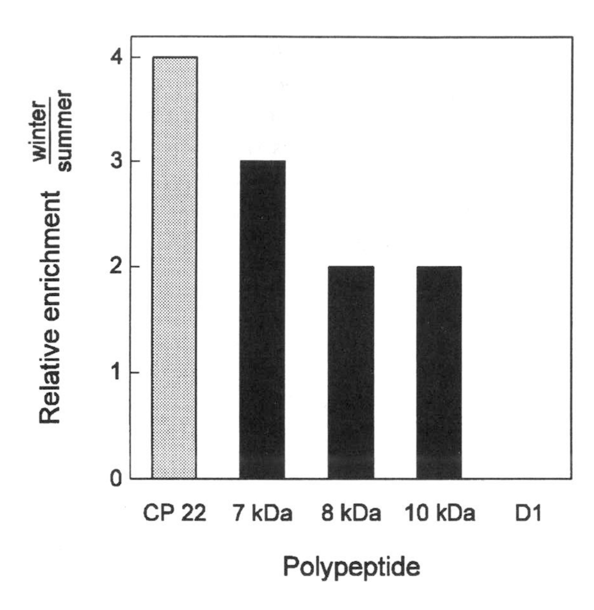


**Figure 18.2.** Winter-induced enhancement of a range of small molecular weight proteins, plus CP22, in LHCII aggregates that partition near the top of mildly denaturing green gels from Scots pine thylakoids. This aggregate band appears in both summer and winter thylakoid preparations, but it increases five-fold in winter and these polypeptides are enriched in the winter aggregates compared with those isolated from summer thylakoids. The polypeptide composition was determined by immunoblotting (stippled columns) or Coomassie staining (filled columns) after SDS-PAGE. Small amounts of other polypeptides were also present in the aggregates but were detected in similar relative amounts in both summer and winter samples. The data show the relative enrichment in winter thylakoids standardised to LHCII content. (From Ottander, Campbell & Öquist, 1995.)

*In vitro* examination of the chlorophyll–protein complexes of cold- and warm-grown rye and pine thylakoids also show a differential stability of LHCII, correlated with a specific change in the fatty acid composition of PG. Winter rye shows a 75% decrease in trans-$\Delta^3$-hexadecenoic acid (trans-16 : 1) associated with PG when grown at 5 °C. The freezing tolerance of wheat and rye cultivars is correlated with the capacity of these plants to reduce trans-16 : 1 content in response to low growth temperatures. However, cold tolerant dicots such as spinach, pea, broadbean and periwinkle do change trans-16 : 1 content upon growth at low temperature. The presence of high levels of trans-16 : 1 is associated with an increase in the stability of the oligomeric form of LHCII in mildly denaturing gel systems, whereas the monomeric form is associated with low levels of trans-16 : 1 PG. Those plants that exhibit a lower level of trans-16 : 1 following growth at low temperatures (pine and monocots) also modulate the organization of LHCII, whereas those that do not change the trans-16 : 1 content of PG (dicots) do not alter the organization of

LHCII. Recent work by Paulsen and co-workers have identified a specific trimerization motif of 16 to 21 amino acids in LHCII that is essential for trimer formation (Hobe *et al.*, 1995). It appears that PG may bind stoichiometrically to this motif to enhance LHCII trimer formation (Hobe *et al.*, 1995). However, we do not know the functional significance of this altered LHCII stability at low temperatures.

## Carbon metabolism

Short-term reductions in leaf temperature lead to the accumulation of soluble carbohydrates. This suppresses photosynthesis by reducing $P_i$ cycling from the cytosol back to the chloroplast, and therefore limits the synthesis of ATP needed in the regeneration of ribulose-1,5-bisphosphate (RuBP). Long-term exposure of evergreen woody perennials such as Scots pine to low growth temperatures also results in a depression of light saturated rates of $CO_2$ assimilation. This is associated with lower light saturated rates of whole chain electron transport and with reductions to the size of the functional plastoquinone pool (see Öquist & Martin, 1986). In contrast, long-term acclimation of herbaceous plants to low growth temperatures, leading to new growth and development, is associated with increased rates of light saturated $CO_2$ assimilation (Table 18.1). This is associated with 1.5-fold higher rates of whole chain and PSI electron transport, and an increase in the size of the functional plastoquinone pool (see Huner *et al.*, 1993). The recovery of photosynthetic capacity in these herbaceous plants is strongly correlated with the development of freezing tolerance and occurs despite the accumulation of soluble carbohydrates (Table 18.1).

The photosynthetic capacity of herbaceous winter annuals recovers following growth and development at low temperature through increases in the activity of several enzymes of the photosynthetic carbon reduction cycle (Table 18.1). Increases in adenylates and phosphorylated intermediates, and an increased capacity for RuBP regeneration, support this increase in enzymatic capacity (Hurry *et al.*, 1994, 1995*b*). The increase in photosynthesis and in soluble carbohydrate content following growth at low temperature also coincides with increases in the activity of cytosolic enzymes associated with sucrose synthesis (Table 18.1). In addition, the activation energy of the photosynthetic carbon reduction cycle enzymes, rubisco and stromal fructose-1,6-bisphosphatase, from cold-tolerant herbaceous annuals decreases following long-term exposure to low growth temperatures (Grafflage & Krause, 1993; Brüggemann, Klaucke & Maas-Kantel, 1994). At this point, we do not know whether the expression of different isoenzymes is involved in these acclimative changes at low temperature.

These data are a clear indication that the observed changes in photosynthetic metabolism and carbon partitioning in cold hardy herbaceous plants are adaptive for growth at low temperatures, and that these changes are an integral component of the cold hardening process. Studies of non-hardy spring cultivars reinforce this conclusion. Prolonged growth at low temperature in these plants is associated with reduced photosynthetic capacity, and in some cases, with patchy loss of chlorophyll (Table 18.1). Non-hardy plants frequently show similar increases in total extractable enzymatic activity per unit chlorophyll to cold-hardy plants, but in contrast to what we observe in plants that are cold tolerant, low growth temperatures can lead to the suppression of the activation state of these enzymes (Table 18.1).

This response of non-hardy herbaceous plants resembles the feed-back regulation of photosynthesis in tobacco that occurs as the leaves mature and go through the transition from sink to source leaves. Studies using transgenic tobacco have also shown that carbohydrate accumulation in leaves suppresses photosynthesis. The loss of photosynthetic capacity is correlated with lower mRNA and protein levels of several photosynthetic carbon reduction cycle enzymes and chlorophyll *a*/*b* binding proteins (see Stitt & Schulze, 1994). Spinach leaves that accumulate soluble carbohydrates following cold-girdling show similar results (Krapp & Stitt, 1995). These studies suggest that sink strength regulates photosynthetic capacity and that sugars may act as the signals to alter photosynthetic gene expression. In contrast, cold-hardy plants show a response that is opposite to that reported from these studies of sugar signalling and sink regulation when they are grown at low, but not warm, temperatures. The data from low growth temperature experiments suggests that there is a developmental requirement to the cold acclimation of hardy cultivars, and that this may include overcoming the regulatory effects of high levels of soluble

Table 18.1. *Effect of low growth temperatures on photosynthesis, carbohydrate content and enzymatic activity in leaves of spring and winter cultivars of herbaceous annuals grown at either 24 °C or 5 °C at irradiances of 250 to 300 μmol $m^{-2}$ $s^{-1}$ PFD.*

| Parameter | Winter rye 24 °C | Winter rye 5 °C | Spring wheat 24 °C | Spring wheat 5 °C | Winter canola 24 °C | Winter canola 5 °C | Spring canola 24 °C | Spring canola 5 °C |
|---|---|---|---|---|---|---|---|---|
| Photosynthesis (μmol $CO_2$ $m^{-2}$ $s^{-1}$) | 7.2 | 9.2 | 6.1 | 5.8 | 7.5 | 9.1 | 8.6 | 0.7 |
| Sucrose (μmol $mg^{-1}$ Chl) | 19.7 | 85.7 | 9.1 | 10.4 | 3.3 | 6.2 | 3.0 | 4.4 |
| Free hexoses (μmol $mg^{-1}$ Chl) | 3.0 | 19.4 | 4.8 | 5.2 | 8.6 | 57.0 | 7.0 | 5.0 |
| Rubisco | | | | | | | | |
| Total activity (μmol $mg^{-1}$ Chl $h^{-1}$) | 291 | 429 | 415 | 408 | 311 | 332 | 334 | 574 |
| Activation state (%) | 68 | 87 | 60 | 84 | 73 | 92 | 79 | 90 |
| stromal Fru-1,6-BPase | | | | | | | | |
| Total activity (μmol $mg^{-1}$ Chl $h^{-1}$) | 72 | 166 | 130 | 245 | 80 | 211 | 85 | 167 |
| Activation state (%) | 91 | 88 | 83 | 82 | 39 | 66 | 47 | 35 |
| Sucrose phosphate synthase | | | | | | | | |
| Total activity (μmol $mg^{-1}$ Chl $h^{-1}$) | 18 | 49 | 30 | 43 | 15 | 31 | 21 | 41 |
| Activation state (%) | 43 | 57 | 52 | 23 | 23 | 38 | 19 | 16 |
| cytosolic Fru-1,6-BPase | | | | | | | | |
| Total activity (μmol $mg^{-1}$ Chl $h^{-1}$) | 33 | 61 | 27 | 54 | 17 | 41 | 18 | 40 |

$CO_2$ assimilation rates were measured at the respective growth temperatures and irradiances, 2 to 3 h into the photoperiod. Data are from (Hurry *et al.*, 1994, 1995*a*).

carbohydrates. This developmental response appears to be lacking in non-hardy plants.

Thus, cold-tolerant plants can recover their photosynthetic capacity at low growth temperatures by increasing the activity of a broad range of photosynthetic and carbon metabolism enzymes. Larger pools of phosphorylated intermediates buttress the increase in enzymatic capacity to support the high rates of flux at low temperatures. The 'cost' of this recovered photosynthetic capacity at low temperatures is lower nitrogen and phosphate use efficiency. This may be one reason for the much lower growth rates of these plants at lower temperatures, despite the similar or higher in situ photosynthetic rates. In addition, we have also shown increases in dark respiration following acclimation to low temperature in cold-tolerant cereals (Hurry *et al.*, 1992), although we have recently shown that this does not necessarily lead to higher rates of oxidase decarboxylation in the light (Hurry *et al.*, 1996). We have shown, however, that mitochondrial activity in the light plays an increased role in supporting photosynthetic oxygen evolution in cold-grown winter rye leaves (Hurry *et al.*, 1995*b*). We interpret these data as an increase in mitochondrial consumption of reducing equivalents generated by the chloroplast electron transport chain, and as an indication of increased metabolic flexibility following growth at low temperatures. Mitochondrial electron transport is important in preventing photoinhibition in pea protoplasts (Saradedevi & Raghavendra, 1992). Such an increase in flexibility in response to growth at low temperatures may 'cost' the plant in terms of higher respiratory losses but the return may be an increased ability to avoid oxidative chilling injury. Despite these higher costs, failure to recover photosynthesis and to regulate respiratory metabolism at low growth temperatures evidently leads either directly to plant death or to higher winter kills. Restoration of the balance between photosynthesis and carbon metabolism is clearly a key element in the low temperature acclimation response of plants.

## WHAT DRIVES ACCLIMATION TO LOW TEMPERATURE?

As mentioned above, the short-term response of *developed* leaves, even from hardy plants, to an accumulation of soluble carbohydrates is to suppress photosynthesis and to down-regulate the expression of photosynthetic enzymes and chlorophyll *a/b* binding proteins. This model of sink regulation has been reviewed elsewhere (Sheen, 1994; Stitt & Schulze, 1994). However, the data presented above is clear evidence that a key aspect of long-term photosynthetic acclimation in cold-hardy versus non-hardy herbaceous annuals is that, contrary to previous studies of sugar signalling, cold-hardy annuals increase all components of the photosynthetic apparatus, from chlorophyll contents to enzymes of carbon metabolism, despite the accumulation of large product pools.

Huner and co-workers have recently proposed an alternative model for low temperature acclimation from studies of both single celled green algae and higher plants (Huner *et al.*, 1996). These studies suggest that photosynthetic adjustment occurs in response to a complex interaction between light and temperature, mediated by the redox poise of intersystem electron transport. However, it is clear from these studies that the acclimation responses of the single-cell algae *Chlorella vulgaris* and *Dunaliella salina* are different in several respects to that of cold-hardy higher plants. *Chlorella* and *Dunaliella* respond to growth at high excitation pressure, no matter whether this is due to high irradiance or to a combination of moderate irradiance and low temperature, by decreasing the light-harvesting antenna and therefore, the capacity to absorb incident radiation (Maxwell *et al.*, 1994; Maxwell, Laudenbach & Huner, 1995). This response is qualitatively very similar to that of *developed* leaves of either non-hardy or hardy plants *shifted* to low temperature. *Chlorella* was also unable to adjust at the level of carbon metabolism and despite continued growth at high excitation pressure, could not overcome the limitation on photosynthesis (Savitch, Maxwell & Huner, 1996). This response is also similar to that reported for leaves of non-hardy plants that develop at low temperature and may reflect restrictions on acclimation imposed by limited export capacity for fixed carbon (Table 18.1). In contrast, winter cereals respond to high excitation pressure, whether induced by high irradiance or moderate irradiance in combination with low temperature, by producing new leaves that have *increased* chlorophyll content and *increased* photosynthetic capacity and higher activity of enzymes involved in carbon metabolism (Gray *et al.*, 1996).

A redox signal may mediate the short-term stress response and sugar accumulation could well generate the trigger through feedback from cytosolic carbon metabolism and $P_i$ cycling. However, recent data from tobacco expressing an antisense construct against the Rieske-FeS protein of the cytochrome $b_6/f$ complex, showed that marked alterations to the redox poise of intersystem electron transport upstream of the cytochrome $b_6/f$ complex did not affect the PSII/PSI ratio or the size of the PSII light-harvesting antennae (Price *et al.*, 1995). This suggests that, if redox poise mediates regulation of gene expression, it may be associated with the thioredoxin system (Levings & Siedow, 1995) and not with the reduction of PSII or the plastoquinone pool. Huner and co-workers also concluded that the redox state of PSII could not be the primary signal for acclimation (Maxwell *et al.*, 1995). Further, the results of Gray *et al.* (1996) show that excitation pressure cannot fully explain the response of cold-hardy plants, as the response to high excitation pressure at low temperature is stronger than that of high excitation pressure alone. This suggests that the long-term response, which clearly has a *developmental* component, involves a response to signals other than redox poise. Huner and co-workers (Huner *et al.*, 1996) have suggested that the redox sensing/signalling mechanisms probably act synergistically with other signal transduction mechanisms to elicit the appropriate physiological response. It seems likely that the developmental response displayed by hardy plants is a different phenomenon to the down-regulation of photosynthesis and photosynthetic gene expression displayed by shifted leaves and by *Dunaliella* and *Chlorella*. We do not know why cold hardy plants require this developmental step. Nor do we know what the consequences of this acclimation are for enzyme function at low temperature. The signals for alteration of gene expression during development may still be sugar accumulation and/or cellular redox status but a different signal transduction path appears to mediate the response. Elucidation of this specific developmental response to acclimation is an exciting area for future research. Understanding this aspect of cold hardening will not only contribute to our knowledge of low temperature responses but also to our knowledge of sugar and redox signalling in plants.

## *IN VIVO* RESPONSE OF PHOTOSYNTHESIS TO HIGH LIGHT AT LOW TEMPERATURE

Photoinhibition, or the light-induced loss of photosynthetic efficiency, occurs when the light-harvesting antennae absorb excitation energy in excess of that used by photosynthesis. Over-wintering evergreens are exposed to the combined stress of high irradiance and low temperature, and thus face a high risk of photoinhibition. Of the four macromolecular complexes found in thylakoid membranes, PSII is the most vulnerable to damage from excess irradiance. For a detailed review of photoinhibition see Chapter 20 of this volume, and for detail on low temperature photoinhibition see (Huner *et al.*, 1993; Ball, 1994; Krause, 1994). We deal here only with aspects of photoinhibition relating to low temperature acclimation of photosynthesis.

## Evergreens at low growth temperatures and high irradiance

Growth of Scots pine at low temperature causes a significant depression in the light-saturated rate of $CO_2$ assimilation, even before exposure to freezing temperatures (see Öquist & Martin, 1986). Further, during prolonged growth under controlled conditions at 4 °C and and 8 h photoperiod, light levels as low as 50 μmol $m^{-2}$ $s^{-1}$ PFD are sufficient to induce a significant reduction in the photon yield of the upper, light exposed side, of the needles (Strand & Öquist, 1988). We have related this low temperature suppression of photosynthesis to the onset of winter dormancy. Photoinhibition of photosynthesis also occurs at low temperatures over-wintering perennial evergreens such as pine, spruce, holly and ivy (see Huner *et al.*, 1993). We view this low temperature-induced photoinhibition as a state of long-term down-regulation of PSII activity to match the rate of photochemistry to the temperature-dependent enzymatic steps of the photosynthetic carbon reduction cycle. Experimental evidence, showing that the photochemical yield of PSII under the high irradiance conditions causing photoinhibition is similar in photoinhibited and non-inhibited leaves, supports this view (Hurry *et al.*, 1992).

Winter stress inhibition of photosynthesis in Scots pine, as induced by frost exposure and light in the field, results from both photoinactivation of PSII and from frost inhibition of $CO_2$ assimilation (Strand & Öquist, 1988; Ottander, Campbell & Öquist, 1995). Winter-stressed Scots pine, growing outdoors and exposed to a combination of light and freezing temperatures, showed almost complete loss of PSII reaction centres, as judged

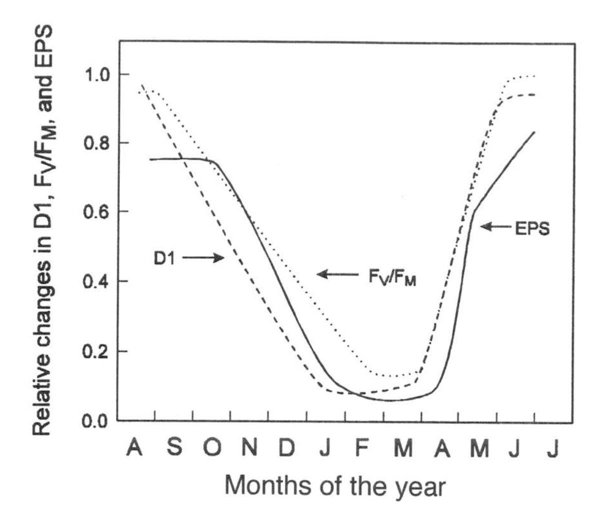


**Figure 18.3.** Relative changes in D1 protein content, photochemical efficiency (assessed by the fluorescence parameter, $F_V/F_M$) and the epoxidation state (EPS) of the xanthophyll cycle pigments (violaxanthin, antheraxanthin and zeaxanthin) in Scots pine needles over the year. (From Ottander, Campbell & Öquist, 1995.)

from D1 degradation, by the end of December (Fig. 18.3). Consequently they also show a severe suppression of photochemical efficiency, as assessed by $F_V/F_M$ (Fig. 18.3). Adams and co-workers have related the loss of $F_V/F_M$ to the accumulation of zeaxanthin in the needles of over-wintering conifers (Adams *et al.*, 1995). However, in Scots pine, all the chlorophyll and LHCII lost over winter, and 50% of the D1 protein lost, occurred before any changes in the total xanthophyll content or in the epoxidation state of the xanthophyll pool. Further, the accumulation of zeaxanthin did not arrest the loss of D1 protein (Fig. 18.3). We have concluded from these studies that the suppression of $F_V/F_M$ in the field in over-wintering Scots pine is due to photoinhibition, as assessed by D1 protein loss, and is not due to sustained energy quenching in the antenna associated with zeaxanthin accumulation. Concomitant with the loss of D1 protein, the light-harvesting complexes of pine appear to become reorganized into large aggregates, as assessed by mildly denaturing green gels. These aggregates consist of the LHCII proteins with enriched amounts of the chlorophyll *a/b* binding 22 kDa protein, encoded by the *psbS* gene, and minor polypeptides of 7, 8 and 10 kDa

(Fig. 18.2). What form such aggregates may take *in vivo* is not clear. However, ultrastructural studies of pine needles collected in the winter have shown that, while the chloroplast envelope and stroma and grana lamellae remain intact, the grana de-stacks during the winter (Öquist & Martin, 1986). The aggregation of LHCII proteins observed by Ottander *et al.* (1995) could account for this de-stacking of the grana over winter.

Besides the reorganization of the pigment/protein complexes in the thylakoid, the amounts and activities of a range of oxygen scavenging enzymes increase in Scots pine exposed to low temperature (Krivosheeva *et al.*, 1996). This may further reduce the risk of photodamaging reactions, despite the severe limitations on photosynthesis in the field during winter. An increased capacity for light induced de-epoxidation of violaxanthin to zeaxanthin upon cold acclimation may also contribute to protection from active oxygen species (Krivosheeva *et al.*, 1996).

We suggest that this late autumn and winter reorganization of the light-harvesting chlorophyll antennae upon degradation of the D1 protein allows pine to maintain large reserves of chlorophyll in a quenched state. In this way, Scots pine can avoid severe photo-destruction of pigments and thylakoid lipids during winter. The enhancement of oxygen radical-scavenging systems, perhaps including the carotenoids of the xanthophyll cycle, may contribute significantly to the preservation of this quenched state during winter. Field-grown Scots pine can fully recover from winter stress over several hours to days if transferred to 20 °C and moderate light conditions in the laboratory. In the field, $F_V/F_M$ and photosynthetic $CO_2$ assimilation recovers in spring before budbreak by deciduous broad-leaved plants (Ottander *et al.*, 1995). This may help explain why, in cold climates, conifers compete so successfully with deciduous trees that shed their leaves every autumn, with concomitant nitrogen and carbon loss.

## Annuals at low growth temperatures and high irradiance

In contrast to evergreens, Somersalo and Krause (1989) presented the first published results showing that cold-grown spinach was less sensitive to photoinhibition at low temperature than warm-grown spinach. Subsequent measurement of 77 K chlorophyll fluorescence, room temperature chlorophyll fluorescence and $CO_2$ saturated $O_2$ evolution have confirmed the novel capacity of spinach and winter cereals to show reduced sensitivity to photoinhibition following growth at low temperature (Huner *et al.*, 1993). In contrast to winter cultivars, we showed that less cold-tolerant cereals had a smaller capacity to reduce their sensitivity to photoinhibition after growth at low temperature (Fig. 18.4). Further,

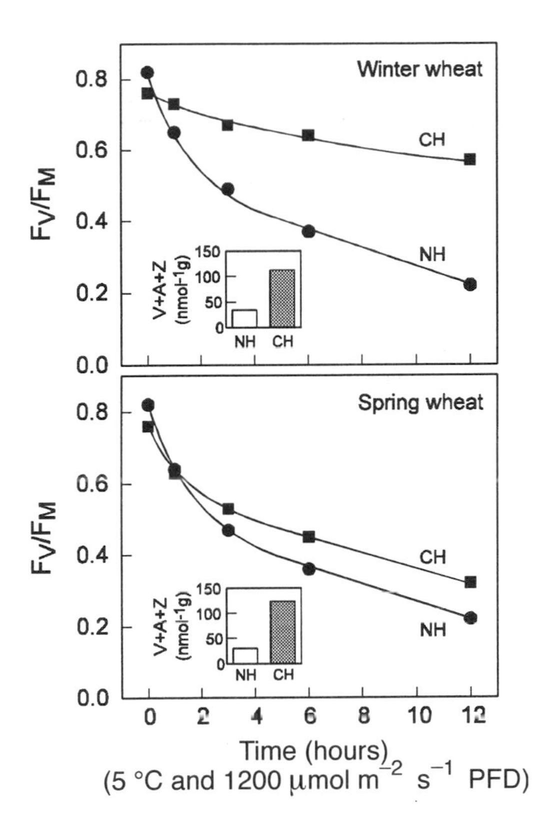


**Figure 18.4.** Sensitivity of warm grown (NH) and cold grown (CH) intact winter and spring wheat leaves to photoinhibition at 5 °C, assessed by the fluorescence parameter, $F_V/F_M$. *Inset* shows the xanthophyll content (V + A + Z = Violaxanthin + Antheraxanthin + Zeaxanthin) of the winter and spring wheat leaves grown under the different temperatures and expressed on the basis of leaf fresh weight. Both winter and spring wheat increase the relative conversion of violaxanthin to zeaxanthin in high light following growth at low temperatures, but no difference was found in the conversion capacity of the two cultivars. (From Hurry & Huner, 1992; Hurry *et al.*, 1992.)

these studies clearly showed that *development* at low temperature was an absolute requirement for the acquisition of resistance to photoinhibition, even in winter rye and spinach (Huner *et al.*, 1993; Gray, Boese & Huner, 1994).

The decrease in sensitivity to photoinhibition in cold-grown winter rye could not be accounted for by changes in the sensitivity of isolated thylakoids. However, isolated intact mesophyll cells from cold-grown rye did show the decrease in sensitivity to photoinhibition. This led to the conclusion that the reduced sensitivity to photoinhibition following growth of winter rye at low temperature was not due to changes in thylakoid composition or in leaf architecture but instead to altered cellular metabolism (Lapointe & Huner, 1993). Analysis of chlorophyll fluorescence quenching showed that cold-grown winter rye leaves required a three-fold increase in irradiance to cause the same degree of PSII closure [$QA_{red}/(QA_{ox}+QA_{red})$] as warm-grown winter rye. When combinations of temperature and light were chosen to set equal redox states of $Q_A$, the differential sensitivity to photoinhibition observed between cold-grown and warm-grown rye was eliminated (Öquist *et al.*, 1993). These findings show a strong correlation between lower sensitivity to photoinhibition and the restoration of the balance between photosynthesis and carbon metabolism in winter rye and wheat (Huner *et al.*, 1993) (Table 18.1). We have concluded from these, and other studies, that modulation of the capacity for photosynthetic carbon assimilation at the low temperature, not an increase in xanthophyll-related dissipative mechanisms (Fig. 18.4), accounts for the ability of winter cereals to reduce their sensitivity to photoinhibition.

In contrast to cereals, the redox state of $Q_A$ during steady-state photosynthesis could not account for resistance to photoinhibition of cold-grown spinach (Gray *et al.*, 1994). Further, spinach does not modulate maximum rates of photosynthesis following growth at low temperature. Increases in xanthophyll pool sizes have been shown for cereals (Fig. 18.4) and dicotyledonous annuals (Koroleva, Thiele & Krause, 1995) grown at low temperatures. These increases in xanthophyll cycle activity may protect cold-grown dicots such as spinach from photoinhibition by increasing the scavenging of active oxygen species (Schöner & Krause, 1990). Alternatively, Adams and co-workers have suggested that the maintenance of a zeaxanthin pool by leaves of evergreen perennials over-wintering in the field may be the explanation for the winter suppression of dark-adapted $F_V/F_M$, and that this may not represent photoinhibition as defined by photoinactivation of the D1 protein (Adams *et al.*, 1995). This notion has been extended even further recently in studies of cold-tolerant herbaceous plants, which suggest zeaxanthin may facilitate the formation of a rapidly reversible 'photoinhibited' quenched (qI) state of PSII, not associated with D1 protein turnover (Thiele, Winter & Krause, 1995). Imagining how either of these mechanisms could function is difficult, given that xanthophyll-related quenching of fluorescence requires both violaxanthin de-epoxidation to antheraxanthin and zeaxanthin and the presence of a transthylakoid pH gradient (Gilmore & Yamamoto, 1993). From the evidence presented above, we believe that such mechanisms are not consistent with our observations of photoinhibition over-wintering Scots pine, or that increased xanthophyll activity is the explanation for reduced sensitivity to photoinhibition in cereals. However, recent evidence suggests that non-photochemical quenching of fluorescence may depend on sequestered proton domains within the thylakoid membrane, rather than the bulk proton gradient (Mohanty & Yamamoto, 1995). This may help explain the slow relaxation of 'quenched dark-adapted $F_V/F_M$' in low temperature photoinhibited leaves that sustain pools of zeaxanthin, if alterations in membrane composition and permeability are shown to affect the equilibration of such proton domains with the bulk proton gradient. This hypothesis, and the possibility that change in the composition and properties of the thylakoid membrane following growth at low temperature may alter protein processing, and assembly of pigment–protein complexes, remain to be tested. Similarly, the possibility that changes in the xanthophyll content may play a more direct role in altering membrane properties at low temperature, perhaps associated with membrane fluidity, also remains to be investigated. Clearly, plant species may use different mechanisms to alter sensitivity to photoinhibition, and the efficacy of these mechanisms will vary depending on the environmental conditions that have led to photoinhibition. It is unlikely that one general mechanism exists to account for the variable sensitivity of all higher plants to photoinhibition at low temperature.

## CONCLUDING REMARKS

When attempting to decide what changes are adaptive for overall plant performance at low growth temperatures, we must consider the growth strategies of the plants in question. Woody evergreens that enter dormancy during autumn and herbaceous winter annuals that continue to grow and develop during autumn, respond very differently to a shift to low growth temperatures. Woody evergreens such as Scots pine down-regulate photosynthesis, specifically inactivating photosystem II. Scots pine also appears to package its light-harvesting complexes into inactivated quenching aggregates during the autumn and winter months. In contrast, the more temperate herbaceous winter annuals exploit the autumn months to establish site occupancy. This requires a different acclimation strategy for herbaceous plants and involves the enhancement of pathways involved in photosynthetic carbon metabolism and coordinated changes in respiratory metabolism. Even if such acclimation is expensive, in terms of lower nitrogen or phosphate use efficiency or high respiratory costs, establishing site occupancy during the autumn may still provide a competitive advantage for over-wintering versus spring germinating annuals. Possible cryoprotective roles for the accumulated sugars, and the function of the altered respiratory capacity in freezing tolerance, remain to be established unequivocally. Changes in the levels of fatty acid desaturation in both plastid and cytosolic membrane lipids are also thought to be important for maintaining membrane stability at low temperature. However, recent results from mutant and transgenic plants with altered lipid metabolism show that membrane composition may also play a key role in protein import and processing at low temperatures. These findings illustrate the importance of viewing adaptation to low temperature as the integrated acclimation of all metabolic processes in the cell, and not as the adaptive change of isolated processes. Alteration in the sensitivity of photosynthesis to photoinhibition at low temperature may relate to changes in lipid biochemistry and cytosolic and respiratory metabolism as much as it does to alterations in the light-harvesting or energy dissipative capacity of the thylakoid pigment-protein complexes. Thus, a full understanding of how herbaceous winter annuals adapt to low growth temperatures will depend on building a more comprehensive picture of the interactions between the different cellular compartments during acclimation. Identifying the signal transduction pathways that mediate the acclimation response of photosynthetic, respiratory and cytosolic metabolism to low temperature will be another challenge. Cellular redox poise is clearly one element in the puzzle and this may be linked, in the short term at least, to the accumulation of sugars. However, long-term acclimation of herbaceous winter annuals to low temperature depends on *development* at the low temperature. Unravelling the basis of this developmental response, and the signals that direct it, is one of the most exciting challenges for the future.

### ACKNOWLEDGEMENTS

We thank Drs Christina Ottander and Douglas Campbell for access to their original data presented in Figs. 18.1 to 18.3, and for helpful discussion and comments on sections of this manuscript. Research grants from the Swedish Natural Science Research Council, the Swedish Council for Forestry and Agricultural Research, the Swedish Engineering Research Council, the Natural Sciences and Engineering Research Council of Canada, and the OECD project for Food Production and Preservation supported this work.

## REFERENCES

Adams, W. W., Demmig-Adams, B., Verhoeven, A. S. & Barker, D. H. (1995). 'Photoinhibition' during winter stress: involvement of sustained xanthophyll cycle-dependent energy dissipation. *Australian Journal of Plant Physiology*, **22**, 261–76.

Ball, M. C. (1994). The role of photoinhibition during tree seedling establishment at low temperatures. In *Photoinhibition of Photosynthesis: from Molecular Mechanisms to the Field*, ed. N. R. Baker & J. R. Bowyer, pp. 365–76. Oxford: βIOS Scientific Publishers.

Browse, J. & Somerville, C. R. (1994). Glycerolipids. In *Arabidopsis*, ed. E. M. Meyerowitz & C. R. Somerville, pp. 881–912. New York: Cold Spring Harbor Laboratory Press.

Brüggemann, W., Klaucke, S. & Maas-Kantel, K. (1994). Long-term chilling of young tomato plants under low light. V. Kinetic and molecular properties of two key enzymes of the Calvin cycle in *Lycopersicon esculentum* Mill. and *L. peruvianum* Mill. *Planta*, **194**, 160–8.

Campbell, D., Zhou, G., Gustafsson, P., Öquist, G. & Clarke, A. K. (1995). Electron transport regulates exchange of two forms of photosystem II D1 protein in the cyanobacterium *Synechococcus. EMBO Journal*, **14**, 5457–66.

Gilmore, A. M. & Yamamoto, H. Y. (1993). Linear models relating xanthophylls and lumen acidity to non-photochemical fluorescence quenching. Evidence that antheraxanthin explains zeaxanthin-independent quenching. *Photosynthesis Research*, **35**, 67–78.

Grafflage, S. & Krause, G. H. (1993). Alterations of properties of ribulose bisphosphate carboxylase related to cold acclimation. In *Advances in Plant Cold Hardiness*, ed. P. L. Li & L. Christersson, pp. 113–24. Boca Raton: CRC Press.

Gray, G. R., Boese, S. R. & Huner, N. P. A. (1994). A comparison of low temperature growth vs low temperature shifts to induce resistance to photoinhibition in spinach (*Spinacia oleracea*). *Physiologia Plantarum*, **90**, 560–6.

Gray, G. R., Savitch, L. V., Ivanov, A. G. & Huner, N. P. A. (1996). Photosystem II excitation pressure and development of resistance to photoinhibition. II. Adjustment of photosynthetic capacity in winter wheat and winter rye. *Plant Physiology*, **110**, 61–71.

Hayden, D. B., Covello, P. S. & Baker, N. R. (1988). Characterization of a 33 kDa polypeptide that accumulates in the light-harvesting apparatus of maize leaves during chilling. *Photosynthesis Research*, **15**, 257–70.

Hobe, S., Kuttkat, A., Förster, R. & Paulsen, H. (1995). Assembly of trimeric light-harvesting chlorophyll a/b complex *in vitro*. In *Photosynthesis: from Light to Biosphere*, vol. 1, ed. P. Mathis, pp. 47–52. Dordrecht: Kluwer Academic Publishers.

Huner, N. P. A., Öquist, G., Hurry, V. M., Krol, M., Falk, S. & Griffith, M. (1993). Photosynthesis, photoinhibition and low temperature acclimation in cold tolerant plants. *Photosynthesis Research*, **37**, 19–39.

Huner, N. P. A., Maxwell, D. P., Gray, G. R., Savitch, L. V., Krol, M., Ivanov, A. V. & Falk, S. (1996). Sensing environmental temperature change through imbalances between energy supply and energy consumption: redox state of photosystem II. *Physiologia Plantarum*, **98**, 358–64.

Hurry, V. M. & Huner, N. P. A. (1992). Effect of cold-hardening on sensitivity of winter and spring wheat leaves to short-term photoinhibition and recovery of photosynthesis. *Plant Physiology*, **100**, 1283–90.

Hurry, V. M., Krol, M., Öquist, G. & Huner, N. P. A. (1992). Effect of long-term photoinhibition on growth and photosynthesis of cold hardened spring and winter wheat. *Planta*, **188**, 369–75.

Hurry, V. M., Malmberg, G., Gardeström, P. & Öquist, G. (1994). Effects of a short-term shift to low temperature and of long-term cold hardening on photosynthesis and ribulose 1,5-bisphosphate carboxylase/oxygenase and sucrose phosphate synthase activity in leaves of winter rye (*Secale cereale* L.). *Plant Physiology*, **106**, 983–90.

Hurry, V. M., Strand, Å., Tobiæson, M., Gardeström, P. & Öquist, G. (1995*a*). Cold hardening of spring and winter wheat and rape results in differential effects on growth, carbon metabolism, and carbohydrate content. *Plant Physiology*, **109**, 697–706.

Hurry, V., Tobiæson, M., Krömer, S., Gardeström, P. & Öquist, G. (1995*b*). Mitochondria contribute to increased photosynthetic capacity of leaves of winter rye (*Secale cereale* L.) following coldhardening. *Plant Cell and Environment*, **18**, 69–76.

Hurry, V. M., Keerberg, O., Pärnik, T., Öquist, G. & Gardeström, P. (1996). Effect of cold hardening on the components of respiratory decarboxylation in the dark and in the light. *Plant Physiology*, **11**, 713–9.

Koroleva, O. Y., Thiele, A. & Krause, G. H. (1995). Increased xanthophyll cycle activity as an important factor in acclimation of the photosynthetic apparatus to high-light stress at low temperature. In *Photosynthesis: From Light to Biosphere*, vol. 4 ed. P. Mathis, pp. 425–8. Dordrecht: Kluwer Academic Publishers.

Krapp, A. & Stitt, M. (1995). An evaluation of direct and indirect mechanisms for the 'sink-regulation' of photosynthesis in spinach: changes in gas exchange, carbohydrates, metabolites, enzyme activities and steady-state transcript levels after cold-girdling source leaves. *Planta*, **195**, 313–23.

Krause, G. H. (1994). Photoinhibition induced by low temperatures. In *Photoinhibition of Photosynthesis: from Molecular Mechanisms to the Field*, ed. N. R. Baker & J. R. Bowyer, pp. 331–48. Oxford: βIOS Scientific Publishers.

Krivosheeva, A., Da-Li, T., Ottander, C., Wingsle, G., Dubé, S. L. & Öquist, G. (1996). Cold acclimation and photoinhibition of photosynthesis in Scots pine. *Planta*, **200**, 296–305.

Lapointe, L. & Huner, N. P. A. (1993). Photoinhibition of isolated mesophyll cells from cold hardened and nonhardened winter rye. *Plant Cell and Environment*, **16**, 249–58.

Levings, C. S. & Siedow, J. N. (1995). Regulation by redox poise in chloroplasts. *Science*, **268**, 695–6.

Maxwell, D. P., Laudenbach, D. E. & Huner, N. P. A. (1995). Redox regulation of light-harvesting complex II and *cab* mRNA abundance in *Dunaliella salina*. *Plant Physiology*, **109**, 787–95.

Maxwell, D. P., Falk, S., Trick, C. G. & Huner, N. P. A. (1994). Growth at low temperature mimics highlight acclimation in *Chlorella vulgaris*. *Plant Physiology*, **105**, 535–43.

Mohanty, N. & Yamamoto, H. Y. (1995). Mechanism of non-photochemical chlorophyll fluorescence quenching. I. The role of de-epoxidised xanthophylls and sequestered thylakoid membrane photons as probed by dibucaine. *Australian Journal of Plant Physiology*, **22**, 231–8.

Moon, B. Y., Higashi, S. I., Gombos, Z. & Murata, N. (1995). Unsaturation of the membrane lipids of chloroplasts stabilizes the photosynthetic machinery against low-temperature photoinhibition in transgenic tobacco plants. *Proceedings of the National Academy of Sciences USA*, **92**, 6219–23.

Öquist, G. & Martin, B. (1986). Cold climates. In *Photosynthesis in Contrasting Environments*, ed. N. R. Baker & S. P. Long, vol. 7, pp. 237–93. New York: Elsevier.

Öquist, G., Hurry, V. M. & Huner, N. P. A. (1993). The temperature dependence of the redox state of photosystem II and susceptibility of winter rye, and spring and winter wheat, to photoinhibition of photosynthesis. *Plant Physiology and Biochemistry*, **31**, 683–91.

Ottander, C., Campbell, D. & Öquist, G. (1995). Seasonal changes in photosystem II organization and pigment composition in *Pinus sylvestris*. *Planta*, **197**, 176–83.

Price, G. D., Yu, J -W., von Caemmerer, S., Evans, J. R., Chow, W. S., Anderson, J. M., Hurry, V. & Badger, M. R. (1995). Chloroplast cytochrome *b/f* and ATP synthase complexes in transgenic tobacco: transformation with antisense RNA against nuclear encoded transcripts for the Rieske FeS and *ATPδ polypeptides*. *Australian Journal of Plant Physiology*, **22**, 285–97.

Saradedevi, K. & Raghavendra, A. S. (1992). Dark respiration protects photosynthesis against photoinhibition in mesophyll protoplasts of pea (*Pisum sativum*). *Plant Physiology*, **99**, 1232–7.

Savitch, L. V., Maxwell, D. P. & Huner, N. P. A. (1996). Photosystem II excitation pressure and photosynthetic carbon metabolism in *Chlorella vulgaris*. *Plant Physiology*, **111**: 127–36.

Schöner, S. & Krause, G. H. (1990). Protective systems against active oxygen species in spinach: response to cold-acclimation in excess light. *Planta*, **180**, 383–9.

Selstam, E. & Widell Wigge, A. (1993). Chloroplast lipids and the assembly of membranes. In *Pigment Protein Complexes in Plastids: Synthesis and Assembly*, ed. C. Sundqvist & M. Ryberg, pp. 241–77. Academic Press, Inc., San Diego.

Sheen, J. (1994). Feedback control of gene expression. *Photosynthesis Research*, **39**, 427–38.

Somersalo, S. & Krause, G. H. (1989). Photoinhibition at chilling temperature. Fluorescence characteristics of unhardened and cold-acclimated spinach leaves. *Planta*, **177**, 409–16.

Somerville, C. (1995). Direct tests of the role of membrane lipid composition in low-temperature-induced photoinhibition and chilling sensitivity in plants and cyanobacteria. *Proceedings of the National Academy of Sciences USA*, **92**, 6215–18.

Stitt, M. & Schulze, E.-D. (1994). Plant growth, storage, and resource allocation: from flux control in a metabolic chain to the whole-plant level. In *Flux Control in Biological Systems*, ed. E -D. Schulze, pp. 57–118. San Diego: Academic Press.

Strand, M. & Öquist, G. (1988). Effects of frost hardening, dehardening and freezing stress on *in vivo* chlorophyll fluorescence of seedlings of Scots pine (*Pinus sylvestris* L.). *Plant Cell and Environment*, **11**, 231–8.

Thiele, A., Winter, K. & Krause, G. H. (1995). Photoinhibition of photosynthesis related to xanthophyll cycle and D1 protein turnover in higher plants. In *Photosynthesis: from Light to Biosphere*, vol. IV. ed. P. Mathis, pp. 437–40 Dordrecht: Kluwer Academic Publishers.

Wu, J. & Browse, J. (1995). Elevated levels of high-melting-point phosphatidylglycerols do not induce chilling sensitivity in an *Arabidopsis* mutant. *Plant Cell*, **7**, 17–27.

# 19 Acclimation to sun and shade

R. W. PEARCY

## INTRODUCTION

The photosynthetic apparatus is remarkable in its capability for adaptation to a wide range of light inputs. A tree seedling may become established in the deep shade of the forest understorey where it will receive only 1% of the photon flux density (PFD) incident on the leaves of its parent in the canopy. During the life of this seedling as it struggles to reach the canopy, it may be exposed to repeated episodes of higher light as nearby overtopping trees fall, creating canopy gaps. The light will then suddenly increase for a period of months to years but will gradually decrease as the gap fills. Within canopies themselves, the growth of new leaves shades older leaves, greatly reducing the PFD they experience. Leaves at the top of the canopy experience large diurnal changes in PFD, while those in the shade may experience brief sunflecks during which the PFD may increase 50–100-fold in less than a second. Seasonal changes and cloudiness patterns are also important sources of variation in the light environment, requiring adjustments in the photosynthetic response for maintenance of optimal efficiency.

These different temporal scales of light change require a great flexibility of the photosynthetic apparatus in adjusting for efficient use of light and for protection against damage during periods of excess light. Short-term changes on the time-scale of a second to a day or so involve regulatory responses of the photosynthetic apparatus such as light activation of photosynthetic enzymes, stomatal regulation and the mechanisms that dissipate excess light energy. These regulatory mechanisms operating on short time-scales of seconds to hours that appear to function to maintain a balance between the supply of light energy and the capacity of the component steps to process this energy. Ultimately, they maintain a high efficiency of light use while minimizing the potential for photoinhibitory damage in the face of large changes of light input. Leaf movements in some species that regulate light capture at the leaf surface also properly fall into this category of regulatory responses to short-term changes in the light environment.

Adjustments to changes in the light environment occurring on longer time-scales involve acclimation of the photosynthetic apparatus. The results of numerous studies indicate that essentially all species have considerable potential for acclimation but the range of light environments over which it is expressed is species specific. Growth in light environments outside of this genetically determined range for a given species may result in a failure to acclimate. Acclimation, as defined here, involves both the dynamic adjustments occurring at the chloroplast and cellular level in extant leaves that are the first line of response to a sustained light changes as well as the developmental changes at the leaf and canopy level that occur over longer time-scales.

Acclimation of leaf photosynthesis to the light climate has been shown by many researchers to involve profound changes in the organization and investment in different components of the photosynthetic apparatus that improve photosynthetic performance in the prevailing light regime. Photosynthetic performance can be measured either as the absolute photosynthetic rate achieved in a given environment or as the return on investment of a resource such as carbon and nitrogen. Since these resources are limited, an investment in one component must result necessarily in reduced investment in another. Therefore, acclimation can be viewed as a series of trade-offs that change dynamically as the light and other conditions change in the environment. These trade-offs operate at several levels, ranging from the partitioning of nitrogen between chloroplast proteins involved in light harvesting, electron transport and carbon metabolism to the partitioning of nitrogen and carbon between shaded and sunlit leaves within a canopy, resulting from the leaf-level acclimation.

In this chapter, the basic state of knowledge will be reviewed regarding the mechanisms by which plants acclimate to changes in their light environment, focusing first on those factors that determine the photosynthetic rate in limiting light and then on those that determine photosynthetic rate at light saturation. The acclimatory mechanisms for avoiding photoinhibition upon an increase in PFD will also be briefly covered. Finally, the scaling of leaf photosynthetic responses to whole plant and canopy carbon gains will be discussed.

## FACTORS DETERMINING LIGHT HARVESTING BY SUN AND SHADE LEAVES

### Pigment composition and concentrations

Leaves are very effective absorbers of photosynthetically active radiation (PAR), in part, because of the high concentrations of chlorophyll (Chl) in them and, in part, because of the organization of this chlorophyll within the profile of the leaf. While other pigments such as anthocyanins and carotenoids make a contribution to the absorption spectrum, and surface properties such as pubescence and waxes increase reflectance, it is clear that Chl is principally responsible for the observed PAR absorption properties. PAR absorption increases with increases in Chl concentration but only at a strongly diminishing return. For example, Evans (1989) reported that a doubling of the Chl content of leaves from 0.3 to 0.6 mmol Chl $m^{-2}$ increased leaf absorbance from about 0.825 to 0.885, an increase of only 6%. A further doubling could bring only a smaller response. Thus, while increases in the chlorophyll content per unit area of shade leaves increases PAR absorption and therefore also increases the assimilation rate at low PFD, this comes at a strongly diminishing return because of the logarithmic relationship between Chl content and light absorption. Shade-acclimated leaves do contain much more Chl per unit mass than sun leaves, which is consistent in an increased allocation of resources to light-harvesting functions as compared to those involved in electron transport and $CO_2$ fixation capacity (Björkman, 1981). However, on a per unit area basis, the results of numerous studies highlight the fact that there is no consistent difference in the Chl contents between sun and shade leaves. In a few cases, Chl contents of obligate sun species have been found to be much lower in shade than sun leaves, probably because of a failure to acclimate successfully to extremely low light conditions of the shaded environment. The more usual similarity of Chl contents of sun- and shade-acclimated leaves is a consequence of the fact that, while shade leaves contain more Chl per unit mass, they have less mass per unit area as compared to sun leaves. In either environment, the diminishing returns probably set a similar upper limit on chlorophyll content per unit area that is realized over a fairly narrow range of Chl contents and is relatively insensitive to the light environment. Increasing Chl is especially costly in terms of the nitrogen required since, while Chl itself contains only 4 mmol N $mol^{-1}$ Chl, the associated Chl-proteins contain 21 to 79 mmol N $mol^{-1}$ Chl (Evans & Seemann, 1989).

The similarity of the Chl contents per unit leaf area in sun- and shade-acclimated leaves is in stark contrast to the substantial differences in the composition and organization of this chlorophyll at the membrane, chloroplast and anatomical level. Acclimation to shade conditions has been shown to result in decreased Chl *a*/*b* ratios that reflect changes in concentrations of light-harvesting Chl complexes relative to reaction centres (Anderson, 1986). The Chl *b* in leaves is located in the light-harvesting Chl *a*/*b* binding protein complexes (LHCII) that constitute the primary antenna complexes for PSII. In contrast, the PSII and PSI reaction centre complexes contain only Chl *a*. With few exceptions, acclimation to shade has been shown to cause an increased investment in LHCII at the expense of a decreased investment in PSII core complexes. In those few shade species that do not show a shift, the ratios remain similar to those of shade species even when these plants are grown in the sun. In all plants studied so far, the concentrations per unit Chl of PSI core complexes by contrast remain much more constant (Anderson, 1986). The shift in LHCPII and PSII has been shown to have several ramifications beyond changes in Chl *a/b* ratios. First, with this model of acclimation, the antenna size for PSII would increase resulting in more efficient utilization of available quanta by PSII. Secondly, LHCII may provide the adhesion between grana necessary for the formation of extensive grana stacks characteristic of chloroplasts of shade leaves (Chow *et al.*, 1988). Individual grana stacks in these shade chloroplasts may contain as many as 100 thylakoids per granum, which may extend completely across the chloroplast. Grana stacks also appear to be randomly orientated within shade chloroplasts. By contrast, chloroplasts of sun leaves typically have no more than ten thylakoids per granum and the grana are all orientated similarly. The function of the increased stacking and random orientation is not fully understood but current evidence suggests that it does not increase either light harvesting or energy transfer efficiency directly (Evans, 1986). Indeed, increasing the local pigment density may actually slightly decrease light absorption because of the sieve effect. The extensive grana stacking may be necessary for proper membrane organization since PSI and the coupling factor $CF_1$ reside primarily in the unappressed regions whereas PSII and LHCII are in the appressed regions.

In an elegant study, Evans (1986) concluded that the actual shifts in Chl *a* and *b* concentrations with sun and shade acclimation have little direct effect on the light absorption *per se* but do have important consequences for the nitrogen economics of acclimation. Increased Chl *b* had been previously proposed to be advantageous in shade light due to an increase in absorbance in the 450–500 mn wavelengths plus smaller differences elsewhere in the spectra. However, through careful comparisons of absorption spectra and the spectrum of sun and shade light, Evans concludes that increased Chl *b* due to the increased concentrations of LHCII does not increase light absorbance in the shade significantly. Comparison of the nitrogen cost per Chl shows that LHCII contains only 25 mmol N $mol^{-1}$ Chl, whereas the PSII core complex contains 83 mmol N $mol^{-1}$ Chl, so that, for a given Chl concentration per unit area, a substantial savings in N cost is accrued by the shift to increased proportion of the chlorophyll being in LHCII. This savings is possible in the shade, but not in the sun, because high capacities of PSII electron transport (ET) are not required in the former conditions.

## Leaf structure

Leaves that have developed under high light typically have a well-developed palisade mesophyll of one to two tiers of columnar cells on the adaxial side and a more loosely organized spongy mesophyll on the abaxial side of the leaf. By contrast, shade leaves generally lack as well organized a palisade layer, and the mesophyll consists primarily of spongy cells with much more air space than found in sun-developed leaves. As a consequence, leaves developing in the shade are generally much thinner with a lower leaf mass per unit area (LMA) than leaves developing in the sun. These anatomical characteristics are, by and large, fixed early in leaf development and well before full expansion of the lamina so that later transfers between light environments have limited effect on leaf structure. The greater number of cells in a cross-section of sun leaves results in more chloroplasts per unit area in sun as compared to shade-developed leaves, although the chloroplasts in shade leaves are typically larger and contain more chlorophyll (Terashima & Hikosaka, 1995).

The differences in leaf structure of sun- and shade-developed leaves have important consequences for the light gradients and light absorption within leaves. As light enters a leaf, it is either absorbed when it is intercepted by a pigment or scattered, especially by air–water interfaces in the cell walls (Vogelmann, 1993). Scattering increases the effective path length and thus the probability of interception by chlorophyll. Green wavelengths therefore, show much less attenuance with depth than red or blue wavelengths that are strongly absorbed by Chl. However, the greater scattering of green wavelengths and therefore, the greater optical path length results in, for the whole leaf, only a small decrease in absorbance of these wavelengths as compared with red or blue wavelengths. Consequently, the absorption spectra of leaves with high Chl contents are relatively flat as compared with an equivalent Chl concentration in solution.

Palisade cells in sun-developed leaves are closely associated with the adaxial epidermis so that much of the incident light can enter the leaf without interruption by intercellular air spaces. Moreover, the chloroplasts are typically appressed along the anticlinal cell walls in the palisade layer in high irradiances so that the palisade layer acts partially as a light pipe, in addition to acting as an absorber of PAR (Terashima & Hikosaka, 1995). Spongy mesophyll is an effective light scatterer, increasing the path length and hence the probability of absorption. The abaxial epidermis also scatters light back into the mesophyll. In sun leaves, the combination of the light piping in the palisade layers and the scattering in the mesophyll mitigates the strong attenuance in PAR across the leaf lamina. Thus, it may be important in providing sufficient light to the lower cell layers. Nevertheless, strong gradients are still present and Terashima and Inoue (1985) have demonstrated, in thin paradermal sections, decreases in Chl *a/b* from the adaxial to abaxial surface strikingly similar to the changes associated with sun versus shade leaves in canopies. Thus, at least some aspects of sun–shade acclimation are evident at the chloroplast level in response to light gradients across leaves. The light piping properties of the palisade function for direct-beam sunlight but not in the diffuse light typical of shaded environments (Vogelmann, 1993). Thus the preponderance of spongy mesophyll in shade plants may favour light absorption in this environment.

## Quantum yields

The quantum yield of photosynthesis is determined by the efficiency with which absorbed photons are used in ATP production and NADP reduction, and the requirements of the metabolic pathways utilizing this ATP and NADPH for $O_2$ evolution or $CO_2$ assimilation. In the absence of the operation of a Q cycle, and under conditions where light is limiting and no photorespiration is occurring (i.e. a 5% $CO_2$ atmosphere), at least nine photons should be required for reduction of one $CO_2$, giving a maximum quantum yield of 0.111 mol $CO_2$ (or mol $O_2$) mol$^{-1}$ photons (Long, Postl & Bolhar-Nordenkampf, 1993). If a Q cycle operates, making ATP synthesis more efficient, then the quantum requirement decreases to 8, giving a quantum yield of 0.125 mol mol$^{-1}$ photons. These values assume that all absorbed photons are utilized photochemically in ET whereas, in reality, losses due to non-specific photon absorption and losses due to double hits or radiationless decay would reduce these quantum yields somewhat. Extensive and careful measurements of the quantum yield of $O_2$ evolution for a variety of species found a mean value for the quantum yield of $O_2$ evolution of $0.106 \pm 0.001$ mol $O_2$ mol$^{-1}$ photons (Björkman & Demmig, 1987). Thus, measured values approach the theoretical maximum expected and show little variation between species adapted to either sun or shade environments, provided that the leaves were not under stressful conditions. Subsequent measurements of the quantum yield of $CO_2$ assimilation under normal $CO_2$ pressures (330 μbar) but also under low $O_2$ pressures (10 mbar) in order to suppress photorespiration found a mean value of $0.093 \pm 0.003$ mol $CO_2$ mol$^{-1}$ photons for a wide variety of species (Long *et al.*, 1993). Again, no difference was found between species from sun or shade environments. Comparisons of the tropical forest understorey species, *Alocasia macrorrhiza*, grown in light environments ranging from 1.7 (deep shade) to 24 (55% of full sun) mol photons m$^{-2}$ day$^{-1}$ found quantum yields of 0.102 mol $O_2$ mol$^{-1}$ photons with no significant decline from the most shaded to the most sunlit environment (Sims & Pearcy, 1989). Thus, at low PFD, the photosynthetic apparatus appears remarkably capable of using the vast majority of absorbed photons for photochemistry, independent of the light environment in which the plants were grown or any genetic adaptation to sun and shade environments. It can be concluded that, in the absence of stress, the maximum quantum yields of sun- and shade-adapted species, or of plants of a species acclimated to different light environments, do not differ significantly. The differences in the quantum yields $CO_2$ and $O_2$ assimilation may be accounted for by utilization of electrons and ATP for other processes such as nitrate reduction which would compete with $CO_2$ reduction. This conclusion is supported by observation of similar values of the fluorescence ratio, $F_v/F_m$, indicating similar maximum quantum yields of PSII photochemistry, in both the studies by Long *et al.* and by Björkman and Demmig.

The nearly maximum possible quantum yield indicates that several potential constraints on the efficiency of light use are overcome under the differing light qualities of sun and shade environments and compositions of the photosynthetic apparatus. First, these high quantum yields indicate that photons are partitioned equally between PSII and PSI, despite the apparent excess of PSII in shade plants and despite differences in light quality. Any other partitioning would reduce the quantum yield. Both light quality and quantity are known to regulate photosystem stoichiometry. Plants grown under one light quality known to preferentially excite either PSII (yellow light) or PSI (red and far red light) and measured in the opposite light quality have reduced quantum yields (Chow, Melis & Anderson, 1990*b*). In sun and shade environments, the changes in excitation of PSII and PSI are not so extreme and calculations of the ratios of PFD absorbed by PSI versus PSII are near unity (Evans, 1987*a*). Comparisons of photosystem stoichiometries in sun and shade plants from natural environments reveal generally lower PSII/PSI reaction centre ratios but also considerable overlap in values (Chow, Anderson & Melis, 1990*a*). Thus, potential differences in excitation must be balanced by adjustments in antenna sizes associated with each photosystem. The interaction between LHCII and PSII cores is known to be regulated, in the short term, by phosphorylation of LHCII, reducing the antenna size (Anderson, 1986), but it is unclear if the result is a greater association of LHCII with PSI, increasing its antenna size. Simply decoupling LHCII from PSII would only reduce the quantum yield.

The high quantum yields of $C_3$ photosynthesis discussed above are, of course, not realized in nature where $CO_2$ lower pressures and higher $O_2$ pressures lead to significant oxygenation of RuBP and photorespiration. This energetic loss reduces the quantum yield at 30 °C by some 40% from the maximum values reported above. In addition, low stomatal conductance can further reduce the quantum yield for much the same reason because of the lower $C_i$. Moreover, because the oxygenation of RuBP increases strongly with temperature, quantum yields for $CO_2$ uptake is strongly temperature dependent (Björkman, 1981), decreasing with increasing temperature. Stress effects such as low nitrogen supply, drought, either too high or too low temperatures may also reduce quantum yields in high light because of photoinhibition. These factors may account for much of the differences in quantum yields reported for sun- and shade-acclimated plants.

## Respiration rates

In order to maintain a positive net carbon balance in highly shaded environments, it is, of course, of utmost importance to minimize carbon losses via respiration. The respiration rate is a primary determinant of the leaf light compensation point, the PFD at which the net photosynthetic rate is zero. For whole plants, the additional respiratory costs of non-photosynthetic tissues and the construction costs associated with growth result in a higher light compensation point than observed for leaves alone (Givnish, 1988). Thus, it comes as no surprise to observe that shade acclimated leaves have much lower respiration rates than sun-acclimated leaves.

Respiration occurs predominantly in response to the growth and maintenance processes necessary for survival. Since light severely limits carbon gain and growth in a shaded environment, the growth costs per unit time are in general low in this environment. Exceptions are obligate high light species where etiolation may cause significant stem elongation and hence extra respiratory costs. These costs will ultimately be fatal. There is little concrete evidence that the construction costs per unit biomass differ between sun- and shade- acclimated plants, so the primary differences in construction costs per unit time will depend on growth rates and allocation. In high light, a large fraction of the leaf respiration observed may be related to the carbohydrate concentration resulting from the high daily photosynthesis and is presumably related to the processing and export of carbohydrates (Pearcy & Sims, 1994). Part may also be related to the operation of an alternative pathway of respiration that functions to remove excess carbohydrates.

Maintenance costs result from the requirement for repair, maintenance of protein levels for those proteins that exhibit turnover, maintenance of ion gradients and other necessary cellular processes. Shade leaves may be characterized by lower maintenance respiration per unit area as compared to sun leaves because of their lower protein concentrations per unit area. However, on a mass basis there is evidence the estimated maintenance respiratory costs may be quite similar (Pearcy & Sims, 1994). There is considerable uncertainty about the true maintenance costs because of the uncertainty of separating maintenance respiration from other respiratory processes. Also, more damage may be occurring in high light, leading to a greater maintenance requirement per unit of protein. Unfortunately, it is difficult to separate this potential component of respiration from the component related to the higher carbohydrate supply.

Another potential component of respiration in leaves is the operation of alternative pathway respiration which utilizes an alternate oxidase and which is energetically much less efficient than respiration that uses the full mitochodrial electron transport chain (Lambers, 1985). This pathway has been suggested to serve as a mechanism for rapid utilization of carbohydrate when it is present in excess. If so, it could be expected to be more prominent in sun than shade leaves. That is the observed pattern in *Piper* species grown in high and low light (Fredeen & Field, 1996). However, as compared to total respiration, the increase in the alternative pathway capacity or the apparent engagement of this pathway was the same in both high light specialist species and shade species.

## FACTORS DETERMINING PHOTOSYNTHETIC CAPACITY IN SUN- AND SHADE-ACCLIMATED LEAVES

In agreement with the initial suggestions of Blackman made in 1905, the photosynthetic capacity ($A_{max}$) of leaves is determined by the capacity of component steps in the photosynthetic apparatus that are not directly dependent on light. Starting with the pioneering work of Björkman (1968), research over the past 25 years has greatly advanced the understanding of the nature of these component steps and how they vary between sun- and shade-acclimated leaves. It is now well established that growth in high versus low light results in increased rubisco activity, PSII ET capacity, capacity of cytochrome F (Cyt F) and chloroplast coupling factor ($CF_1$) per unit leaf area (Anderson & Osmond, 1987). Strong linear relationships between $A_{max}$ and either Cyt F or PSII content have been found for plants grown at different nitrogen or light levels while the relationship between $A_{max}$ and rubisco often shows a slight curve linearity at high rubisco concentrations (Evans, 1987*b*). It is likely that most, if not all, of the enzymes of the Calvin cycle and ET carriers will show a linear or nearly linear increase, suggesting that there is a tight co-ordination between the investments in these enzymes so that the capacity of each component step is in balance. The exception to this rule is the PSI core complex content, which is relatively insensitive to growth at different irradiances. The capacity of PSI is known to exceed that of PSII by a substantial margin so that it is unlikely that it ever becomes limiting. The reasons for the apparent over-capacity of PSI relative to PSII and its insensitivity to growing conditions are unclear.

In addition to the changes in the contents of ET carriers and Calvin cycle enzymes, high light-grown plants are characterized by increased stomatal conductance ($g_s$) as compared to low light-grown plants. When light is not limiting, the capacity of the biochemical steps sets the demand for $CO_2$ whereas $g_s$ sets the capacity for supply of $CO_2$. A maximum efficiency of resource use, such as in this case, nitrogen use, occurs when there is a balanced co-limitation by all components in the system and not just a single rate-limiting step. As shown by the elegant studies of von Caemmerer and Farquhar (1981), carboxylation capacity and the capacity for RuBP regeneration via electron transport and the Calvin cycle often appear to co-limit $CO_2$ assimilation at the intercellular $CO_2$ levels established at normal ambient $CO_2$ pressures and values of $g_s$ of a leaf in saturating PFDs.

Changes in photosynthetic capacity in sun versus shade acclimation can therefore be unequivocally ascribed to balanced and co-ordinated changes in the contents per unit area of the enzymes and carriers responsible for the carboxylation and ET capacities of the leaf, coupled with a concomitant change in $g_s$. It is worth asking how these changes are brought about. With respect to $g_s$, a higher stomatal density in sun versus shade leaves may account for the difference. Additionally, there is usually considerable scope for increasing $g_s$ above those values normally occurring in a light-saturated leaf, as evidenced by increased conductance when $CO_2$ pressures are lowered or humidities are increased. Thus, for leaves that have matured in a low light environment and have low stomatal densities, there may be no great constraint on increasing $g_s$ at least moderately. Increases in the concentrations of enzymes or carriers per unit leaf area could result from one or all of three different factors. First, the chloroplast concentrations of these components could increase with the numbers of chloroplasts per unit area remaining relatively constant. Secondly, the concentration of chloroplasts per cell could increase. And thirdly, the number (or volume) of cells per unit leaf area could increase. Although there are few studies that have explored these factors, some indication can be gained by comparing the photosynthetic capacities of sun- and shade-developed leaves on a per unit leaf area basis and a per unit mass basis. For the species that have been examined, large changes in $A_{max}$ per unit area evident with acclimation become much smaller or even disappear when comparisons are made on the basis of leaf mass. For some species from shaded habitats, photosynthetic rates per unit mass may decline when plants are grown in high light because of photoinhibitory damage. It would appear that, in plants where photoinhibitory damage is not occurring, owing either to the genetic potentials or to the range of growth light environments, photosynthetic capacity per unit cell volume is relatively constant and increased photosynthetic capacity per unit leaf area results from the increased cell volume per unit area. This type of pattern underscores the importance of the anatomical differences occurring during development in sun and shade. So far, there is no quantitative understanding of the roles of leaf structure versus cellular or chloroplast level acclimation. Since Chl *a*/*b* and the concentrations per unit Chl of rubisco, PSII core complexes, etc. all change, then clearly acclimation at these levels, reflecting shifts in resource allocation between light harvesting and carboxylation and ET capacity, is also occurring. These shifts should operate in the same direction as the effects of leaf thickness with respect to photosynthetic capacity resulting in a higher photosynthetic capacity per unit cell volume. That they do not bring about this result may indicate that there are other limitations operating. These may relate to changes in the gradients of cellular and chloroplast properties or increased intercellular air space diffusion resistance in the thicker sun leaves. Further research will be required to elucidate these important relationships.

Dynamic measurements of acclimation following transfer of plants between light environments have revealed a range of responses of light-saturated photosynthetic capacity depending on the species. For some species, mature leaves exhibit no, or at best only a modest, acclimation response (Pearcy & Sims, 1994). Thus, all of the potential for acclimation from low to high light depended on development, and specifically on the elaboration of cell layers that increase leaf thickness. In other species, such as the tropical forest canopy tree *Bischofia javanica*, whose seedlings initially establish in the deep shade of the understorey, mature leaves exhibit some capacity for acclimation upon transfer from low to high light, but this capacity is considerably less than that

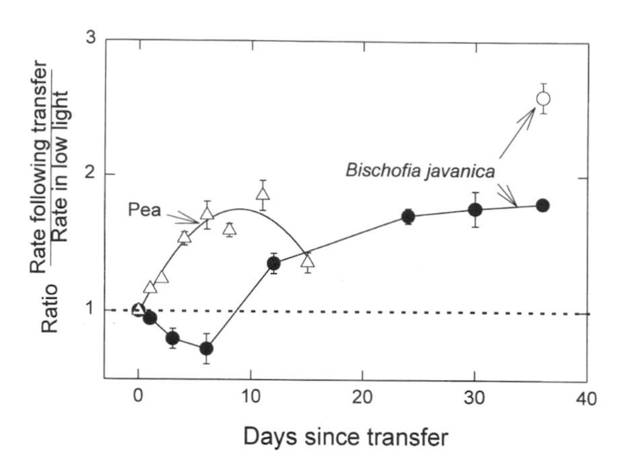


**Figure 19.1.** The dynamics of acclimation in a herbaceous annual, Pea (Δ) and a tropical forest tree, *Bischofia javanica* (●), illustrating the difference in the time scales of the response of leaf photosynthetic capacity to an increase in PFD. In each case the plants were transferred to high PFD at day 0. Peas responded with an acclimation response with no lag and a $\frac{1}{2}$ time of about 4 days whereas *B. javanica* exhibited an initial photoinhibition followed by a slow acclimation giving an overall half-time for acclimation of approximately 15 days. In contrast to pea leaves, which were able to reach the photosynthetic capacity of leaves that developed in high PFD the Amax achieved during acclimation of *B. javanica* leaves was still considerably less than that of leaves that had developed continuously in high PFD. (Redrawn from data in Chow & Anderson, 1987 and Kamaluddin & Grace, 1992.)

expressed by leaves developing in different light environments (Kamaluddin & Grace, 1992). For *B. javanica*, leaves that had developed in the shade but were then transferred to high light exhibited a 1.8-fold increase in $A_{max}$ per unit leaf area after 30 days (Fig. 19.1). By contrast, leaves that developed in high light exhibited a photosynthetic capacity per unit leaf area higher by a factor of 2.5 than those of shade developed leaves. Leaf mass per unit area also differed by a factor of 3.2 and 3.9 for the high light developed and transferred leaves, respectively, as compared to the shade leaves. Thus, light-saturated photosynthetic capacity per unit leaf mass actually declined with either acclimation or development in high light. Since considerable starch may have accumulated in high light, adding to the leaf mass per unit area, this may, at least in part, account for the decline in photosynthetic rate per unit mass. In addition, the possibility that there was some residual photoinhibition, which was quite prominent immediately following the transfer, cannot be ruled out. The transferred leaves, however, also exhibited an increase in thickness that was due to an elongation of the palisade cells. Thus, much of the acclimation response observed for photosynthetic capacity per unit area in this case may be related as well to increased cell volume.

Studies with peas in which the dynamics of acclimation were followed in a cohort of leaves show a small transient decrease immediately after transfer followed by a substantial and co-ordinated increase in rubisco, Cyt f and light saturated $O_2$ evolution rates and decreases in PSII core complexes and Chl *a*/*b* ratios (Anderson & Osmond, 1987). These adjustments are rapid with a half-time of about 4 days (see Fig. 19.1), illustrating the capacity of these leaves to respond in a rapid and co-ordinated way to changes in the light environment. Thus, when measured in terms of chlorophyll or leaf area, a substantial capacity for acclimation in apparently fully developed leaves is present in peas and other herbaceous species that contrasts with the limited capacity exhibited by already developed leaves of the herbaceous shade plant *Alocasia macrorrhiza* and of some woody species. There is, unfortunately, no information on which to judge whether these changes may also be related to changes in leaf mass per unit area due to cellular expansion. Chloroplast biologists typically report concentrations per unit chlorophyll, whereas gas exchange physiologists typically report photosynthetic rates per unit leaf area. Reporting of simple measurements of Chl per unit leaf area and leaf mass per unit area in each study would allow ready conversion of rates and concentrations to common units for comparison. Also, morphometric studies are needed to understand how cell volumes and chloroplast numbers are changing so that a better understanding of the relationships between chloroplast-based acclimation and whole-leaf acclimation can be gained.

## PHOTOINHIBITION IN SUN- AND SHADE-ACCLIMATED LEAVES

Unless a leaf has an unusually high photosynthetic capacity in high light, there will always be photons in excess of those that can be utilized, and therefore also the potential for photoinhibitory damage. The increase in photosynthetic capacity with acclimation to high light may partially alleviate some of the excess but, even after acclimation, much of it remains. Thus, mechanisms that dissipate this excess energy or alternatively repair the damage must be present.

It is now known that, in high light, PSII can be down-regulated so that excess energy is dissipated as heat, avoiding excess excitation of the reaction centre (Demmig-Adams & Adams, 1992). The xanthophyll cycle located in the thylakoid membranes is intimately involved in this dissipation. In high light, violaxanthin and antheraxanthin are converted to zeaxanthin via successive de-epoxidation steps. In the presence of a $\Delta pH$ across the thylakoid, as occurs in high light, potentates zeaxanthin and, to a lesser extent, antheraxanthin in some way facilitates conversion of the excited state of Chl to heat which can be harmlessly dissipated. Sun-acclimated leaves possess a greater total pool size of xanthophyll pigments as well as a greater capacity for rapid conversion of these pools to antheraxanthin and zeaxanthin than shade leaves when the light is suddenly increased. Sun-acclimated leaves are therefore much more capable of thermal dissipation of excess radiation than shade leaves. Shade plants are thought to be more vulnerable to photoinhibition in high light because of the larger PSII antenna associated with the higher LHCII concentrations. Indeed, shade acclimated leaves exposed to full sun show inactivation of PSII specifically associated with the D1 protein of the reaction centre, indicating that the capacity for photoprotection is exceeded (Öquist, Chow & Anderson, 1992). Recovery from this photoinhibitory damage requires repair of the reaction centre. Recovery from photoinhibitory damage is slower in shade than sun plants because repair occurs in the stroma lamellae, which, due to the extensive grana stacking, are of limited extent in shade chloroplasts (Tyystjarvi *et al.*, 1992). Although damaged D1 centres accumulate in shade-acclimated leaves exposed to high light because of this limitation, there is evidence that the accumulation of damaged centres themselves may serve an important photoprotective role. The damaged centres continue to dissipate energy and may therefore serve to protect the remaining undamaged PSII centres, presumably through some remaining interconnectivity of damaged and undamaged PSII centres.

The interactions between photoinhibition and acclimation are likely to be of great importance as plants respond to changes in light brought about by the formation of canopy gaps or by alteration of canopies due to branch falls. Transfer of a shade leaf to high light conditions, for example 2 hours of high PFD in the middle of a day (similar to a canopy gap), results in an initial photoinhibition evident for the first few days followed by a recovery and acclimation to the new conditions. Environmental stresses such as low N availability or high temperatures can exacerbate photoinhibition and delay the recovery, provided that conditions are not so extreme that the leaf dies. The recovery of the leaves is important in maintaining leaf area which then allows for photosynthate production and development of new, fully acclimated leaves.

## ACCLIMATION AT THE WHOLE-PLANT/ CANOPY LEVEL

Ultimately, to understand the role of acclimation, insight into how the processes occurring at the leaf level integrate to whole plant photosynthesis and growth must be developed. The outlines for this problem have been visible for some time but, surprisingly, little attention has been given to the integrative processes linking whole plant behaviour with leaf behaviour, and especially how these processes differ in response to sun and shade conditions or other environmental factors. If the carbon concentration of biomass is constant, then the growth rate must be proportional to the whole-plant net photosynthetic rate. The whole-plant photosynthetic rate is equal to the net photosynthetic rate of the leaves as determined by their leaf-specific environment and physiological properties, multiplied by the leaf area of the plant. The plant leaf area is determined by the growth of leaves as influenced by the resource allocation patterns, which are also set by the environment and the genetic make-up of the plant, as well as the leaf longevity. Finally, the leaf area may be displayed in different ways depending on the architecture of the plant and how it responds to the growth environment and ontogenetic factors.

An important linkage in this process is clearly the LMA. This single factor is highly correlated with changes in photosynthetic capacity but is also critically important in the resource allocation patterns of the plant. The leaf area ratio (LAR) of a plant is the ratio of total leaf area to total plant biomass including roots. The leaf weight ratio (LWR) is the ratio of the mass of leaves to the total plant mass. Relative growth rates of plants have been shown to be much more highly correlated with LAR of plants than with the $A_{max}$ of this leaf area. Most studies (see Björkman, 1981) have found that growth in low light tends to increase LWR but this effect, with a few exceptions, is generally rather small. However, the large changes in LMA accompanying leaf development in high and low light cause large changes in LAR between sun- and shade-acclimated plants. In low light, the low LMA of shade-acclimated leaves causes the leaf area ratio to increase substantially, effectively compensating for the lower photosynthetic rates per unit area due to the low PFDs. In the sun, however, the higher $A_{max}$ associated with the acclimation to the high light is offset by the lower LAR due to the greater LMA of these leaves. To the extent that changes in $A_{max}$ are linked either directly or indirectly to LMA, there clearly exists a trade-off for a plant of a given size between the $A_{max}$ per unit leaf area that can be realized and the leaf area that can be produced. These arguments apply to acclimation of a given species, since, in the absence of stress effects, there appears to be in many cases relatively little capacity to alter the $A_{max}$ per unit leaf biomass or perhaps more appropriately per unit mesophyll cell volume. They do not apply in cross-species comparisons where there is essentially no correlation, or even a negative correlation, between $A_{max}$ and LMA.

Sims, Gebauer and Pearcy (1994) investigated the trade-offs involved in whole plant acclimation of *Alocasia macrorrhiza* using both a measurement and a modelling approach. This species, which occurs predominately in tropical forest understories but can survive in high light, is a rhizomatous rosette-former with five to six leaves borne on the end of long petioles. Sun plants had a 2.5-fold higher $A_{max}$ per unit area (Fig. 19.2) but also had a 60% smaller LAR than shade plants of the identical size. Sun and shade plants had about equal allocation to stems and petioles but sun plants allocated more to roots (20% versus 10% of total biomass) and slightly less to leaves (38% versus 42%) as compared to shade plants. In terms of the costs required to construct the plant (grams of photosynthate as glucose equivalents per gram

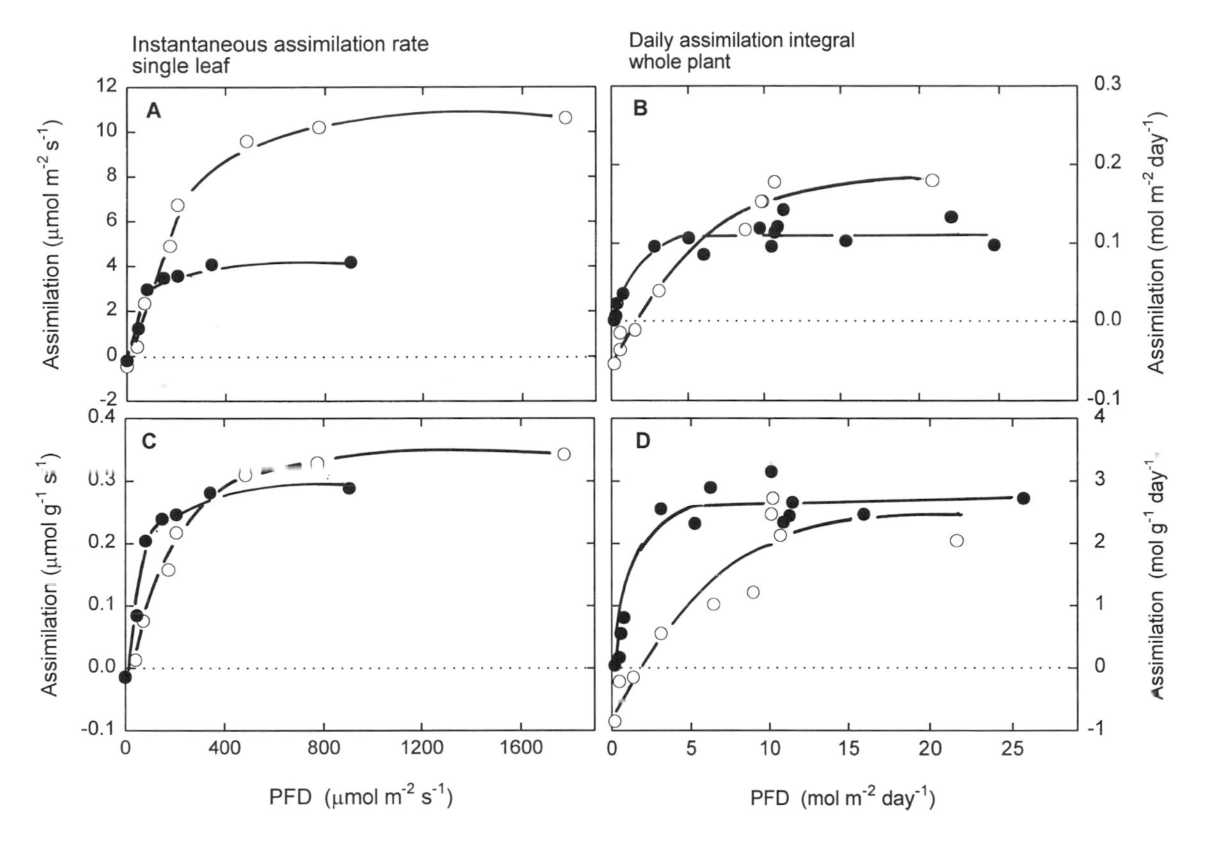


**Figure 19.2.** Comparison of acclimation of *Alocasia macrorrhiza* at the leaf and whole plant level. A and C show the instantaneous response to PFD on a leaf area and leaf mass basis, respectively for plants grown at 1.7 (●) and 25 (○) mol $m^{-2}$ $day^{-1}$ (peak PFD at midday of 45 and 1100 µmol photons $m^{-2}$ $s^{-1}$). B and D show the whole plant gas photosynthetic response to variation in the daily PFD for plants acclimated to these regimes on both a leaf area and leaf mass basis, respectively. Symbols are the same as for A and C. Plants were grown in their respective light regimes and then transferred to a whole plant assimilation chamber for a 24-hour measurement period under a light regime created with differing layers of shade cloth in a glasshouse during a period of cloudless skies. Each point represents the 24-hour assimilation measured for a single plant, with different plants being used for each. The large difference in instantaneous light saturated assimilation rate per unit leaf area (A) essentially disappears when rates are expressed per unit leaf mass but differences in assimilation at low PFD are amplified. At the whole plant level, the higher leaf area ratio of the shade plants effectively compensates for their lower photosynthetic rate per unit leaf area. Thus, on a per unit mass basis shade plants had a large advantage in shade but even in the sun had no disadvantage relative to sun plants. (Redrawn from data in Sims & Pearcy, 1989, 1994.)

of biomass produced), leaves accounted for about 50% of the cost of a shade plant and 44% of the cost of a sun plant. However, of these costs, those attributable to synthesis of Chl and the ET Calvin cycle components represented less than 20% of the cost of the leaf. The remainder, consisting of other enzymes and especially the cell wall were clearly a much larger cost than the photosynthetic apparatus itself. Nitrogen allocation to leaves represented 68% and 60% to the total plant N in sun and shade leaves, respectively. About 35% of the leaf N in shade leaves was in photosynthetic components whereas the remainder was in 'other' components. In sun leaves, 30% was in photosynthetic components. Thus, both in terms of the costs of construction and N, the photosynthetic apparatus is but a small part of the overall costs. Those items needed to support the photosynthetic apparatus, both structurally and metabolically, are a much bigger component of the costs.

The consequences of these patterns of allocation can be observed either via whole plant gas exchange experiments or in simulations to test their effects on growth. When the daily gas exchange of whole plants acclimated to either sun or shade but measured under different daily light doses were compared, the shade plants exhibited a large advantage at low daily doses (Fig. 19.2). This advantage was apparent when either 'gross' (daily 'day' respiration added back) or net photosynthesis was considered and can be attributed to the greater LAR of the shade plants due to their lower LMA (Pearcy & Sims, 1994). At high daily light doses, however, sun-acclimated plants exhibited no advantage and possibly even a slight disadvantage as compared to shade plants. Thus the higher photosynthetic rate per unit leaf area of sun plants in the high light environment was offset, and perhaps even more than offset by their low LAR. Simulation modelling of growth in sun- and shade-type plants in different environments also revealed that in a high light environment shade-type plants out-performed sun-type plants (Sims *et al.*, 1994). This is in agreement with an observation made more than 50 years ago that, immediately following a transfer to a high light environment, the relative growth rate of shade-grown sunflower plants was greater that that of sunflower plants grown continuously in the high light environment (see Björkman, 1981). It can be concluded that, for *Alocasia* and possibly also for sunflower, acclimation to high light does not serve to directly increase performance in high light. It may, in concert with other changes in leaf angles, $g_s$ and photoprotection act to increase resistances to stresses associated with high light (photoinhibition, high leaf temperatures) and therefore enhance long-term performance in this manner. Investigations with other species are needed. Of particular interest would be those herbaceous species like peas that show a large potential for acclimation of mature leaves where the mechanisms might be quite different.

At the level of the individual plant or plant stand, acclimation to the gradient of PFD through the canopy also needs to be understood in the context of how properties at the chloroplast and leaf level scale to the whole canopy performance. For a given level of a resource such as N, the daily canopy photosynthesis will be maximized with respect to investments of that resource when it is equally limiting for photosynthesis of all leaves. That is to say that the ratio of the marginal gain in daily assimilation with the addition of an increment of N (increase in carboxylation and ET capacity) to the cost of that N is equal in all leaves (Field, 1983). Following similar reasoning, the maximum return of canopy photosynthesis will be achieved if all leaves reach light saturation simultaneously. This could be achieved partially by architectural adjustments in leaf angles throughout the canopy so that steeper upper leaves intercept less light and have a lower PFD in their surface but more PFD is transmitted to the lower leaves. It certainly also involves sun/shade acclimation responses that adjust the light response curve giving differing light saturation points. Because the acclimation response involves changes in leaf N per unit area ($N_L$), optimal use of nitrogen and optimal light utilization are intimately linked.

Studies of photosynthesis and nitrogen distributions in canopies reveal that, in general, the gradient of $N_L$ with depth is in the same direction as the optimal gradient but that it is somewhat shallower than the optimum gradient. In a dense canopy, the optimal $N_L$ gradient has been shown to yield a 20–35% increase in canopy photosynthesis as compared to a uniform N distribution. The advantage accrued via an optimal gradient as compared to a uniform distribution depends on canopy density and N availability. For real canopies, the enhancement appears to be typically more of the order of 8–20% (Hirose & Werger, 1987). The gradient is due to the acclimation of leaves, leaf senescence, and consequently the redistribution of N made available by these processes to newly developing leaves. The gradient achieved also depends on the N supply. The optimum may not be reached because these processes lag behind or there are upper and lower limits on the $N_L$ that constrain the gradient that can actually be achieved. Additionally, the light gradient changes over the day and sun-fleck penetration and utilization has not been considered. Experiments in which light levels on different leaves were manipulated independent of leaf age gradients indicate that sun–shade acclimation is a major determinant of the gradient but leaf ageing and nitrogen nutrition are also important (Hikosaka, Terashima & Katoh, 1994).

Acclimation is perhaps best viewed as an economic

process in which the pattern of investment of resources is optimized so that the daily carbon gain or resource use efficiencies can be maximized. Hikosaka and Terashima (1995) have shown that shifts in the partitioning of nitrogen among components of the photosynthetic apparatus as occurs during acclimation can enhance daily leaf carbon gain. Moreover, the optimal partitioning of N at different PFDs corresponded rather well with observed patterns of partitioning for leaves acclimated to different light environments. Evans (1993) used a similar but less detailed model of leaf photosynthesis and examined both how leaf carbon gain and canopy carbon gain was maximized. Changes in the partitioning of N between light harvesting and carbon fixation/electron transport components within leaves that had become shaded accounted for about a 4% improvement in canopy carbon gain whereas changes in distribution between leaves accounted for a 20% improvement. Thus, it is necessary to consider both the dynamics of the canopy as well as the dynamic re-allocation of N within the leaf and within the chloroplast in order to understand the response of photosynthesis to changing light environments.

## CONCLUDING REMARKS

Acclimation to sun and shade conditions has been demonstrated to involve a remarkably well co-ordinated series of adjustments in the structure and composition of the photosynthetic apparatus. At the chloroplast level, this appears to involve mostly a trade-off between maximizing light absorption via investment in chlorophyll–protein complexes and investment in maintenance of high photosynthetic capacity versus investment in rubisco and electron transport components. Simulation models show that, for any given environment, there is an optimal partitioning of these investments between light harvesting and photosynthetic capacity that will maximize the daily carbon gain. Indeed, the actual partitioning achieved by leaves in a canopy appears to be remarkably close to the predicted optimal levels. In addition, other co-ordinated adjustments in, for example, the xanthophyll cycle pool may help to minimize the risk of photoinhibition. It still remains to be shown how specifically the light environment is detected and allocation of nitrogen and other resources is allocated and re-allocated to achieve this optimal partitioning, but progress is being made in understanding the nature of the putative signal transduction pathways (Anderson, Chow & Park, 1995).

One of the conclusions that can be drawn is that, to fully understand the adaptive significance of acclimation, it is necessary to understand its role in whole plant carbon gain and productivity in different environments. There is clearly a trade-off in, at least some species, between the development of thicker leaves in high light having a higher photosynthetic capacity and production of leaf area for light interception. Indeed, in *Alocasia*, this trade-off is sufficient that the large advantage seen in photosynthetic capacity per unit leaf area observed in leaves developing in the sun as compared to the shade essentially disappears when whole plant photosynthetic rates per unit mass are compared. Although a greater leaf thickness in sun- versus shade-developed leaves is commonly observed, it is not yet clear how far the results for *Alocasia* can be generalized. In species such as pea, which exhibit a greater potential for acclimation of extant leaves, it is possible that the trade-off is less severe. Further research is required to understand the differences in response of species that depend more on development of new leaves for acclimation versus those with a greater capacity for acclimation of already developed leaves.

## REFERENCES


Anderson, J. M. (1986). Photoregulation of the composition, function and structure of thylakoid membranes. *Annual Review of Plant. Physiology*, **37**, 93–136.

Anderson, J. M. & Osmond, C. B. (1987). Shade–sun responses: compromises between acclimation and photoinhibition. In '*Photoinhibition*.' ed. D. J. Kyle; C. B. Osmond & C. J. Arntzen. pp. 1–38. Amsterdam: Elsevier.

Anderson, J. M., Chow, W. S. & Park, Y. I. (1995). The grand design of photosynthesis: acclimation of the photosynthetic apparatus to environmental cues. *Photosynthesis Research*, **46**, 129–39.

Björkman, O. (1968). Carboxydismutase activity in shade-adapted and sun- adapted species of higher plants. *Physiologia Plantarium*, **21** 1–10.

Björkman, O. (1981). Responses to different quantum flux densities. In *Physiological Plant Ecology. I. Physical Environment*, ed. O. L. Lange, P. S. Nobel, C. B. Osmond & H. Ziegler, pp. 57–107. Berlin: Springer-Verlag.

Björkman, O. & Demmig, B. (1987). Photon yield of $O_2$

evolution and chlorophyll fluorescence characteristics at 77 K among vascular plants of diverse origins. *Planta*, **170**, 489–504.

Chow, W. S. & Anderson, J. M. (1987). Photosynthetic responses of *Pisum sativum* to an increase in irradiance during growth. I. Photosynthetic activities. *Australian Journal of Plant Physiology*, **14**, 1–8.

Chow, W. S., Qian, L., Goodchild, D. J. & Anderson, J. M. (1988). Photosynthetic acclimation of *Alocasia macrorrhiza* (L.) G. Don to growth irradiance: structure, function and composition of chloroplasts. *Australian Journal of Plant Physiology*, **15**, 107–22.

Chow, W. S., Anderson, J. M. & Melis, A. (1990*a*). The photosystem stoichiometry in thylakoids of some Australian shade adapted plant species. *Australian Journal of Plant Physiology* **17**, 665–74.

Chow, W. S., Melis, A. & Anderson, J. M. (1990*b*). Adjustments of photosystem stoichiometry in chloroplasts improve the quantum efficiency of photosynthesis. *Proceeding of the National Academy of Sciences USA*, **87**, 7502–6.

Demmig-Adams, B. & Adams, W. W. (1992). Photoprotection and other responses of plants to high light stress. *Annual Review of Plant Physiology and Plant Molecular Biology*, **43**, 599–626.

Evans, J. R. (1986). A quantitative analysis of light distribution between the two photosystems, considering variation in both the relative amounts of the chlorophyll–protein complexes and the spectral quality of light. *Photochemistry and Photobiophysics*, **10**, 135–47.

Evans, J. R. (1987*a*). The dependence of quantum yield on wavelength and growth irradiance. *Australian Journal of Plant Physiology*, **14**, 68–79.

Evans, J. R. (1987*b*). The relationship between electron transport components and photosynthetic capacity in pea leaves grown at different irradiances. *Australian Journal of Plant Physiology*, **14**, 157–70.

Evans, J. R. (1989). Photosynthesis – the dependence on nitrogen partitioning. In *Causes and Consequences of Variation in Growth Rate and Productivity of Higher Plants*. ed. H. Lambers, *et al.*, pp. 159–74. Netherlands: SPB *Academic Publishing*.

Evans, J. R. (1993). Photosynthetic acclimation and nitrogen partitioning within a lucerne canopy II. stability through time and comparison with a theoretical optimum. *Australian Journal of Plant Physiology*, **20**, 69–82.

Evans, J. R. & Seemann, J. R. (1989). The allocation of protein nitrogen in the photosynthetic apparatus: costs, consequences and control. *In Photosynthesis*. ed. W. R. Briggs, pp. 183–205. New York: Alan R. Liss Inc.

Field, C. (1983). Allocating leaf nitrogen for the maximization of carbon gain: leaf age as a control on the allocation program. *Oecologia (Berl.)*, **56**, 341–7.

Fredeen, A. L. & Field, C. B. (1996). Ecophysiological constraints on the distribution of *Piper* species. In *Tropical Forest Plant Ecophysiology*, ed. S. S. Mulkey, R. L. Chazdon & A. P. Smith, pp. 597–618. New York: Chapman and Hall.

Givnish, T. J. (1988). Adaptation to sun and shade: a whole-plant perspective. *Australian Journal of Plant Physiology*, **15**, 63–92.

Hikosaka, K. & Terashima, I. (1995). A model of the acclimation of photosynthesis in the leaves of $C_3$ plants to sun and shade with respect to nitrogen use. *Plant Cell and Environment*, **18**, 605–18.

Hikosaka, K., Terashima, I. & Katoh, S. (1994). Effects of leaf age, nitrogen nutrition and photon flux density on the distribution of nitrogen among leaves of a vine (*Ipomoea tricolor* Cav) grown horizontally to avoid mutual shading of leaves. *Oecologia*, **97**, 451–7.

Hirose, T. & Werger, M. A. (1987). Maximizing daily canopy photosynthesis with respect to the leaf nitrogen pattern in the canopy. *Oecologia*, **72**, 520–6.

Kamaluddin, M. & Grace, J. (1992). Photoinhibition and light acclimation in seedlings of *Bischofia javanica*, a tropical forest tree from Asia. *Annals of Botany*, **69**, 47–52.

Lambers, H. (1985). Respiration in intact plants and tissues: Its regulation and dependence on environmental factors, metabolism and invaded organisms. *In Encyclopedia of Plant Physiology*, New Series, vol. 18. ed. R. Douce & D. A. Day, pp. 418–73. Berlin: Springer-Verlag.

Long, S. P., Postl, W. F. & Bolhar-Nordenkampf, H. R. (1993). Quantum yields for uptake of carbon dioxide in $C_3$ vascular plants of contrasting habitats and taxonomic groupings. *Planta*, **189**, 226–34.

Öquist, G., Chow, W. S. & Anderson, J. M. (1992). Photoinhibition of photosynthesis represents a mechanism for the long-term regulation of Photosystem-II. *Planta*, **186**, 450–60.

Pearcy, R. W. & Sims, D. A. (1994). Photosynthetic acclimation to changing light environments: scaling from the leaf to the whole plant. In *Exploitation of Environmental Heterogeneity by Plants: Ecophysiological Processes Above and Below Ground*, ed. M. M. Caldwell & R. W. Pearcy, pp. 223–34. San Diego: Academic Press.

Sims, D. A. & Pearcy, R. W. (1989). Photosynthetic characteristics of a tropical forest understory herb, *Alocasia macrorrhiza*, and a related crop species, *Colocasia esculenta* grown in contrasting light environments. *Oecologia*, **79**, 53–9.

Sims, D. A. & Pearcy, R. W. (1994). Scaling sun and shade photosynthetic acclimation to whole plant performance. I. Carbon balance and allocation at different daily photon flux densities. *Plant, Cell and Environment*, **17**, 881–7.

Sims, D. A., Gebauer, R. & Pearcy, R. W. (1994). Scaling sun and shade photosynthetic acclimation to whole plant performance. II. Simulation of carbon balance and growth at different daily photon flux densities. *Plant, Cell and Environment*, **17**, 889–900.

Terashima, I. & Hikosaka, K. (1995). Comparative ecophysiology of leaf and canopy photosynthesis. *Plant, Cell and Environment*, **18**, 1111–28.

Terashima, I. & Inoue, Y. (1985). Vertical gradient in photosynthetic properties of spinach chloroplasts dependent on intra-leaf light environment. *Plant and Cell Physiology*, **26**, 781–5.

Tyystjarvi, E., Aliyrkko, K., Kettunen, R. & Aro, E. M. (1992). Slow degradation of the D1 protein is related to the susceptibility of low-light-grown pumpkin plants to photoinhibition. *Plant Physiology*, **100**, 1310–17.

Vogelmann, T. C. (1993). Plant tissue optics. *Annual Review of Plant Physiology and Plant Molecular Biology*, **44**, 231–51.

von Caemmerer, S. & Farquhar, G. D. (1981). Some relationships between the biochemistry of photosynthesis and the gas exchange of leaves. *Planta*, **153**, 376–87.

# 20 Photoinhibition

C. CRITCHLEY

## INTRODUCTION

For many years now the term 'photoinhibition' has been used to signify a suboptimal condition of the photosynthetic system in a leaf or an alga that is induced when irradiance exceeds levels required for saturation. This condition is always characterized by a lowering in quantum efficiency but is not always accompanied by a reduction in maximal, light-saturated (and $CO_2$-saturated) photosynthesis rate. In some species, significantly reduced quantum efficiency has been measured at the same time as maximum, saturated rates were actually increased (Walker, 1992). It is specifically the quantum efficiency of photosystem II (PSII) that is affected and not, in most cases studied, photosystem I.

Much argument and terminology have developed in the search for the mechanism of photoinhibition. Recently, the concepts of dynamic and chronic photoinhibition (Osmond, 1994), and photoprotection (Demmig-Adams & Adams III, 1992) have been introduced, although the mechanisms underlying these processes, and how they might relate to the original concept of photoinhibition, have not been clearly identified. Whether an 'inhibition' of PSII actually occurs has also been questioned (Critchley, Russell & Bolhàr-Nordenkampf, 1992). In the remainder of this chapter, these concepts will be discussed and, as far as possible, clarified and reconciled with our current knowledge of photoinhibition and PSII structure and function.

## HETEROGENEITY OF PHOTOSYSTEM II

While we have a reasonable understanding of the overall structure and function of the photosystems, particularly PSII, and the other complexes involved in photosynthetic electron transport, we know very little about their interaction with each other in the thylakoid membrane. We also do not understand the dynamics of structural changes in the complexes and/or the thylakoid membranes in constant light, in changing light or in light–dark and dark–light transitions. Furthermore, because of structural and functional heterogeneity, we have difficulties with the biochemical and biophysical measurement of complexes such as PSII. This heterogeneity is likely to be not only lateral, i.e. related to the location and relative abundances of the complexes in the non-appressed or the appressed parts of the thylakoid membrane in an individual chloroplast, but also vertical, i.e. in the distribution within the grana stacks and among the different chloroplasts in the different tissues of the leaf.

Each leaf with more than one cell layer has a light gradient and hence photosystems which are exposed to different quantities and qualities of irradiance (Nishio, Sun & Vogelmann, 1994). In practical terms, this means that bifacial leaves illuminated on their upper side will contain photosystems fully irradiated in the upper cell layer, with much less light reaching those in the lower leaf tissues.

While this variability of the irradiance conditions across the leaf must affect all electron transport complexes directly or indirectly, structural heterogeneity within one type of complex seems to be restricted to PSII. Interestingly, no structural and/or functional heterogeneity has been reported for the cytochrome $b_6/f$ complex, the ATP synthase or PSI, although the latter shows variations in antenna size. It is clear now that at least two types of PSII can be found in chloroplasts exposed to irradiance at, or in excess of, saturation: active complexes engaged in linear electron transport from water to plastoquinone, and dissipative complexes from which the incoming excitation energy is lost as heat.

This chapter will explore distributional and structural heterogeneity of PSII and their relevance for PSII functionality. It will also discuss the significance of this heterogeneity, and of grana stacking, for photoinhibition.

## FUNCTIONALITY OF PHOTOSYSTEM II

Photosystem II is a large multisubunit complex and an integral part of the photosynthetic electron transport chain. It is localized in the thylakoid membrane and associated with the other electron transport complexes, particularly the cytochrome $b_6/f$ complex. Its monomeric structure consists of a peripheral trimeric light harvesting complex (LHCII), an inner antennae system or core, and a reaction centre. The water-oxidizing Mn-complex is bound to the reaction centre. The complete structure may not, as was assumed until recently, be arranged in the layered onion-type models often seen in the literature. Instead the monomer may be wedge-shaped with the reaction centre proteins D1/D2 at its tip (Fig. 20.1, ACTIVE). It has also been suggested that PSII functions as a dimer (Boekema *et al.*, 1995).

Of all the chloroplast proteins, including the electron transport complexes as well as the enzymes of the Calvin cycle, only the D1 protein of the PSII reaction centre has the most unusual property of being turned over much faster than any of the other proteins. This property was recognized long ago before the identity of the protein and its functional significance were known (Bottomley, Spencer & Whitfeld, 1974). Since D1 turnover is light dependent, and under some circumstances shows some light intensity dependence, it was originally suggested that this turnover is really a damage–repair cycle necessary because excess irradiance damages the protein and incapacitates PSII (Barber & Andersson, 1992).

D1 turnover rates, however, do not correlate at all with rates of photoinhibition. Many studies have shown that D1 turnover rates (i) saturate at relatively low light intensities and do not change over a more than ten-fold increase in irradiance (Geiken, Critchley & Renger, 1992); (ii) are maximal at growth light intensity and not in photoinhibitory conditions (Sundby, McCaffery & Anderson, 1993; Russell *et al.*, 1995); and (iii) are significant in very low light (Keren, Gong & Ohad, 1995). The relationship between D1 turnover and photoinhibition is therefore far from straightforward and it is perhaps necessary to consider models for mechanisms of the two phenomena independently.

## PHOTOINHIBITION AND CHLOROPHYLL FLUORESCENCE

Because PSII emits, in special conditions, a characteristic and complex fluorescence signal, one of the favourite measures of photoinhibition is the $F_v/F_m$ ratio. This chlorophyll fluorescence parameter has been shown to correlate with the number of functional PSII complexes as determined, for example, by $O_2$ flash yield measurements (Russell *et al.*, 1995). This is presumably because measurements of the two parameters involve the same population of photosystems, i.e. mostly those in the cells of the palisade parenchyma. Because of the leaf light gradient, photoinhibited PSII complexes are to be expected mainly in the upper palisade parenchyma cells. This can be shown by measuring the $F_v/F_m$ ratio of photoinhibited leaves on both the irradiated and the non-irradiated side. While significant decreases in $F_v/F_m$ occur on the irradiated side, none or very little, is seen on the non-irradiated side (Syme, Bolhàr-Nordenkampf & Critchley, 1992).

The $F_v/F_m$ ratio is only measurable after 10–15 min dark adaptation and so reflects maximal fluorescence emission after all significant and short-term quenching, such as photochemical quenching (qP) and high energy quenching related to the establishment of the ΔpH (qE) has relaxed. Therefore, this parameter really measures the total photochemical capacity, as well as the overall efficiency of the population of PSII in a leaf that can be reached with the actinic beam. A significant decrease in the value of this parameter is considered to constitute photoinhibition and equivalent to the decrease in quantum efficiency of $O_2$ evolution or $CO_2$ fixation. Typical $F_v/F_m$ values of sun leaves are around 0.8 or just below 0.8. Shade or low light leaves, on the other hand, always have values of around 0.83–0.85, a difference of about 4%. Any value below 0.725 would certainly be considered photoinhibition.

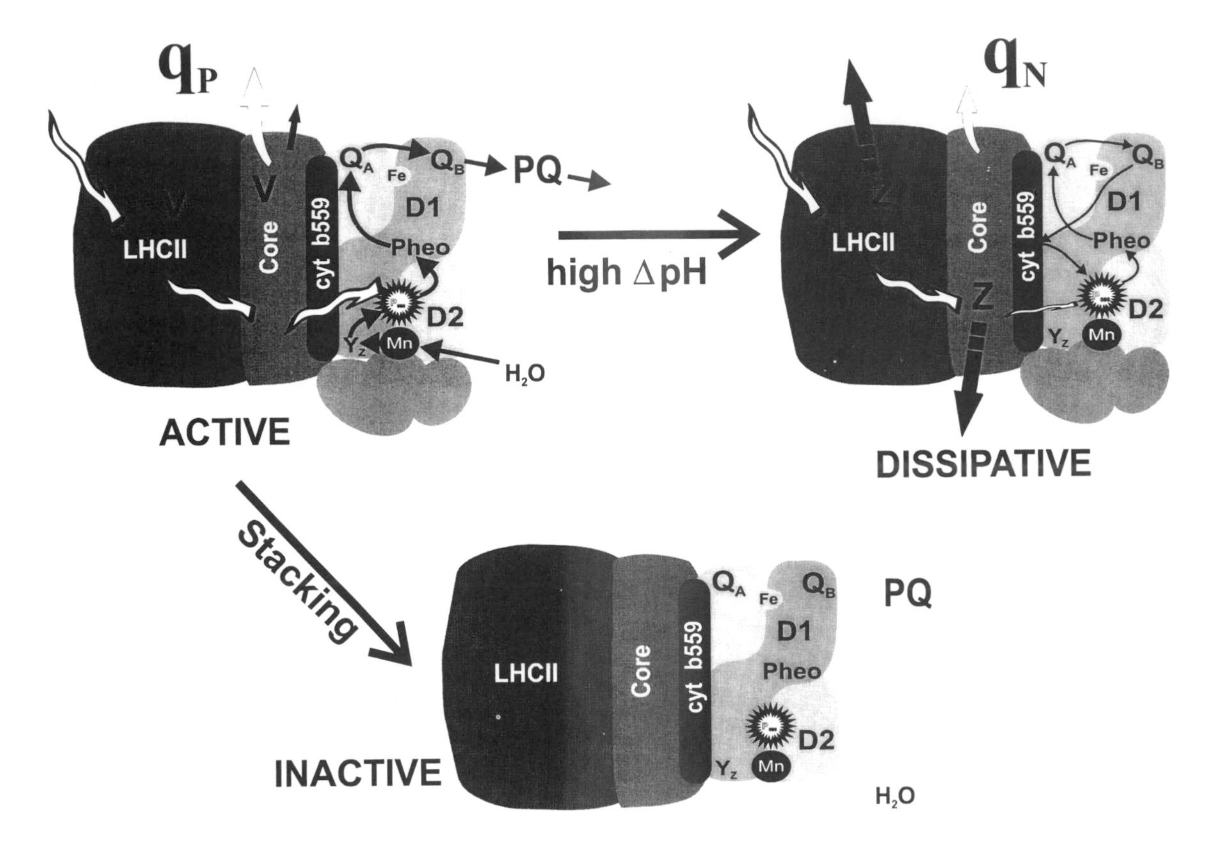


**Figure 20.1.** Structure and function of different forms of PSII. Curved, open arrows indicate excitation energy transfer from the antennae to the reaction centre, thin arrows show the direction of electron flow, open arrow indicates fluorescence, broken arrow indicates heat dissipation. LHCII, light harvesting chlorophyll protein complex of PSII; core, proximal antennae proteins of 47 and 43 kDa carrying chlorophyll a molecules; LHCII and core proteins also bind violaxanthin (V) or zeaxanthin (Z); D1/D2, heterodimer of the reaction centre (RC) proteins carrying the trap $P_{680}$, pheophytin (pheo), the permanently bound quinone $Q_A$, and $Q_B$ which dissociates from the RC as $PQH_2$. Electrons are extracted from water by a 4-manganese cluster (Mn) and passed on to $Y_Z$, a tyrosine which is part of the amino acid chain of the D1 protein. Cytochrome $b_{559}$ (cytb559) whose function is unknown, cannot be removed from the PSII RC without destroying its functionality. Electron-transport active (ACTIVE), dissipative (DISSIPATIVE) and inactive (INACTIVE) conformations of PSII: The active centre is shown in a low fluorescent state due to photochemical quenching, qP, but little nonphotochemical quenching, qN; the dissipative centre, formed under the influence of a high ΔpH, has converted its V into Z, dissipates most of the energy as heat and contributes to qN because little fluorescence is seen from these centres; other mechanisms may contribute to the dissipative PSII conformation; inactive centres are mainly formed in shade chloroplasts with large grana stacks where they do not receive any or very few photons. Water oxidation is turned off in these centres but can be rapidly photoactivated. Little or no xanthophylls may be bound to these centres.

## DYNAMIC AND CHRONIC PHOTOINHIBITION

Osmond (1994) recently suggested that there is a multitude of research cultures that all seek to explain the reduced efficiency in PSII and that all seem to claim exclusivity. For example, the concept of dynamic photoinhibition is based on the assessment of the rapidly relaxing, ΔpH dependent fluorescence quenching (qE), a component of qN (non-photochemical quenching), which Weis & Berry claim is reaction centre based (Weis & Berry, 1987), while Horton and his collaborators are convinced that qE is due to antenna quenching through zeaxanthin-mediated chlorophyll aggregation in LHCII (Horton & Ruban, 1994). Demmig-Adams and Adams (1996) suggest that all non-photochemical quenching is photoprotective and occurs through the formation of dissipating centres in which zeaxanthin quenches chlorophyll triplets directly. The only difference, they propose, in terms of whether relaxation of quenching occurs in minutes or hours or even days, is the means by which PSII and the thylakoid membrane can hold on to a high ΔpH and/or zeaxanthin, i.e. xanthophylls in the de-epoxidized state and able to quench. Recent experiments by Ruban & Horton (1995) indicate that qI, the slowly reversible component of non-photochemical quenching, is not related to the bulk phase pH gradient. They suggested that more localized protonation of LHCII proteins is involved in this quenching.

Whatever its mechanism, dynamic photoinhibition, according to Osmond, is the province of the sun chloroplast (or leaf): an instantaneous efficiency decline as the light intensity approaches saturation, which does not involve a significant reduction in maximal light saturated rate. It is also rapidly reversible (within minutes) and non-damaging. Perhaps an example of such dynamic photoinhibition is seen in sun leaves (maximal light intensity at midday 1800–2200 μmoles $m^{-2}$ $s^{-1}$) of *Schefflera arboricola*, exposed to various irradiances in the laboratory, showing rapid $F_v/F_m$ decline over a large range in conditions (Fig. 20.2 (*a*)). Recovery in these leaves is also rapid. Chronic photoinhibition, on the other hand, occurs mostly in shade chloroplasts (or plants) and is chronic because the photoinhibition levels reached in shade leaves are often only slowly reversible. Shade leaves of *Schefflera arboricola* (maximal light intensity at midday 30–100 μmole $m^{-2}$ $s^{-1}$) provide perhaps an example of this type of photoinhibition (Fig. 20.2(*b*)). However, chronic photoinhibition is, according to Osmond, also associated with reduced maximal, light-saturated rates. This is not easily explained mainly because it has not yet been shown conclusively that whenever maximum, light (and $CO_2$), saturated rates are not affected, in both sun and shade chloroplasts, we have rapidly reversible, dynamic photoinhibition. Conversely, it also has not been demonstrated that, whenever apparent quantum yields and maximum rates are both reduced,

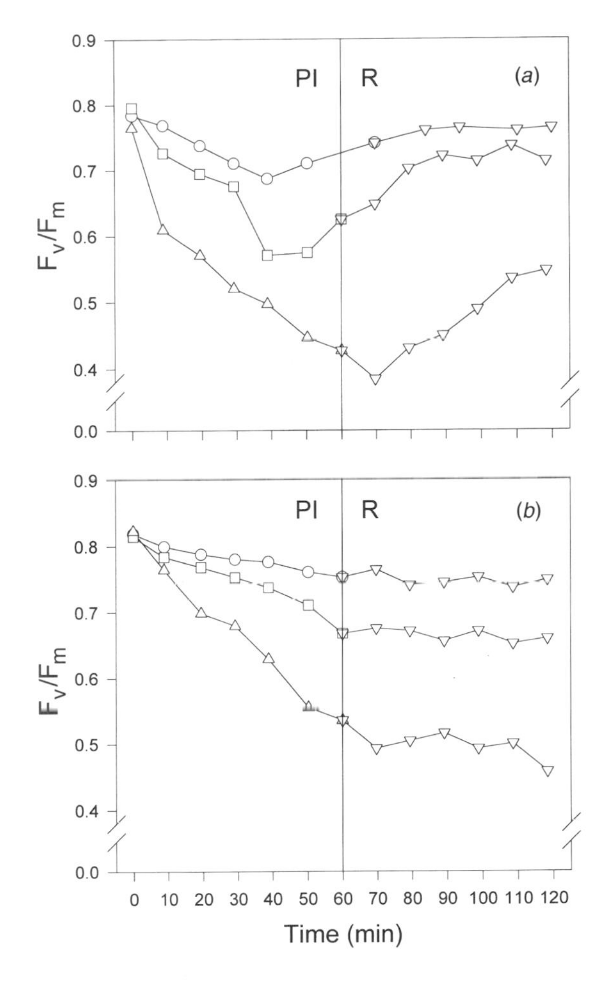


**Figure 20.2.** $F_v/F_m$ values (measured after 10 min dark adaptation) (*a*) in sun and (*b*) in shade leaves of *Schefflera arboricola* during irradiance treatment (PI) at 500 (○), 1500 (□) or 2500 (Δ) μmoles $m^{-2}$ $s^{-1}$ followed by recovery (R) in 100 μmoles $m^{-2}$ $s^{-1}$. Note the difference in $F_v/F_m$ before treatment.

chronic photoinhibition occurs which only recovers very slowly and may require protein synthesis.

It has also been suggested that photon damage occurs particularly in shade leaves when both photochemical efficiency and maximal rates are reduced. If this is the case, the interesting questions then are whether different mechanisms underlie chronic and dynamic photoinhibition which have very similar characteristics, if not time courses, of $F_v/F_m$ decline (see also

Fig. 20.2), why this should be so, and how these mechanisms differ.

## PHOTOPROTECTION

Decreases in PSII efficiency, lasting many hours or even days, have been reported, particularly in evergreen species when they are subjected to very large excesses of irradiance during bright sunny days in the winter months (Bolhàr-Nordenkampf *et al.*, 1993; Adams & Demmig-Adams, 1995). Many other studies have been published which show a correlation between excess irradiance induced formation of the xanthophyll pigments, zeaxanthin (Z) and antheraxanthin (A) at the expense of violaxanthin (V) and the decrease in dark adapted $F_v/F_m$ values. Taking many such measurements and plotting the level of A+Z against $F_v/F_m$ yielded a curvilinear relationship. From this it was concluded that decreases in $F_v/F_m$ reflect a photoprotective down-regulation of PSII based on increased energy dissipation in the antennae through the operation of the xanthophyll cycle. However, there seem to be many questions that need to be asked in this context. For example, which chlorophylls in LHCII, the core, or the reaction centre are likely to form triplets and, following from this, how many zeaxanthin molecules per photosystem could be effective in direct energy quenching via such triplets? Sun chloroplasts, in particular, seem to contain large amounts of xanthophylls compared to shade chloroplasts, even though they do not contain any more PSII per leaf area. So, what are all those xanthophylls doing in sun chloroplasts if they are largely converted to Z but perhaps not needed in dissipation? Conversely, why do shade chloroplasts contain so little xanthophyll when they have so much PSII? Does it mean that shade PSII complexes are structurally different from the sun complexes, introducing more heterogeneity? Perhaps the xanthophyll pool size is related to the amounts of active and dissipative centres. These are only some of the questions that should be answered before we are able to decide what roles the xanthophylls play in the structure and function of PSII and the thylakoid membrane.

## THE ROLE OF D1 TURNOVER IN DYNAMIC AND CHRONIC PHOTOINHIBITION AND PHOTOPROTECTION

Like the rates of photosynthesis, D1 turnover is high in sun leaves and often very low in shade leaves (Geiken, Critchley & Renger, 1992). From the data that show maximal D1 turnover rates in leaves operating at maximal rates of photosynthesis, optimal apparent quantum yields and optimal photochemical efficiency, it must be concluded that the D1 protein is exchanged in active as well as dissipative PSII. If, as suggested by Keren *et al.* (1995), the likelihood of charge recombination between $Q_B^-$ and the $S_{2,3}$ states may be involved in the D1 degradation process and, if this state is likely to occur more frequently in active rather than dissipative centres, the different D1 turnover rates may be explained. In fact, if the $Q_B^-/S_{2,3}$ charge recombinations generate chlorophyll triplets which are potentially dangerous, then relatively less triplet-induced damage and D1 turnover would be expected in dissipative centres because the chlorophyll triplets would be quenched by zeaxanthin.

From recent studies (van Wijk *et al.*, 1995) comes some support for this suggestion of D1 turnover in both active and dissipative centres. They found, in biochemical studies of *in vitro* protein synthesis in isolated intact chloroplasts and isolated thylakoids, that newly synthesized D1 protein becomes incorporated into different kinds of PSII complexes. Whether these different kinds of PSII complexes represent sequential stages of PSII reassembly, as the authors seem to think, or whether they actually represent heterogeneous forms of PSII, remains to be seen. One way of deciding this question would be to repeat these *in vitro* incorporation experiments with chloroplasts and thylakoids isolated from different kinds of photoinhibited leaves.

As pointed out above, D1 turnover may occur in both electron transport active and dissipative centres, since the same turnover rates were observed whether there were a lot of dissipative centres or none. Furthermore, since there is no significant change in D1 pool

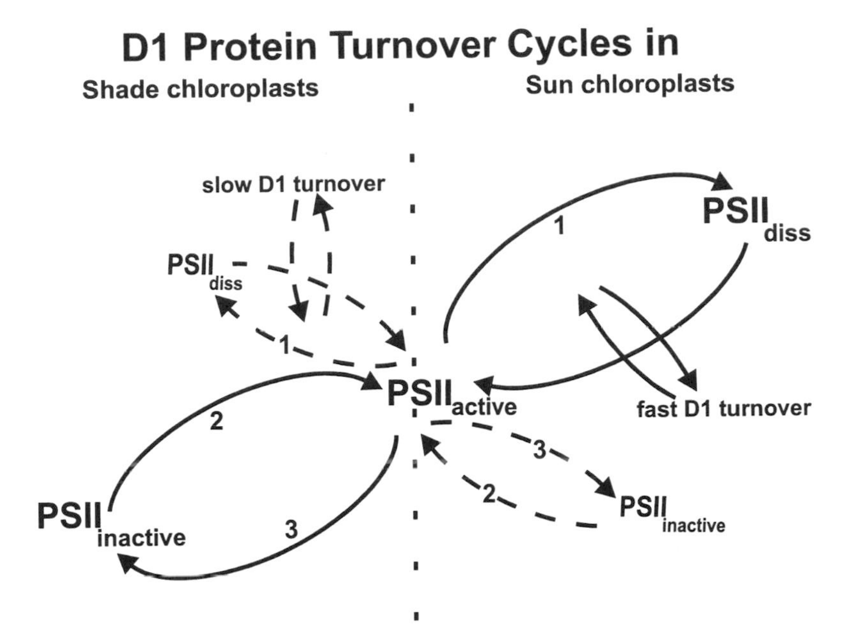


**Figure 20.3.** The relationship between the formation of electron transport active ($PSII_{active}$), dissipative ($PSII_{diss}$) and inactive ($PSII_{inactive}$) PSII complexes (see also Fig. 20.1) and D1 protein turnover in sun and shade chloroplasts. In sun chloroplasts with small grana stacks and large stroma exposed thylakoid membrane surfaces, rapid interconversion of active and dissipativc PSII complexes is the major metabolic activity, accompanied by high D1 turnover activity in both types. (1) signifies a possibly irreversible change in D1 conformation causing formation of dissipative complexes. In sun chloroplasts only few PSII complexes would be in the inactive form, so that there is only little (2) light activation or (3) grana stacking. In shade chloroplasts with large grana stacks and relatively little stroma exposed thylakoid membrane surfaces, interconversion of active into inactive PSII complexes is the major metabolic activity so that formation of inactive complexes leads to stacking. Partial unstacking may occur when inactive complexes are light activated. Interconversion of active and dissipative complexes and D1 turnover would be slow.

size, the location of degradation events may determine which D1 proteins are exchanged. It has been suggested previously that D1 protein removal is the limiting factor in turnover, not its intrinsic rate of synthesis (Vasilikiotis & Melis, 1994). If we further assume that new synthesis, insertion and processing provide for the form of the protein appropriate to the type of centre, i.e. whether active or dissipative, to be inserted, then the following scenarios can be envisaged (Fig. 20.3).

In shade chloroplasts, perhaps only half of all PSII receives photons and is active in electron transport (Fig. 20.1 ACTIVE); the other half is inactive (Fig. 20.1 INACTIVE), highly concentrated and isolated from the other electron transport complexes in the large grana stacks. This would account for the significant amounts of inactive PSII found in most leaves and it would, to some extent, explain the low rates of electron transport (photosynthesis) and D1 turnover measured in shade leaves under low intensity conditions. Exposure to suprasaturating intensities would cause the formation of dissipative centres (Fig. 20.1 DISSIPATIVE) at the expense of electron transport active centres, but it would also activate inactive centres. This would explain the relatively slow decrease in $F_v/F_m$ that is often measured in shade leaves (see Fig. 20.2(*b*)). Recovery in low light may be so slow that it does not register for several hours because of little D1 turnover (and hence slow regeneration of active PSII) and cessation of light activation of inactive PSII because of low light intensity. In these shade chloroplasts, and this includes chloroplasts in the spongy mesophyll cells of typical bifacial sun leaves, the main metabolic activity is not D1 turnover or the conversion of active to dissipative PSII, but the formation of large grana stacks into which most of the PSII complexes are concentrated in an inactive form. Only if light intensity exceeds saturation, the normally very slow D1 turnover may accelerate somewhat (Geiken, Critchley & Renger, 1992). A second, significant, effect of excess irradiance impinging on shade chloroplasts, however, would be the light activation of inactive PSII complexes. These additional, active PSII would balance to some extent the formation of dissipative PSII and therefore, only slow decreases in apparent quantum yield (or $F_v/F_m$) would be measured (Fig. 20.2(*b*)). Because of low D1 turnover in these chloroplasts, particularly at irradiances below saturation, recovery light intensities have very little effect on the photoinhibited status of the system, even several hours after photoinhibitory irradiance ceases.

D1 turnover, photosynthetic carbon fixation reactions and even electron transport rates will also be slow at low temperatures. A drop in temperature therefore, will also trap the photosynthetic system in a photoinhibited state. Moreover, cold-acclimated species would have a lower temperature of significant reduction in D1 turnover than warm-climate species. A very interesting question to study is how D1 turnover affects the xanthophyll pigments and their function in dissipation or, conversely, whether the operation of the xanthophyll cycle affects D1 turnover.

High light leaves with abundant sun chloroplasts in the upper cell layers, would have high rates of photosynthesis and D1 turnover and little inactive PSII. Many dissipative PSII with a longer life-time would be formed and photoinhibition would be more rapid than in shade leaves at similar irradiances (Fig. 20.2(*a*)). As soon as the active to dissipative complex conversion rate becomes faster than the D1 turnover rate, the ratio of active PSII to dissipative PSII would steadily decrease until the light intensity is reduced. Significantly, it has been shown in many systems that D1 turnover rates saturate at quite low intensities. Hence, with D1 turnover remaining fast even at low recovery light intensity, the ratio of active to dissipative PSII would increase, particularly if PSII complexes in active conformation were mostly turned over, and recovery would be significant. This proposition can be tested in experiments where the saturating intensity for D1 turnover can be determined and where recovery should be slow below that saturating intensity, but rapid and almost independent of light intensity above it. The main metabolic activity would be fast D1 turnover in the light and, in excess irradiance, the interconversion of active and dissipative PSII. Very little inactive PSII would be formed in sun chloroplasts.

## FUTURE OUTLOOK

Many issues, for example, PSII functional dimers, photosystem co-operativity and connectivity and state transitions and how they would affect excitation energy distribution, photoinhibition and D1 protein turnover have not been considered in this chapter. This is because we have too little knowledge of photosystem and electron transport complex interaction in the membrane and their dynamics of change, for them to be considered in models of the mechanisms of photoinhibition and D1 turnover.

The role of the xanthophylls in PSII structure and function, both of the de-epoxidized and the epoxidized form, needs to be clarified. This may come from studies with mutants which are deficient in one or the other of the conversion enzymes. It may also come from analysis of sun and shade photosystems which would include their pigment binding properties.

The mechanism of rapid recovery from photoinhibition, the nature of any possible differences between dynamic and chronic photoinhibition and the nature of the $F_v/F_m$ decline need to be systematically investigated. This could be approached with experiments on sun and shade leaves with quite different rates of photoinhibition, measured as the decline in $F_v/F_m$, in which, at the same time, only a decrease in apparent quantum yield, not in maximum rate, occurs. Does this actually ever happen in shade leaves? Such measurements could then be compared with photoinhibition in which both the apparent quantum yield and the maximum rate are reduced. The third system that should be investigated is that of leaves which, after a period of excess irradiation, show reduced apparent quantum yields and increased maximum saturated rates.

Identification and biochemical characterisation of the different types of PSII may come from studies in which different types of chloroplasts, isolated from a leaf with a significant light gradient can be separated, and where they can be compared with chloroplasts from more homogeneous tissues. Further experiments using the *in vitro* translation systems recently described (van Wijk *et al.*, 1995) with chloroplasts isolated from differentially photoinhibited leaves should yield some interesting results and shed some light on PSII structural and functional heterogeneity and how it may be related to photoinhibition and D1 turnover.

## ACKNOWLEDGEMENTS

I thank most sincerely Dr W. Bottomley for his help with the Figures and constructive comments on the manuscript. I am very grateful to Professor H. R. Bolhàr-Nordenkampf for many fruitful discussions and criticisms and to his group at the University of Vienna, where I am currently on Special Study Leave. They provided the environment conducive to doing experiments as well as write and think. I am grateful to the University of Queensland for the Special Study Leave.

## REFERENCES

Adams, W. W. & Demmig-Adams, B. (1995). The xanthophyll cycle and sustained thermal energy dissipation activity in *Vinca minor* and *Euonymus kiautschovicus* in winter. *Plant, Cell and Environment*, **18**, 117–27.

Barber, J. & Andersson, B. (1992). Too much of a good thing: light can be bad for photosynthesis. *Trends in Plant Science*, **17**, 61–6.

Boekema, E. J., Hankamer, B., Bald, D., Kruip, J., Nield, J., Boonstra, A. F., Barber, J. & Rögner, M. (1995). Supramolecular structure of the photosystem II complex from green plants and cyanobacteria. *Proceedings of the National Academy of Sciences, USA*, **92**, 175–9.

Bolhàr-Nordenkampf, H. R., Haumann, J., Lechner, E. G., Postl, W. F. & Schreier, V. (1993). Seasonal changes in photochemical capacity, quantum yield, P700-absorbance and carboxylation efficiency in needles from Norway spruce. In *Current Topics in Plant Physiology: Photosynthetic Responses to the Environment*, vol. 8, ed. H. Y. Yamamoto & C. M. Smith, pp. 193–200. An American Society of Plant Physiologists Series.

Bottomley, W., Spencer, D. & Whitfeld, P. R. (1974). Protein synthesis in isolated chloroplasts: comparison of light-driven and ATP-driven synthesis. *Archives of Biochemistry and Biophysics*, **164**, 106–17.

Critchley, C., Russell, A. W. & Bolhàr-Nordenkampf, H. R. (1992). Rapid protein turnover in photosystem II. Fundamental flaw or regulatory mechanism? *Photosynthetica*, **27**, 183–90.

Demmig-Adams, B. & Adams III, W. W. (1992). Photoprotection and other responses of plants to high light stress. *Annual Review of Plant Physiology and Plant Molecular Biology*, **43**, 599–626.

Demmig-Adams, B & Adams III, W. (1996). The role of xanthophyll cycle carotenoids in the protection of photosynthesis. *Trends in Plant Science*, **1**, 21–6.

Geiken, B., Critchley, C. & Renger, G. (1992). The turnover of photosystem II reaction centre proteins as a function of light. In *Research in Photosynthesis*, vol. IV, ed. N. Murata, pp. 634–46. Dordrecht: Kluwer Academic Publishers.

Horton, P. & Ruban, A. (1994). The role of light-harvesting complex II in energy quenching. In *Photoinhibition of Photosynthesis – From Molecular Mechanisms to the Field*, ed. N. R. Baker & J. R. Bowyer, pp. 111–28. Oxford: Bios Scientific Publishers Ltd.

Keren, N., Gong, H. & Ohad, I. (1995). Oscillations of reaction centre II-D1 protein degradation *in vivo* induced by repetitive light flashes. Correlation between the level of RCII-$Q_B$- and protein degradation in low light. *Journal of Biological Chemistry*, **270**, 806–14.

Nishio, J. N., Sun, J. & Vogelmann, T. C. (1994). Photoinhibition and the light environment within leaves. In *Photoinhibition of Photosynthesis – From Molecular Mechanisms to the Field*, ed. N. R. Baker & J. R. Bowyer, pp. 221–37. Oxford: Bios Scientific Publishers Ltd.

Osmond, C. B. (1994). What is photoinhibition? Some insights from comparisons of shade and sun plants. In *Photoinhibition of Photosynthesis – From Molecular Mechanisms to the Field*, ed. N. R. Baker & J. R. Bowyer, pp. 1–24. Oxford: Blos Scientific Publishers Ltd.

Ruban, A.V. & Horton, P. (1995). An investigation of the sustained component of nonphotochemical quenching of chlorophyll fluorescence in isolated chloroplasts and leaves of spinach. *Plant Physiology*, **108**, 721–6.

Russell, A. W., Critchley, C., Robinson, S. A., Franklin, L. A., Seaton, G. G. R., Chow, W. S., Anderson, J. M. & Osmond, C. B. (1995). Photosystem II regulation and dynamics of the chloroplast D1 protein in *Arabidopsis* leaves during photosynthesis and photoinhibition. *Plant Physiology*, **107**, 943–52.

Sundby, C., McCaffery, S. & Anderson, J. M. (1993). Turnover of the photosystem II D1 protein in higher plants under photoinhibitory and nonphotoinhibitory irradiance. *Journal of Biological Chemistry*, **268**, 25476–82.

Syme, A. J., Bolhàr-Nordenkampf, H. R. & Critchley, C. (1992). Light-induced D1 protein degradation and photosynthesis in sun and shade leaves. In *Research in Photosynthesis*, vol. IV, ed. N. Murata, pp. 337–40. Dordrecht: Kluwer Academic Publishers.

van Wijk, K. J., Bingsmark, S., Aro, E -M. & Andersson, B. (1995). *In vitro* synthesis and assembly of photosystem II core proteins. *Journal of Biological Chemistry*, **270**, 25685–95.

Vasilikiotis, C. & Melis, A. (1994). Photosystem-II reaction center damage and repair cycle – chloroplast acclimation strategy to irradiance stress. *Proceedings of the National Academy of Sciences, USA*, **91**, 7222–6.

Walker, D. (1992). Tansley review No. 36: excited leaves. *New Phytologist*, **121**, 325–45.

Weis, E. & Berry, J. A. (1987). Quantum efficiency of Photosystem II in relation to 'energy'-dependent quenching of chlorophyll fluorescence. *Biochimica et Biophysica Acta*, **894**, 198–208.

# 21 Photosynthesis, respiration and global climate change

B. G. DRAKE, J. JACOB AND M. A. GONZÀLEZ-MELER

## INTRODUCTION

Atmospheric carbon dioxide concentration ($C_a$) has been increasing for the past 150 years from a level of about 270 ppm where it remained for the entire history of human civilization to the present concentration of nearly 365 ppm. Deforestation, the major source for the increase in $C_a$ in the nineteenth century, currently accounts for about a quarter to a third of the total anthropogenic emission. Since the end of World War II, the rapid expansion in the use of fossil fuels and nitrogen fertilizers has resulted in a 30% increase. The most certain aspect of the anthropogenic effect on the composition of the atmosphere is that $C_a$ continues to rise and that a future world with high $C_a$ is inevitable.

For the climate, both the level and the rapid rise in $C_a$ are significant. Climatologists worry that this rapid rise will disrupt the climate system. The occurrence of higher temperatures seems to be an article of faith in the scientific community. Understanding the mechanisms regulating carbon flow in ecosystems in response to rising $C_a$ and climate change is therefore essential for predicting future levels of $C_a$.

Carbon enters the food chain of the biosphere through photosynthesis and it is this process by which plants sense global climate change. The expectation that rising $C_a$ and rising global temperature will be accompanied by increasing production of plant biomass is based on the well-known responses of photosynthesis, growth and water use efficiency to elevated $C_a$ concentration (Long & Drake, 1992). The primary physiological mechanisms responsible for these responses are the stimulation of photosynthesis, reduction of stomatal aperture and the inhibition of mitochondrial respiration. In this chapter, we will consider the direct effects of rising $C_a$ and temperature on photosynthesis, the modification of this effect by acclimation of the process to high $C_a$, and the consequences of acclimation for other processes including mitochondrial respiration, nitrogen use, and ecosystem carbon balance.

## DIRECT EFFECTS OF $C_a$ ON PHOTOSYNTHESIS

Plants sense and respond to $C_a$ primarily through photosynthesis, because rubisco, the primary carboxylase enzyme in $C_3$ plants, responds directly to alterations in $C_a$ (Bowes, 1993; Stitt, 1991). At the current levels of $C_a$, rubisco is normally operating at sub-saturating levels of leaf internal space concentration of $CO_2$ ($C_i$) in $C_3$ plants. This enzyme is also an oxygenase catalysing a reaction between RuBP and $O_2$ leading to an eventual loss of carbon through photorespiration (Long & Drake, 1992). An increase in $C_a$ will increase $C_i$ in the vicinity of Rubisco, increasing carboxylation while reducing oxygenation of RuBP. These effects on the primary carboxylating enzyme will produce a higher rate of photosynthetic $CO_2$ assimilation (A). Long and Drake (1992) discuss these effects in detail.

The interaction between $CO_2$ and temperature is a consequence of the properties of rubisco. As temperature increases, the affinity of the enzyme for $CO_2$ declines, while the affinity for $O_2$ remains more or less constant. This effect reduces the specificity ratio ($V_{cmax}\ K_c/V_{omax}\ K_o$) 86% between 0 and 50 °C. The solubility of $CO_2$ also declines by about 32% over this temperature range. Since $CO_2$ competes with $O_2$ for the same sites on the enzyme, the relative stimulation by a unit increase in the

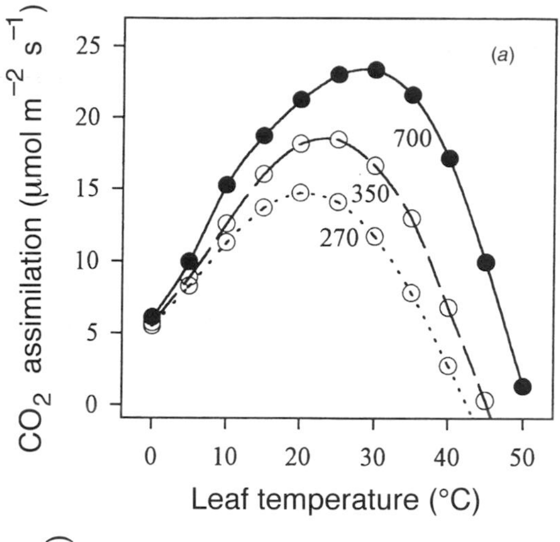


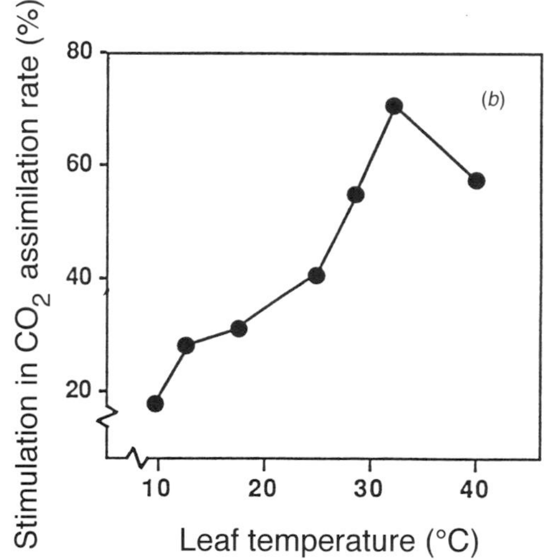


**Figure 21.1.** (*a*) The effect of $C_a$ on the rate of photosynthesis in a leaf as a function of temperature as predicted by a model for leaf photosynthesis. Data redrawn from Figure 4.2 of Long and Drake (1992). (*b*) The temperature dependence of the stimulation of photosynthesis by elevated $C_a$ (700 ppm) from experiments with the $C_3$ sedge, *Scirpus olneyi*.

concentration of $CO_2$ also increases with temperature (Fig. 21.1). The properties of photosynthesis are sufficiently well understood that they can be modelled (Long & Drake, 1992). In Fig. 21.1(*a*), the modelled response of photosynthesis to different levels of $C_a$ as a function of temperature is illustrated. In Fig. 21.1(*b*), the relative stimulation of photosynthesis by a doubling of normal ambient concentration of $C_a$ in *Scirpus olneyi* is shown for comparison with the modelled responses.

One of the major results of this effect is that the temperature optimum increases from near 20 °C at 270 ppm to near 30 °C at 700 ppm. For most species growing in a $C_a$ of twice the present level, an increase in temperature of 1–4 °C would increase photosynthesis and the added stimulation by the elevated $C_a$ contributes another large fraction of the increase.

Will these effects occur in plants exposed to elevated $C_a$ in low light? Long and Drake (1992) point out that elevated $C_a$ increases quantum yield and this increase would be greatest at high temperatures. In addition, the high $C_a$ will reduce the light compensation point. A doubling of the $C_a$ concentration would cause the LCP to decline by about 40% at 25 °C.

Three points can be made which summarize the direct effects of elevated $C_a$ on photosynthesis. (i) Photosynthesis is markedly increased by a doubling of the present ambient level of $C_a$. (ii) This is true in both light-limiting and light-saturating conditions. (iii) Rising temperature will enhance, rather than offset, this effect.

## ACCLIMATION OF PHOTOSYNTHESIS TO ELEVATED $CO_2$

Plants acclimate to a wide range of many environmental conditions. Would they not adjust their responses to an increase in $C_a$? If so, how will this alter the basic patterns in the responses? It has long been noted that, if a plant grown in normal ambient $C_a$ is transferred to a higher concentration, adjustments to elevated $C_a$ occur through reduction in photosynthetic capacity. This is manifested as a reduction in the photosynthetic capacity at high $C_a$ as well as a reduction in the carboxylation efficiency of the leaf (obtained from the initial slope of the $A/C_i$ relationship, e.g. Jacob, Greitner & Drake, 1995). All combinations of changes in the initial slope and maximum capacity have been observed. Sometimes, there is no downward adjustment or even an increase in capacity (Arp, 1991; for review see Long & Drake, 1992). In order to model the effects of rising $C_a$ and temperature on photosynthesis, it is crucial to understand the basic patterns in the acclimation response. This need has stimulated considerable work to clarify the mechanism for these responses, what elicits them, how the response interacts with environmental conditions and which species are subject to them. Acclimation often results in a reduction in photosynthetic capacity. This response is often observed in plants grown in small pots or under conditions in which the supply of some critical resource, such as nitrogen, limits growth (Arp, 1991; Long & Drake, 1992).

Acclimation responses may be thought of in terms of the balance between the activity of the source, i.e. the capacity of photosynthesis, and the strength of the sink, which in this case is meant all of the possible uses to which additional carbon may be put in the plant. One of the most commonly observed responses to elevated $C_a$ is an accumulation of carbohydrate and starch during the day. This may persist through the night or the levels may be similar in plants from both high and low $C_a$. Long and Drake (1992) indicate that, under growth-limiting conditions (i.e. small rooting volume) in 16 studies, starch increased an average of more than three times while sucrose increased about 1.5 times the concentration in plants grown in normal ambient $C_a$. In addition to the significant increase in the concentration of soluble carbohydrates and starch, there was an average reduction in rubisco activity of about 20% (four studies, six species).

Many of these studies included crop plants grown in well-watered, high-nutrient culture conditions. Jacob *et al.* (1995) present data for acclimation of photosynthesis of a native species, the $C_3$ sedge, *Scirpus olneyi*, which had been exposed to twice ambient $C_a$ for six years in its native habitat. Here, the plant had limited nitrogen available. Table 21.1 summarizes the responses of acclimation of photosynthesis (measured on excised stems) and carbon reduction. Increased concentration of carbohydrates and a decline in soluble protein and rubisco were accompanied by a reduction in the rate of photosynthesis measured at normal ambient $C_a$ of 46%. Nevertheless, when compared at their respective growth concentrations, plants in elevated $C_a$ had higher rates of photosynthesis by about 50%. And, when measurements were made on intact stems in the field, the rates of photosynthesis were 100% greater in plants grown and tested in elevated $C_a$ suggesting that elevated $C_a$ relieves stress and that this acts to increase the relative effects of

Table 21.1. *Acclimation of photosynthesis, respiration and tissue composition in* Scirpus olneyi *exposed to elevated $C_a$ in open top chambers in the native environment.*

| | A (350 ppm) | E (680 ppm) | (E-A)/A % |
|---|---|---|---|
| CE | 0.14 | 0.11 | −21.4 |
| $A_o$ at growth $C_a$ | 18.0 | 27.5 | 52.7 |
| A at 350 ppm | 18.0 | 9.7 | −46.0 |
| Soluble protein | 2.1 | 1.3 | −38.0 |
| Rubisco | 0.95 | 0.55 | −42.1 |
| Sugars | 19.7 | 29.8 | 51.3 |
| Starch | 51.3 | 79.0 | 54.0 |
| $R_d$ | 2.39 | 1.52 | −36.4 |
| LCP | 47 | 19 | −59.6 |

Data on photosynthesis and tissue composition are from Jacob *et al.*, 1995, Fig. 2, and Table 1, 1995. Respiration data from Gonzàlez-Meler, 1995. Carboxylation efficiency (CE, mol $m^{-2}$ $s^{-1}$); Assimilation (A) and respiration ($R_d$) at the growth $C_a$ ($A_o$,$R_d$; μmol $m^{-2}$ $s^{-1}$), total soluble protein, rubisco, sugars and starch (g $m^{-2}$). All differences significant at $P<0.05$ except in soluble protein and rubisco ($P<0.07$).

$C_a$ (Jacob *et al.*, 1995). Long and Drake (1992) reviewed 11 studies utilizing 20 species and the rate of photosynthesis in elevated $C_a$ was 50% higher than in normal ambient $C_a$. This result is similar to that found by several other reviews (Arp, 1991; Bowes, 1993). Modelled responses of photosynthesis show that, above 25 °C, even with a reduction of 30% in the rubisco activity, there is an increase in the rate of photosynthesis in doubled $C_a$ compared with normal ambient (Long & Drake, 1992).

Stitt (1991) proposed that accumulation of nonstructural carbohydrates was the signal to which the plant in elevated $C_a$ responds. Biochemical and genetic approaches have been used to demonstrate how accumulation of certain nonstructural carbohydrates can mediate acclimation. In tomato plants exposed to high $C_a$, the concentrations of glucose, sucrose and fructose increased and the level of total *rbcS* mRNA decreased compared with their ambient $C_a$ grown counterparts. Thus end product (carbohydrate) accumulation can mediate the expression of photosynthetic genes, and this may represent a basic mechanism for acclimation of photosynthesis to elevated $C_a$. Accumulation of carbohydrates in transgenic tobacco plants (Stitt, von Schaewen & Willmitzer, 1991) and feeding glucose to excised spinach leaves decreased the activities of rubisco, chloroplastic fructose 1,6-bisphosphatase and NADP-glyceraldehyde 3-phosphate dehydrogenase.

The basic problem for the plant is to adjust all aspects of carbon metabolism to the availability of photosynthates. The rate of photosynthesis is regulated in the long term of days to weeks by adjusting the level of Rubisco, an enzyme which is so abundant (possibly the most abundant protein in the biosphere) that it may require as much as 25% of nitrogen in the leaf. If there is sufficient sink capacity, as may be the case for a crop plant, such as wheat, with a large capacity to create new tillers, and to exploit the availability of nutrient, there may be little need to reduce the complement of rubisco because there is sufficient sink to accommodate all additional carbohydrate. Nie *et al.* (1995) report little effect of $C_a$ on the acclimation of wheat exposed to elevated $C_a$ in a free air carbon enrichment (FACE) study in Arizona. Only late in the season, and in combination with a water-stress treatment, was there any indication of reduction of the photosynthetic capacity and rubisco. Rowland-Bamford *et al.* (1991) show that the reduction in rubisco across a wide range of $C_a$ treatments occurs in concert with the rubisco activity in the leaf.

The increase in carbohydrates is also accompanied by a decrease in soluble protein and rubisco. Rowland-Bamford *et al.* (1991) reported decreased activity and content of rubisco in rice exposed to a range of ambient $C_a$ concentrations between 190 and 900 ppm. But, this response is complex because others report no decrease in the activity of rubisco in elevated $C_a$. Often, the reduction in rubisco occurs under nitrogen-limiting conditions, suggesting that the enzyme may be a pool of nitrogen which the plant can adjust depending upon need. There are also wide variations in the effects of age and species on acclimation of the concentration of rubisco. In any case, there are enough exceptions to make us cautious about assigning a rule to the observation that high $C_a$ often reduces rubisco.

While there is considerable evidence that acclimation results in major changes in the carbon metabolism, there is very little information concerning the possible reduction in RuBP regeneration and the capacity for electron transport. Webber, Nie and Long (1994) proposed two hypothetical scenarios of acclimation of photosynthesis in plants that are not sink limited but are limited by N. (i) Re-allocation of N away from rubisco into RuBP regeneration apparatus of the Calvin cycle and proteins of thylakoid at the expense of rubisco. (ii) Re-allocation of N from rubisco into growth of other organs that act as sinks. Overall, these two scenarios are applicable to native species which have large sink capacity (roots, wood, etc) and are N limited. But, while it is possible that N could be reallocated from rubisco to more limiting processes in order to optimize its use, there is no evidence that such a reallocation actually occurs in native species acclimated to elevated $C_a$ in the field.

In summary, elevated $C_a$ often results in acclimation of the rate of photosynthesis accompanied by a reduction in the concentration of rubisco. The reduction in the protein content of plants in elevated $C_a$, sufficient to reduce total leaf nitrogen by about 20%, is a result of the loss of protein. The reduction in tissue N is a very important aspect of the response of plants to elevated $C_a$ because it changes the maintenance components of respiration, the palatability for insects and the tissue quality for decomposition by microbes. All of these have major effects on ecosystem level responses to rising $C_a$ and may have much to do with the regulation of ecosystem carbon balance (discussed later). While acclimation results in a reduction in the photosynthetic capacity in many cases, the evidence that plants grown in elevated $C_a$ have higher rates of photosynthesis than those grown in normal ambient $C_a$ is overwhelming.

## ELEVATED $CO_2$ AND STOMATA

The short-term response of stomata to elevated $C_a$ is a well-described phenomenon. Moreover, there appears to be little, if any, adjustment of the basic response pattern after long-term exposure of plants to elevated $C_a$. However, plants grown in a greenhouse in high humidity showed little responsiveness of stomata to $CO_2$ until given an exposure to a stress; chilling or even a single drying cycle is sufficient to induce the response to $C_a$. The effects of elevated $C_a$ on stomata reduces transpiration and increases water-use efficiency. This response occurs irrespective of whether or not the plants were grown in conditions which resulted in acclimation of photosynthesis (Long & Drake, 1992). In fact, the reduction in stomatal conductance was greatest in plants in which there were growth limitations and acclimation of photosynthesis. Thus, there are certain to be effects of rising $C_a$ irrespective of whether or not there is acclimation of photosynthesis. Long and Drake (1992) point out that, in order for the plant to maintain an internal $CO_2$ concentration of approximately 70% of the external $C_a$ concentration, there would have to be a reduction of stomatal conductance of about 40% for a doubling of $C_a$. Because of the increase in leaf area of plants grown in elevated $C_a$, canopies exposed to elevated $C_a$ through the season show that the overall reduction of evapotranspiration is somewhat less than this. There are a number of reports of a reduction in evapotranspiration of both $C_3$ and $C_4$ canopies averaged over the season of about 20%. The effects of elevated $C_a$ on stomata do not translate directly to proportional reductions in transpiration or evapotranspiration because there are complex feedbacks between the effects of $C_a$ on leaf temperature and, when considering long-term effects, on regional humidity. Nevertheless, and contrary to speculations based on theoretical analyses of water and energy balance, gas flux measurements and soil water measurements in ecosystems exposed to elevated $C_a$ show that there are real savings in water. These results must be considered preliminary since it is thought that the microenvironment of the open top chamber in which most of these measurements have been made is sufficiently different from the real environment of plants that no solid conclusions on this question are yet possible.

A very interesting effect of the improved water balance of plants grown in elevated $C_a$ is the relief of water stress which has secondary effects on growth, photosynthesis and respiration. If this leads to improved soil water balance, then there are other benefits for soil processes such as improved nutrient cycling.

In summary, improved water balance has a number of consequences for plants exposed to environmental stress including relief of dehydration, improved salt tolerance, protection from exposure to toxic atmospheric pollutants, etc. Some of the effects of elevated $C_a$ on primary production in native species may be due to the effects on stomatal conductance which is reduced 20–40% in plants grown in elevated $C_a$. These effects will occur whether or not there is acclimation of photosynthesis or other processes to elevated $C_a$.

## RESPONSES OF MITOCHONDRIAL RESPIRATION TO ELEVATED $CO_2$

It would be expected that respiration would increase in elevated $C_a$ as a result of increased photosynthesis and the higher growth implied by this and, at least in the short term, as a result of increased carbohydrate content as reported, for example, by Azcón-Bieto and Osmond (1983). But, it has been known for many years that plant respiration is reduced by exposure to high concentrations of $C_a$ (Kidd, 1916; Nilovskaya & Razoryonova, 1968). Thus, it must be said that the inhibition of respiration by elevated $C_a$ has recently been rediscovered (Reuveni & Gale, 1985; Ziska & Bunce, 1993, 1994; Azcón-Bieto *et al.* 1994). The effects of rising $C_a$ and climate change on respiration have been reviewed many times (Amthor, 1991; Wullschleger, Ziska & Bunce, 1994). Although each reviewer emphasizes the potential for direct effects of $CO_2$ on different aspects of plant physiology, collectively they indicate the considerable potential for rising $C_a$ to reduce the respiration of vegetation. Two classes of effects can be distinguished experimentally: a direct effect, in which the rate of respiration is rapidly and reversibly reduced following a rapid increase in $C_a$ (Amthor, Koch & Bloom, 1992), and a reduction of the rate of respiration in tissues from plants grown in elevated $C_a$ compared with the rate in those grown in normal ambient $C_a$ when both are tested at a common background of $C_a$ (Azcón-Bieto *et al.*, 1994). Some experimental evidence suggests that inhibition of respiration of foliage is sufficient to increase plant growth over and above what might be expected strictly from the stimulation of photosynthesis (Reuveni & Gale, 1985).

At the most fundamental level, respiration can be considered to have a dual nature: it is the source of metabolic intermediates used in the synthesis of cellular constituents as well as the source of ATP and reduced nucleotide (particularly pyridine nucleotide), the driving force of these syntheses. A variety of substrate (carbohydrates, lipids, and proteins) can be respired through pathways occurring in several cell compartments (e.g. cytosol, mitochondria, chloroplast) with the products converging on the tricarboxylic acid cycle and hence on mitochondrial oxidative phosphorylation. This duality, source of substrate and source of the energy to drive the reactions, is probably reflected in the regulation of the rate by respiratory substrate and by ADP availability, although it remains a goal of plant physiologists to establish fully the mechanism behind this control.

Phytomass and carbohydrate content generally increase in $C_3$ plants grown in elevated $C_a$ levels as a result of increased photosynthetic activity. In some studies, particularly those using young, rapidly growing tissues, increased dark respiration has been shown to accompany the stimulation of photosynthesis by elevated $C_a$ (Azcón-Bieto & Osmond, 1983) and, in these cases, it seems likely that the use of young, rapidly expanding leaves complicates the interpretation of the results. Many studies clearly indicate that long-term acclimation of plants to elevated $C_a$ significantly reduces dark respiration rates (see above references). Mechanisms for this effect have been sought in various aspects of plant growth. The primary effect seems to be the short-term effect as it appears to alter the kinetics of the responses of several key enzymes of mitochondrial respiration, including cytochrome *c* oxidase, and is therefore likely to be present irrespective of any acclimation which may take place in the plant after long-term growth in elevated $C_a$ (Fig. 21.2; Table 21.2).

Although little effort has been made to distinguish between the direct, reversible effect of elevated $C_a$ from the long-term residual effect, this distinction almost certainly points to different aspects of the phenomenon. Response to rapid changes in $C_a$ has been observed. In these 'short-term effects', the rate of $CO_2$ emission from the plant tissue is decreased within minutes after $C_a$ was increased and is reversed upon returning the $C_a$ to a lower level (Fig. 21.2). Short-term effects have been reported by Kidd (1916), Nilovskaya and Razoryonova (1968), Reuveni and Gale (1985) and Amthor *et al.*, (1992).

A plausible mechanism underlying the short-term effect is the inhibition of activity of enzymes of the mitochondrial electron transport system. It has been shown that elevated $C_a$ reduces the activity of cytochrome *c* oxidase and succinate dehydrogenase (Gonzàlez-Meler, 1995). Recently, it has been shown that increasing the $C_a$ in solution, through addition of bicarbonate to reaction media sufficiently to be equivalent to a doubling of the present atmospheric $C_a$, reduces *in vivo* activity of cytochrome *c* oxidase from beef-heart (Table 21.2), from soybean mitochondria, and in crude extracts from the sedge *Scirpus olneyi* (Gonzàlez-Meler, 1995).

Acclimation of respiration to elevated $C_a$ can be measured as reduction in $CO_2$ emission (or $O_2$ assimilation) from tissues at a common $C_a$ background (Fig. 21.2). This class of effects involves secondary effects of elevated $C_a$ on tissue composition, especially carbohydrate concentration and/or protein concentration (which is reflected in the tissue nitrogen concentration). But the reduction in tissue nitrogen, through reduction of soluble proteins (mainly rubisco) explains only part of the $CO_2$-dependent variations in respiratory rates of photosynthetic tissues of *Scirpus olneyi* and *Lindera benzoin* (Azcón-Bieto *et al.*, 1994). Azcón-Bieto *et al.* (1994) concluded that 'inhibition of the rate of respiration in *Scirpus olneyi* and *Lindera benzoin* was shown to be due largely to reduction in some enzymatic complexes of the mitochondrial electron transport chain' (cytochrome *c* oxidase and complex III), resulting in the reduction in the capacity of tissue respiration. Long-term effects have been reported by Bunce (1990), Reuveni and Gale (1985) and Wullschleger, Norby and Gunderson (1992).

In summary, there now appears to be evidence that elevated $C_a$ reduces dark respiration in foliage and possibly in other plant tissues and microbes. Doubling of present $C_a$ leads to a significant reduction in dark respiration of 20–30% of the rate obtained for tissues grown in present normal ambient $C_a$. The primary effect is immediate and reversible although this direct effect of elevated $C_a$ appears to be modulated by changes in tissue composition in some species through reduction of tissue nitrogen or increased tissue carbohydrate concentrations. Direct or acclimation effects of elevated $C_a$ on the alternative respiratory pathway have not been demonstrated and so we assume that this pathway is not involved directly in the overall inhibition of respiration. There is some evidence that this direct effect of elevated $C_a$ on respiration extends to microbes involved in decomposition. A mechanism that will explain the direct effect of $C_a$ on respiration includes the inhibition of enzyme activity. The candidates for possible attention in

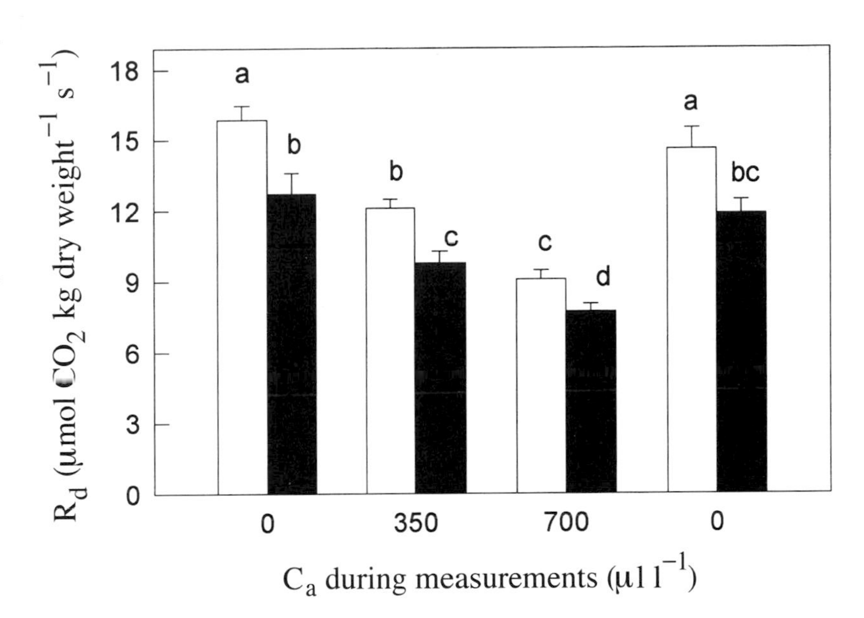


**Figure 21.2.** Respiration of photosynthetic tissues of the $C_3$ sedge *Scirpus olneyi* measured at different concentrations of atmospheric $C_a$ for plants grown in ambient $C_a$ (open bars) or in elevated $C_a$ (closed bars) in open top chambers in the field after 7 years of exposure to elevated $C_a$ (data from Gonzàlez-Meler, 1995). Measurements were done at 25 °C in closed cuvettes and wcrc started with $CO_2$-free air (for experimental protocols see Gonzàlez-Meler, 1995). a, b, c and d indicate significant differences of the means ($p$<0.05).

Table 21.2. *The effect of increasing $C_a$ by the equivalent of twice normal ambient $C_a$ on the activity of bovine cytochrome* c *oxidase*

| $C_a$ | $O_2$ uptake (nmol mg$^{-1}$ prot min$^{-1}$) |
|---|---|
| 360 | 379 (10) |
| 680 | 267 (13) |
| Inhibition % | 29 |

The enzyme was incubated for 10 minutes with or without increased $C_a$ at pH 7.2 as described in Gonzàlez-Meler, 1995. Values are mean and standard errors for six replicates.

this regard are the succinate dehydrogenase and cytochrome *c* oxidase. This effect has the potential to greatly alter the expected effect of rising $C_a$ on the carbon cycle in terrestrial ecosystems.

## RISING $CO_2$ AND ECOSYSTEMS

Our field studies in three ecosystems during the past eight years have shown that elevated $C_a$ caused higher rates of photosynthesis and lower rates of respiration (Drake & Leadley, 1991; Azcón-Bieto *et al.*, 1994; Jacob *et al.* 1995; Gonzàlez-Meler, 1995; see also Table 21.1 and Fig. 21.2). Figure 21.3 and Table 21.3 demonstrate how these effects alter the carbon balance of a wetland ecosystem. Figure 21.3 shows how elevated $C_a$ increases the efficiency of daily photosynthetic carbon assimilation. These measurements were made using ecosystem gas exchange and incident photosynthetic photon flux (PPF) measured periodically throughout the season (Drake & Leadley, 1991). Then, using the function describing the daily photosynthetic efficiency (total carbon assimilation/total PPF) with the total PPF, we estimated total daily carbon uptake. Table 21.3 shows the annual sums for gross primary production (GPP), ecosystem respiration ($R_e$) and net ecosystem production (NEP). The accumulation of carbon and the direct effects of elevated $C_a$ on specific processes resulted in the following effects. Increased canopy photosynthesis, increased net primary production, reduced tissue nitrogen concentration, reduced dark respiration and increased soil carbon content in communities dominated by $C_3$ photosynthetic-type species. The increase in soil carbon stimulated soil microbial production of methane (Dacey, Drake & Klug,

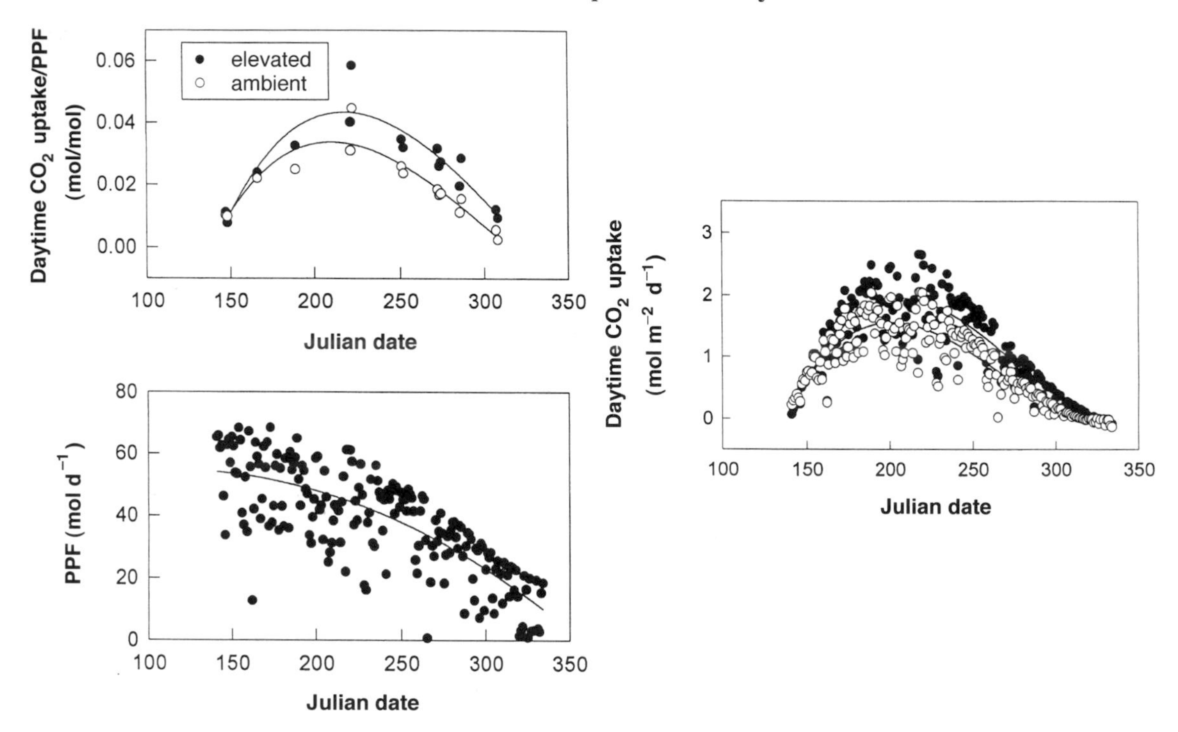


Figure 21.3. The effect of long-term exposure to elevated $C_a$ on ecosystem $CO_2$ flux in a $C_3$ wetland. Data for daytime $CO_2$ uptake (gross primary production, GPP) are in panel on the right. These data derived from the product of the data in the two panels on the left. Integrated values for total seasonal carbon assimilation determined from the sums of daily carbon uptake are given in Table 21.3 under GPP.

Table 21.3. *Annual ecosystem carbon fluxes over a wetland exposed to normal ambient (350 ppm) and elevated (680 ppm)* $C_a$ *for 8 years*

| $C_a$ | GPP | $R_e$ (kg C $m^{-2}$) | NEP |
|---|---|---|---|
| 350 | 1.89 | 0.44 | 1.45 |
| 680 | 2.45 | 0.28 | 2.17 |
| (E-A)/(A)% | 30 | −36 | 50 |

Data are for 1994 (kg C $m^{-2}$) in a community dominated by the $C_3$ sedge, *Scirpus olneyi*. Treatment differences for values of carbon fluxes, gross primary production (GPP), net ecosystem respiration ($R_e$), and net ecosystem production (NEP = GPP-$R_e$) significant at $p<0.05$.

1994), increased numbers of foraminifera and nematodes, increased nitrogen fixation, and increased soil respiration (unpublished data). Reduction in tissue N concentration (resulting from reduction of the protein concentration during acclimation of photosynthesis) caused increased C : N in plant tissues, increased N use efficiency and reduced insect grazing and the severity of fungal infection (Thompson & Drake, 1994). In each of the three ecosystems, the most important effects were a consequence of stimulation of $CO_2$ assimilation or a reduction in the tissue N concentration. Photosynthesis and acclimation of photosynthesis and respiration to rising atmospheric $C_a$ thus plays a pivotal role in ecosystem carbon balance.

## RISING $CO_2$ AND THE GLOBAL CARBON BUDGET

There is an analogous relationship between the role of physics in the climate system and the role of physiological processes in ecosystem functioning. In both cases, fundamental processes, about which we require a lot more basic science, will be used to model overall system responses to rising $C_a$. Nevertheless, models of the impacts of rising $C_a$ on physiological processes have been created and are being used to project these effects into the future. Used in conjunction with climate models, plant and ecosystem models are being employed to estimate the levels of future atmospheric $C_a$. Before projections can be made with certainty, these models must capture recent findings from research on fundamental aspects of physiological studies. The direct effects on photosynthesis are very well known but the mechanisms involved in acclimation of plants to higher $C_a$ concentrations are just now being elaborated. The effects of $C_a$ on stomatal aperture and on the inhibition of mitochondrial respiration are not understood well enough to model them.

We lack fundamental information on each of the key physiological processes. Our results do show that there are major responses of photosynthesis and respiration to $C_a$ and these are sufficiently large that they would have major consequences for global carbon budgets. Gross primary production, the amount of carbon assimilated by photosynthesis of the terrestrial biosphere, has been estimated to be between 100 and 125 GtC/yr (Gifford, 1994). Stimulation of global photosynthesis and inhibition of respiration of about 10% would accommodate all anthropogenic carbon. Estimates put the stimulation of photosynthesis and net ecosystem production by the present level of $C_a$ as being about 0.5–4.0 GtC/annum which is roughly equivalent to the loss of carbon by deforestation.

The consumption of fossil fuel and burning of wood products introduces about 6 GtC into the atmosphere annually. About half of this goes into the oceans and terrestrial ecosystems and about half remains in the atmosphere. With estimates of the various sources and sinks, we cannot balance the carbon account and there is somewhere between 0.5 and 4.3 GtC/yr unaccounted for (Gifford, 1994). This missing carbon is really a measure of our ignorance. In 1990, this amounted to about 2.0 GtC but in 1991, there was no increase in $C_a$ and so the missing carbon that year was about 5 GtC. Since we can only carry our predictions into the future with confidence if the budget balances in the short term, a gap in knowledge which places the missing carbon at almost equivalent to the total anthropogenic input is unacceptable. In order to predict future $C_a$, it is essential to understand how the biosphere responds to ever-increasing $C_a$.

## CONCLUSIONS AND OUTLOOK FOR THE FUTURE

Much research during the past ten years has shown that rising $C_a$ will improve the efficiency of key physiological processes including photosynthesis, respiration, and nitrogen and water use. The evidence that rising $C_a$ will stimulate photosynthesis and growth is overwhelming and there is also persuasive evidence that this effect translates into carbon accumulation in native ecosystems and improved yield of crops. The direct effect of elevated $C_a$ on photosynthesis and the acclimation of both photosynthesis and respiration to prolonged exposure to elevated $C_a$ are being intensively studied. But, the mechanism by which elevated $C_a$ is sensed and the details of acclimation are not sufficiently understood to be modelled. The effect of elevated $C_a$ on respiration is particularly important since it has the potential to influence the global carbon budget independent of the effects on photosynthesis.

### REFERENCES

Amthor, J. S. (1991). Respiration in a future, higher-$CO_2$ world: opinion. *Plant, Cell and Environment*, **14**, 13–20.

Amthor, J. S., Koch, G. W. & Bloom, A. J. (1992). $CO_2$ inhibits respiration in leaves of *Rumex crispus L. Plant Physiology*, **98**, 757–60.

Arp, W. J. (1991). Effects of source-sink relations on photosynthetic acclimation to elevated $CO_2$. *Plant, Cell and Environment*, **14**, 869–75.

Azcón-Bieto, J. & Osmond, C. B. (1983). Relationship between photosynthesis and respiration. The effect of carbohydrate status on the rate of $CO_2$ production by respiration in darkened and illuminated wheat leaves. *Plant Physiology*, **71**, 574–81.

Azcón-Bieto, J., Gonzàlez-Meler, M. A., Doherty, W. &

Drake, B. G. (1994). Acclimation of respiratory $O_2$ uptake in green tissues of field-grown native species after long-term exposure to elevated atmospheric $CO_2$. *Plant Physiology*, **106**, 1163–8.

Bowes, G. (1993). Facing the inevitable: plants and increasing atmospheric $CO_2$. *Annual Review Plant Physiology and Plant Molecular Biology*, **44**, 309–32.

Bunce, J. A. (1990). Short-term and long-term inhibition of respiratory carbon dioxide efflux by elevated carbon dioxide. *Annals of Botany*, **65**, 637–42.

Dacey, J. W. H., Drake, B. G. & Klug, M. J. (1994). Stimulation of methane emission by carbon dioxide enrichment of marsh vegetation. *Nature*, **370**, 47–9.

Drake, B. G. (1992). A field study of the effects of elevated $C_a$ on ecosystem processes in a Chesapeake Bay wetland. *Australian Journal of Botany*, **40**, 579–95.

Drake, B. G. & Leadley, P. W. (1991). Canopy photosynthesis of crops and native plant communities exposed to long-term elevated $CO_2$ treatment. *Plant, Cell and Environment*, **14**, 853–60.

Gifford, R. M. (1994). The global carbon cycle: a viewpoint on the missing sink. *Australian Journal of Plant Physiology*, **21**, 1–15.

Gonzàlez-Meler, M. A. (1995). Effect of increasing the atmospheric concentration of carbon dioxide on plant respiration. PhD thesis, Universitat de Barcelona, Barcelona, Spain.

Jacob, J., Greitner, C. & Drake, B. G. (1995). Acclimation of photosynthesis in relation to rubisco and non-structural carbohydrate contents and *in situ* carboxylase activity in *Scirpus olneyi* grown at elevated $CO_2$ in the field. *Plant, Cell and Environment*, **18**, 875–84.

Kidd, F. (1916). The controlling influence of carbon dioxide. Part III. The retarding effect of carbon dioxide on respiration. *Proceedings of the Royal Society of London, Series B*, **89**, 136–56.

Krapp, A., Quick, W. P. & Stitt, M. (1991). Ribulose-1,5-bisphosphate carboxylase-oxygenase, other Calvin cycle enzymes and chlorophyll decrease when glucose is supplied to mature spinach leaves via transpiration stream. *Planta*, **186**, 58–69.

Long, S. P. & Drake, B. G. (1991). The effect of the long-term $CO_2$ fertilization in the field on the quantum yield of photosynthesis in the $C_3$ sedge *Scirpus olneyi*. *Plant Physiology*, **96**, 221–6.

Long, S. P. & Drake, B. G. (1992). Photosynthetic $CO_2$ assimilation and rising atmospheric $CO_2$ concentrations. In *Crop Photosynthesis: Spatial and Temporal Determinants*, ed. N. R. Baker & H. Thomas, pp. 69–103. Amsterdam: Elsevier Science Publishers BV.

Nie, G. Y., Long, S. P., Kimball, B. A., Lamorte, R. A., Pinter, P. A., Walls, G. W. A. & Webber, A. N. (1995). Effects of free-air $CO_2$ enrichment on the development of the photosynthetic apparatus in wheat, as indicated by changes in leaf proteins. *Plant, Cell and Environment*, **18**, 855–64.

Nilovskaya, N. T. & Razoryonova, T. A. (1968). Respiration rate of vegetable plants at various partial pressures of carbon dioxide. *Fiziologiya Rastenig*, **5**, 873–6.

Reuveni, J. & Gale, J. (1985). The effect of high levels of carbon dioxide on dark respiration and growth of plants. *Plant, Cell and Environment*, **8**, 623–8.

Rowland-Bamford, A. J., Baker, J. T., Allen, L. H. Jr & Bowes, G. (1991). Acclimation of rice to changing atmospheric carbon dioxide concentration. *Plant, Cell and Environment*, **14**, 577–83.

Stitt, M. (1991). Rising $CO_2$ levels and their potential significance for carbon flow in photosynthetic cells. *Plant, Cell and Environment*, **14**, 741–62.

Stitt, M., von Schaewen, A. & Willmitzer, L. (1991). Sink regulation of photosynthetic metabolism in transgenic tobacco plants expressing yeast derived invertase in their cell wall involves a decrease of the Calvin cycle enzymes and an increase in glycolytic enzymes. *Planta*, **183**, 40–50.

Thompson, G. & Drake, B. G. (1994). Insects and fungi on a $C_3$ sedge and a $C_4$ grass exposed to elevated atmospheric $CO_2$ concentrations in open top chambers in the field. *Plant, Cell and Environment*, **17**, 1161–7.

Webber, A. N., Nie, G.-Y. & Long, S. P. (1994). Acclimation of photosynthetic proteins to rising $CO_2$. *Photosynthesis Research*, **39**, 413–25.

Wullschleger, S. D., Norby, R. J. & Gunderson, C. A. (1992). Growth and maintenance respiration in leaves of *Liriodendron tulipifera* L. saplings exposed to long-term carbon dioxide enrichment in the field. *New Phytologist*, **121**, 515–23.

Wullschleger, S. D., Ziska, L. H. & Bunce, J. A. (1994). Respiratory responses of higher plants to atmospheric $CO_2$ enrichment. *Physiologia Plantarum*, **90**, 221–9.

Ziska, L. H. & Bunce, J. A. (1993). Inhibition of whole plant respiration by elevated $CO_2$ as modified by growth temperature. *Physiologia plantarum*, **87**, 459–66.

Ziska, L. H. & Bunce, J. A. (1994). Direct and indirect inhibition of single leaf respiration by elevated $CO_2$ concentrations: interaction with temperature. *Physiologia Plantarum*, **90**, 130–8.

PART IV

# Special topics and applications

# 22 Evolution

W. NITSCHKE, U. MÜHLENHOFF AND U. LIEBL

## THE MAJOR HISTORICAL LANDMARKS

All data presently available suggest that photosynthesis is a rather ancient principle of energy conversion. In former years, several authors therefore, suggested a close if not indispensable link between photosynthesis and the origin of life, i.e. the so-called 'photosynthesis-first'-hypothesis. New insights gathered during recent years, however, suggest that photosynthesis actually represents a 'relatively' late invention of life in the sense that ancient organisms thrived already quite happily using well-developed non-photosynthetic ways of energy conversion when photosynthesis was added to the stock of strategies for survival. Although photosynthesis was therefore, probably not rocking the cradle of life, it is doubtlessly a major (if not the most important) innovation during the evolution of organisms. A specific form of photosynthetic energy conversion, oxygenic photosynthesis, even managed to induce global changes in earth's atmosphere radically altering living conditions on this planet. Today, photosynthetic organisms represent the vast majority of world's biomass not only in the form of plants but also of phototrophic microorganisms, such as algae and cyanobacteria.

Some of the presently known historical landmarks of life and more specifically photosynthetic life on earth are depicted in Fig. 22.1. There is common agreement on the fact that, immediately after the formation of the earth about 4.5 Gyrs ago, continuous meteorite bombardment resulted in frequent atmosphere heating and ocean boiling events thereby creating at times, at least locally, rather hostile environments to the evolution of organisms. Geological evidence dates the end of this bombardment to about 3.8 Gyrs ago. Roughly at the same time, first evidence for biological carbon fixation, i.e. for life, is provided by isotope compositions in Archaean rock. Photosynthetic microorganisms clearly inhabited the earth about 1 Gyr later. The fossil records show the presence of oxygen-evolving cyanobacteria at about 2.8 Gyrs ago. Cyanobacterial-type microfossils have even been found as early as −3.5 Gyrs, i.e. only 0.3 Gyrs after the first records of life itself. However, the question whether these organisms were really phototrophs is still under debate.

By about 2.5 Gyrs ago, the cyanobacteria had proven the most successful organisms on earth by becoming the dominant form of life on this planet. Concomitantly, their oxygenic mode of photosynthesis initiated the first 'life-made' global climate catastrophe in earth's history. Whereas in earth's primordial atmosphere, oxygen was present at best in trace amounts, cyanobacterial oxygen evolution did away with this relatively reducing environment by increasing the oxygen content of the atmosphere to roughly one-fourth of the present-day value in the span of about 1 Gyr (see Fig. 22.1).

The great success of the cyanobacteria is certainly only in part due to the 'superiority' of a photosynthetic metabolism. The global changes in oxygen content of the atmosphere brought about by cyanobacterial photosynthesis constitute an aggressive element in the competition between organisms existing at that time since oxygen was almost certainly extremely toxic to the vast majority of species on earth. One might even be tempted to speculate that the invention of oxygen evolution in photosynthesis was driven less by the search for an ubiquitous electron donor than by the intention to create a powerful toxic agent in order to get rid of most competitors in one shot.

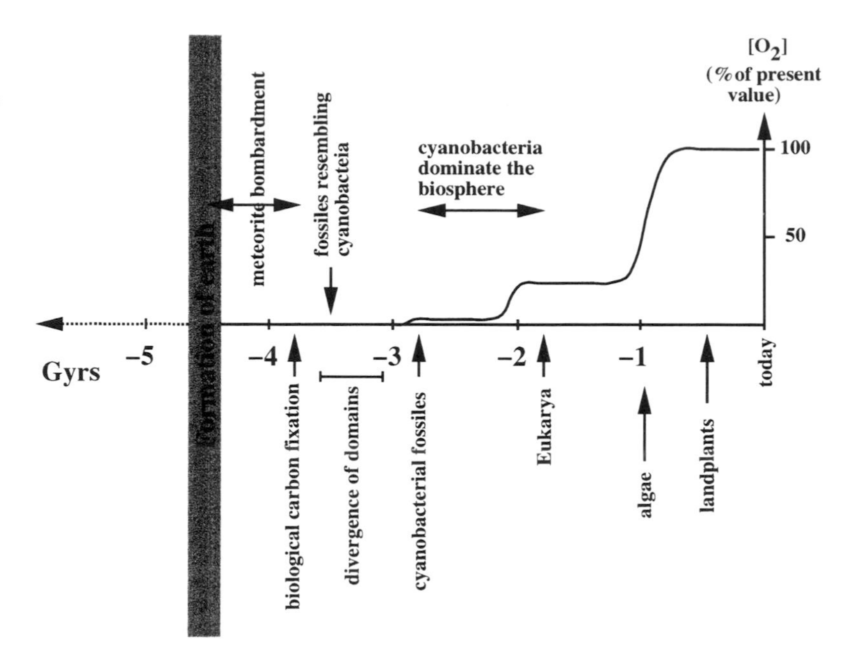


**Figure 22.1.** Schematic representation of the major geological and evolutionary events having an impact on the origin and evolution of phototrophic organisms on earth.

At about 1.8 Gyrs ago, the first Eukarya appeared on our planet and it took less than one Gyr to convince many Eukarya of the benefits provided by phototrophic metabolism. Hence, at about 1 Gyr ago, eukaryotic organisms, i.e. the algae, had begun to 'ingest' the principle of photosynthesis via endosymbiotic uptake of photosynthetic Bacteria. The descendants of these algae, the landplants, appeared on the earth's surface at 0.5 Gyrs ago. The combined efforts of the algae and the land plants apparently further lifted the oxygen concentration of the atmosphere up to its present-day level.

## WHAT HAD HAPPENED BILLIONS OF YEARS AGO?

### The basic theories

A truly empirical approach to the study of evolution needs some kind of 'written' record rescuing information concerning specific 'historical' events into future times. Until the end of the 1960s, only rudimentary pieces of such written records were available with respect to the evolution of microorganisms. Therefore, a number of theories have been proposed starting out from basic reasoning and the few data available at that time.

The so-called heterotrophic theory for the origin of life, first suggested by Horowitz in 1947, assumes that a reducing atmosphere of ammonia, methane and hydrogen, in combination with a source of free energy, would give rise to a prebiotic 'soup' of monomers such as amino acids, sugars, lipids and nucleotides in the hydrosphere. These monomers would self-assemble to form ribonucleic acids, proteins and membranes which would eventually evolve into cells able to replicate and mutate. Once the heterotrophic cells had exhausted the prebiotic soup, there was a strong selective pressure favouring cells with a photo-autotrophic metabolism that were able to obtain their energy from sunlight. In this theory, the invention of photosynthesis took place significantly later than the origin of life and metabolic pathways were developed when resources became scarce, i.e. from the final products backwards to the fixation of carbon dioxide and nitrogen.

The autotrophic theory for the origin of life, in contrast, assumes that life did not originate in a substrate-rich prebiotic soup but in water and an atmosphere of nitrogen and carbon dioxide, making the origin of life and the origin of photosynthesis dependent on each other. In this theory, the first organized metabolic unit would be a primitive energy conversion system that could perform the basic processes of photosynthesis and respiration. Around this basic unit, increasingly complex organic molecules would be formed and the metal catalysts eventually would become metalloenzymes. The biosynthetic pathways would have started with carbon fixation and would subsequently have evolved gradually to higher complexity.

The most recent concept elaborated by Wächtershäuser proposes an autotrophic origin of life and does not require any photosynthetic mechanisms. Instead, its basic concept relies on redox reactions taking place on the surfaces of solid state catalysts (e.g. pyrite) allowing the spontaneous development of archaic versions of central metabolic pathways such as the citric acid cycle.

During the last decade, research on evolution has, to some extent, lost its speculative nature by exploiting fossil evidence and comparative sequence data. Parts of the presented theories may now be re-evaluated in the light of this new information.

## Fossils

The 'Darwinian' way to study the evolution of photosynthesis consists of retrieving information from comparisons of traits in extant species. The obvious macroscopic photosynthetic organisms, i.e. green plants, however, yield little information since almost all essential steps in the evolution of photosynthetic processes have already occurred in the 'microbial period' (see below and Fig. 22.1). An obvious task therefore, consists in the uncovering of fossilized microorganisms of the Archaean period of the earth. The oldest extant rock formations suitable for this approach date back to 3.8 Gyrs ago. There may well have been microorganisms before that date; all fossil evidence, however, has very likely been wiped out by erosion. Furthermore, there are two obvious problems with fossilized microorganisms. First, it is not always clear whether the imbedded microfossils are of the same age as the parent rock. Secondly, the morphological appearance of microorganisms is of limited help for their classification, and morphological comparisons with extant organisms can therefore be misleading. Despite these problems, such microfossils represent the only direct source of evidence of life from the Archaean era.

## Protein- and nucleotide sequences; universal books of history

A second, more indirect way of reconstructing evolution was provided when a hitherto cryptic language, i.e. the code of amino acid sequences was deciphered. Zuckerkandl and Pauling proposed in 1965 that phylogenetic relationships between species may be deduced by comparing primary sequences of selected proteins. In more recent years, this approach was also applied to nucleotide sequences of DNA and RNA molecules. Small subunit (SSU) rRNA became the most widely used entity for determination of phylogenetic relationships between species. Mathematical models and computer programs were developed in order to retrieve the respective information from the sequences under consideration. The most frequently used algorithms for the construction of phylogenetic trees are currently based either on 'maximum likelihood' between sequences or on the principle of 'parsimony' in building the respective trees.

Based on small subunit r-RNA sequence comparisons, it is now generally accepted that the world of all living organisms consists of three 'kingdoms' or domains: the Eukarya, Archaea and Bacteria (Fig. 22.2(a)). Photosynthesis is found in the domains of the Bacteria and the Eukarya. In Eukarya, the photosynthetic processes occur within specialized organelles, the chloroplasts, that have been recognized as endosymbiotic descendants of the cyanobacteria, i.e. of species that belong to the bacterial kingdom (see also p. 299). Phototrophic ability has therefore been 'imported' into the Eukarya (as represented by the arrow in Fig. 22.2(a)) and does not represent an evolutionary achievement of this domain. No chlorophyll-based photosynthesis has so far been found within the Archaea. The light-driven ion-pumps of the archaeal halobacteria rely on photoisomerization phenomena rather than on photo-ionization and therefore, represent an independent way of using solar energy for metabolic needs. The 'birthplace' of photosynthesis therefore, seems to be located within the domain of the

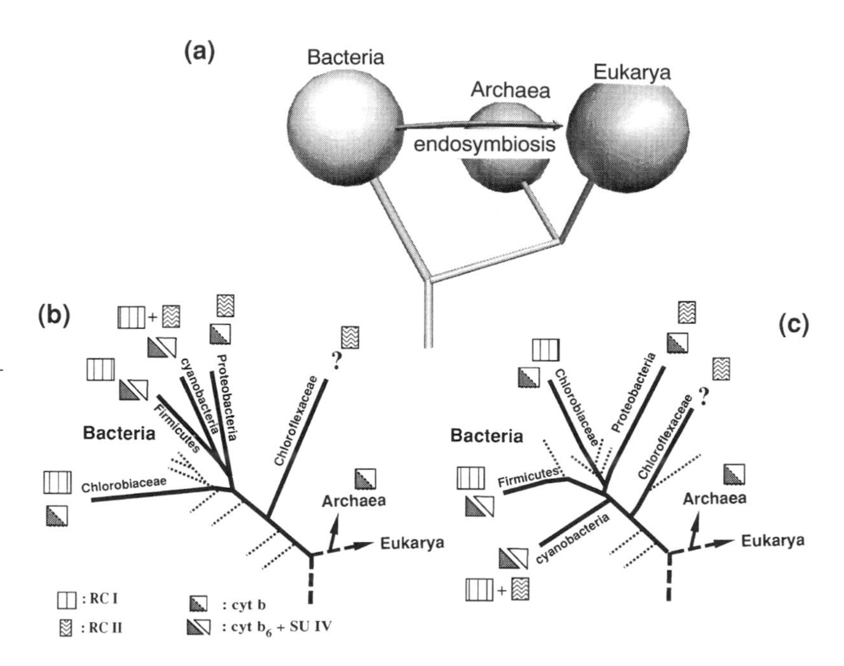


**Figure 22.2.** Phylogenetic relationships based on SSU-rRNA (a) between the three major domains, i.e. the Bacteria, the Archaea and the Eukarya; (b) and (c) for phyla containing phototrophic species within the domain of the Bacteria as reported (b) by Woese (1987) or (c) by Olsen *et al.* (1994). The types of photosynthetic reaction centres and cytochrome bc complexes found in the respective branches and as discussed in the text are indicated. Dashed lines denote branches that do not contain photosynthetic representatives.

Bacteria. In addition, when restricting one's interest to the bioenergetic 'heart' of the photosynthetic apparatus, its present-day state employed by plants was already reached within the domain of the Bacteria with the invention of oxygen-evolving photosynthesis by the ancestors of contemporary cyanobacteria. This bioenergetic 'heart' of cyanobacterial electron transport was subsequentially (i.e. during the evolution of chloroplasts within the eukaryotic lineages) only marginally modified.

At the end of the 1980s, deducing evolutionary relationships from sequence comparisons appeared as the ultimate tool able to answer all questions concerning the evolution of species. Very recently, however, a slight disenchantment concerning these methods is starting to emerge. It was shown that many of the proposed trees are actually not 'robust' and that several different tree topologies can be obtained upon slight modifications of parameters. One prominent example for varying tree topologies consists in the phylogenetic tree of the domain of the Bacteria itself. As is obvious from what was described above, this phylogeny is a crucial parameter for the study of the evolution of photosynthesis. Unfortunately, two topologically rather divergent phylogenetic trees for the domain of the Bacteria are presently in the literature (cf. Fig. 22.2(b) vs. (c)). The main difference between these two trees consists in the positioning of the Chlorobiaceae and the cyanobacteria. Whereas the older phylogenetic tree shown in Fig. 22.2(b) considered the Chlorobiaceae as a branch diverging rather early preceded only by the Chloroflexaceae, the more recent tree (Fig. 22.2(c)) reserves this position for the cyanobacteria which were formerly seen as a relatively young radiation. The repositioning of the cyanobacteria is paralleled by a similar shift of the position of the Firmicutes (containing Gram-positive and the phototrophic heliobacteria) keeping cyanobacteria and Firmicutes closely related in both trees.

## THE THREE PRINCIPLES OF PHOTOSYNTHETIC ELECTRON TRANSPORT CHAINS

Photosynthetic organisms are found dispersed over the entire domain of the Bacteria rather than being clustered together in a confined subset (Fig. 22.2(b),(c)). This suggests that photosynthesis was invented by an early representative of the Bacteria or that possibly the whole domain diverged from a photosynthetic species.

Presently, five different branches (that is roughly half of the 'Bacterial' phylogenetic tree) are known to contain photosynthetic organisms. These are the green filamentous bacteria (Chloroflexaceae), the cyanobacteria, the heliobacteria, the green sulphur bacteria (Chlorobiaceae) and the purple bacteria. Detailed functional studies have demonstrated that, within these five branches, only three clearly distinguishable basic principles of photosynthetic electron transport are operative (Fig. 22.3, schemes [1] to [3]).

Whereas cyanobacteria use two functionally quite distinct photosynthetic reaction centres, i.e. PSI and PSII, together with a third membrane–integral redox complex, the cytochrome $b_6f$ complex (Fig. 22.3, scheme [3]), all the other photosynthetic species content themselves with only one light-driven reaction centre, related to either PSI or PSII from cyanobacteria, and a cytochrome bc complex homologous to the cytochrome $b_6f$ complex in the cyanobacterial system. The photosystems of Chloroflexaceae and purple bacteria belong to the 'RCII-family' (i.e. are related to PSII) and electron transport is mainly cyclic around the cytochrome bc complex and the photosystem (Fig. 22.3, scheme [2]). The cytochrome bc complex seems to play a comparable role in the electron transport chains of Chlorobiaceae and heliobacteria. The photosystem employed by these bacteria, however, is of the RCI-type, i.e. homologous to PSI (Fig. 22.3, scheme [1]). Non-cyclic electron transfer is predominantly found in oxygen-evolving organisms like cyanobacteria and higher plants where water gets oxidized, and $CO_2$ acts as the ultimate electron acceptor and probably as well (to a still ill-determined extent) in the electron transport chain of Chlorobiaceae and heliobacteria (see Fig. 22.3, scheme [1]). However, cyclic electron flow around PSI is also operational in cyanobacteria and chloroplasts and, in addition, a special group of cyanobacteria, the Oscillatoriae, can also perform 'true' anoxygenic photosynthesis using $H_2S$ as the ultimate electron donor.

As is evident from Fig. 22.3, the more complex electron transport chain of cyanobacteria (scheme [3]) looks suggestively like the sum of the electron transport pathways in [1] and [2] (i.e. '[1] + [2] = [3]').

The photosynthetic membranes of phototrophic bacteria and chloroplasts therefore, appear like a construction kit using only three basic building blocks, i.e. cytochrome bc complexes, RCI- and RCII-type photosystems joined together by somewhat variable bolts and screws like quinones and soluble electron carrier proteins.

Two different evolutionary questions arise from such a representation of photosynthetic electron transport chains:

(a) How are [1], [2] and [3] evolutionarily related?
(b) How did the individual building blocks evolve?

Concerning (b), i.e. the evolutionary pathways of the enzymes involved in photosynthetic electron transport, 'written information' has been 'excavated' in the form of primary sequences of the respective proteins from the majority of relevant species. In contrast, present scenarios with respect to (a) are still rather speculative. In the following we will therefore first focus on the evolution of the enzymes trying to collect, at the same time, pieces of information helpful for assessing the validity of the more global evolutionary hypotheses.

## EVOLUTION OF THE MULTIPROTEIN COMPLEXES OF THE PHOTOSYNTHETIC ELECTRON TRANSPORT CHAINS

### Cytochrome bc complexes

Unlike the photosynthetic reaction centres, cytochrome bc complexes are not uniquely associated with photosynthesis but also play a pivotal role in respiration. In facultatively respiring purple bacteria, the same cytochrome bc complex integrates either in the respiratory (under aerobic conditions) or in the photosynthetic (under anaerobic conditions) electron transport chains. The vast majority of non-photosynthetic bacteria contain cytochrome bc complexes as part of their aerobic or anaerobic (using terminal electron acceptors different from oxygen) respiratory chains (see Fig. 22.3, scheme [*]).

The failure to discover photosynthetic archaebacteria indicates that the invention of photosynthesis occurred in the domain of the Bacteria. In contrast, this seems not to be the case for the enzyme ancestral to extant cytochrome bc complexes. Two essential parts of cytochrome bc complexes, i.e. cytochrome b and the Rieske-2Fe2S-protein, have been detected and characterized in Archaea and it therefore seems likely that the

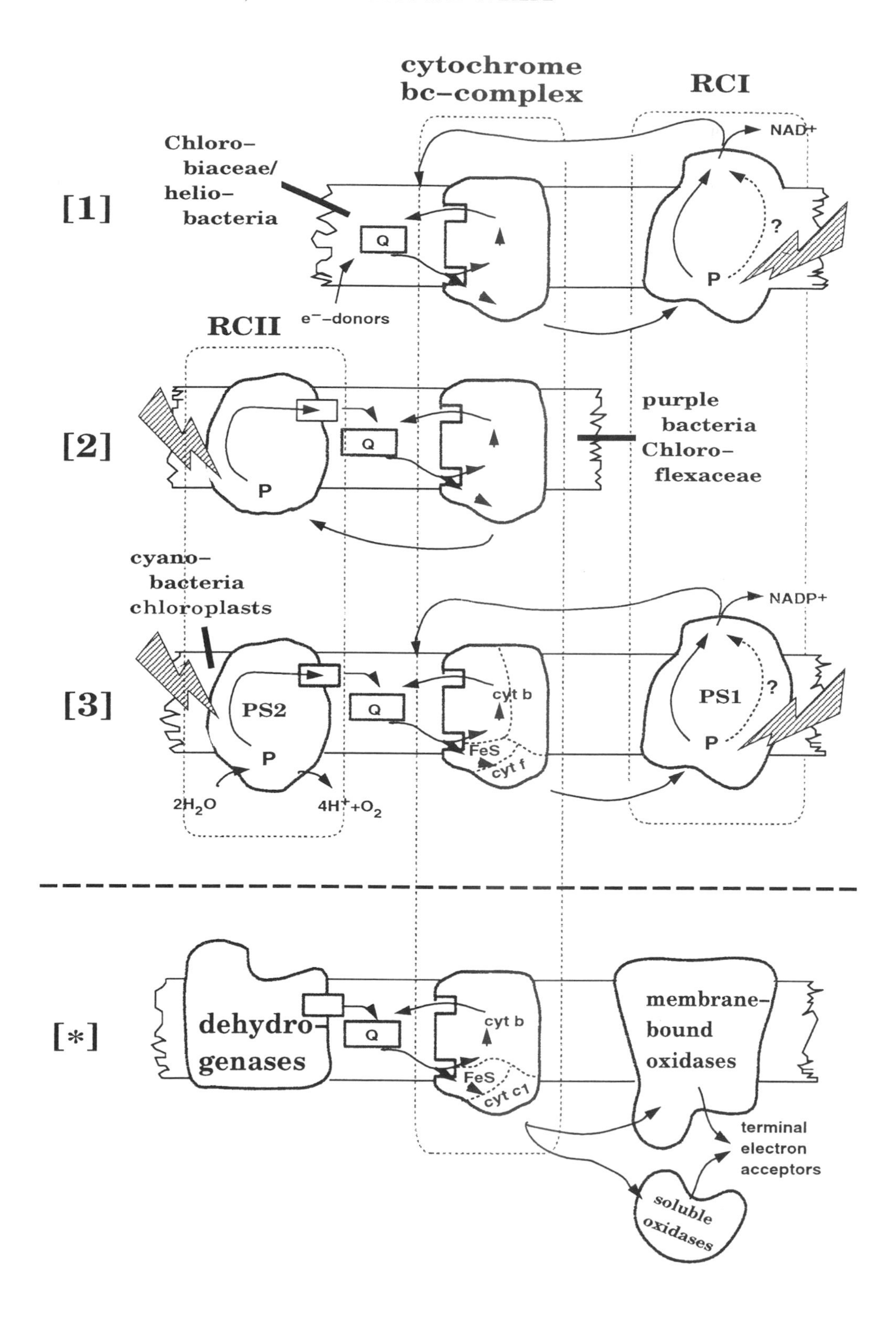
cytochrome
bc–complex
RCI
Chloro-
biaceae/
helio-
bacteria
[1]
NAD+
Q
?
P
e⁻-donors
RCII
[2]
Q
P
purple
bacteria
Chloro-
flexaceae
cyano-
bacteria
chloroplasts
NADP+
[3]
PS2
Q
cyt b
PS1
?
P
FeS
cyt f
P
2H2O
4H++O2
[*]
dehydro-
genases
Q
cyt b
FeS
cyt c1
membrane-
bound
oxidases
terminal
electron
acceptors
soluble
oxidases

evolution of cytochrome bc complexes and their function largely precedes the development of photosynthesis.

As is evident from a comparison of the photosynthetic schemes [1]–[3] to that of respiratory electron transport [*], the basic enzymatic mechanism of the cytochrome bc complexes is similar in photosynthetic and non-photosynthetic systems. The cytochrome bc complex therefore, appears as the central pole of photosynthetic electron transport, inherited from the ancestral non-photosynthetic energy-conserving chains. The remaining components (e.g. the reaction centres) have evolved around this functional core into the three basic photosynthetic mechanisms shown in Fig. 22.3.

### *The divergence towards cytochrome $bc_1$ and $b_6f$ complexes*

The cytochrome bc complexes in oxygenic photosynthesis (see Fig. 22.3), scheme [3]) are referred to as cytochrome $b_6f$ complexes whereas those involved in the anoxygenic photosynthesis of purple bacteria are generally called cytochrome $bc_1$ complexes. The dichotomy of $b_6f$-type and $bc_1$-type enzymes goes beyond mere nomenclature. The major criterion distinguishing between the two types consists in the arrangement of the cytochrome b-part of the complex. Cytochrome $bc_1$ complexes contain a continuous polypeptide for cytochrome b whereas cytochrome $b_6f$ complexes are characterized by a cytochrome b which is present as two separate proteins, corresponding to the N- and C-terminal parts of the 'long' b-polypeptide, respectively.

Recently obtained data concerning cytochrome b in hitherto poorly characterized species, i.e. the Chlorobiaceae, several Gram-positive bacilli and an archaebacterium have provided first information helpful in addressing the question of the evolutionary pathways towards $b_6f$- and $bc_1$-type enzymes. Unfortunately, however, the two different phylogenetic trees proposed so far, yield contradictory answers (Fig. 22.2(b) vs. (c)). According to the 'older' tree (Fig. 22.2(b)), the 'long' cytochrome b would be ancestral with a *gene fission* event having taken place on the line towards Firmicutes and cyanobacteria. The more recent tree (Fig. 22.2(c)), however, suggests a reverse scenario. Against the background of this tree, the 'two-protein'-cytochrome b appears phylogenetically older and the 'long' cytochrome b would have been created by *gene fusion* early on in the branch leading towards proteobacteria and Chlorobiaceae. One might argue that the more recent 16 S r-RNA tree will be the more reliable one. However, the 'outgroup'-protein, i.e. the cytochrome b detected in Archaea, corresponds to a long, unsplit haem protein. Therefore, either the unsplit cytochrome b was indeed the ancestral protein, which would be in conflict with the newer phylogenetic tree, or the gene fusion event had occurred independently twice in evolution.

Helpful information for solving this evolutionary riddle can be expected from sequence data on the cytochrome bc complex from Chloroflexaceae. In both versions of the phylogenetic tree, Chloroflexaceae are close to the root of the Bacterial tree and it is precisely the lack of data at this position that precludes at the present time less ambiguous scenarios concerning the evolutionary relationships between the different types of cytochrome bc complexes.

### *Cytochrome b and the 2Fe2S-protein*

Both cytochrome b and the Rieske 2Fe2S-protein have been detected in the majority of branches of the domain of the Bacteria and have also been described in a representative of the Archaea. Reported evolutionary trees based on sequences of cytochrome b and the 2Fe2S-protein and including the archaeal species differ slightly from each other and from the phylogenetic trees of species as shown in Fig. 22.2. Whereas such differences are frequently interpreted to indicate lateral gene transfer, we tend to recommend some caution in drawing such kind of conclusions. As will be discussed below, the presently emerging controversies concerning several of the

---

**Figure 22.3.** Schematic representation of the three basic mechanisms of photosynthetic electron transport found so far. [1] and [2] show the electron transport pathways employed by Chlorobiaceae/heliobacteria and by Chloroflexaceae/purple bacteria, respectively. [3] represents the oxygenic photosynthesis of cyanobacteria and chloroplasts. The scheme marked [*] shows the general principle of respiratory-type (aerobic as well as anaerobic) electron transport chains.

previously accepted trees show that the detailed methodology of the mathematical approaches is still 'evolving'.

With respect to global structure and function, however, both cytochrome b and the 2Fe2S-protein are strikingly conserved over the huge evolutionary span from Archaea to chloroplasts.

### *Cytochrome $c_1$/f and the 'cytochrome b/FeS-complex'*

Such a high conservation is completely absent for the case of the c-type haem proteins contained in these enzymes. The level of sequence homology between cytochrome f and cytochrome $c_1$ is close to that of unrelated proteins and it has even been speculated that cytochrome f and cytochrome $c_1$ might represent an example of convergent evolution. According to the results of algorithms predicting secondary structural elements such a $\alpha$-helices, turns and $\beta$-sheets, there may nevertheless be significant structural similarities between cytochrome $c_1$ and cytochrome f and the imminent 3D-structure of a cytochrome $bc_1$ complex will once and for all put an end to speculations concerning the relatedness of these proteins.

On moving towards species that are phylogenetically more distant from cyano- and proteobacteria, the cytochrome c-part of 'cytochrome bc'-complexes becomes increasingly elusive. Recent sequence data on the complex from Gram-positive bacilli suggest the presence of a c-type haem protein that is unrelated to cytochrome $c_1$ or cytochrome f. The operon coding for cytochrome b and the Rieske 2Fe2S-protein in Chlorobiaceae is even devoid of such a c-type haem protein and, whereas both cytochrome b and the Rieske protein have been identified biochemically in Chlorobiaceae, there is no hint of a c-type haem protein associated with the complex so far. In the Archaeon *Sulfolobus acidocaldarius*, c-type haems even seem to be fully absent but cytochrome b and the Rieske protein are nevertheless involved in the respiratory electron transport chain of this bacterium.

Taking all these facts together suggests that the functional core of 'cytochrome bc' complexes may actually be a 'cytochrome b/FeS'-complex. This core complex would then be the evolutionary precursor of cytochrome $bc_1$ and cytochrome $b_6f$ complexes and is apparently still present in a number of extant species. In this scenario the acquisition of the 'additional' c-type cytochrome would correspond to the trapping of a previously soluble electron carrier protein followed by structural and functional incorporation as a new subunit of the former cytochrome b/FeS complex. A comparable event is well documented for the case of the RCI-type photosystem where a soluble 8Fe8S ferredoxin has most probably evolved into a bound subunit of the RC (see p. 294).

## The two extant types of photosynthetic reaction centres

All photosynthetic reaction centres studied so far can be classed as either RCI- or RCII-type reaction centres (Fig. 22.3), a classification based on both structural and functional criteria. The architecture of these photosynthetic RCs is based on the dimerization of two structurally rather similar subunits (as in the case of the heterodimeric RCs of PSI, PSII and the RCs from Chloroflexaceae and purple bacteria) or of two copies of the same protein (i.e. yielding the homodimeric RCs of Chlorobiaceae and heliobacteria). No evidence for the existence of a monomeric RC has been obtained so far. In fact, the structures of the photosynthetic RCs known to date even suggest that photosynthetic function may have been brought about by the very event of dimerization. The functional heart of photosynthesis, i.e. the photooxidizable pigment, actually consists in almost all RCs of two chlorophyll-type pigments symmetrically stacked against each other in the contact region between the two subunits of the dimeric complex. The specific structure (and hence the electronic properties) of this 'special pair' of pigments significantly lowers the electrochemical potential of the pair with respect to that of individual chlorophyll cations and brings it into the range of those covered by the redox co-factors of terminal oxidases. Therefore, in species able to perform both photosynthesis and (aerobic or anaerobic) respiration, these photosynthetic RCs can substitute terminal oxidases as ultimate electron sinks in their respective electron transport chains without modifying any of the 'upstream' redox proteins. The only exception to this rule is the 'water-splitting' PSII with a redox potential for its photooxidizable pigment that is sufficiently oxidizing to extract electrons from water. However, the identity of the photooxidizable pigment in PSII is still under debate and it seems possible that this role is played by a chlorophyll molecule other than the 'special pair'.

In all photosynthetic RCs, the photoinduced charge pair is stabilized by a series of ultrafast electron transfer reactions which spatially separate the electron from the special pair cation. Comparable electron transfer chains are found in enzymes involved in other redox mechanisms and it is indeed only the initial event of photooxidation of a special pigment that is specific for photosynthesis.

### *RCII-type photosystems*

RCII- (or quinone-) type photosystems are characterized by a dimeric polypeptide core with molecular masses of around 30 kDa each and by the presence of a two-electron-gating quinone as terminal electron acceptor (see Chapter 3, this volume).

Presently known representatives of the RCII-group comprise the reaction centres of Chloroflexaceae and purple bacteria as well as the water-splitting PSII operating in the oxygenic photosynthesis of cyanobacteria and chloroplasts.

Whereas the RCs from Chloroflexaceae and purple bacteria look astonishingly similar to each other considering the enormous phylogenetic distance between the two branches (see Fig 22.2 (b),(c)), the ability of PSII to oxidize water is paralleled by the presence of several additional cofactors involved in rereduction of the photooxidized pigments that are absent from the RCs from purple bacteria and Chloroflexaceae. The arrangement of co-factors on the electron-accepting side of PS2 as well as the sequence of electron transfer steps through these electron acceptors, however, betray the membership of PSII to the RCII-group of photosystems. This is especially true for the unidirectionality of electron transfer through the chain of electron acceptors using only one of the two structurally quite symmetrical branches. This unidirectionality of electron transfer is a common feature of all RCII-type photosystems examined so far. It is obviously a consequence of one of the principal tasks of RCII which is to reduce quinones, i.e. two-electron redox centres. Therefore, the electrons from two subsequent turnovers have to be collected to complete the catalytic cycle of the RC's reducing side. This is achieved by inducing functional asymmetry between the two quinone acceptors $Q_A$ and $Q_B$, entailing the requirement for asymmetry of the whole electron transfer chain.

Whereas all these functional similarities are evident, the overall sequence homology between the RCs from purple bacteria and Chloroflexaceae on one side, and PSII on the other side, is quite low. Nevertheless, important amino acid residues are conserved as well as the overall structure based on five membrane-spanning helices in each half of the dimer.

Evolutionary trees have been constructed from the available primary sequences of the two core subunits of RCII-type photosystems. Unexpectedly, these trees suggested that the divergence into PSII on one side, and *Chloroflexus* and purple bacteria on the other side, happened on the level of a homodimeric ancestor. This would imply that subsequent gene duplications occurring independently in the two lineages had created heterodimeric RCs performing asymmetric electron transfer in an astonishingly similar way. Such an extreme case of 'convergent evolution' is somewhat difficult to believe, and several authors have therefore attempted to provide alternative solutions for this paradox.

Recently, however, the validity itself of the detailed mathematical approach for constructing the evolutionary tree of the RCII-type photosystems was challenged and it was shown that a slight alteration of the employed parameters yields a tree supporting heterodimerization of the RCII-type photosystems prior to the divergence into Chloroflexus/purple bacteria and PSII-lineages.

Thus, the precise evolutionary pathway leading to the different RCII-type photosystems is presently as controversial as it possibly can be. One might argue that, given this lack of agreement, it was not worthwhile to go into details concerning this question. However, we feel that this specific example (just as well as the divergence of the phylogenetic trees in Fig. 22.2(b) and (c)) should serve as a warning against exaggerated confidence in the results of sequence-based evolutionary trees. The computer programs performing the numerous calculations needed to produce these trees (e.g. multiple sequence alignments, parsimony searches, etc.) are freely available on the Internet and are usually characterized by a high user-friendliness. Therefore, it is tempting to consider the output obtained on running these programs as plain 'truth' forgetting about the inherent mathematical problems and the large number of adjustable parameters (some of the authors of this contribution feel guilty of this fault). A prominent example for controversies concerning this field is provided by the heated debate on the origin of the genus *Homo*, deduced from parsimony analyses of the mitochondrial genome. Nevertheless, when the limits of the methods are kept in mind, sequence comparisons are certainly the most powerful tool presently at hand for the study of evolution.

### *RCI-type photosystems*

The reaction centres found in Chlorobiaceae and heliobacteria differ significantly from those described above. Together with PSI from the oxygenic electron transport chain of cyanobacteria and chloroplasts, they constitute the class of RCI-type photosystems characterized by (a) a dimeric core formed by two proteins with molecular masses of 70–80 kDa, (b) by the direct association of a significant number of light-harvesting pigments to this dimeric core and (c) the presence of a small, extrinsic protein carrying two 4Fe4S-clusters, the terminal electron acceptors of the RCs (see also Chapter 3, this volume). Several additional subunits may be present in the different RCI-type photosystems but are not common to the whole class.

The dimeric core of both the chlorobial and the heliobacterial RC is made up of two copies of a single protein, meaning that these cores are *homo*dimeric. PSI is a *hetero*dimer, being formed by the psaA and psaB gene products. In their C-terminal halves, i.e. in the part presumably harbouring all the essential co-factors involved in charge separation, however, PsaA and PsaB are almost identical. We therefore tend to add as a fourth characteristic feature distinguishing between the two classes of RCs the fact that RCII-type photosystems are heterodimeric whereas the RCI-group contains homo- or quasi-homodimeric structures.

This divergence into the homo- (RCI-) and the heterodimeric (RCII-) groups has certainly been favoured by functional constraints. By contrast to the quinone-reducing RCIIs, electron transfer in RCIs results in the eventual reduction of FeS centres, i.e. one-electron redox centres. Therefore, a strong functional pressure favouring the evolution towards a heterodimer was probably lacking for the case of RCIs.

All RCI-type photosystems examined to date seem to contain a third subunit which is attached to the cytoplasmic/cytosolic side of the RC and carries two 4Fe4S centres. Its amino acid sequence as well as structural and functional data clearly identify this subunit as a member of the family of 8Fe8S-ferredoxins. The respective protein in Chlorobiaceae contains a long N-terminal extension of unknown function fused onto the FeS-subunit. Since 8Fe8S-ferredoxins serve as soluble cytoplasmic electron carriers in many photosynthetic and non-photosynthetic microorganisms, it seems highly likely that the FeS-subunit of extant RCIs represents a 'captured' form of a soluble electron transfer partner of the ancestral RCI-type photosystem.

### *A common ancestor to all RCs?*

Despite the presence of numerous hints that could be drawn from functional properties, the presence of two classes of RCs became generally acceptable only when characteristic sequence homologies for each of the classes had been revealed. The approach of comparing primary sequences can, in principle, also be used when looking for a link between the two classes of photosystems. The results of such attempts, however, are rather discouraging. So far, no unambiguous sequence homologies between RCII- and RCI-type photosystems have been reported.

Does this mean that, at present, we cannot decide whether RCI- and RCII-type photosystems have a common ancestor or whether they were invented independently? Despite the lack of sequence homologies, in our opinion the descent of RCI and RCII from a common ancestor is conspicuous. The electron acceptor chains of both types of RC are similar with respect to number and chemical identity of cofactors. The emerging 3D-structure of a cyanobacterial PSI shows that the arrangement of these co-factors is suggestively similar to that seen in the purple bacterial RC and that even the arrangement of the inner transmembrane helices harbouring these co-factors in PSI is strongly reminiscent of the geometry of the purple bacterial RC-core. The actual difference between RCI and RCII consists in the presence of several additional helices in RCI (or absence thereof in RCII) which ligate the chlorophyll-molecules serving as the inner antenna of the RCI-type photosystems. Therefore, the main question concerning the structure of the common ancestor of both types of RCs is whether this ancestral RC resembled RCI, subsequently

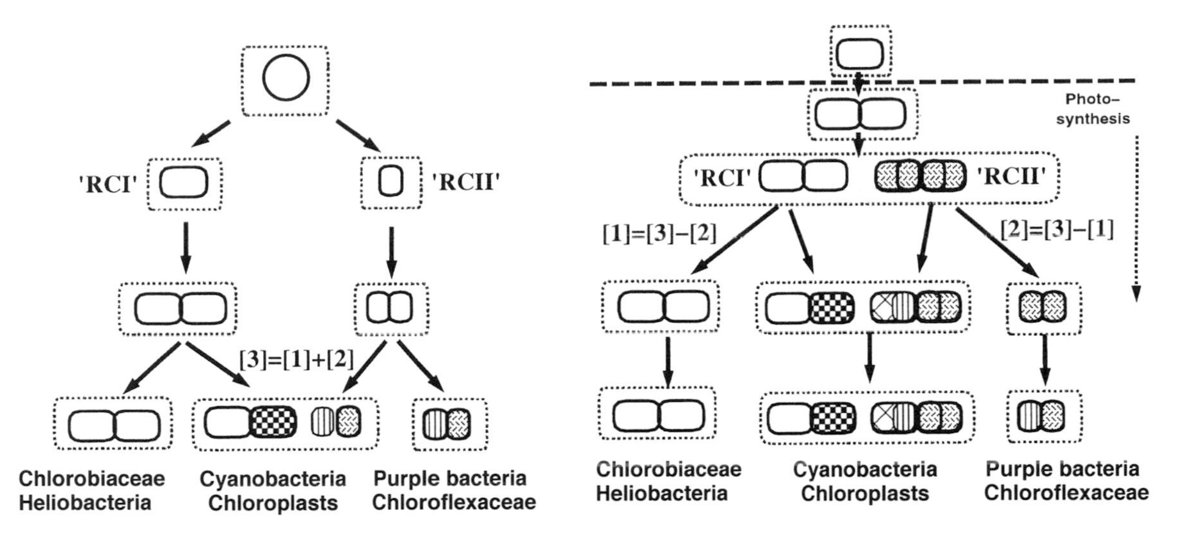


**Figure 22.4.** Two extreme scenarios for the evolutionary pathways of photosynthetic RCs from their invention to their appearance in extant organisms. The left scheme assumes early divergence into monomeric ancestors of RCI – and RCII-type photosystems within different species evolving via gene duplications into homo- and heterodimeric forms. In this scenario, RCI- and RCII-based electron transport chains from two different species would have been fused together to yield the oxygen-evolving electron transport chain of cyanobacteria. The right scheme, by contrast, suggests (a) that photosynthetic RCs were always dimeric, (b) that early in the evolution of photosynthesis, species containing two (or more) different types of RC had appeared and (c) that the extant 'single RC'-electron transport chains result from selective loss of one or the other of the two types of RC.

getting rid of the antenna-part of the protein to become an RCII-type photosystem or whether it rather resembled the 2 × five-helices RCII fusing antenna proteins onto its core in order to evolve towards RCI. At present, there is no data giving hints with respect to this question. The determination of 3D-structures of antenna proteins from diverse photosynthetic species, however, might provide some clues, especially since it has been proposed that the close antenna of PSII, i.e. the CP47/43-proteins might be evolutionarily related to the additional 'antenna-helices' of RCI.

In any case, the scarcity of data with respect to the evolution from the common ancestor towards RCI and RCII leaves room for a large number of possible evolutionary scenarios, of which two rather divergent examples are depicted in Fig. 22.4.

## THE ORIGIN OF PHOTOSYNTHESIS

In former years, the evolution of photosynthetic electron transport was frequently thought to have nucleated in a primordial photosystem performing transmembrane charge separation. The ejected electron was supposed to be returned to the photooxidised pigment via some membrane-diffusible carrier (e.g. quinones, porphyrins) with occasional build-up of membrane-potential due to redox linked protonation/deprotonation of this carrier. The subsequent development of a cytochrome bc complex was considered a necessary step in order to improve the efficiency of membrane-potential build-up.

As explained on pp. 289–92, such 'photosynthocentric' scenarios are not in line with presently available data.

In contrast to the older scenarios, we therefore consider it quite likely that the invention of photosynthetic reaction centres occurred in an environment already containing a fully functional cytochrome bc complex. This complex would operate in a 'respiration-type' electron transport chain delivering electrons to some terminal electron acceptors (almost certainly different from oxygen) using a 'Mitchellian' way of ATP production.

In such an environment, a nascent photosynthetic RC would have been of use once it managed to accept electrons from the cytochrome bc (b/FeS?) complex and to photoreduce quinones. Every quinone reduced by such an RC would result in an extra turnover of the cytochrome bc complex, hence would contribute to maintain the membrane potential and increase ATP production. Since such an organism would be able to live using only its respiratory pathway, a respective increase in ATP production will provide an evolutionary advantage over other species even if the RC worked very inefficiently, i.e. with a low quantum-yield. On the other side, since the RC is not indispensable for survival, it is free to mutate rapidly exploring many different evolutionary directions. Such an RC might even have evolved into a water-splitting PSII-like photosystem (provided that the organism found ways to cope with the produced oxygen) without obligatorily needing PSI to complement the linear electron transport chain. Gene duplications may have resulted in a number of differing RCs in the same energy-conserving membrane, two of which managed to survive and to evolve into RCI and RCII (see below).

It is unnecessary to say that the details of these scenarios are pure speculation and should be taken as such. However, the underlying assumption that photosynthesis originated after the development of respiratory electron transport chains seems well founded at present. This takes away many of the constraints previously considered as important during the origin of photosynthesis. It therefore seems likely that photosynthesis was born in a playground free of far-going responsibilities for the survival of its parent organisms. This fact may explain the extraordinarily rapid evolution and diversification of photosynthesis in the domain of the Bacteria (Figs 22.1 and 22.2).

## EVOLUTION OF THE THREE BASIC PRINCIPLES

In the beginning of the 1990s, most authors were favouring evolutionary scenarios in which, after divergence into RCI- and RCII-lines of photosystems, the two principles of photosynthetic electron transport [1] and [2] (see Fig. 22.3) were evolving independently until in some ill-defined event they were fused together ('[1] + [2] = [3]') in an organisms that became the ancestor of the cyanobacteria (Fig. 22.4, left scenario). It is furthermore tacitly assumed that it is the very fusing of the two principles that made water-splitting/oxygen evolution possible. Such evolutionary scenarios were most probably animated by the prevalence of the phylogenetic tree shown in Fig. 22.2(b), which locates cyanobacteria high up in the tree with almost all other photosynthetic branches having diverged before the appearance of the cyanobacteria. This tree topology comforted a view that considers the oxygen-evolving Z-scheme as the summit of the evolution of photosynthesis, concluding that all 'inferior' mechanisms of photosynthetic energy conversion must necessarily be more ancient.

Whereas this evolutionary scenario may well be correct, it is neither more plausible nor more in agreement with available data than a second group of hypotheses that envisage a rather inverted sequence of events. In fact, the topology of the more recent phylogenetic tree shown in Fig. 22.2(c) suggests a very early appearance of organisms containing both types of RC, i.e. the ancestors of the cyanobacteria. The divergence of branches containing phototrophs that rely on either RCI (Chlorobiaceae and heliobacteria) or RCII (purple bacteria) *after* the divergence of the cyanobacterial line appears a paradox at first sight. Therefore, a rather far-going uncoupling of the phylogenetic relationships between species from the evolution of the photosynthetic apparatus brought about by frequent events of lateral gene transfer has been evoked. However, already in 1989, Pierson and Olson proposed an alternative scenario that did not have to recur to extensive lateral exchange of genes in order to rationalize the distribution of photosynthetic mechanisms across the phylogenetic tree. In this scenario, organisms containing both types of RC emerged very early in evolution. Subsequently, several lineages lost either RCI or RCII due to adaptation to specific environments. As already mentioned above, the appearance of two or more RCs in one species by way of gene duplication seems quite likely. Thus, gene duplication events, together with subsequent mutation of the individual genes, may well have led to the formation of two independent populations of homodimeric RCs. Further gene duplication events could subsequently have transformed these two different ancestral homodimeric photosystems into heterodimeric forms.

The extant photosynthetic electron transport chains found in Chloroflexaceae/purple bacteria on one side (scheme [2]) and in heliobacteria/Chlorobiaceae on the other side (scheme [1]) would then represent 'amputated' versions of the original RCI+RCII forms having lost either RCI ('[2] = [3] − [1]') or RCII ('[1] = [3] − [2]'). Only those organisms that managed to specialize the function of each photosystem would have found an advantage in keeping them both. The decisive act of specialization may well have been the transformation of RCII into a water-splitting photosystem. A specific example from this group of scenarios is depicted in Fig. 22.4 (left scheme).

## OXYGENIC PHOTOSYNTHESIS

### Invention of oxygenic photosynthesis

#### *The cyanobacteria*

Today, there is little doubt that 'plant-like' photosynthesis was invented by the representatives of the Bacteria (Fig. 22.1) and more specifically that ancient cyanobacteria were the ancestors of present-day chloroplasts. However, these insights were obtained only fairly recently since the ability of the cyanobacteria to perform oxygenic photosynthesis has, for a long time, disguised their bacterial nature and their ancient origin, and had confined them to the backwaters of botany as 'blue-green algae' until the 1960s. Proper classifications based on microbiological criteria were not performed before the late 1970s.

Based on 16S r-RNA sequences, cyanobacteria are part of a rather coherent group that also includes the eukaryotic plastids and the chlorophyll *a*/*b*-containing prochlorophytes (Fig. 22.5). This obvious close relationship between the photosynthetic eukaryotes and the cyanobacteria is also reflected in their photosynthetic apparatus, which has evolved only marginally on the way from cyanobacteria to the chloroplasts. All protein components of the photosynthetic electron transfer chain of cyanobacteria are present within the chloroplasts and only few accessory protein subunits have been added since. In addition, the enzymes of the bioenergetic chain which are essential for electron transport are highly conserved between phototrophic eukaryotes and cyanobacteria and identities of up to 80% are not unusual. Major differences only exist in the antenna systems. While most cyanobacteria and the red algae (the rhodophytes) contain phycobilin antenna pigments assembled in water-soluble phycobilisomes (PB), green chloroplasts use membrane-intrinsic chlorophyll (Chl) *a*/*b*-binding proteins for light harvesting. As discussed below, antenna systems are of limited use as evolutionary markers. The principal theme of the evolution of phototrophic Eukarya lies in the evolution of the biological species rather than in further optimization of the photosynthetic apparatus.

#### *The LHC-confusion: phycobilisomes and prochlorophytes*

A major, although transient, boost for the hypothesis of an endosymbiotic origin of the chloroplasts came with the discovery of the prochlorophytes, the only cyanobacteria that, like chloroplasts, contain Chl *a* and *b* and lack phycobilin. Consequently, it was suggested that these 'prochlorales' were the ancestors of green plastids while the phycobilin-containing organelles of rhodophytes would have originated from cyanobacteria proper. The photosynthetic membranes of the prochlorophytes show membrane appression, moderate grana formation and a certain degree of lateral heterogeneity of the reaction centres, all characteristics of green chloroplasts. However, the chloroplast-like morphology of the photosynthetic membrane is not particularly obvious and was alternatively rationalized as a secondary effect caused by the loss of phycobilisomes. All other morphological characteristics clearly indicate that prochlorophytes are cyanobacteria.

The presumed evolutionary importance of the prochlorophytes declined once it became apparent that their major Chl *a*/*b* light harvesting proteins were not related to the 'cab' (chlorophyll *a*/*b*-binding) proteins constituting the light harvesting complexes (LHCs) of chloroplasts but are, instead, homologous to a cyanobacterial chlorophyll-binding subunit of PSII, which is expressed under iron deprivation only. In addition, several sequence-derived phylogenetic studies uniformly showed that prochlorophytes are typical members of the cyanobacteria without any particular relationship to the chloroplasts (Fig. 22.5(a)). Furthermore, the prochlorophytes themselves do not form a coherent group. Instead, the individual members are scattered over the cyanobacterial radiation and are in fact only distantly related to each other. Thus, the substitution of the phycobilisomes by Chl *a*/*b* antenna systems must have

**Figure 22.5.** Phylogenetic trees of the cyanobacteria, the prochlorophytes and the phototrophic eukaryotes based on 16 S rRNA sequence comparisons. (*a*) Phylogenetic relationships among the cyanobacteria. (○) prochlorophytes, (□) algae and land plants; (*b*) Phylogenetic relationships within the plastids of algae and land plants. (□) chl a/b algae and land plants, (■) chl a-phycobilisome algae, (●) chl a/c algae.

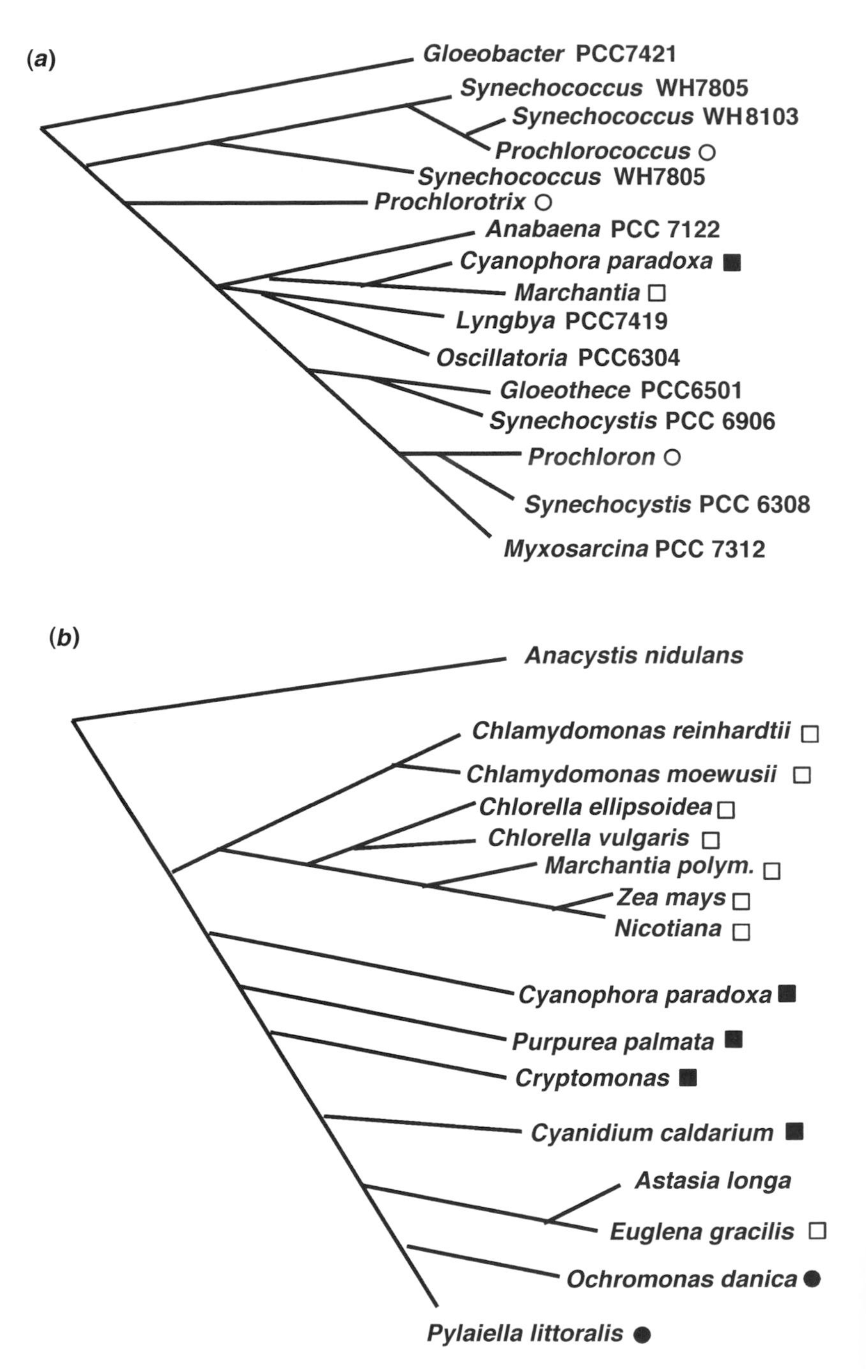

occurred independently several times, once in the ancestry of the green plastid lineage and several times within the cyanobacteria themselves. It is not known why some cyanobacteria have opted for Chl *a*/*b* antenna systems, especially since phycobilisomes largely surpass their Chl *a*/*b* counterparts as light harvesting complexes for visible light. Phycobilisome-deficient mutants of cyanobacteria which survived by inventing an alternative antenna system (or retrieving a dormant one from the genetic repertoire) are therefore likely candidates for the ancestors of the prochlorophytes. Therefore, the lesson to be learnt from the prochlorophytes is that pigmentation constitutes very obvious characteristics of photosynthesis, but should be handled with care as an evolutionary marker.

## The evolution of phototrophic eukaryotes

### *The endosymbiont theory*

According to the endosymbiont theory, first formulated in 1882, mitochondria and plastids and probably other organelles of today's Eukarya are descendents of formerly free-living bacterial ancestors, which entered an endosymbiotic relationship with, and were ultimately incorporated into, a eukaryotic host cell. The large amount of molecular data accumulated during the last 15 years has clearly shown that 'the nuclear genome of photosynthetic eukaryotes descended from a lineage other than that from which organellar genomes descend', a criterium formulated in 1982 as a future experimental test for the endosymbiont theory. Today, there is no doubt any longer that the plastids of extant phototrophic Eukarya are related to the cyanobacteria and that the mitochondria descend from a purple bacterium. Since the oldest known Eukaryotes, the archaezoa, do not yet contain mitochondria nor plastids, the event of their acquisition must have occurred only after the formation of the domain of the Eukarya. Except in the archaezoa, however, mitochondria are almost ubiquitous among eukaryotes suggesting that they were acquired relatively early in the evolutionary history of the eukaryotic domain. In contrast, plastids are significantly less widespread and were therefore most probably incorporated later than the mitochondria. Thus, phototrophic eukaryotes are phylogenetic chimeras of at least three different evolutionary traits, those of the eukaryotic nucleus, the plastid and the mitochondrion.

### *The phylogeny of plastids*

*A single ancestor?* The question concerning the number of primary endosymbiotic events eventually leading to extant plastids and mitochondria is still unsettled. In view of the enormous and often surprising variety of present-day symbiotic relationships of cyanobacteria and algae with non-photosynthetic and even photosynthetic eukaryotes, it seems difficult to imagine that extant algal and land plant plastids descended from a single endosymbiont. Nevertheless, most studies based on sequences of molecular markers for phylogenetic analyses are consistent with a monophyletic origin of today's chloroplasts. According to SSU rRNA (16S)-derived phylogenies, all plastids root within the cyanobacterial phylum but they are more closely related to each other than to any of the cyanobacteria (Fig. 22.5(*a*)). Thus, the line of descent of the plastids looks like a cyanobacterial subline deriving from a single cyanobacterial ancestor. This view is corroborated by several different sequence-derived phylogenetic analyses that are uniformly consistent with a monophyletic origin of plastids. Nevertheless, the relationships between plastids reported in the respective studies are not fully consistent with each other. Some unexpected findings are observed, the branching orders differ between trees and the respective trees frequently are not robust. In addition, it has been noted that, among other sources of artefact inherent to phylogenetic algorithms, substitutional bias may be encountered with sequences derived from different cellular compartments, i.e. nucleus and plastids, which tend to render the A/T-rich plastid sequences related to each other more closely than they might actually be. In summary, phylogenetic studies currently seem to indicate a monophyletic origin of the plastids, but the respective conclusions are still somewhat preliminary.

***Phylogenetic relationships among the algae*** Algae are usually classified according to the accessory pigments of their plastids into three groups, i.e. (i) the red algae (rhodophytes) and glaucophytes, which possess phycobilisomes and Chl *a*, (ii) the Chl *a*/*b*-containing green algae (chlorophytes, gamophytes and euglenophytes) and

(iii) the Chl *a/c*-containing algae (see Fig. 22.5(*b*)). The latter are a diverse group of nine taxa, of which the cryptophytes and eustigmatophytes also contain phycobiline. While this classification is still of some use, the underlying assumption that these three types of plastids with different pigmentations have different evolutionary ancestors is no longer valid. Investigations of the PS1 complex of red algae have shown that it is associated with LHC complexes which are immunologically related to both the Chl *a/b* LHC proteins of green plastids and to the fucoxantin Chl *a/c* antenna of chromophytes. Therefore, the conclusion must be drawn that the ancestor of today's plastids contained phycobilisomes as well as the precursors of the Chl *a/b*- and the Chl *a/c*-LHC complexes, and that one or the other of these antenna complexes was lost during the evolution of the different types of algae. The decision of the ancestors of present-day algae to choose a Chl *a/b* antenna instead of phycobilisomes must have occurred very early in plastid evolution since the phylogenetic tree of the algae diverges early into a 'green' and a 'red' branch (Fig. 22.5(b)).

Today's green algae and land plants descend from a single Chl *a/b* containing algal ancestor since there is a clear co-evolution between the chloroplast and its host. Branching off first from the green branch are the chlamydia and chlorellae, as the most distant relatives of today's plants. Based on ultrastructural characteristics, the algal group with the closest relationship to land plants are the charophyces, a finding supported by sequence evidence. Two introns in plastidic tRNAs of land plants were also detected in this algal subgroup and are absent in other chlorophytes. Apparently, this insertion of introns occurred fairly recently after the separation between the chlorophyte and the charophyte/land plant lineages. The development of flowering plants from non-flowering bryophytes is also reflected in the phylogeny of their plastids, with *Marchantia* plastids being more closely related to algal organelles than the plastids of the flowering plants.

The plastid type with the closest relationship to the putative ancestor of the plastids is the cyanelle of the glaucophytes, of which *Cyanophora paradoxa* has been extensively investigated. The unique feature of these plastids, which closely resemble cyanobacteria in overall morphology, consists in the fact that, although imbedded in the host cytoplasm, they are surrounded by a peptidoglycan cell wall. Typical murein components have been identified in these walls and these algae are killed by penicillins just like Gram-negative bacteria. Cyanelles are currently believed to be evolutionary relics of a different kind of plastid establishment which came to an early dead end and they are rarely encountered in nature. SSU rRNA-derived analysis of organellar rRNA shows a close relationship of *Cyanophora* to rhodophytes and cryptophytes (Fig. 22.5(*b*)). Other studies place cyanelles at an intermediate position between plastids and cyanobacteria, indicating that they may have diverged from the main plastidic line of descent even before radiation into the remaining types of plastids occurred.

The next lineages after the cyanelles to diverge from the 'red' branch of the 16S rRNA-derived phylogenetic tree are the rhodophytes and the bilin-containing plastids of cryptophytes. The common progenitor of this branch was thus most likely to be a phycobilin-containing red plastid and the Chl *c*-containing algae are likely to be the descendants of rhodophyte ancestors. It is generally accepted that today's Chl *a/c*-containing plastids evolved from rhodophyte algae by secondary endosymbiotic events (as discussed below).

### *The (cyano)bacterial heritage of plastid genomes*

Plastid chromosomes are highly condensed versions of bacterial genomes. Land plant plastomes contain about 120 genes of bacterial origin, the remaining 90% of plastid proteins are encoded in the nucleus and imported into the organelle. Chloroplast DNA from different land plants is very similar in structure and gene content and encodes all rRNA and tRNA needed for their translation system and carries genes encoding several subunits of the ribosomes, a plastidic RNA polymerase and several components of the bioenergetic apparatus of the chloroplast membrane. Remarkably, chlorophyte plastomes almost completely lack genes encoding enzymes of anabolic pathways. In contrast, the plastomes of cyanelles, rhodophytes and chromophytes encode about twice as many genes as their chlorophyte counterparts. For example, the plastid genomes of *Cyanophora paradoxa* and the rhodophyte *P. purpurea* contain significantly more genes for ribosomal subunits and bioenergetic membrane complexes and, in addition, they carry several genes encoding enzymes involved in anabolic pathways and regulation of transcription. Especially this fact that rhodophyte plastomes therefore, have retained some control over their own gene expression shows that they resemble more closely the ancestral form of the plastome than do the plastid DNAs of chlorophytes, which apparently were exposed to high pressure to condense their gene content. A further characteristic of plastomes of phycobilin-containing algae is that they almost completely lack introns in gene-coding regions, a feature shared with cyanobacteria. In contrast, chloroplast DNA is notorious for containing introns, which, for instance, in *Euglena gracilis*, can make up close to 40% of the organellar DNA. The almost complete absence of introns in rhodophytes, however, indicates that the common progenitor of today's plastids was likely almost intron-free and the majority of introns in chlorophyte plastids were acquired later.

Although it is safe to assume that the condensation of plastid DNA was, to a large part, already accomplished by the progenitor of present-day chloroplasts, this process is still operational today. Examples of fairly recent events are found in the plastomes of non-photosynthetic protists and parasitic plants. The plastome of *Epifagus virginiana*, a non-photosynthetic plant, has been reduced to about half the size of its plant relatives and has consequently deleted all genes required for photosynthesis. Similarly, the colourless euglenoid *Astasia longa* has lost all photosynthetic genes but has retained most of the plastid DNA structure of photosynthetic euglenoid algae. Most intriguingly, a crippled version of a plastome has recently been detected in the malaria-inducing parasite *Plasmodium*, which is probably descended from an alga.

A number of characteristics of the plastidic transcription/translation system clearly show the close relatedness between plastids and cyanobacteria. (i) Plastidic genes are mostly assembled into bacterial-type operons and are cotranscribed as polycystronic mRNAs from a single promotor equipped with typical bacterial consensus motifs. The RNA polymerase of plastids provides striking evidence for the descendence of algal plastids from cyanobacteria. In the plastidic version its subunit $\beta'$ is split into two parts encoded by two adjected genes, a feature shared only by cyanobacteria. (ii) mRNAs are translated by 70S ribosomes of the bacterial type. (iii) The genes encoding ribosomal rRNAs are arranged and processed in the same way as in bacteria. (iv) Genes for the ribosomal subunits are often assembled into typical bacterial operons. For instance, the ribosomal operons str and S10, which are very similar with respect to gene content and arrangement in all bacteria studied, are fully retained in the plastid DNA of rhodophytes. In the course of the evolution of the plastids, they were sometimes fused with other operons, rearranged and their gene content reduced. Rearrangements and size reduction follow different pathways during the development of the individual algal lineages, but the original composition of these operons is always recognizable, thus making these operons valuable molecular markers for tracing developmental pathways leading to the different plastid types. Besides the ribosomal operons, other examples for these rearrangement/condensation processes can be found, some of them starting with typical cyanobacterial operons, for instance, the ATPase gene cluster.

On the other hand, several operon structures are found in plastids whereas their individual genes are not linked in cyanobacteria. These operons must therefore, have been assembled after the event of endosymbiosis. A striking and evolutionarily important example is a plastidic operon related to PS2 (the psbB-ORF31-psbN-psbH operon). In present-day cyanobacteria, these genes are not linked. Remarkably, in all plastidic DNA the composition of this operon is strictly conserved, arguing against a polyphyletic origin of the plastids.

### *First- and second-hand plastids: the phylogeny of the eukaryotic host*

The plastids of red and green algae, with the exception of the euglenophytes, are surrounded by a double membrane, which probably arose directly from the double membrane system of the cyanobacteria-like progenitor. The plastids of all other major algal classes are enveloped by three or four membranes. The presence of the additional membranes of these 'complex plastids' can be rationalized by assuming that they are remnants of a secondary endosymbiotic event in which a non-photosynthetic protist engulfed a photosynthetic alga, which was then reduced to little more than its plastid. The additional two outer membranes would then represent the plasma membrane of the endosymbiont and a phagosomal membrane of the host. In algae containing three plastidic membranes, the phagosomal membrane has been lost. Especially for the cases of the dinoflagellates and the euglenoid algae, the origin of these complex plastids was probably that of a predator and a prey, in which the predator thought that he could get more out of his meal when retaining the plastid of its prey.

Cryptophyte algae are the living archaeopteryxes witnessing the secondary origin of complex plastids. The plastid of these Chl *a*/*c*-algae is closely related to those of red algae, contains phycobilin and is surrounded by four membranes. The two outer membranes, the chloroplast endoplasmic reticulum, are separated from the inner membranes by a distinct compartment, the periplastidal space which contains a 'bonsai' version of a nucleus, the so-called nucleomorph. This organelle, the remnant of a proper nucleus of an alga, is surrounded by a double membrane envelope with nuclear pores and nucleolus-like structures can be distinguished in its interior. Its genome contains three miniature linear chromosomes and is transcriptionally active. Comparisons between the nuclear and the nucleomorph-encoded SSU-rRNAs of cryptomonad algae showed that the two types of rRNA are phylogenetically distinct (Fig. 22.6). While the nucleomorph rRNA is specifically related to the nuclear rRNA of rhodophytes, the nuclear rRNA roots within the fungal and land plant lineage of the Eukarya. Thus, cryptomonad algae evolved from a secondary endosymbiotic event involving a phagocytotic protist and a red alga. Nucleomorph-like structures have also been detected in a few other algae. In most algae with complex plastids, no DNA of the engulfed alga has been retained and consequently, the phylogenetic heritage of their plastids has been lost.

The concept of plastid acquisition by secondary endosymbiosis is required in order to rationalize the phylogeny of the algae, scattered over the domain of the Eukarya (Fig. 22.6). Coherent groupings are only found for the chlorophytes and the rhodophytes, both likely descendants from a primary endosymbiotic event involving a cyanobacterium and a non-photosynthetic eukaryote. Both the chlorophytes and the rhodophytes form distinct lineages within the Eukarya. In agreement with the phylogenetic analysis of their plastids, the lineage of the land plants is a sister group of the chlorophytes with a common green ancestor. By contrast, the taxonomic positions of the Chl *a*/*c*-algae cannot be explained without assuming secondary endosymbiotic events, and it is not surprising that the number of these events leading to the various distantly related Chl *a*/*c*-algae is still a matter of debate. A Chl *a*/*c*-containing subgroup, the chromophytes apparently forms a single, separate lineage, which may well be descended from a single ancestor. This subline, however, did not derive from a cryptophyte-like alga, a finding that strongly favours the polyphyletic origin of algae with complex plastids. Especially difficult to identify are the ancestors of the dinoflagellates and the euglenophytes which are particularly unrelated to any other algal group. Euglenophytes represent a difficult case since they appear to be a fairly ancient eukaryotic subgroup whereas their plastids, although probably related to those of the rhodophytes and the chromophytes, contain Chl *a*/*b*. Since many euglenophytes are aplastidic and phagotrophic, it was suggested that they originated from an endosymbiotic event that was more recent than the origin of their eukaryotic lineage. For similar reasons, it is not clear whether the dinoflagellates evolved from a photosynthetic or from a non-photosynthetic ancestor. Thus, for many of the algae with complex plastids, the particular origin of the primary algal host still remains largely elusive.

## CONCLUDING REMARKS

As the reader of this chapter certainly has noticed, our knowledge of the evolution of photosynthesis from its origin up to the very recent developments of extant species as depicted above is full of question marks, ambiguities and controversies. We consider this to reflect the strengths of present day's research on evolution. Whereas it is relatively straightforward to build flawless and fully self-consistent theories into empty space, empirical science is characterized by constant confrontation between models and experimental data resulting in the absence of 'absolute truth' and the menace that any glorious theory can be sent to the waste-paper bin by the outcome of a single new experiment. During the last decade, the study of the evolution of photosynthesis clearly has 'evolved' into such an empirical science. An example of a data-inflicted change of paradigm is provided by the scenarios outlined above viewing photosynthesis as an offspring of respiration-related principles of energy conversion.

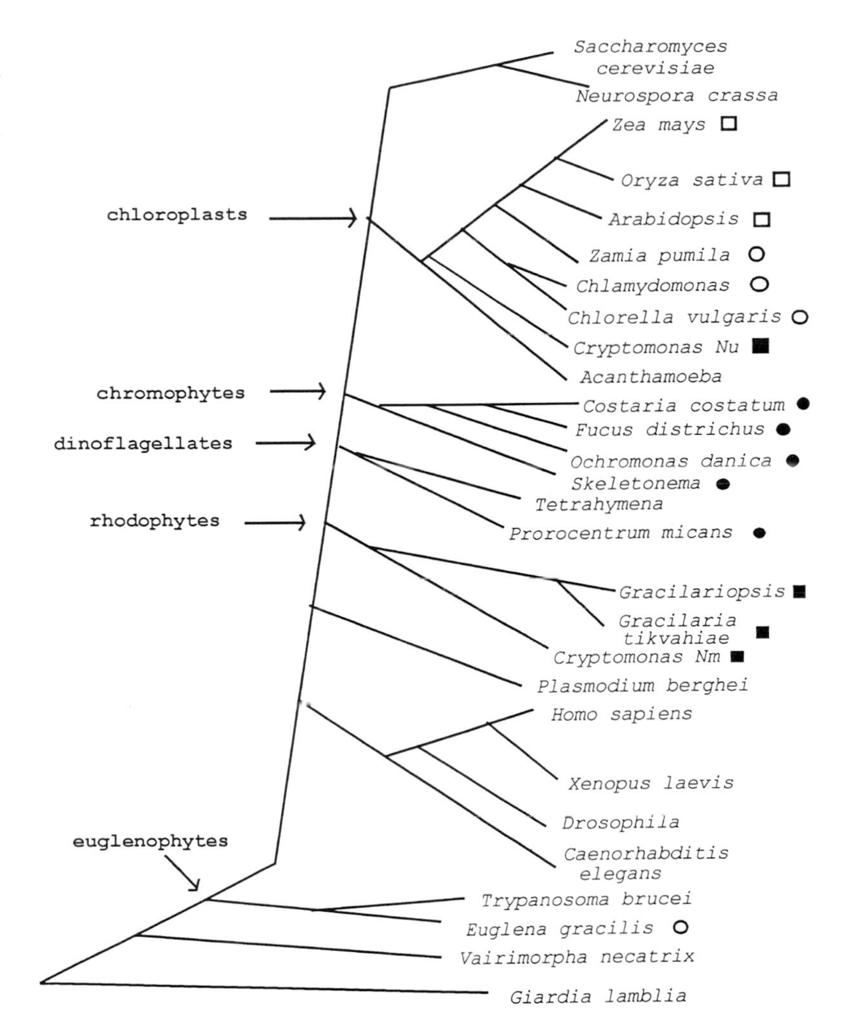


**Figure 22.6.** Distribution of photosynthetic organisms within the Eukarya. The phylogenetic relationships were derived from sequence comparisons of nuclear SSU rRNA and are not drawn to scale. (○) Chl *a/b* algae, (□) land plants, (●) Chl *a/c* algae, (■) Chl *a*-PB algae.

Therefore, in contrast to former years, when experimental scientists had a tendency to look down disparagingly on 'evolutionary' theories, research on evolution has turned not only into an empirical discipline but its results have also influenced and sometimes redirected developments of other disciplines more directly involved in the study of the photosynthetic mechanisms. We therefore hope to convince as many young scientists as possible to include thinking in evolutionary terms in their research on photosynthesis.

## ACKNOWLEDGEMENTS

This contribution has benefited enormously from suggestions or communications of results by D. Albouy (Paris/France), R. E. Blankenship (Tempe/USA), M. Bruschi (Marseille/France), D. A. Bryant (University Park/USA), G. S. Bullerjan (Bowling Green/USA), E. Gantt (College Park/USA), G. Hauska (Regensburg/FRG), D. M. Kramer (Pullman/USA), P. Joliot (Paris/France), A. W. D. Larkum (Christchurch/New Zealand), N. Nelson (Nutley/USA), A. W. Rutherford (Paris/France), A. Schricker, (Freiburg/FRG), N. Sone (Tochigi, Japan) and H. Strotmann (Düsseldorf/FRG). Furthermore, we would like to thank the group of F. Guerlesquin (Marseille/France) for extensive access to the SG-workstations on which most of the Figures were produced. W. N. and U. L. acknowledge support from European

Commission (grants ERBCHB6CT930253 and B102-CT93-00760.

FURTHER READING

Baltscheffsky, H., (ed.). (1996). *Origin and Evolution of Biological Energy Conservation*. New York: VCH Publ. Inc.

Blankenship, R. E. (1992). Origin and early evolution of photosynthesis. *Photosynthesis Research*, **33**, 91–111.

Blankenship, R. E., Madigan, M. T. & Bauer, C. W. (eds). (1995). *Anoxygenic Photosynthetic Bacteria*. Dordrecht: Kluwer Academic Publishers.

Bryant, D. A. (ed.). (1994). *Molecular Biology of the Cyanobacteria*. Dordrecht: Kluwer Academic Publishers.

Castresana, J., Lübben, M. & Saraste, M. (1995). New archaebacterial genes coding for redox proteins: implications for the evolution of aerobic metabolism. *Journal of Molecular Biology*, **250**, 202–10.

Douglas, S. E. (1992). Eukaryote–eukaryote endosymbioses: insights from the studies of cryptomonad algae. *BioSystems*, **28**, 57–68.

Felsenstein, J. (1988). Phylogenies from molecular sequences: inferences and reliability. *Annual Review of Genetics*, **22**, 521–65.

Fitch, W. M. & Bruschi, M. (1987). The evolution of prokaryotic ferredoxins – with a general method correcting for unobserved substitutions in less branched lineages. *Molecular Biology and Evolution*, **4**, 381–94.

Gray, M. W. (1992). The endosymbiont hypothesis revisited. *International Review of Cytology*, **141**, 233–357.

Knoll, A. H. (1992). The early evolution of eukaryotes: a geological perspective. *Science*, **256**, 622–7.

Lockhart, P. J., Steel, M. A. & Larkum, A. W. D. (1996). Gene duplication and the evolution of photosynthetic reaction center proteins. *FEBS Letters*, **385**, 193–6.

Lockhart, P. J., Larkum, A. W. D., Steel, M. A., Waddel, P. J. & Penny, D. (1996). Evolution of chlorophyll and bacteriochlorophyll: the problem of invariant sites in sequence analysis. *Proceedings of the National Academy of Sciences* USA, **93**, 1930–4.

Löffelhart, W. & Bohnert, H. J. (1994). Structure and function of the cyanelle genome. *International Review of Cytology*, **151**, 29–65.

Malden, B. E. H. (1995). No soup for starters? Autotrophy and the origin of metabolism. *Trends in Biochemical Sciences*, **20**, 337–42.

Matthijs, H. C. P., van der Staay, G. W. M. & Mur, L. R. (1994). Prochlorophytes: the 'other' cyanobacteria? In *The Molecular Biology of the Cyanobacteria*, ed. D. A. Bryant, Dordrecht: Kluwer Academic Publ.

McFadden, G.I. (1993). Second-hand chloroplasts: evolution of cryptomonad algae. *Advances in Botanical Research*, **19**, 189–230.

Morden, C. W., Delwiche, C. F., Kuhsel, M. & Palmer, J. D. (1992). Gene phylogenies and the endosymbiotic origin of plastids. *BioSystems*, **28**, 75–90.

Olsen, G. J., Woese, C. R. & Overbeek, R. (1994). *Journal of Bacteriology*, **176**, 1–6.

Pierson, B. K. & Olson, J. M. (1989). Evolution of photosynthesis in anoxygenic photosynthetic prokaryotes. In *Microbial Mats: Physiological Ecology of Benthic Microbial Communities*, ed. Y. Cohen, & E. Rosenberg, pp. 402–27. Washington, DC: American Society of Microbiology.

Post, A. F. & Bullerjan, G. S. (1994). The photosynthetic machinery in Prochlorophytes: structural properties and ecological significance. *FEMS Microbiology Reviews*, **13**, 393–414.

Schopf, J. W. & Walter, M. R. (1982). Origin and early evolution of cyanobacteria: the geological evidence. In *The Biology of Cyanobacteria*. eds. N. G. Carr & B. A. Whitton, Oxford: Blackwell Scientific Publishers.

Schopf, J. W. & Packer, B. M. (1987). Early Archean (3.5-billion-year-old) microfossils from Warrawoona group, Australia. *Science*, **237**, 70–3.

Schütz, M., Zirngibl, S., le Coutre, J., Büttner, M., Xie, D.-L., Nelson, N., Deutzmann, R. & Hauska, G. (1994). A transcription unit for the Rieske FeS-protein and cytochrome b in *Chlorobium limicola*. *Photosynthesis Research*, **39**, 163–74.

van der Oost, J., de Boer, A. P. N., de Gier, J.-W. L., Zumft, W. G., Stouthamer, A. H. & van Spanning, R. J. M. (1994). The heme-copper oxidase family consists of three distinct types of terminal oxidases and is related to nitric oxide reductase. *FEMS Microbiology Letters*, **121**, 1–10.

Wächtershäuser, G. (1994). Life in a ligand sphere. *Proceedings of the National Academy of Sciences* USA, **91**, 4283–7.

Whatley, J. M. (1993). The endosymbiotic origin of chloroplasts. *International Review of Cytology*, **144**, 259–99.

Woese, C. R. (1987). Bacterial evolution. *Microbiology Reviews*, **51**, 221–71.

Yu, J., Hederstedt, L. & Piggot, P. J. (1995). The cytochrome bc complex (menaquinone: cytochrome c reductase) in *Bacillus subtilis* has a nontraditional subunit organisation., *Journal of Bacteriology*, **177**, 6751–60.

# 23 Modelling leaf/canopy photosynthesis

W. R. BEYSCHLAG AND R. J. RYEL

## INTRODUCTION

Measuring the primary production of whole canopies has become an increasingly important aspect of ecological research. Questions pertaining to plant competition for light at the community level, to concern over changes in canopy flux rates resulting from global warming or increasing atmospheric $CO_2$, can in part be addressed with measurements of whole canopy photosynthesis. Since photosynthesis measurements of individual foliage elements will generally not represent the behaviour of the whole plant (due to differences in age, physiology and exposure to microclimatic conditions), an important method for estimating these fluxes has involved the use of whole-canopy photosynthesis models. In this chapter, we will present a class of these models that scale up from single-leaf estimates to the whole canopy (Fig. 23.1).

Canopy photosynthesis models generally consist of two interconnected components: (i) a single-leaf photosynthesis submodel, and (ii) a microclimatic submodel that incorporates the interactions of structure with the physical environment. The structure of the canopy is typically divided into subregions of similar foliage characteristics, and the microclimatic conditions at defined time-steps are determined with the microclimatic submodel. Rates of photosynthesis are then calculated for representative foliage elements within each subregion, weighted appropriately for foliage density and summed to generate whole plant or canopy rates.

Models discussed in this chapter range from relatively simple for homogeneous single-species canopies, to complex for diverse multispecies canopies, and are frequently used formulations suitable for addressing a range of ecological questions. Sufficient detail is provided to guide the reader in understanding and perhaps developing similar models. We limit our discussion of the microclimatic submodel to the light climate and energy balance of leaves, recognizing that addressing some questions may require consideration of additional interactions of microclimatic factors (e.g carbon dioxide concentration, air humidity, or pollutant concentrations) and foliage structure.

## SINGLE-LEAF PHOTOSYNTHESIS

The basic unit of canopy photosynthesis models is single-leaf photosynthesis where assimilation rate organ is calculated based on the physiological characteristics and surrounding microclimatic conditions. Single-leaf photosynthesis models are of two basic types: (i) empirical models, where mathematical descriptions are formulated directly between measured variables, and (ii) mechanistic models, which attempt to describe the process of photosynthesis using mathematical formulations of the underlying natural laws. Questions concerning effects of environmental conditions on the process of photosynthesis itself are best addressed with mechanistic models, while empirical models are useful when questions concern changes in canopy structure or competition for light as parametrization is often easier. Single-leaf photosynthesis models can be further divided into $C_3$ and $C_4$ models based on the two light-based metabolic pathways.

### $C_3$ photosynthesis

#### *Mechanistic models*

The most widely used predominantly mechanistic model for single-leaf $C_3$ photosynthesis is that of Farquhar, von

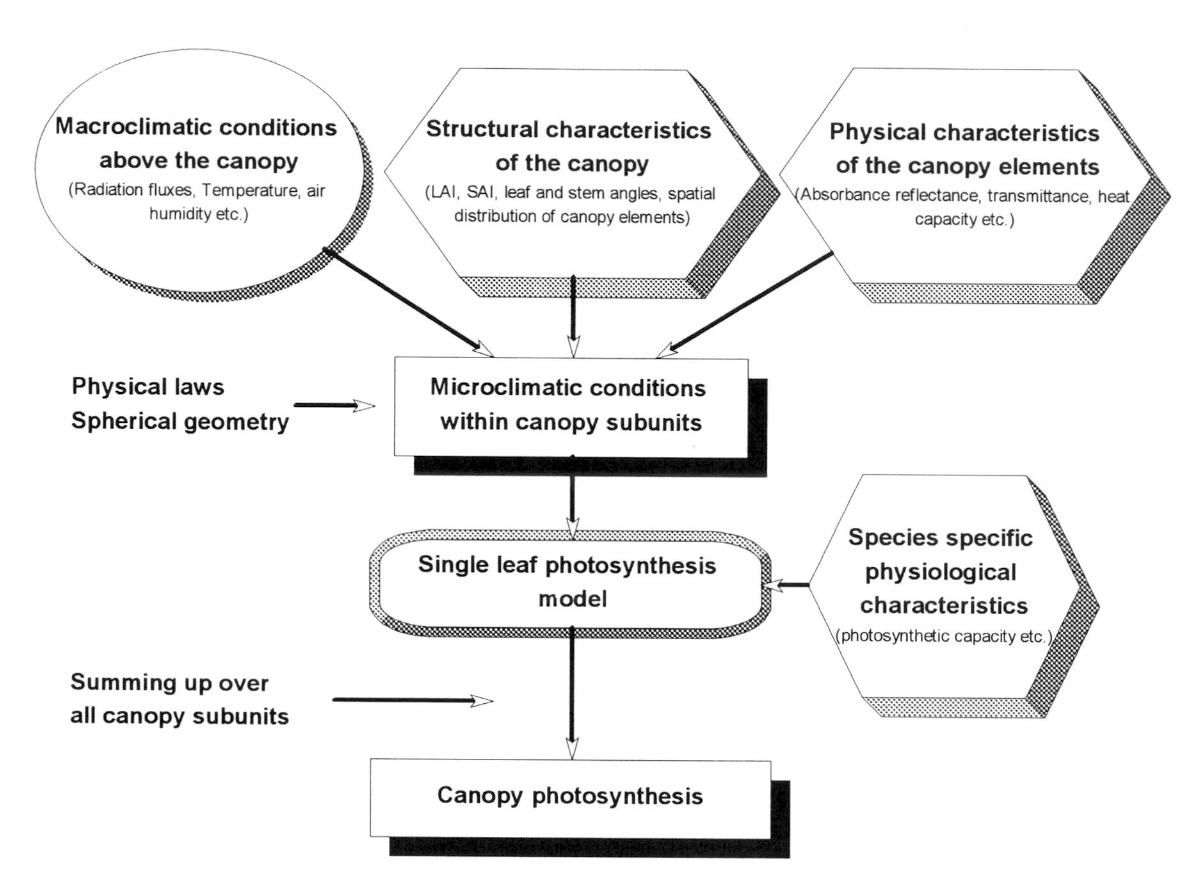


**Figure 23.1.** Generic flow chart of a canopy photosynthesis model. Parameters in rectangular boxes are the results calculated by the model. Parameters in oval and hexagonal boxes are input parameters. Parameters in hexagons have to be provided separately for each canopy subunit.

Caemmerer and Berry (1980) and Farquhar and von Caemmerer (1982). This model assumes that photosynthesis is primarily limited by activity of the $CO_2$ binding enzyme RuBP-1,5-carboxydismutase (rubisco), and by the RuBP regeneration capacity of the Calvin cycle as mediated by electron transport. We follow the development of Harley *et al.* (1992) to present the basic model structure as this formulation focuses on field parametrization.

Since 0.5 mol of $CO_2$ is released in the photorespiratory carbon oxidation cycle for each mol of $O_2$ reduced, net assimilation (A) may be expressed as:

$$A = V_c - 0.5V_o - R_d = V_c\left(1 - \frac{0.5O}{\tau C_i}\right) - R_d \qquad (1)$$

where $V_c$ and $V_o$ are the rates of carboxylation and oxygenation at rubisco, $C_i$ and O are the partial pressures of $CO_2$ and $O_2$ in the intercellular air space, $R_d$ is the rate of $CO_2$ evolution in the light resulting from processes other than photorespiration, and $\tau$ is the specificity factor for rubisco.

The rate of carboxylation, $V_c$, is assumed to be limited by three factors and is set as the minimum of:

$$V_c = \min.\{W_c, W_j, W_p\}, \qquad (2)$$

where $W_c$ is the carboxylation rate as limited solely by the quantity, activation state and kinetic properties of rubisco, $W_j$ is the rate of carboxylation limited solely by the rate of RuBP regeneration in the Calvin cycle, and $W_p$ is the carboxylation rate as limited solely by available inorganic phosphate. A can then be rewritten as:

$$A = \left(1 - \frac{0.5O}{\tau C}\right) \min.\{W_c, W_j, W_p\} - R_d. \quad (3)$$

$W_c$ is assumed to have Michaelis–Menten kinetics, with competitive inhibition by $O_2$ and is expressed as:

$$W_c = \frac{V_{c_{max}} C_i}{C_i + K_c\left(1 + \frac{O}{K_o}\right)} \quad (4)$$

where $V_{c_{max}}$ is the maximum rate of carboxylation and $K_c$ and $K_o$ are Michaelis constants for carboxylation and oxygenation, respectively.

$W_j$ is proportional to the rate of electron transport J and is expressed as:

$$W_j = \frac{JC}{4\left(C_i + \frac{O}{\tau}\right)} \quad (5)$$

under the assumption that four electrons generate sufficient ATP and NADPH for the regeneration of RuBP in the Calvin cycle (Farquhar & von Caemmerer, 1982). The rate of electron transport J is dependent on incident photosynthetic photon flux density (PFD) and is expressed empirically by Harley *et al.* (1992) using the equation of Smith (1937) as:

$$J = \frac{\alpha I}{\left(1 + \frac{\alpha^2 I^2}{J_{max}^{\;2}}\right)^{1/2}} \quad (6)$$

where $\alpha$ is the efficiency of converting PFD energy, I is incident PFD and $J_{max}$ is the light saturated rate of electron transport. The third carboxylation rate of eqn. 2, $W_p$, can be expressed as:

$$W_p = 3TPU + \frac{V_o}{2} = 3 \cdot TPU + \frac{V_c\, 0.5O}{C_i \tau} \quad (7)$$

where TPU is the rate of phosphate release in triose-phosphate utilization during starch and sucrose production.

Model factors $K_c$, $K_o$, $R_d$ and $\tau$ have temperature dependencies expressed empirically by Harley *et al.* (1992) as:

$$K_c, K_o, R_d, \text{and } \tau = \exp\left[c - \frac{\Delta H_a}{RT_k}\right] \quad (8)$$

where c is a parameter-specific scaling constant, $\Delta H_a$ is the activation energy, R is the gas constant, and $T_k$ is leaf temperature.

The temperature dependency of A is also mediated through the temperature dependency of $J_{max}$ and $V_{c_{max}}$, and several formulations are possible. The formulation of Harley *et al.* (1992) involving activation and deactivation energies based on Johnson, Eyring and Williams (1942) is:

$$J_{max} \text{ and } V_{c_{max}} = \frac{\exp\left[c - \frac{\Delta H_a}{RT_k}\right]}{1 + \exp\left[\frac{\Delta S T_k - \Delta H_d}{RT_k}\right]} \quad (9)$$

where $\Delta H_d$ is the deactivation energy, and $\Delta S$ is an entropy term.

### *Empirical Models*

Numerous empirical models for leaf photosynthesis appear in the literature, and their popularity is based on the general ease of parametrization. A popular empirical model which gives excellent fits to a wide range of leaf photosynthesis data is that of Thornley and Johnson (1990) with gross photosynthesis (P) expressed as:

$$P = \frac{\alpha I P_m}{\alpha I + P_m} \quad (10)$$

where $\alpha$ is the photochemical efficiency, I is incident PFD, and $P_m$ is the maximum gross photosynthesis rate at saturating PFD. Incorporating a factor ($\theta$) for resistance between the $CO_2$ source and site of photosynthesis, P can be expressed as:

$$P = \frac{1}{2\theta}\left\{\alpha I + P_m - \left[(\alpha I + P_m)^2 - 4\theta\alpha I P_m\right]^{1/2}\right\} \quad (11)$$

for $0 < \theta < 1$. The factor $\theta$ can be determined by the best fit to parametrization data.

Another rather easily parametrized empirical model is that of Tenhunen *et al.* (1987). This model describes PFD and $CO_2$ dependency of net photosynthesis with formulations from Smith (1937). The dependency of A on PFD is calculated from $\alpha$ and incident PFD (I) as in eqn. 6, and $CO_2$ dependency is calculated similarly using the leaf internal $CO_2$ partial pressure ($c_i$) in place of I and carboxylation efficiency (initial slope of A-$c_i$ response curve) in place of $\alpha$. A light dependency of carboxylation efficiency is also formulated using the equation of Smith (1937). The maximum capacity of photosynthesis is calculated using the formulation of Johnson *et al.* (1942) (eqn. 8).

## $C_4$ photosynthesis

In $C_4$ photosynthesis, carbon dioxide passes through the $C_4$ cycle involving carboxylation of phospho*enol*pyruvate (PEP) in the mesophyll cells, transfer of $C_4$ compounds to the bundle sheath cells and decarboxylation of these compounds before entering the $C_3$ photosynthesis cycle (Peisker & Henderson, 1992). Thus, $C_3$ photosynthesis models can be used in modelling $C_4$ photosynthesis with an initial submodel describing the $C_4$ processes and calculating the pool of inorganic carbon in the bundle sheath cells.

### *Mechanistic models*

A simple model for $C_4$ photosynthesis on the leaf level has been presented by Collatz, Ribas-Carbo and Berry (1992), and assumes that the carboxylation of inorganic carbon catalysed by PEP carboxylase is related linearly to $CO_2$ concentration of the mesophyll internal air space. A more complete model is that of Chen *et al.* (1994) where the $C_4$ cycle is assumed controlled solely by PEP carboxylase and the rate of this cycle ($V_4$) is assumed dependent on $CO_2$ concentration and described by Michaelis–Menten kinetics as:

$$V_4 = \frac{V_{4m} C_m}{C_m + k_p} \quad (13)$$

where $C_m$ is the $CO_2$ concentration in the mesophyll space, $k_p$ is the Michaelis constant, and $V_{4m}$ is the maximum velocity related to incident PFD (I) by:

$$V_{4m} = \frac{\eta_p I}{\left(1 + \frac{\eta_p^2 I^2}{V_{pm}^{\ 2}}\right)^{1/2}} \quad (14)$$

where $\eta_p$ is a fitted parameter and $V_{pm}$ is the potential activity of PEP carboxylase. The $C_3$ and $C_4$ parts of the model are linked through the $CO_2$ and $O_2$ concentration in the bundle sheath cells by characterizing the diffusion flux ($V_b$) of $CO_2$ between the mesophyll and the bundle sheath (i.e. rate of $C_3$ photosynthesis), and the net $CO_2$ exchange rate ($A_n$) between the mesophyll intercellular air space and the atmosphere, (i.e. rate of $CO_2$ fixation by PEP carboxylase). Given these relationships, $V_4$ is assumed to also be expressed as:

$$V_4 = V_b + A_n \quad (15)$$

and calculated jointly with eqn. 13 by iteratively solving for the $CO_2$ and $O_2$ concentrations in the bundle sheath cells.

### *Empirical models*

While there are many fewer empirical models for single-leaf $C_4$ photosynthesis than for $C_3$, reduced numbers of parameters over mechanistic formulations is again characteristic of these models. Dougherty *et al.* (1994) used the minimum of two photosynthetic capacities. $A_1$ is assumed to be limited by incident PFD (I) and maximum photosynthetic capacity ($A_m$) and was expressed as a non-rectangular hyperbola as:

$$A_1 = \frac{A_m + \alpha I^2 \sqrt{A_m^2 - 2A_m \alpha I(2\beta - 1) + \alpha^2 I^2}}{2\beta} \quad (16)$$

where $\alpha$ is quantum efficiency and $\beta$ is an empirical shape parameter. $A_2$ is assumed limited by intracellular $CO_2$ ($c_i$) and $A_m$ and expressed as:

$$A_2 = \frac{A_m c_i}{c_i + \frac{1}{E_c}} \quad (17)$$

where $E_c$ is an empirical index of leaf $CO_2$ efficiency. Equation 9 was used for temperature dependence of $A_m$.

Thornley and Johnson (1990) extended their empirical $C_3$ photosynthesis model (see above) to create two formulations of $C_4$ models. A distinction is made between a model with uncoupled mesophyll, where energy available to the mesophyll for pumping $CO_2$ into the bundle sheath is assumed to be independent of the energy supply to the bundle sheath for $C_3$ photosynthesis and photorespiration, and a model with coupled mesophyll where a common supply of energy to both mesophyll and bundle sheath is assumed. The reader is referred to Thornley and Johnson (1990) for details of these lengthy formulations.

## Conductance

$CO_2$ is supplied to the mesophyll through the stomata, and stomatal diffusive conductance ($g_s$), and affected by irradiance, temperature, air humidity, $CO_2$ partial pressure, water status of the plant and endogenic rhythms. Since mechanisms behind stomatal regulation are not fully understood, there are presently no true mechanistic models for stomatal conductance. Existing models can be divided up into uncoupled models in which $g_s$ is calculated solely as a function of physical conditions, and coupled models where $g_s$ is also a function of the rate of leaf photosynthesis.

### *Coupled models*

Ball, Woodrow and Berry (1987) developed a relatively simple coupled model for stomatal conductance which often has good correspondence with experimental data and is widely used. $g_s$ is assumed linearly related to net photosynthesis (A) and relative humidity ($h_s$), and is expressed as:

$$g_s = \frac{kAh_s}{c_s} \quad (18)$$

where $c_s$ is the mole fraction of $CO_2$ at the leaf surface, and k a constant.

Leuning *et al.* (1995) modified this model to account for interactions between gas molecules leaving and entering stomata, and the observations that stomata respond to the rate of transpiration. Stomatal conductance is expressed as:

$$g_s = g_0 + \frac{a_1 A}{(c_r - \Gamma)\left(1 + \frac{D_s}{D_0}\right)} \quad (19)$$

where $a_1$ and $D_0$ are an empirical coefficients, $\Gamma$ is the $CO_2$ compensation point, $g_o$ is the stomatal conductance when A approaches zero, and $D_s$ and $c_s$ are humidity deficit and $CO_2$ concentration at the leaf surface, respectively.

### *Uncoupled models*

In these models $g_s$ is calculated strictly as a function of environmental factors, and is often related linearly to incident PFD, leaf temperature, water vapour mole fraction difference, air humidity and/or leaf water potential. While uncoupled conductance models may be easier to parametrize and often give good correspondence to measured data, coupled models are recommended when more physiologically based questions are addressed.

## WHOLE PLANT/CANOPY PHOTOSYNTHESIS

Predicting whole-plant or whole-canopy gas-exchange rates requires the simultaneous calculation of fluxes at multiple locations within the plant canopy. Because of the strong dependence of photosynthesis on incident light and influences of light on stomatal conductance, variability in incident light intensity is usually the overwhelming contributor to variability in rates of net photosynthesis and transpiration in the vegetative crown. In this section, model structures are reviewed that define light–foliage interactions within individual plants and plant canopies.

## Radiation–foliage interactions

Radiation intercepted by plant foliage can be divided into long- and short-wave radiation, with short-wave radiation classified as total short-wave and photosynthetically active photon flux (PFD). Total short-wave radiation includes the visible light and near infra-red portions of the electromagnetic spectrum (400–3000 nm), while PFD is the range of visible light where absorption by chlorophyll *a* and *b* is high. Long-wave radiation includes the infra-red spectrum > 3000 nm. Interception of PFD is important as a driving factor in the process of photosynthesis, while total short-wave and long-wave radiation affect leaf temperature by affecting the leaf energy balance.

Short-wave radiation incident on a leaf surface is the sum of fluxes from three sources: direct solar beam, diffuse radiation from the sky, or diffuse radiation reflected and transmitted by other foliage elements (leaf diffuse). Direct solar beam radiation is affected by position of the sun and cloud conditions, and has altitude and azimuth components that depend on latitude, date and time of day. Diffuse radiation from the sky emanates from the hemisphere of sky above the horizon. Under clear or uniformly overcast skies, the radiation from all portions of the hemisphere may be quite similar, but under patchy clouds, the intensity may vary depending on location and density of clouds. Leaf diffuse radiation results from the reflection and transmission of direct beam and sky diffuse radiation within the canopy, and flux depends on the proximity, and optical properties of adjacent foliage.

## Leaf energy balance

Absorbed short-wave and long-wave radiation, convection, leaf transpiration and leaf long-wave emittance affect leaf temperature. Long-wave radiation reaches the leaf surface from the sky, soil surface and from surrounding foliage and is dependent on the temperature and emissivity of the radiation surfaces. Convective heat transfer occurs between the leaf and the surrounding air, and is affected by air and leaf temperature and wind speed across the leaf surface, while latent heat loss from the leaf is affected by transpiration rate. Leaf temperature is determined by equilibrating energy gains and losses:

$$I_S a_S + I_L a_L = C_I + H_I + L_I \qquad (20)$$

where $I_S$ and $I_L$ are the intercepted short-wave and long-wave radiation, respectively, $a_S$ and $a_L$ are fractions of radiation absorbed, $C_I$ is the leaf convective loss (or gain), $H_I$ is latent heat loss from transpiration, and $L_I$ is long-wave radiation emittance from the leaf surface. Norman (1979) and Gates (1980) provide formulations for convective and latent heat transfer and leaf emittance.

Calculations of leaf temperature require iterative calculation procedures to balance energy gains and losses, and model formulations are more complex when leaf temperature is included. When leaves are small or narrow in stature, the assumption that leaf and air temperature are identical is often made (e.g. Ryel *et al.*, 1990; Ryel, Beyschlag & Caldwell, 1993; Wang & Jarvis, 1990), while models used for plants with wider leaves (e.g. Caldwell *et al.*, 1986) or in situations of minimal wind and low convective loss (e.g. Ryel & Beyschlag, 1995) have incorporated energy balance routines for calculating leaf temperature.

## Monotypic plant stands

The simplest canopy photosynthesis models are for single species plant communities with relatively homogeneous distribution of foliage (e.g. lawns, pastures, crop canopies and some forest canopies). The approach is to divide the canopy into layers of approximately uniform foliage density and orientation, and calculate radiation interception and gas flux rates for points located within the centre of these layers (Fig. 23.2, *upper*). These models are one-dimensional since differences in radiation interception will only occur in the vertical dimension.

Thickness of model layers is determined by the foliage density distribution in the canopy, and layers ideally have leaf area index (LAI) less than 0.5 $m^2$ foliage area per $m^2$ ground area to facilitate calculation of leaf diffuse (Norman, 1979). Foliage is typically divided into leaves and stems (or branches), as these units have different optical and photosynthetic properties, and often different orientations. Foliage elements are typically assumed to be randomly distributed throughout the canopy layer, but provisions can be made for more clumped foliage (Caldwell *et al.*, 1986).

The orientation of foliage is described by the angle of

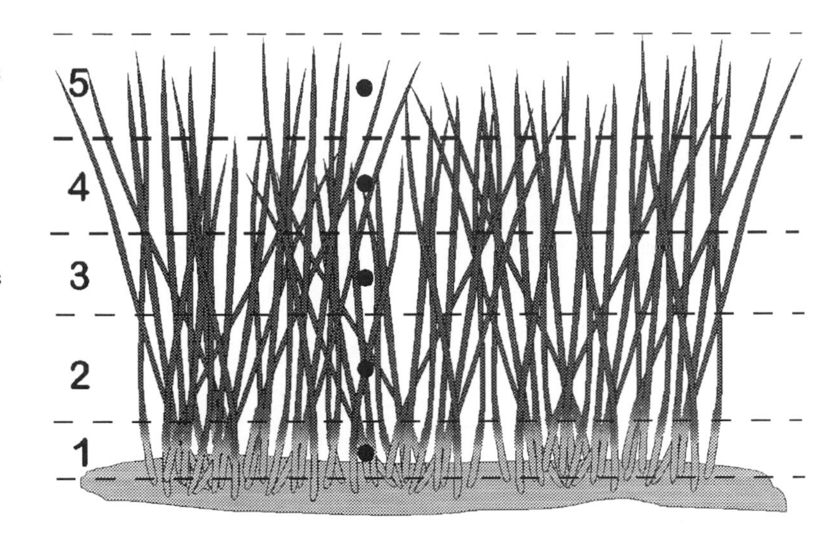


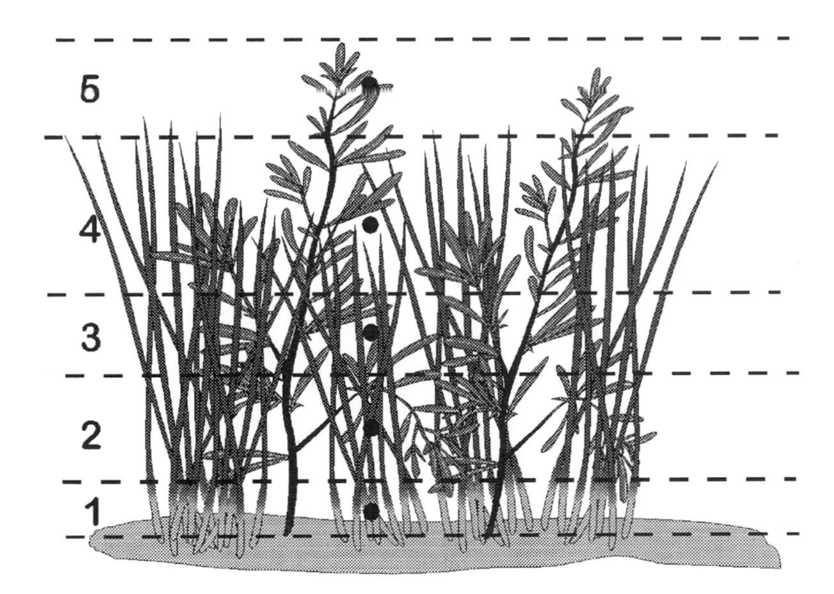


**Figure 23.2.** *Upper:* Schematic depiction of a uniform single-species grass stand divided into five layers of similar foliage density and orientation. Points represent locations within layer for which calculations of light interception and net photosynthesis are performed. *Lower:* Schematic depiction of a two species mixed canopy divided into five layers of similar foliage density and orientation. Points represent locations within layer for which calculations of light interception and net photosynthesis are performed.

the element from horizontal (leaf angle), and directional alignment (azimuth angle). Leaf angle is usually assumed to be a constant within layers, estimated as the average of foliage elements in the layer, but various distributions have also been proposed (e.g. Norman, 1979; Campbell, 1986). Foliage elements are usually assumed to have random azimuth orientations, but models have considered non-random distributions of leaf azimuth angles (e.g. Caldwell *et al.*, 1986). We present the case where foliage angles are assumed constant within layers, and elements have random azimuth orientation.

Interception of PFD is calculated for sunlit and shaded foliage within each layer. As direct beam radiation penetrates a canopy, it is assumed to decline in area

**Figure 23.3.** Schematic representation of direct beam PFD flux (*left*) and diffuse PFD flux (*right*) through a portion of a plant canopy. PFD flux for direct beam is intercepted by leaves to reduce the area sunlit in lower foliage, while the flux remains constant. Flux for diffuse PFD is reduced as it passes through foliage, but all foliage remains illuminated.

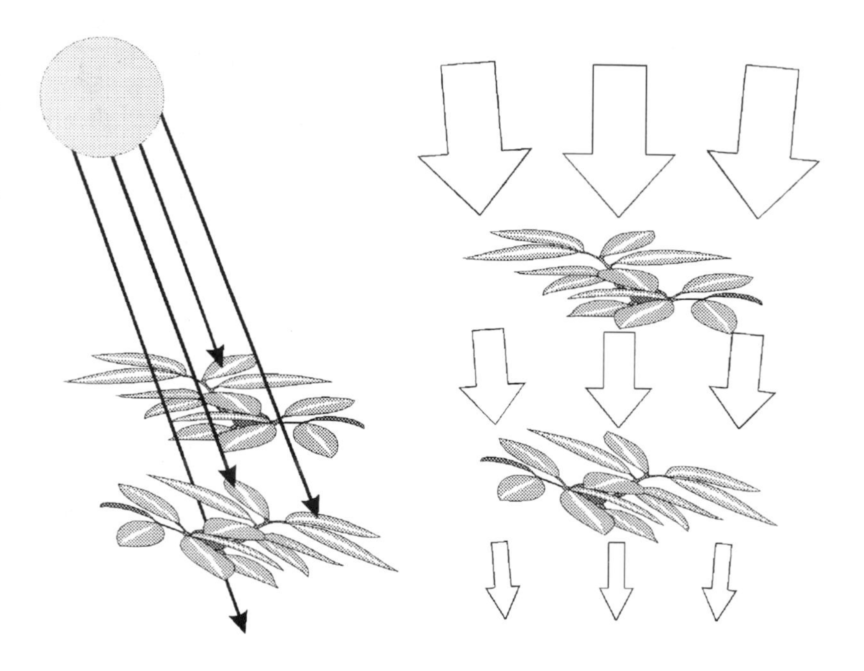

but not in intensity (Fig. 23.3, *left*), the area of direct beam PFD can be assumed to decline exponentially with passage through increasing quantities of randomly distributed foliage. The fraction ($P_I$) of foliage sunlit (area of direct beam) at a point in layer I (I = 1 for bottom layer; I = n for top layer) can be expressed as:

$$P_I = \exp\left(\sum_{i=I}^{n}(L_i\,Kl_i + S_i\,Ks_i)\,l_i\right) \quad (21)$$

where $L_i$ and $S_i$ represent the leaf and stem area densities ($m^2\ m^{-3}$) in layer i, respectively, and $l_i$ is the path length (m) of PFD through layer i. Light extinction coefficients for leaves and stems ($Kl_i$, and $Ks_i$, respectively) are calculated according to Duncan *et al.* (1967) for foliage angle $\phi_i$, from horizontal and angle of sun $\gamma$ above horizon as:

$$K_i = \cos\phi_i\,\sin\lambda \quad (22)$$

when $\phi_i$, is less than or equal to $\lambda$, and as:

$$K_i = \frac{2}{\pi}\sin\phi_i\,\cos\lambda\,\sin\gamma_i + \left(1-\frac{\gamma_i}{90}\right)\cos\phi_i\,\sin\lambda \quad (23)$$

when $\phi_i$, is greater than $\lambda$ where $\gamma_i$ is the angle (0–90°) that satisfies:

$$\cos\gamma_i = \cot\phi_i \cdot \tan\lambda \quad (24)$$

The incident flux of direct beam radiation on a sunlit leaf surface depends on the relative orientations of the leaf surface, and beam direction and intensity of the sun. Flux on the upper leaf surface can be expressed as:

$$Qu_i = B(\cos\phi_i\,\sin\lambda - \sin\phi_i\,\cos\lambda\cos\delta_i) \quad (25)$$

and on the lower surface as:

$$Ql_i = B(\cos(180-\phi_i)\sin\lambda - \sin(180-\phi_i)\cos\lambda\cos\delta_i) \quad (26)$$

where B is the flux of PFD on a surface normal to the solar beam, and $\delta_i$ is the azimuth of the axis of the leaf angle relative to the sun azimuth.

In contrast to direct beam, sky-diffuse radiation emanates from a hemisphere above the canopy, not a point. Calculation of attenuation is most readily accomplished by dividing this hemisphere into bands of equal width or area, and calculating the flux as if it originates as a point from the centre of the band. Equations are analogous to that for direct beam (eqn. 21) except that flux, not area, is reduced (Fig. 23.3. right) and may be expressed for layer I as:

$$D_{I,w} = D_{sky} A_w \exp\left(\sum_{i=I}^{n}(L_i Kl_i + S_i Ks_i) l_i\right) \quad (27)$$

where w is the skyband (e.g w = 1, 2, . . ., 9 for bands = 0–10°, 10–20°, . . ., 80–90°), $D_{sky}$ is the flux of sky diffuse on a horizontal surface and $A_w$ is the fraction of the total hemisphere contained in band w visible to the leaf surface (see Duncan *et al.*, 1967).

Diffuse radiation reflected and transmitted by foliage is the most difficult aspect of radiation interception to model. Leaf optical properties (reflectance and transmittance), foliage density and proximity of adjacent foliage all contribute to variation in scattering of this radiation. The simplest approach for modelling the attenuation of leaf diffuse radiation is to assume that radiation striking the upper surface of a leaf is reflected upwards or transmitted downwards, while the reverse would hold for radiation striking the lower leaf surface. Downward flux of PFD leaving the midpoint of layer I would be:

$$\begin{aligned} Td_i = {} & t_i L_i (Qu_i P_i + Du_i) + r_i L_i (Ql_i P_i + Dl_i) \\ & + Tu_{i-1} r_i L_i + Td_{i+1} t_i L_i + Td_{i+1}(1 - L_i - S_i) \end{aligned} \quad (28)$$

while the upward flux leaving layer I would be:

$$\begin{aligned} Tu_i = {} & r_i L_i (Qu_i P_i + Du_i) + t_i L_i (Ql_i P_i + Dl_i) \\ & + Td_{j+1} r_i L_i + Tu_{i-1} t_i L_i + Tu_{i-1}(1 - L_i - S_i) \end{aligned} \quad (29)$$

where $r_i$ and $t_i$ are leaf reflectance and transmittance, respectively, for PFD. The total PFD incident on both sides of a sunlit leaf in layer I is then:

$$I_i = Qu_i + Ql_i + Du_i + Dl_i + Tu_{i-1} + Td_{i+1} \quad (30)$$

where $Du_i$ and $Dl_i$ are sum of sky diffuse for all sky bands for upper and lower leaf surfaces, respectively. Shaded leaves would be without the contributions of the direct beam components $Qu_i$ and $Ql_i$.

## Multispecies plant stands

The model for uniform monotypic plant stands can be extended rather easily to include multiple species. Foliage components of each species are assumed to be relatively uniformly distributed horizontally, but the vertical mixing of crowns may vary. The simplest case has the canopy divided horizontally into monotypic layers by species, the situation when one species overtops another, while the more general case has foliage elements mixed within canopy layers (Fig. 23.2, lower) such as in grassland or crop–weed canopies (Ryel *et al.*, 1990).

Equations describing interception of PFD are very similar to those for the monotypic plant stands. The analogous equation to eqn. 21 for the fraction of foliage lit by direct beam radiation in layer I for species x is:

$$P_{x,I} = \exp\left(\sum_{i=I}^{n}\left(\sum_{y=1}^{N}(L_{y,i} Kl_{y,i} + S_{y,i} Ks_{y,i})\right) l_i\right) \quad (31)$$

where N is the number of species in the canopy. Similarly, eqns. 27–30 can be extended to apply to multiple species.

## Canopies without homogeneous structure

Many plant canopies contain clumped or discontinuous vegetation and are not adequately represented with canopy models that have only vertically varying foliage density. Interception of light by plants in these canopies is affected by surrounding vegetation positioned at varying distances and compass orientation, and light may reach plants from both the sides and top. Because of canopy complexity, a three-dimensional model is necessary.

The basic approach to modelling these heterogeneous canopies is to fit individual plants or clumps of vegetation with suitable three-dimensional geometric shapes such as cubes, cones, ellipses, and cylinders. We illustrate this approach using the cylinder model of Ryel *et al.* (1993). Analogous to the layer model, an individual plant is divided into concentric cylinders and horizontal layers of relatively uniform foliage density and orientation (Fig. 23.4). Instead of a single vertical array of points, light interception and gas exchange are calculated

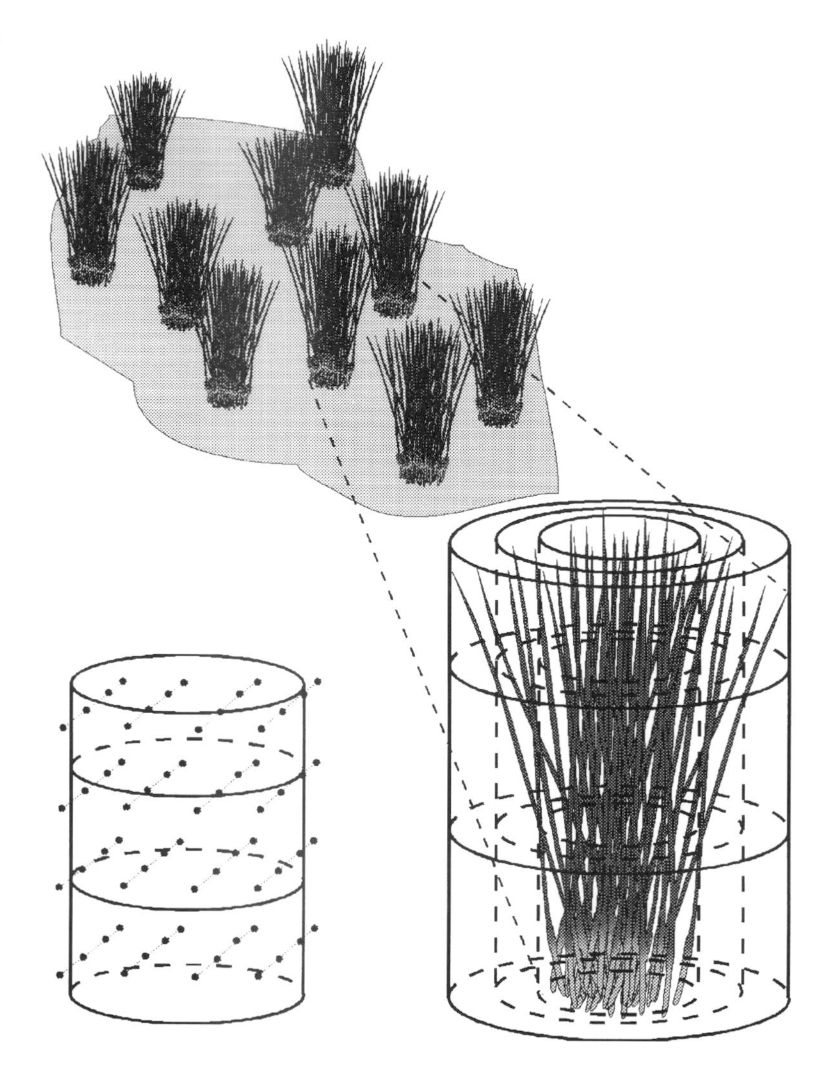

**Figure 23.4.** Schematic depiction of a grass tussock divided into three layers and three subcylinders of similar foliage density and orientation. Cylindrical model structure would be applied to all tussocks within the stand for calculation of whole stand net photosynthesis. Figure in lower left corner depicts matrix of points at which calculations are performed by the model.

for a three-dimensional array of points within the cylinder (Fig. 23.4). Averages of gas-flux rates calculated for points within each canopy subsection are weighted by green foliage area and summed to generate whole plant flux rates. Calculations can be made for all plants within a desired vegetation canopy, for representative individuals of plant classes of similar structure and stature, or for single plants when all plants are assumed to be identical and equal. Whole canopy flux rates are simply sums of rates for the individual plants. As many as 1000 or more points may be necessary to 'sample' an individual plant to obtain consistent estimates of flux rates (Ryel *et al.*, 1993).

The area of direct beam radiation (fraction of sunlit foliage) for point k in plant x is:

$$P_{x,k} = \exp\left(\sum_{y=1}^{N_p}\left(\sum_{i=1}^{n_{y,l}}\sum_{m=1}^{n_{y,c}}(L_{y,i,m}Kl_{y,i,m} + S_{y,i,m}Ks_{y,i,m})l_{y,i,m}\right)\right) \quad (32)$$

where $N_p$ is the total number of plants in the canopy, and $n_{y,l}$ and $n_{y,c}$ are the number of layers and subcylinders in plant y, respectively.

The flux of sky diffuse radiation is calculated for azimuth directions within sky bands, as neighbouring plants usually do not form a uniform border of foliage in all compass directions. The analogous equation to eqn. 27 is:

$$D_{x,p,w,a} = D_{sky} A_{w,a} \exp\left(\sum_{y=1}^{N_p}\left(\sum_{i=1}^{n_{y,l}}\sum_{m=1}^{n_{y,c}}(L_{y,i,m} K_{l\,y,i,m} + S_{y,i,m} K_{s\,y,i,m}) l_{y,i,m}\right)\right) \quad (33)$$

where $A_w$ is extended ($A_{w,a}$) to include azimuth orientation (a) of the skyband subsection.

Reflected and transmitted diffuse within a complex canopy is difficult to model. Norman and Welles (1983) and Ryel *et al.* (1993) calculated the average reflected and transmitted diffuse within layers of points within the canopy. This is most easily done within individual plants where Q and D in eqn. 28 and 29 for matrix layer i are replaced with the average values for all points in the matrix layer.

### 'Big leaf' models

'Big-leaf' models (Sellers *et al.*, 1992; Amthor, 1994) for calculating whole-canopy net photosynthesis assume gas fluxes of whole canopies can be approximated by a single leaf. Model development involves reducing properties of the whole canopy to that of a single leaf and using modified equations for single-leaf net photosynthesis and conductance (e.g. eqn. 1 and 18 above), resulting in greatly reduced model complexity and fewer parameters to estimate. They have appeal when flux rates are modelled at the landscape scale for several vegetation communities.

Despite this simplicity, 'big-leaf' models have some significant drawbacks. Parameters for these models cannot be measured directly, and because of non-linearities of the functions involved, simple arithmetic means of parameters for individual leaves should not be used (Leuning *et al.*, 1995). Calculation of flux rates for the 'big leaf' is also affected by the non-linearity of functions, clearly illustrated by McNaughton (1994) who found that the average canopy conductance that preserves whole-canopy transpiration flux was different from that preserving whole-canopy $CO_2$ assimilation. However, 'big-leaf' models may prove useful for long-term simulations (e.g. monthly, seasonal, annual) if calculated fluxes are comparable to those from more detailed canopy models for a suitable range of meteorological conditions.

## ESTIMATION OF MODEL PARAMETERS

While a detailed discussion of methods for parametrizing canopy photosynthesis models is beyond the scope of this chapter, a brief overview of data collection methods is presented. Obtaining suitable parameters for canopy photosynthesis models is often a challenging endeavour due to the difficulty of measuring flux rates and structural parameters. In addition, parameter estimates can significantly affect model outputs and study conclusions. Care must be taken to ensure reasonable parameter estimation, and model sensitivity analyses to parameter estimates should be used to assess the robustness of conclusions based on model simulations.

### Single-leaf photosynthesis

Many parameters for the model of Farquhar *et al.* (1980) can be derived directly from gas-exchange measurements, as the initial slope of the A vs. $c_i$ relationship reflects the activity status of rubisco, and A at saturating light and $CO_2$ characterizes the maximum RuBP regeneration rate of the Calvin cycle. Further, the initial slope of the relationship A vs. incident (or absorbed) PFD at saturating $CO_2$ is an estimate of the maximum light-use efficiency (or quantum-use efficiency) of the light reaction of photosynthesis. These light-use efficiency parameters can also be estimated through measurements of chlorophyll fluorescence (Schreiber, Bilger & Neubauer, 1994; see Chapter 24 of this volume). Estimates of biochemical parameters can often be obtained from the literature, as the biochemical properties of specific enzymes should be independent of the plant species. Nevertheless, if responses at the physiological level are desired, it may be important to obtain *in vitro* parameter estimates, since some of these biochemical parameters may vary between species or varieties (Evans & Seemann, 1984). For stomatal models, relationships between apparent stomatal conductance and physical variables (e.g. air humidity, incident PFD) can be calculated from gas-exchange measurements.

### Canopy structure

Density and distribution of foliage within a canopy can be obtained using destructive or non-destructive methods, and Norman and Campbell (1989) provide a comprehensive review. Harvesting foliage from subsets of the plant canopy is the most common destructive method, and analytical devices such as a leaf-area meter can aid in this quantification. Non-destructive methods often use within canopy light measurements or linear probes. Inversion of eqns. 21 and 27 combined with within canopy light measurements can be used to estimate foliage area in many types of canopies. The LAI-2000 plant canopy analyser (LI-COR, Lincoln, NE, USA) is a portable canopy light analyser that automatically calculates foliage area and orientation with some limitations. High contrast fish-eye photography is another light-based method used to estimate foliage area. Linear probe methods, commonly called inclined-point quadrats, involve repeatedly moving a rod orientated at a fixed angle through the canopy, using the frequency of contacts of foliage in calculating foliage area.

Foliage orientation can be estimated *in situ*, with the LAI-2000 analyser or with linear-probe methods. Leaf reflectance and transmittance of PFD, total shortwave and longwave radiation can be measured using a spectroradiometer, but suitable values can often be obtained from the literature.

## MODEL VALIDATION

Model validation is designed to enhance model performance by determining whether the model output approximates to the simulated system. Validation is typically conducted by comparing model outputs and independent estimates of selected parameters. These comparisons may be made visually or statistically. Single-leaf photosynthesis and conductance models are often validated by comparing model and measured values of net photosynthesis, transpiration, intercellular $CO_2$ and conductance throughout the course of one or more days (Tenhunen *et al.*, 1987; Beyschlag *et al.*, 1990; *Ryel et al.*, 1993). Model routines for within, and below, canopy light interception can be validated by comparing light sensor readings with model predictions for incident light in similar canopy locations (Norman & Welles, 1983; Ryel *et al.*, 1990; Beyschlag, Ryel & Dietsch, 1994). Diurnal courses of whole-plant gas exchange measured within large gas-exchange cuvettes can be compared with similar model outputs (Ryel *et al.*, 1993). Sap-flow measurements of a suitable sample of plants within a canopy are useful in the validation of whole-canopy transpiration rates, while eddy-correlation measurements of gases within and above the canopy can also be useful in whole-canopy flux-rate validation.

## PERSPECTIVES

Canopy photosynthesis models link measurable single-leaf photosynthesis with estimates of whole-canopy photosynthesis which are difficult to measure. As presented in this chapter, models are now available that range from formulation for single-species, homogeneous canopies to complex, non-homogeneous mixed species canopies. They are modular in format allowing relatively easy replacement of components with other routines more suitable for study objectives or available data sets.

While important in addressing questions concerning basic plant ecophysiology (e.g. Beyschlag, Ryel & Caldwell, 1994), environmental change (e.g. Ryel *et al.*, 1990; Reynolds *et al.*, 1992) and crop management (e.g. Grace, Jarvis & Norman, 1987), outputs from these models also provide carbon-gain inputs to allocation and growth models (e.g. Johnson & Thornley, 1985; Charles-Edwards, Doley & Rimmington, 1986; Reynolds *et al.*, 1987). Linking canopy photosynthesis and growth/allocation models necessitates a feedback mechanism since carbon allocation and growth change the structural parameters of plant canopies, which in turn affects canopy photosynthesis.

## LIMITATIONS

While there are numerous possibilities to apply canopy photosynthesis models, there are also limitations. Time and cost expenditure for parametrization may be high (particularly for structural parameters), and increases with the number of species and complexity of canopy structure. Model complexity can also increase computational time to impractical levels. Model structure, however detailed, may not fit the system modelled, and model validation is important in reducing this risk.

Extrapolation of model outputs beyond the range of validation data can produce erroneous results, but may be the only way to gain insights into phenomena that cannot be measured. Care must be exercised in interpreting results.

## FUTURE DEVELOPMENTS

New developments within the area of canopy photosynthesis models will likely point in opposite directions. Additional phenomena may be added, primarily at the leaf level, including stomatal patchiness and effects of sun-flecks or leaf flutter on short-term photosynthesis. Linkages between macroclimate above and microclimate within the canopy may also be improved, including better representation of air turbulence and concentration gradients. These additions will increase the complexity of existing models.

Reduced complexity of canopy photosynthesis models will also be necessary with the need to estimate flux rates for the patchwork of communities at the landscape level. Further developments of big-leaf models may be important in this quest, but problems with model structure and parametrization need to be addressed. In addition, canopy-level models of photosynthesis based upon remote sensing data are likely to see more development in the future (Field, Gamon & Peñelas, 1994).

Further developments in linking plant and canopy photosynthesis models with growth models are also expected, particularly in assessing how competition for light affects growth patterns. As growth models become more mechanistic, the feedback linkage between carbon gain and structural changes within the canopy can be better defined. Another area where canopy photosynthesis estimation is useful is in the analysis of plant succession and the formation of stable patterns, where the development of cellular automaton models (Silvertown *et al.*, 1992) becomes increasingly popular.

## CONCLUSIONS

Mathematical simulation models are valuable tools in quantifying photosynthetic primary production at the level of the whole plant or canopy, where measurements are often difficult. Models have been developed to apply to a wide spectrum of canopies. The modular approach outlined here allows for the development and expansion of these models for application to new and varied questions. Validation techniques continue to be developed and refined to aid in the improvement and parametrization of these models. New ideas and concepts will be invaluable in the future application, development and role in research of these models.

### ACKNOWLEDGEMENTS

Part of this work has been funded by the Deutsche Forschungsgemeinschaft, Bonn (SFB 251, University of Würzburg, FRG).

### REFERENCES

Amthor, J. S. (1994). Scaling $CO_2$-photosynthesis relationships from the leaf to the canopy. *Photosynthesis Research*, **39**, 321–50.

Ball, J. T., Woodrow, I. E. & Berry, J. A. (1987). A model predicting stomatal conductance and its contribution to the control of photosynthesis under different environmental conditions. In *Progress in Photosynthesis Research*, vol. IV, ed. J. Biggens, pp. 221–4. Dordrecht: Martinus Nijhoff Publishers.

Beyschlag, W., Barnes, P. W., Ryel, R. J., Caldwell, M. M. & Flint, S. D. (1990). Plant competition for light analyzed with a multispecies canopy model. II. Influence of photosynthetic characteristics on mixtures of wheat and wild oat. *Oecologia*, **82**, 374–80.

Beyschlag, W., Ryel, R. J. & Caldwell, M. M. (1994). Photosynthesis of vascular plants: assessing canopy photosynthesis by means of simulation models. In *Ecophysiology of Photosynthesis. Ecological Studies*, vol. 100, ed. E-D. Schulze & M. M. Caldwell. pp. 409–30, Berlin, Heidelberg, New York: Springer-Verlag.

Beyschlag, W., Ryel, R. J. & Dietsch, C. (1994). Shedding of older needle age classes does not necessarily reduce photosynthetic primary production of Norway spruce. Analysis with a three-dimensional canopy photosynthesis model. *Trees, Structure and Function*, **9**, 51–9.

Caldwell, M. M., Meister, H. P., Tenhunen, J. D. & Lange, O. L. (1986). Canopy structure, light microclimate and leaf gas exchange of *Quercus coccifera* L. in a Portuguese macchia: measurements in different canopy layers and simulations with a canopy model. *Trees, Structure and Function*, **1**, 25–41.

Campbell, G. (1986). Extinction coefficients for radiation in plant canopies calculated using an ellipsoidal inclination angle distribution. *Agricultural and Forest Meteorology*, **36**, 317–21.

Charles-Edwards, D. A., Doley, D. & Rimmington, G. M. (1986). *Modelling Plant Growth and Development*. Sydney: Academic Press.

Chen, D-X., Coughenour, M. B., Knapp, A. K. & Owensby, C. E. (1994). Mathematical simulation of $C_4$ grass photosynthesis in ambient and elevated $CO_2$. *Ecological Modelling*, **73**, 63–80.

Collatz, G. J., Ribas-Carbo, M. & Berry, J. A. (1992). Coupled photosynthesis-stomatal conductance model for leaves of $C_4$ plants. *Australian Journal of Plant Physiology*, **19**, 519–38.

Dougherty, R. L., Bradford, J. A., Coyne, P. I. & Sims, P. L. (1994). Applying an empirical model of stomatal conductance to three C-4 grasses. *Agricultural and Forest Meteorology*, **67**, 269–90.

Duncan, W. C., Loomis, R. S., Williams, W. A. & Hanau, R. (1967). A model for simulating photosynthesis in plant communities. *Hilgardia*, **38**, 181–205.

Evans, J. R. & Seemann, J. R. (1984). Differences between wheat genotypes in specific activity of ribulose-1,5-bisphosphate carboxylase and the relationship to photosynthesis. *Plant Physiology*, **74**, 759–65.

Farquhar, G. D., von Caemmerer, S. & Berry, J. A. (1980). A biochemical model of photosynthetic $CO_2$ assimilation in leaves of $C_3$ species. *Planta*, **149**, 78–90.

Farqhuar, G. D. & von Caemmerer, S. (1982). Modelling of photosynthetic response to environmental conditions. In *Encyclopedia of Plant Physiology, New Series*, Vol. 12B, ed. O. L. Lange, P. S. Nobel, C. B. Osmond & H. Ziegler, pp. 549–87. Berlin, Heidelberg, New York: Springer-Verlag.

Field, C. B., Gamon, J. A. & Peñelas, J. (1994). Remote sensing of terrestrial photosynthesis. In *Ecophysiology of Photosynthesis. Ecological Studies*, vol. 100. ed. E-D. Schulze & M. M. Caldwell. pp. 511–27, Berlin, Heidelberg, New York: Springer-Verlag.

Gates, D. M. (1980). *Biophysical Ecology*. Berlin, Heidelberg, New York: Springer-Verlag.

Grace, J. C., Jarvis, P. G. & Norman, J. M. (1987). Modelling the interception of solar radiant energy in intensively managed stands. *New Zealand Journal of Forestry Science*, **17**, 193–209.

Harley, P. C., Thomas, R. B., Reynolds, J. F. & Strain, B. R. (1992). Modelling photosynthesis of cotton grown in elevated $CO_2$. *Plant, Cell and Environment*, **15**, 271–82.

Johnson, F., Eyring, H. & Williams, R. (1942). The nature of enzyme inhibitions in bacterial luminescence; sulfanilamide, urethane, temperature, and pressure. *Journal of Cell Comparative Physiology*, **20**, 247–68.

Johnson, I. R. & Thornley, J. H. M. (1985). Dynamic model of the response of a vegetative Grass crop to light temperature and nitrogen. *Plant, Cell and Environment*, **8**, 485–99.

Leuning, R. (1983). Transport of gases into leaves. *Plant, Cell and Environment*, **6**, 181–94.

Leuning, R., Kelliher, F. M., dePury, D. G. G. & Schulze, E-D. (1995). Leaf nitrogen, photosynthesis, conductance and transpiration: scaling from leaves to canopies. *Plant, Cell and Environment*, **18**, 1183–1200.

McNaughton, K. G. (1994). Effective stomatal and boundary-layer resistances of heterogeneous surfaces. *Plant, Cell and Environment*, **17**, 1061–8.

Norman, J. M. (1979). Modeling the complete crop canopy. In *Modification of the Aerial Environment of Crops*. ed. B. J. Barfield & J. F. Gerber, pp. 249–77. St Joseph, MI: American Society of Agricultural Engineers.

Norman, J. M. & Campbell, G. S. (1989). Canopy structure. In *Plant Physiological Ecology. Field Methods and Instrumentation*. ed. R. W. Pearcy, J. R. Ehleringer, H. A. Mooney & P. W. Rundel, pp. 301–25. London, New York: Chapman and Hall.

Norman, J. M. & Welles, J. M. (1983). Radiative transfer in an array of canopies. *Agronomy Journal*, **75**, 481–4.

Peisker, M. & Henderson, S. A. (1992). Carbon: terrestrial $C_4$ plants. *Plant Cell and Environment*, **15**, 987–1004.

Reynolds, J. F., Acock, B., Dougherty, R. L. & Tenhunen, J. D. (1987). A modular structure for plant growth simulation models. In *Biomass Production by Fast Growing Trees*. ed. J. S. Pereira & J. J. Landsberg, pp. 123–4. Dordrecht: Kluwer Academic Publishers.

Reynolds, J. F., Chen, J. L., Harley, P. C., Hilbert, D. W., Dougherty, R. L. & Tenhunen, J. D. (1992). Modeling the effects of elevated $CO_2$ on plants: extrapolating leaf response to a canopy. *Agricultural and Forest Meteorology*, **61**, 69–94.

Ryel, R. J., Barnes, P. W., Beyschlag, W., Caldwell, M. M. & Flint, S. D. (1990). Plant competition for light analyzed with a multispecies canopy model. I. Model development and influence of enhanced UV-B conditions on photosynthesis in mixed wheat and wild oat canopies. *Oecologia*, **82**, 304–10.

Ryel, R. J. & Beyschlag, W. (1995). Benefits associated with steep foliage orientation in two tussock grasses of the American Intermountain West. A look at water-use-efficiency and photoinhibition. *Flora*, **190**, 251–60.

Ryel, R. J., Beyschlag, W. & Caldwell, M. M. (1993). Foliage orientation and carbon gain in two tussock grasses as assessed with a new whole-plant gas-exchange model. *Functional Ecology*, **7**, 115–24.

Schreiber, U, Bilger, W. & Neubauer, C. (1994). Chlorophyll fluorescence as a nonintrusive indicator for rapid assessment in *in vivo* photosynthesis. In *Ecophysiology of Photosynthesis. Ecological Studies* vol. 100. ed. E -D. Schulze & M. M. Caldwell, pp. 49–70, Berlin, Heidelberg, New York: Springer-Verlag.

Sellers, P. J., Berry, J. A., Collatz, G. J., Field, C. B. & Hall, F. G. (1992). Canopy reflectance, photosynthesis and transpiration. III. A reanalysis using improved leaf models and a new canopy integration scheme. *Remote Sensing Environment*, **42**, 187–216.

Silvertown, J., Holtier, S., Johnson, J. & Dale, P. (1992). Cellular automaton models of interspecific competition for space – the effect of pattern on process. *Journal of Ecology*, **80**, 527–34.

Smith, E. (1937). The influence of light and carbon dioxide on photosynthesis. *General Physiology*, **20**, 807–30.

Tenhunen, J. D., Harley, P. C., Beyschlag, W. & Lange, O. L. (1987). A model of net photosynthesis for leaves of the sclerophyll *Quercus coccifera*. In *Plant Response to Stress – Functional Analysis in Mediterranean Ecosystems*, ed. J. D. Tenhunen, F. Catarino, O. L. Lange & W. C. Oechel. pp. 339–54. Berlin, Heidelberg, New York: Springer-Verlag.

Thornley, J. H. M. & Johnson, I. R. (1990). *Plant and Crop Modelling. A Mathematical Approach to Plant and Crop Physiology*. Oxford: Clarendon Press.

Wang, Y. P. & Jarvis, P. G. (1990). Description and validation of an array model – MAESTRO. *Agricultural and Forest Meteorology*, **51**, 257–80.

# 24 Chlorophyll fluorescence as a diagnostic tool: basics and some aspects of practical relevance

U. SCHREIBER, W. BILGER, H. HORMANN AND C. NEUBAUER

## INTRODUCTION

Chlorophyll (Chl) fluorescence has many facets, reflecting the complexity of the photosynthetic process, from the primary photochemical charge separation at the picosecond time scale to activation of $CO_2$ fixation in minutes and adaptational changes in hours or even days. It is primarily a matter of appropriate instrumentation and methodology to collect the relevant information contained in chlorophyll fluorescence, in order to assess specific aspects of photosynthesis. In this respect, there has been considerable progress during the past ten years. In order to make optimal use of Chl fluorescence, it is important to understand the principal mechanisms controlling this signal *in vivo* and to choose the appropriate instrumentation and methodology for selective assessment of specific aspects to be studied. For example, picosecond relaxation measurements are ideally suited for the study of exciton transfer between pigment–protein complexes in the antenna system, whereas there are better ways to assess photosynthetic performance of a leaf in its natural environment in full sunlight. This chapter deals primarily with aspects of practical relevance. A brief introduction to the principles governing fluorescence yield *in vivo* will be presented and the major methods of measurements will be outlined, with some emphasis on the most frequently used techniques of pulse-amplitude-modulation and pump-and-probe. It is not intended to present a comprehensive review with numerous citations of original papers, but rather a brief outline of the present state-of-the-art and discussion of some selected topics which unavoidably are somewhat biased by the research interests of the authors.

In the past, a number of reviews on Chl fluorescence have been published, with emphasis on different aspects, covering a large amount of literature. Current knowledge on the relationship between fluorescence emission and the physiology of photosynthesis is summarized in the reviews of Briantais *et al.* (1986), Horton and Bowyer (1990), Krause and Weis (1991), Schreiber and Bilger (1993), Dau (1994*a*,*b*), Govindjee (1995) and Joshi and Mohanty (1995). The literature on fluorescence lifetime measurements and theoretical aspects was reviewed by Moya, Sebban and Haehnel (1986), Holzwarth (1991) and Dau (1994a). Of the numerous articles surveying the wide field of Chl fluorescence applications in plant and stress physiology, the following reviews should be mentioned: Karukstis (1991), Björkman and Demmig-Adams (1994), Schreiber, Bilger and Neubauer (1994); Mohammed, Binder and Gillies (1995).

## PRINCIPLES GOVERNING CHLOROPHYLL FLUORESCENCE YIELD *IN VIVO*

Chl *a* in solution displays a fluorescence yield of 20–35%, which does not change substantially upon a dark–light transient. In contrast, Chl *a* fluorescence originating from plants is characterized by a large variable yield (2–10%), changes of which can be readily induced by a dark–light transition (Kautsky effect). Contrary to the monomeric Chl *a* in solution, the *in vivo* chlorophyll is embedded in form of protein complexes in the thylakoid membrane, and the excited singlet state (Chl *a**) is subjected to a number of competing de-excitation reactions, including energy transfer, photochemical trapping, radiationless de-excitation and fluorescence emission. Any change in the probability of one of these pathways will result in corresponding changes of fluorescence yield. Hence, in view of the known dynamics of the thylakoid membrane organization and conformation, as well as of

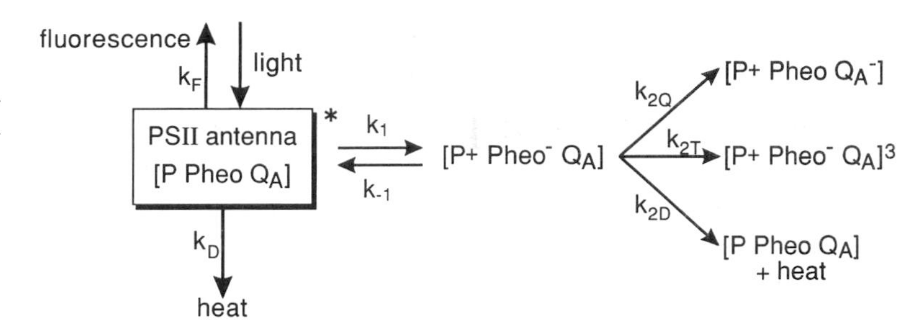


**Figure 24.1.** Exciton dynamics of PSII controlling the yields of fluorescence, photochemistry and nonradiative dissipation, based on the reversible radical pair model of Schatz, Brock and Holzwarth (1988).

the complex responses of photosynthetic electron flow to light and other environmental factors, *a priori* an appalling complexity of fluorescence phenomenology may be expected. Fortunately, it has turned out that after all, in practice, quite reliable information can be obtained by steady-state fluorescence measurements (Weis & Berry, 1987; Genty, Briantais & Baker, 1989) and that a simple empirical expression for the relationship between effective photochemical yield and fluorescence parameters (see Fig. 24.3) comes very close to an expression derived on the basis of a sophisticated theoretical model of exciton dynamics in PSII (Lavergne & Trissl, 1995).

Light quanta are absorbed mainly by the light-harvesting pigment-protein complexes (LHCI and LHCII) of the two photosystems (PSI and PSII). The interpretation of Chl *a* fluorescence changes is greatly facilitated by the fact, that at room temperature variable fluorescence originates almost exclusively from PSII. Hence, from the viewpoint of fluorescence, it is the excitation equilibration and the various competing de-excitation processes *within* PSII which are of primary importance. For a quantitative description of the PSII exciton dynamics, a number of models have been proposed (for review see Dau 1994*a*), which differ primarily in the assumed rates of excitation equilibration between distinct pigment–protein pools, the relative rate of energy trapping by charge separation, and the extent of reversibility of the trapping process. Present evidence from time-resolved fluorescence spectroscopy and picosecond absorption measurements argues strongly for the reversible radical pair model of Schatz, Brock & Holzwarth (1988) (for review see Dau 1994a). The essence of this model, which can be summarized by the reaction steps depicted in Fig. 24.1, consists of the following assumptions:

(i) Excitation equilibration between all pigment pools, including the reaction centres, is very rapid and influences neither the fluorescence life-time nor the trapping time.

(ii) Charge separation at the PSII reaction centres is reversible, i.e. the radical pair, $P_{680}^+$ $Pheo^-$ can recombine to the singlet state, and an exciton can visit an open centre several times before it is trapped ('shallow trap').

(iii) In principle, reversible radical pair formation can also take place when $Q_A$ is reduced.

The main factor determining Chl *a* fluorescence yield *in vivo* is the redox state of the first stable acceptor $Q_A$. When $Q_A$ is oxidized, photochemical charge separation and stabilization of the electron on $Q_A$ (rate constant $k_{2Q}$) can take place, i.e. excitation trapping essentially becomes irreversible and the probability for fluorescence emission is low. When $Q_A$ is reduced, the probability for fluorescence emission is high for two reasons, namely a decrease in the rate constant of primary charge separation ($k_1$) (see Fig. 24.1), and an increase in the rate constant for charge recombination ($k_{-1}$). Both changes can be explained by electrostatic interaction between $Q_A^-$ and $Pheo^-$ (Schatz *et al.*, 1988; Dau, 1994*a*). It should be noted that the two types of variable fluorescence will be affected differently by changes in the rates of alternative pathways of radical pair decay ($k_{2T}$, spin dephasing to triplet state, and $k_{2D}$, non-radiative decay to ground state) (see also pp. 327–8).

In practice, the ratio between maximal fluorescence yield (all $Q_A$ reduced) and minimal fluorescence yield (all $Q_A$ oxidized) is approximately 5 to 6 in dark-adapted, healthy leaves and good-quality isolated chloroplasts. However, this ratio can vary largely, depending on the illumination state and on a number of more or less physiological treatments, mostly involving a decrease of the maximal fluorescence yield. As photochemistry at

PSII is prevented, a decrease in maximal fluorescence yield normally is indicative of an increase in heat formation either at the reaction centres or in the antenna (see Fig 24.1).

From the foregoing remarks, it is clear that changes in chlorophyll fluorescence yield may be divided into two basic categories. Assuming that a physiologically intact photosynthetic organism can display a certain maximal fluorescence yield, any lowering of fluorescence with respect to this yield may be caused either by photochemical energy conversion in PSII (photochemical quenching) or by an increase of nonradiative deexcitation, involving heat formation or energy transfer to non-fluorescing PSI (non-photochemical quenching). Normally, both types of fluorescence changes overlap and, without appropriate deconvolution, a straightforward interpretation is not possible. In this context, the following points are relevant:

(i) The formation and disappearance of most types of non-photochemical quenching are relatively slow and, hence, can be separated in time from the potentially much faster changes in photochemical quenching. This is a prerequisite of the so-called saturation pulse method which will be outlined below (see pp. 325–9). Rapid changes in fluorescence yield, as can be induced by a dark–light transient, in first approximation, may be considered to reflect changes in the rate of photochemical charge separation of PSII.

(ii) The light-induced increase of most forms of non-photochemical quenching requires the formation of a transthylakoidal ΔpH, which, in isolated chloroplasts, can be dissipated by small amounts of uncouplers.

## WAYS OF MEASURING FLUORESCENCE

With a number of commercial instruments being available, it has become rather easy to measure chlorophyll fluorescence (for reviews see Schreiber & Bilger 1987; Bolhar-Nordenkampf *et al.*, 1989; Horton & Bowyer 1990; Mohammed *et al.*, 1995). On the other hand, it should be realized that there are many different ways of measuring fluorescence and that it takes some judgement to choose the appropriate instrument for solving a given problem. Therefore, some of the most commonly used methods of measuring chlorophyll fluorescence will be outlined briefly.

### Conventional fluorometers

For a long time, starting with Kautsky and co-workers, until *ca.* 10 years ago, the most common way of measuring chlorophyll fluorescence was based on the following 'conventional' principle. Fluorescence is excited by continuous light which is strongly absorbed by chlorophyll (normally blue light) and fluorescence emission is detected at wavelengths beyond 660 nm (red long-pass filter) with the help of a suitable photosensor (photomultiplier or photodiode). Such conventional systems display the following properties:

(i) Separation of the weak fluorescence signal from much stronger excitation light relies exclusively on the quality of optical filtering.

(ii) The excitation beam serves at the same time as the actinic beam.

(iii) With different actinic intensities, vastly different signal levels will be obtained.

(iv) At high actinic intensity, very fast shutter and recording systems are required to resolve the minimal fluorescence yield, Fo.

(v) The measured parameter is fluorescence *intensity*, which must be divided by the excitation intensity in order to obtain a measure of fluorescence yield.

(vi) The signal is disturbed by ambient light, with the consequence that, preferentially, experiments should be carried out in darkened rooms or the apparatus must be absolutely light-tight.

(vii) The obtained information is derived primarily from dark–light induction curves, which are strongly influenced by a given dark-adaptation state of the samples.

### Modulation fluorometers

In modulation fluorometers a modulated excitation light is used, which is created either mechanically (chopper) or electronically, and the modulated fluorescence signal is processed by a selective amplifier which rejects all non-modulated signals. Hence, modulated fluorometers are characterized by the following advantages:

(i) In addition to optical filtering electronical filtering is also employed.
(ii) The fluorescence excited by actinic light does not contribute to the signal.
(iii) Even with vast changes in actinic intensity, the measured signal remains within a fixed range.
(iv) At sufficiently low intensities of modulated measuring light, this does not cause substantial closure of PSII reaction centres, such that the Fo-level can be continuously monitored, and upon application of actinic light the induction kinetics can be recorded.
(v) The measured parameter represents the relative fluorescence *yield*, as the excitation intensity normally is constant and the applied actinic illumination affects the signal only by changing the state of the sample, i.e. by inducing changes of photochemical and non-photochemical quenching.
(vi) The signal is not disturbed by ambient light, such that experiments can be carried out in daylight and even full sunlight in the field.
(vii) Essential information can be collected not only from dark–light induction curves but also from light–dark relaxation kinetics and under steady-state light conditions in conjunction with the saturation pulse method (see pp. 325–9).

Modulation fluorometry, and with it also the use of chlorophyll fluorescence as a diagnostic tool in photosynthesis research, have profited greatly from recent advances in solid-state- and opto-electronics. Presently, the most advanced instruments are based on the so-called pulse amplitude modulation (PAM) technique (Schreiber, Schliwa & Bilger 1986), which will be briefly described. Fluorescence is excited repetitively by 1 μs pulses of light from a light-emitting diode (LED) passing through a short-pass filter ($\lambda < 670$ nm). A fast responding photodiode detector is protected by a long-pass filter ($\lambda > 700$ nm), which cuts off the excitation light and passes the lower wavelength part of fluorescence. A highly selective pulse amplification system ignores all signals except for the fluorescence pulse excited during the 1 μs measuring pulse. Due to the large linearity range of the photodiode (up to $10^9$), this system tolerates extreme changes in background light intensity (up to several times the intensity of full sunlight) at a low integrated measuring light intensity which, by itself, does not induce fluorescence changes and monitors the Fo-level in the absence of background light.

## Pump-and-probe technique

The 'pump-and-probe' or 'double-flash' technique measures fluorescence yield with the help of a relatively weak 'probe' flash following a 'pump' flash which normally is made saturating in terms of PSII charge separation. By varying the delay time between the two flashes, the relaxation kinetics of fluorescence yield can be followed. Furthermore, by varying the intensity of the pump flash, apparent cross-sections can be measured. This technique has contributed considerably to the elucidation of primary reactions in photosynthesis. Examples of application were described by Falkowski *et al.* (1986), Horton and Bowyer (1990), Kolber and Falkowski (1993) and Govindjee (1995).

A main advantage of the pump-and-probe technique is its high time resolution with which rapid electron transfer steps at the PSII reaction centre can be monitored. This relates primarily to $Q_A$ re-oxidation by the secondary acceptor $Q_B$, which is blocked by PSII inhibitors, like the common herbicides atrazin and diuron (DCMU), and is characteristically slowed down in atrazin-resistant mutants. This leads to important practical applications of this technique in agriculture and weed control. Another advantage is its high sensitivity, which has opened the way for applications in limnology and oceanography. The exceptional sensitivity is due to the fact that the intensity of single probe pulses, although much weaker than the pump pulses, can be made orders of magnitude higher than that of continuous excitation light, without inducing any significant closure of PSII centres. Hence, it has become possible to determine key parameters of photosynthetic performance, like PSII photochemical quantum yield and absorption cross-section of the natural phytoplankton contained in ocean water samples (Kolber & Falkowski, 1993).

## Time-resolved fluorescence measurements

Time-resolved fluorescence measurements provide the most detailed information on the various deexcitation pathways competing with fluorescence emission. A very short pulse of light is applied to create a population of excited singlet states (within femto- or picoseconds) and then the fluorescence decay is followed by sophisticated high-speed detection techniques, such as single photon counting (for reviews, see Moya *et al.*, 1986; Holzwarth, 1991).

Following a picosecond excitation pulse, the initial number of excited states ($N_o$) in a simple one-component model system may be assumed to decay in parallel by four first-order processes:

$$-\frac{dN}{dT} = N_o(k_F + k_P + k_D + k_T)$$

with $k_F$, $k_P$, $k_D$ and $k_T$ representing the rate constants of fluorescence emission, photochemical energy conversion, non-radiative dissipation and excitation energy transfer, respectively. The excited states decay according to:

$$N(t) = N_o e^{-t/\tau}$$

where the lifetime, $\tau$, represents the time required for N to drop to $N_o/e$, and is defined as:

$$\tau = \frac{1}{k_F + k_P + k_D + k_T}$$

The natural or intrinsic lifetime ($\tau_o = 1/k_F$) is given when fluorescence is the only de-excitation pathway. The fluorescence yield is related to these lifetimes by the equation:

$$\Phi_F = \frac{\tau}{\tau_o}$$

Chl *a* fluorescence in solution is characterized by a $\tau_o$ of *ca.* 15 ns, whereas fluorescence *in vivo* displays overall lifetimes ranging from fractions of a ns to several ns, depending on the conditions, i.e. primarily on the state of PSII. The overall decay is composed of several components, the exact number of which, with associated lifetimes, amplitudes and assignments of origin, has been a matter of debate over the past 15 years (for reviews see Moya *et al.*, 1986; Holzwarth 1991; Dau 1994*a*). The complexity is mainly due to dynamic changes in the concentration of open reaction centres and of non-radiative excitation sinks. A straightforward analysis has been complicated by the fact that deconvolution depends to a considerable extent on the assumed model of PSII organization. Further progress will rely strongly on the elucidation of PSII heterogeneity (for review see Lavergne & Briantais, 1996) and of regulatory changes in PSII organization (for review see Allen, 1995).

## 77 K fluorescence spectroscopy

When photosynthesizing organisms or isolated membranes are cooled to liquid nitrogen temperature, the total fluorescence yield increases, and fluorescence emission and excitation spectra change dramatically (for reviews see Briantais *et al.*, 1986; Krause & Weis, 1991). At 77 K major emission peaks in green plants are at 685, 695, and 720–740 nm. The emission at 685 and 695 nm is attributed to PSII, whereas the emission at 720–740 nm originates from PSI (peak at 735 nm in higher plants). The various bands can be split into various components by deconvolution techniques, thus revealing the high complexity of Chl protein complexes in the thylakoid membrane. Obviously, contrary to room temperature, rapid equilibration of excitation energy between the various components of antenna and core complexes is not possible at 77 K.

77 K spectroscopy has played an essential role in the study of antenna and reaction centre organization and in its application for assessment of PSII quantum yield in leaves (for review see Björkman, 1987). Much fundamental work on the relationship between Fv/Fm and quantum yield, as well as on photoinhibition, was based on 77 K fluorescence measurements before the advent of selective modulation fluorometers and the saturation pulse method. In the meantime, it has been mostly replaced by the latter method, as this has the advantage of being non-invasive. An advantage of the 77 K method is the fact that relatively low light intensities can be applied to close PSII centres, avoiding the potential complication of non-photochemical quenching induced by strong light (see p. 329). No appreciable re-oxidation of $Q_A^-$ occurs at 77 K, such that changes caused by removal of PQ-quenching (Vernotte, Etienne & Briantais 1979) are avoided. The fact that maximal values of Fv/Fm determined at 77 K and by the saturation pulse method

are identical (0.83–0.87), supports the reliability of the latter method.

## The polyphasic rise of fluorescence in saturating light

For a long time, starting with the work of Kautsky and co-workers, fluorescence induction curves (Kautsky effect) have been the main basis of information (for reviews see Briantais *et al.*, 1986; Horton & Bowyer, 1990; Dau, 1994*b*; Govindjee, 1995; Joshi & Mohanty, 1995). Such induction curves are characterized by a biphasic rise from an initial level O (= Fo) via an intermediate level I to a peak level P, sometimes with a dip D preceding the I–P rise (nomenclature according to Govindjee, 1995). The various transients reflect the interplay between PSII driven reduction of the primary stable acceptor $Q_A$ and its re-oxidation by the secondary acceptor pool (mainly plastoquinone) and PSII. For a discussion of the significance of these 'conventional' transients the reader is referred to the above-mentioned reviews.

Additional features of the fluorescence rise are revealed at saturating light intensity, which will be briefly discussed here because they provide a new type of information and are also relevant with respect to the widespread quenching analysis by the saturation pulse method. Figure 24.2 shows typical recordings of the polyphasic fluorescence rise in saturating white light observed with dark-adapted chloroplasts in the absence and presence of the PSII inhibitor DCMU. In the control, the more than five-fold increase in fluorescence yield occurs in distinct phases, which have been analysed in detail by Neubauer and Schreiber (1987), Schreiber and Neubauer (1987) and more recently, using a somewhat different technique and nomenclature, also by Strasser, Srivastava and Govindjee (1995). The first fluorescence rise from Fo to the $I_1$-level is called the 'photochemical phase' because its slope (not resolved in Fig. 24.2) is proportional to actinic intensity. On the contrary, the following rise phases (from $I_1$ via $I_2$ to Fm) cannot be further speeded up by increasing intensity and are called 'thermal phases'. A small dip, D, can be distinguished, which is emphasized by treatments which weaken the PSII donor side. A minimal time of *ca.* 25 ms is required to reach the $I_2$-level and *ca.* 200 ms to reach Fm. The polyphasic rise characteristics can be observed in a large variety of photosynthetic organisms and, hence, appear to reflect very fundamental properties of PSII. While the O–$I_1$ rise can be understood in a conventional way as reflecting the kinetics of $Q_A^-$ accumulation, an interpretation of the thermal phase is more difficult. In this context, the effect of the PSII inhibitor DCMU is important, which raises the $I_1$-level to the original $I_2$-level and eliminates the subsequent rise to Fm see Fig. 24.2). For an interpretation, the following observations also appear relevant:

(i) Hill reagents, as ferricyanide and methylviologen, like DCMU, suppress the $I_2$-Fm phase.
(ii) Preillumination by saturating single turnover flashes causes lowering of $I_1$, following a period–4 oscillation (S-state-dependent quenching).
(iii) All treatments which are known to slow down electron donation to $P_{680}^+$ preferentially suppress the thermal phases ($I_1$–$I_2$-Fm).
(iv) A light intensity dependence shows clear cut saturation of $I_1$ and $I_2$ levels.

On the basis of these observations, it must be concluded that $Q_A$ is fully reduced at $I_1$ and that fluorescence yield at this point is lowered by a non-radiative decay process with respect to Fm. This process, which is favoured by a weak PSII donor side and is mostly prevented by DCMU, obviously is associated with energy dissipation at PSII reaction centres ($k_{2D}$ and $k_{2T}$ in Fig. 24.1). This is of considerable practical relevance in view of the fact that determination of maximal fluorescence yield, involving the application of saturating light pulses, is essential for the assessment of photochemical yield by the saturation pulse method (see below). Possible mechanisms of quenching at $I_1$ will be discussed on p. 331.

## SATURATION PULSE METHOD AND QUENCHING ANALYSIS

### Principle of the method

The rationale of the saturation pulse method is simple. Upon application of a sufficiently strong light pulse, $Q_A$ rapidly becomes fully reduced, and photochemical fluorescence quenching is suppressed, such that the remaining quenching may be considered non-

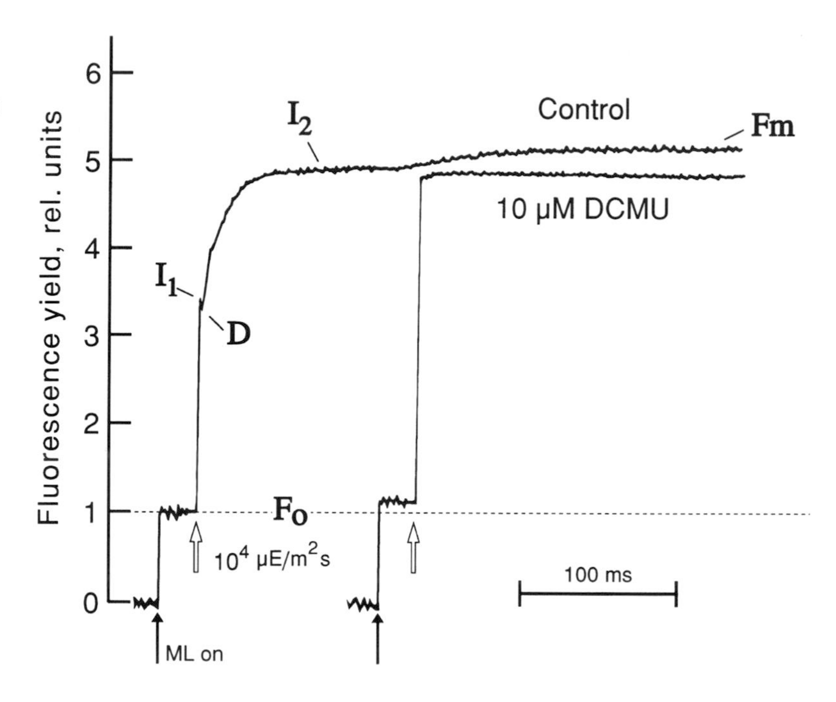


**Figure 24.2.** Polyphasic rise of fluorescence yield in intact spinach chloroplasts upon onset of saturating continuous light in the absence and presence of the PSII inhibitor DCMU. See text for significance of the various phases and typical fluorescence levels.

photochemical quenching (Bradbury & Baker, 1981). In Fig. 24.3, schematic traces of a measurement of modulated fluorescence are shown, with the characteristic fluorescence levels and quenching coefficients defined according to a proposal for standard nomenclature (van Kooten & Snel, 1990). While the sample is in a dark-adapted state, the minimal and maximal yields, Fo and Fm, are determined by application of a saturating pulse of light. The ratio (Fm–Fo)/Fm = Fv/Fm is a measure of the potential maximal PSII quantum yield (see p. 329; for review see Björkman, 1987). During illumination, the fluorescence yield, F, undergoes complex changes until eventually it reaches a steady-state level (the Kautsky effect). Before the introduction of the saturation pulse

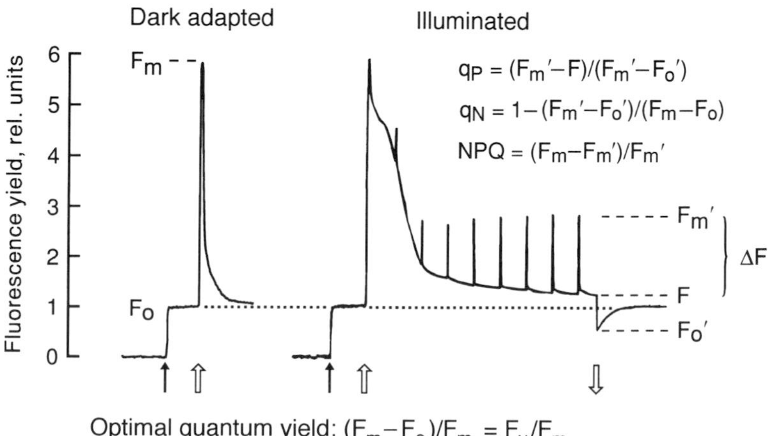


**Figure 24.3.** Fluorescence quenching analysis by the saturation pulse method. Definition of quenching coefficients, basic fluorescence parameters and typical fluorescence levels.

method, such induction curves, with complex phenomenology, were the main source of information. However, at illumination times exceeding a few seconds, this information is ambiguous, as, e.g. a low yield could be caused either by high photochemical efficiency or by a high yield of radiationless dissipation, which not only lowers fluorescence but lowers photochemical yield as well. The relevant information can be obtained quasi-continuously by saturation pulses, which may be applied repetitively to monitor the changing level of maximal fluorescence yield, Fm′. To a first approximation it may be assumed that Fm–Fm′ represents non-photochemical quenching, whereas Fm′-F represents photochemical quenching. So-called quenching coefficients, qP and qN, have been defined (see Fig. 24.3). It was first assumed that only variable fluorescence is quenched non-photochemically. However, later it was shown that also, when all PSII centres are open, the Fo-level can be lowered by so-called 'energy-dependent' non-photochemical quenching. Therefore, correct calculation of qP and qN requires previous determination of Fo′, which is not always a simple matter. Under steady-state conditions Fo′ can be determined simply by darkening the sample, as $Q_A$ will be oxidized within a few seconds when the PQ-pool is partially oxidized. Simultaneous application of weak far-red background light can enhance the oxidation rate. Knowledge of Fo′ is indispensable in order to obtain information on the extent of PSII 'openness' via qP-calculation. On the other hand, the extent of non-radiative energy dissipation, relative to a standard dark state, can also be assessed conveniently by the NPQ parameter (Bilger & Björkman, 1990) which does not contain Fo′ (definition in Fig. 24.3).

Another fluorescence parameter, which can be derived by the saturation pulse method and which does not require knowledge of Fo′, is ΔF/Fm′ (definition in Fig. 24.3), the so-called Genty-parameter (Genty *et al.*, 1989). This parameter closely reflects the effective quantum yield of PSII (see p. 329). The extraordinary usefulness and 'popularity' of this parameter is due to the fact that, on one hand, it can be easily measured with available portable instruments (for reviews see, e.g. Bolhar-Nordenkampf *et al.*, 1989; Mohammed *et al.*, 1995) and, on the other hand, provides very fundamental information. This will be further elaborated on pp. 332–4.

## Different forms of non-photochemical quenching

Non-photochemical quenching of Chl fluorescence, as defined by the saturation pulse method, can have a number of causes. In the framework of the reversible radical pair model (see Fig. 24.1), maximal fluorescence yield is characterized by low values of $k_D$ (heat dissipation in the antenna and the reaction centre in its ground state) and of $k_{2T} + k_{2D}$ (non-radiative decays of radical pair). The potential complexity results from the fact that these rate constants, in principle, may be affected by numerous factors. However, under *in vivo* conditions, the factors determining pigment organization and energy dissipation are quite stable and under physiological control (for review see Allen, 1995). After dark adaptation, leaves from non-stressed plants show little variation in Fv/Fm, which is a measure of the potential photochemical yield (see Björkman, 1987 and Fig. 24.3). Illumination generally results in more or less suppression of Fm, which has been studied extensively in the past (for reviews see Krause & Weis, 1991; Demmig Adams & Adams, 1992; Horton & Ruban, 1992; Walters & Horton, 1993; Schreiber & Bilger, 1993). The mechanisms of various types of non-photochemical quenching are still under debate, and no attempt will be made here to discuss all proposals. In the following subsections, just the three major types will be briefly described and, for more detailed information, the reader is referred to the above-mentioned reviews and to the original literature cited therein.

### *Energy-dependent quenching (qE)*

With isolated chloroplasts, energy-dependent quenching can be defined as that part of non-photochemical quenching which is quickly reversed by addition of an uncoupler, as the transthylakoidal ΔpH is a *sine qua non* condition for this type of quenching. With leaves or algae, a distinction from other types of non-photochemical quenching is less straightforward and is complemented favourably by 77 K measurements (Walters & Horton, 1993). When the dark-relaxation of qN (coefficient of overall non-photochemical quenching) is studied by repetitive saturation pulses, the most rapid phase ($t_{1/2} < 1$ min) generally is attributed to the energy-dependent part (qE), as the ΔpH is supposed to be rapidly dissipated. However, hydrolysis of ATP accumulated during excessive illumination in principle could maintain a $\Delta pH$ for some time in the dark. In algae and cyanobacteria, pathways of $O_2$-dependent electron flow do exist (chlororespiration), which can maintain a $\Delta pH$ in the dark.

Energy-dependent quenching, as the main expression of non-radiative energy dissipation, has been proposed to reflect down-regulation of PSII in excess light (Weis & Berry, 1987). As this process is of considerable importance for survival of plants in rapidly changing natural environments, it has been the topic of numerous fluorescence studies, mainly with the aim of revealing the molecular mechanisms (see above-mentioned reviews). From this work the following findings will be mentioned:

1. A large part of qE is closely correlated with zeaxanthin formation, which has been viewed as a dissipative trap in the antenna (increased $k_D$ in Fig. 24.1) (for reviews see Demmig-Adams & Adams, 1992; Pfündel & Bilger, 1994).
2. Horton and co-workers proposed that zeaxanthin facilitates aggregation of LHCII and, as a consequence, increases non-radiative dissipation (increased $k_D$ in Fig. 24.1) (Horton & Ruban, 1992).
3. Part of qE was suggested to result from stimulation of non-radiative decay of the reversible radical pair (increase of $k_{2D}$ or $k_{2T}$ in Fig. 24.1), associated with a weakened PSII donor side at low pH (Schreiber & Neubauer, 1990). Krieger & Weis (1993) proposed that internal acidification leads to the release of $Ca^{2+}$ from a site close to water splitting, causing formation of 'inactive centres' which are characterized by rapid charge recombination involving $P_{680}^+$ and a modified, high-potential $Q_A^-$.

## *State 2 quenching (qT)*

Distribution of excitation energy between the two photosystems is regulated by phosphorylation and dephosphorylation of LHCII, characterized by state 2 and state 1, respectively (for reviews see Williams & Allen, 1987; Allen, 1995). Phosphorylated LHCII detaches from PSII in appressed membranes and supposedly associates with PSI in non-appressed membranes, resulting in decreased PSII excitation and increased PSI excitation. Phosphorylation is under redox control of a component, X, with $E_m \sim +40$ mV, which is located between the two photosystems and closely associated with the PQ-pool, but not identical to it. Dark adapted leaves are in state 1 (oxidized X, LHCII, high Fm) and upon illumination partially change to state 2 (reduced X, LHCII-P, low Fm). In algae and cyanobacteria, dark reduction of X (via NDH-activity) can induce state 2. Pronounced reversible state 1 – state 2 changes can be observed upon alternating application of PSII and PSI light.

State 2 with respect to state 1 is characterized by a lowered fluorescence yield, in particular of Fm. The corresponding component of overall non-photochemical quenching has been denoted by the coefficient qT (Horton & Hague, 1988). In higher plant chloroplasts, upon a light–dark transition, qT relaxes slowly (half-time *ca.* 5 min). Relaxation is prevented by NaF which inhibits dephosphorylation. When samples are frozen at 77 K, in state 2 the PSII/PSI fluorescence ratio is 10 – 30% lower than in state 1, with a lowering of PSII emission being mostly responsible. In leaves, optimal state 2 is observed at rather low light intensities (50 – 100 μmol quanta $m^{-2}$ $s^{-1}$) (Walters & Horton, 1993).

In practice, detection of reversible state changes has been facilitated by the saturation pulse method. However, it is important to realize that frequent application of saturation pulses will induce state 2, such that when qN-relaxation is studied by repetitive pulses, the qT-component will be over-estimated.

The mechanism of state 2 quenching may be less clear than is suggested by the simple rationale given above. Contrary to expectations, the Fo-level is not quenched significantly (Walters & Horton, 1993).

## *Photoinhibitory quenching (qI)*

Photoinhibitory quenching, as characterized by the coefficient qI, is defined as that part of overall light induced qN which does not relax within 30 – 40 min (Schreiber & Bilger, 1987; Horton & Hague, 1988). It results in suppression of Fv/Fm, which in numerous studies was found to be correlated linearly with a suppression of the measured optimal quantum yield of photosynthesis (for reviews see Björkman 1987; Demmig-Adams & Adams, 1992). Fv/Fm measurements have been widely used as a tool for elucidating the mechanism of photoinhibition and the adaptational mechanisms for protection against photodamage. A discussion of the very extended literature on this topic is out of scope and the reader is referred to the reviews of Krause and Weis (1991), Demmig-Adams and Adams (1992) and Schreiber and Bilger (1993).

The mechanisms causing qI are still controversial. Arguments have been put forward for non-radiative dissipation in the antenna (Björkman, 1987; Demmig-Adams & Adams, 1992) and formation of inhibited PSII centres which convert excitation energy to heat (Cleland, 1988; Giersch & Krause, 1991). In this context, it may be mentioned that, in intact chloroplasts, qI correlates with preferential suppression of the $I_1$–$I_2$ phase in the polyphasic fluorescence rise in saturating light (Schreiber & Neubauer, unpublished observations). This may argue for a stimulation of $k_{2D}$ in the framework of the reversible radical pair model (Fig. 24.1).

## RELATIONSHIP BETWEEN FLUORESCENCE YIELD AND THE QUANTUM YIELD OF PSII

### Theoretical considerations

Understanding the relationship between fluorescence yield and the quantum yield of PSII is very fundamental for all practical applications of chlorophyll fluorescence. There are more or less sophisticated ways to approach this question. Theoretical derivations all are based on the competition of fluorescence emission with other pathways of exciton de-excitation, the details of which depend on the assumed model of exciton dynamics in the antenna system and of trapping at the reaction centres (for review see Dau, 1994*a*). For a plant physiologist, such theoretical derivations are often difficult to follow, with the consequence that uncertainty and even confusion may remain. Therefore, here a very general derivation will be presented, which does not require any specific model and is just based on the assumption that the ratio of the rate constants of fluorescence emission and non-radiative deexcitation does not change during a brief saturation pulse.

1. It can be assumed that the sum of quantum yields of photochemistry, fluorescence and non-radiative de-excitation is unity:

$$\Phi_P + \Phi_F + \Phi_D = 1 \quad \text{or} \quad \Phi_P = 1 - \Phi_F - \Phi_D$$

2. With the application of a saturating light pulse, the photochemical quantum yield, $\Phi_P$, becomes zero while the remaining yields are maximal:

$$(\Phi_F)_m + (\Phi_D)_m = 1 \quad \text{or} \quad (\Phi_D)_m = 1 - (\Phi_F)_m$$

3. The ratio between dissipation and fluorescence is assumed to be constant during a brief saturation pulse:

$$\frac{(\Phi_D)_m}{(\Phi_F)_m} = \frac{\Phi_D}{\Phi_F}$$

and considering eqn. 2, an expression is derived which describes the yield of dissipation on the basis of measured fluorescence yields:

$$\frac{1-(\Phi_F)_m}{(\Phi_F)_m} = \frac{\Phi_D}{\Phi_F} \quad \text{or} \quad \Phi_D = \frac{\Phi_F}{(\Phi_F)_m} - \Phi_F$$

4. Using this expression for $\Phi_D$, it is now possible to express $\Phi_P$, as defined in eqn 1, by fluorescence parameters alone:

$$\Phi_P = 1 - \Phi_F - \left(\frac{\Phi_F}{(\Phi_F)_m} - \Phi_F\right) = 1 - \frac{\Phi_F}{(\Phi_F)_m} = \frac{(\Phi_F)_m - \Phi_F}{(\Phi_F)_m}$$

5. In practice, the photochemical quantum yield of PSII (with the definition $\Phi_{II} = \Phi_P$) can be assessed either after dark adaptation, when the fluorescence yield before application of a saturation pulse is defined as Fo and the maximal yield during the pulse is Fm, or during illumination at a given fluorescence yield F, when the maximal yield during the pulse is Fm, (see Fig. 24.2). Then, the following fluorescence expressions for determination of PSII photochemical quantum yield are obtained:

$$\text{Maximal quantum yield:} \quad \Phi_{II} = \frac{\text{Fm} - \text{Fo}}{\text{Fm}} = \frac{\text{Fv}}{\text{Fm}}$$

$$\text{Effective quantum yield:} \quad \Phi_{II} = \frac{\text{Fm} - \text{F}}{\text{Fm}'} = \frac{\Delta\text{F}}{\text{Fm}'}$$

These are the well-known expressions which have been widely confirmed experimentally by simultaneous

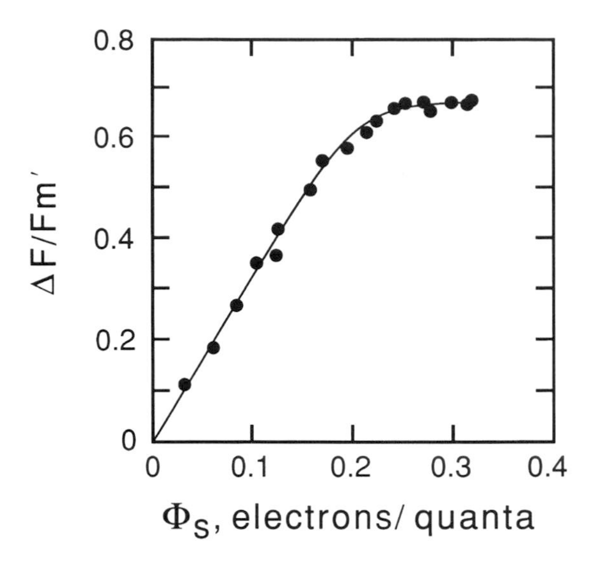


**Figure 24.4.** Relationship between the effective quantum yield of PSII, $\Delta F/Fm'$, and the quantum yield of ferricyanide reduction, $\Phi_S$, in spinach thylakoids. The varied parameter is actinic light intensity. Further details are described in Hormann *et al.*, 1994.

measurements of fluorescence and the yield of photosynthetic electron transport (Genty *et al.*, 1989, see reviews by Krall & Edwards, 1992; Schreiber & Bilger, 1993). Obviously, most of PSII photochemistry and fluorescence emission *in vivo* are related to each other in a very straightforward way, understanding of which is possible by common-sense assumptions.

When a thorough theoretical derivation is carried out, based on a sophisticated model of exciton dynamics derived from picosecond lifetime measurements, expressions result which are not identical but similar to the above equations, with the possible deviations being rather small (~14%) (Lavergne & Trissl, 1995). Actually, experimentally observed deviations are not trivial and their clarification is important to gain new insights into the mechanisms involved. For example, several laboratories have observed that the linear correlation between $\Delta F/Fm'$ and $\Phi_s$ (quantum yield of electron transport) breaks down at low light intensities, i.e. high $\Phi_{II}$-values (for review Schreiber *et al.*, 1995*a*). As shown in Fig. 24.4, this phenomenon is also observed in isolated thylakoids in the presence of an excellent electron acceptor. It appears to be related to a fundamental heterogeneity of PSII, involving approximately one-third of all $O_2$-evolving centres, which can be closed at relatively low light intensities without much loss in variable fluorescence yield. This aspect of PSII heterogeneity (for a review see Lavergne & Briantais, 1996) must be taken very seriously, particularly when low light intensities are applied.

## Practical aspects of $\Phi_{II}$-determinations

When differences between the $\Phi_{II}$-values are observed, which are determined via fluorescence or electron transport measurements, this may have at least three different reasons: (i) fluorescence parameters (Fo, F, Fm, Fm′: for definition see Fig. 24.2) were not properly determined, (ii) the measured electron transport rate is not identical to the rate of charge separation at PSII, (iii) the observed differences are caused by PSII heterogeneity (as discussed above). Aspects relating to the first mentioned two points will be briefly discussed below.

### *Contribution of PSI fluorescence*

In most fluorescence studies, the tacit assumption is made that the measured fluorescence at room temperature exclusively originates from PSII. However, while this normally is true for *variable* fluorescence, there is a significant contribution of PSI fluorescence to Fo, which recently was determined to amount to 5 and 11% of *maximal* fluorescence in $C_3$ and $C_4$ plants, respectively (Pfündel, 1995). While, in principle, this contribution can be minimized by measuring only short-wavelength fluorescence (below 700 nm), where PSII emission is dominating, this means a considerable loss in signal amplitude. If no correction for PSI fluorescence is made, $\Phi_{II}$ will be underestimated. For example, with maximal quantum yields determined by the PAM fluorometer ranging around 0.82–0.85 in $C_3$ plants, after appropriate correction (Pfündel, 1995), these will be increased to values between 0.86 and 0.90, equivalent to Fm/Fo ratios of 7–10.

### *Determination of the relevant* Fm

According to the rationale of the saturation pulse method (see pp. 325–7), very intense light is applied in order to saturate PSII photochemistry within less than a second, before any secondary changes will be induced which would cause a change in non-photochemical quenching (or of $\Phi_D/\Phi_F$, see p. 329). In practice, there is some uncertainty and even controversy, as to the proper way of measuring the Fm or Fm′-values required for Fv/Fm or ΔF/Fm′-determination. In this context, the following empirical facts are important.

(i) It is not possible to induce the maximal fluorescence yield within less than 100 ms, even by several-fold saturating light intensities. This is clearly apparent from the saturation characteristics of the polyphasic fluorescence rise upon onset of strong continuous illumination, which was already discussed above (see p. 325 and Fig. 24.2)

(ii) The maximal fluorescence yield observed following a saturating single turnover flash is distinctly lower than the maximal fluorescence reached after completion of the 'thermal phases' in a longer pulse of saturating light (Schreiber *et al.*, 1995*a*; Kramer, Di Marco & Loreto, 1995).

(iii) Oxidized plastoquinone (PQ) can quench maximal fluorescence yield non-photochemically (Vernotte *et al.*, 1979), with the consequence that, contrary to the original rationale (see p. 325 and p. 329) the rise from Fo to Fm will also include a phase-paralleling PQ reduction and, hence, representing the removal of non-photochemical quenching.

(iv) Chl fluorescence yield depends also on the state of the PSII donor side. So-called S-state dependent non-photochemical quenching has been described (see, e.g. Govindjee, 1995).

(v) When $Q_A$ is reduced, formation of the radical pair $P_{680}^+$ $Pheo^-$ can still occur, although with a lower rate constant (Schatz *et al.*, 1988). Recombination to the singlet state will constitute only one of three possible deexcitation pathways (see Fig. 24.2). If spin dephasing to the triplet state or non-radiative decay to the ground state take place, this will result in non-photochemical quenching of Fm. It has been hypothesized that spin dephasing may be favoured by membrane energization (Schreiber & Neubauer, 1990).

Researchers using the PAM fluorometer for saturation pulse quenching analysis usually apply relatively long pulses of strong light, with the aim of fully reducing the PSII acceptor side and of empirically reaching a maximal plateau of fluorescence yield. Recently, it has been pointed out that, at least with green algae and cyanobacteria, shorter pulses of saturating light are advantageous (20–50 ms), because in these organisms there is a rather rapid decline of fluorescence yield after the peak level, P (Schreiber *et al.*, 1995*a*). Such short pulses may be also best suited for quenching analysis in leaves, because they lead to the $I_2$ level and avoid the $I_2$–P rise (see Fig. 24.2), which most likely is caused by the disappearance of PQ quenching (Vernotte *et al.*, 1979).

With the pump-and-probe technique (see p. 323), saturating single turnover flashes are used to cause total reduction of the primary acceptor and to induce maximal fluorescence yield. However, as this is distinctly lower than the Fm determined by the saturation pulse method, the question arises, which of the two Fm values should be taken for calculation of Fv/Fm or ΔF/Fm′ (Kramer *et al.* 1995). At present, no definite answer to this question can be given, but a close look at the arguments put forward by Kramer *et al.* (1995) may be instructive. In the original work of Vernotte *et al.* (1979) the PQ quenching amounted to 15–20%, which can be correlated with the $I_2$–P phase of fluorescence induction in saturating light (Neubauer & Schreiber, 1987). On the other hand, the maximal fluorescence yield following a single turnover saturating flash corresponds to the $I_1$ level and in intact chloroplasts can be raised to the $I_2$ level by DCMU. Hence, if PQ quenching were directly involved, this would have to be bound $Q_B$, which can be removed by DCMU from its binding site. An involvement of PQ is suggested by the fact that reduction of the PQ pool by illumination raises $I_1$ and also F(50 µs). However, there is also the surprising finding that the $I_1$–$I_2$ rise is the mirror image of the simultaneously measured P515 decay transient (Schreiber & Neubauer, 1990). Hence, there is the alternative possibility that quenching at $I_1$ or of F(50 µs) is associated with a strong local field affecting the properties of the primary radical pair. For example, the rate constant ($k_{2T}$) for formation of the triplet state could be increased (see Fig. 24.1). For judging this point, it should be realized that a measurement of F(50 µs) by the pump-and-probe method involves charge separation by the probe pulse at centres almost exclusively in the state P Pheo $Q_A^-$ in the presence of a strong local field.

Whatever eventually will turn out to be the cause of the lowering of $I_1$ and of F(50 µs) with respect to Fm, from a practical point of view, it appears most essential to avoid uncontrolled variations of it. This aspect favours long saturation pulses, which make sure that this non-photochemical quenching is eliminated. If, in this way, as suggested by Kramer *et al.* (1995), photochemical yield were over-estimated, at least it would be so consistently. On the other hand, F(50 μs) would be more or less non-photochemically quenched, depending on the conditions (affecting the redox state of PSII acceptor and donor sides). It appears unlikely that the true quantum yield is really substantially lowered by PQ quenching, (i) because maximal quantum yields determined by gas exchange measurements are rather higher than the mean value of 0.83 determined by the saturation pulse method, and (ii) because identical values are also determined by 77 K fluorescence spectroscopy (see p. 324 and it is known that $Q_A^-$ is not re-oxidized at 77 K, such that PQ reduction should be prevented.

### *Cyclic flow around PSII*

The parameter ΔF/Fm′ relates to the effective quantum efficiency of charge separation at PSII centres, $\Phi_{II}$, irrespective of whether charge separation leads to secondary stabilization (and linear electron transport) or not. The possibility of a cyclic electron flow at PSII is well established. In particular, it can be strongly stimulated by low concentrations of so-called ADRY-reagents like CCCP or ANT-2p, catalysing the reduction of the $S_2$ and $S_3$ states of the water-splitting system by cyt b559 which is re-reduced via the PSII acceptor side. As expected, in the presence of CCCP, the relationship between $\Phi_{II}$ and $\Phi_S$ is distorted (Schreiber *et al.*, 1995*a*), as at increasing concentrations $\Phi_S$ can be suppressed substantially without substantial loss in $\Phi_{II}$.

The significance of cyclic PSII under physiological conditions is under debate. Evidence for pronounced cyclic flow at high light intensities has been claimed by Falkowski *et al.* (1986), on the basis of a correlation between pump-and-probe fluorescence measurements and $O_2$ yield measurements, with the indication of a substantial fraction of PSII turnover (as judged from fluorescence) which is not correlated with $O_2$ evolution at light saturation. Such behaviour so far has not been observed using the saturation pulse method with leaves (Genty *et al.*, 1989), isolated chloroplast (Hormann, Neubauer & Schreiber, 1994) and green algae (Schreiber *et al.*, 1995*a*; Heinze, Dau & Senger, 1996). This apparent discrepancy could result from the different methods of Fm determination outlined above.

### *$O_2$-dependent electron flow*

$O_2$ can be reduced at various sites of the photosynthetic electron transport chain, with the electrons originating from PSII activity. Hence, $O_2$-dependent electron flow will be expressed as a component of overall ΔF/Fm′, like any other type of electron transport involving PSII. However, like cyclic electron transport around PSII, $O_2$-dependent electron flow *in vivo* cannot be detected via gas exchange measurements and, hence will affect the relationship between $\Phi_{II}$ and $\Phi_s$. This is due to the fact that the stoichiometry of the so-called Mehler ascorbate peroxidase (MAP) cycle in intact chloroplasts results in zero net $O_2$ exchange (Asada & Badger, 1984). The significant contribution of the MAP cycle to overall electron transport in intact systems and its important regulatory role have been elucidated mainly with the help of fluorescence quenching analysis (for recent review see Schreiber *et al.*, 1995*b*). The most relevant aspects will be briefly summarized.

1. $O_2$ reduction at the PSI acceptor side (Mehler reaction) is followed by light-driven reduction of the monodehydroascorbate radical formed in close vicinity due to the action of the membrane-bound enzyme complexes of superoxide dismutase, ascorbate peroxidase and monodehydroascorbate reductase.
2. $O_2$ reduction displays a pH 5 optimum and the MAP cycle is stimulated by the transthylakoidal ΔpH.
3. The MAP cycle contributes decisively to the overall ΔpH which causes 'energy-dependent' non-photochemical quenching (see pp. 327–9) and which has important functions in the regulation of photosynthesis in excess light (see below).

## APPARENT ELECTRON TRANSPORT RATE AND CONCEPT OF EXCESS PFD

From a practical point of view, probably the most important application of chlorophyll fluorescence in plant science is the assessment of effective PSII quantum yield *during steady-state illumination* in the laboratory, greenhouse or under natural field conditions. In this way, with knowledge of the incident flux density of photosynthetically active radiation (PFD) the relative electron transport rate can be determined:

$$\text{relative rate} = \Phi_{II} \times \text{PFD}$$

Assuming that an average leaf absorbs 84% of the incident PFD, and that the quanta are evenly distributed between the two photosystems, even an estimate of the absolute rate can be obtained:

$$\text{rate} = \Phi_{II} \times 0.42 \times \text{PFD}$$

Fundamental information on the photosynthetic performance of a plant is available from light response curves (plots of apparent electron transport rates versus PFD). In analogy to corresponding curves obtained from gas exchange measurements, the slope at low PFD reflects maximal quantum yield and the plateau reached at light saturation is a measure of capacity. Typical examples of such light response curves are shown in Fig. 24.5(*a*) where differences in the photosynthetic performance of wild-type and transgenic potato plants are assessed (Bilger *et al.*, 1995). The transgenic plants are characterized by expression of an antisense mRNA to cp-FBPase, a key enzyme of the reductive pentose phosphate cycle. When grown at moderate PFD, wild-type and antisense plants have almost indistinguishable phenotypes. Without going into the physiologically relevant details, it is obvious from Fig. 24.5(*a*) that both plants show identical rates at low light intensities, whereas the transgenic plant saturates at much lower PFD and displays a much lower saturated rate than the wild-type. The initial slopes of both curves fall on the same line ($\Phi_{II} \times \text{PFD} = 0.8 \times \text{PFD}$), which predicts the expected ideal rate, if there were no limitations. Hence, the difference between the observed rates at a given PFD (e.g. 500 µmol $m^{-2}$ $s^{-1}$ in Fig. 24.5) and the ideal rate can be taken as a measure of limitation. Obviously, for a rapid screening of the overall photosynthetic performance of a particular plant, it is not necessary to record a full light response curve. However, a PFD should be chosen, at which under the given conditions of other environmental factors (as temperature, water stress, $CO_2$ concentration) a significant limitation in control plants is observed (500 µmol $m^{-2}$ $s^{-1}$ in the example of Fig. 24.5).

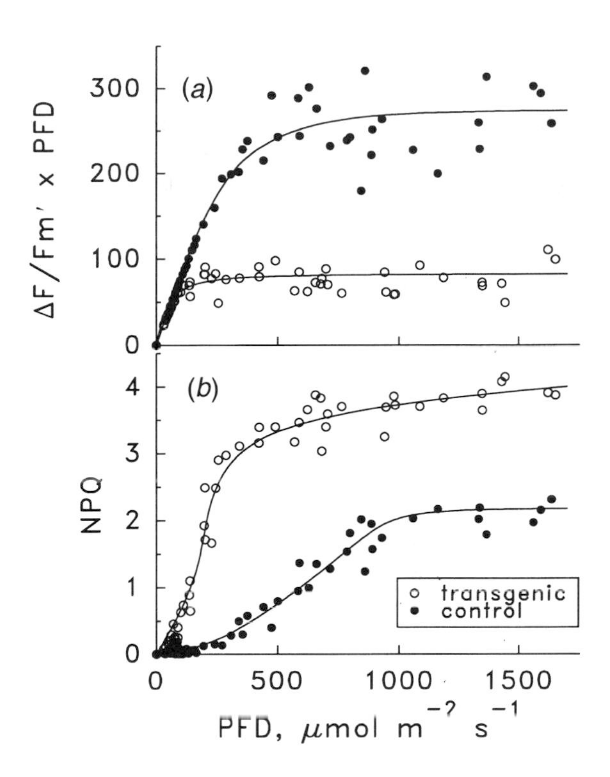


**Figure 24.5.** Light response curves of apparent electron transport rates. (*a*) $\Delta F/Fm' \times$ PFD, and (*b*) of non-photochemical quenching, NPQ, measured with leaves from control and transgenic plants of *Solanum tuberosum* L. Transgenic plants were characterised by expression of antisense mRNA to cp-FBPase. For further details, see text and Bilger *et al.*, 1995.

When, at sufficiently high PFD, a rate limitation becomes apparent, the 'excess PFD' may be defined as the difference between the ideal line and the measured curve (Björkman & Demmig-Adams, 1994). For plant survival, it is essential that the energy corresponding to excess PFD is dissipated by special mechanisms which prevent photodamage. Non-photochemical quenching of fluorescence allows some insight into these mechanisms (see pp. 327–9). As shown in Fig. 24.5(*b*), non-radiative energy dissipation (expressed by the non-photochemical quenching parameter NPQ, for definition see Fig. 24.3) increases at much lower PFD in the transgenic plant, where it also reaches distinctly higher values than in the wild type. Actually, closer inspection reveals that the relationship between NPQ and the excess PFD is almost linear (not shown) and that there is also a linear relationship between NPQ and the zeaxanthin content in both plants (Bilger *et al.*, 1995). Obviously, as has been confirmed by numerous studies (for reviews see Demmig-Adams & Adams, 1992; Horton & Ruban, 1992; Björkman & Demmig-Adams, 1994; Pfündel & Bilger, 1994), the epoxidation state of the violaxanthin cycle plays a decisive role in regulated dissipation of excess PFD. As discussed (p. 328), the actual molecular mechanism is still controversial.

## CONCLUSIONS AND OUTLOOK

In conclusion, it may be stated that Chl fluorescence measurements have not only contributed decisively to our present knowledge of photosynthesis but have also emerged as the most frequently used tool for assessment of photosynthetic performance and the general physiological state of plants. Outstanding advantages of this tool are its non-invasiveness, speed of data acquisition and high sensitivity. Hence, for example, the photosynthetic performance of a leaf in its natural position on a tree can be analysed from some distance, without even touching the leaf; information on the overall quantum yield can be obtained within seconds; and the most advanced instruments now even allow assessment of photosynthesis in single cells or in water samples from lakes, rivers and oceans. This remarkable development, which has been the result of a very fruitful interaction between basic and applied research, is still going on, as is revealed by a look into the Proceedings of the 10th International Photosynthesis Congress (held in Montpellier in 1995). Very promising fields of further progress are Chl fluorescence imaging and microfibre fluorometry, as well as phytoplankton fluorescence analysis and remote sensing.

Despite the considerable progress which has been made during the past ten years, there are still a number of nagging open questions concerning details of fluorescence analysis, some of which were also addressed in this chapter. These questions are closely linked with basic unsolved aspects of photosynthesis, like PSII heterogeneity and dynamic PSI/PSII interactions. In order to solve these questions, other measuring techniques have to complement Chl fluorescence, and fluorescence analysis has to be further refined. In particular, it will be important to reach the point where detailed knowledge of PSII organization allows an unequivocal analysis of time resolved fluorescence decays. While only few specialized laboratories can afford the considerable investments required for such analysis, it may be expected to be of great help in the interpretation of steady-state fluorescence.

### ACKNOWLEDGEMENTS

We wish to thank J. Lavergne and H. Dau for helpful literature. H. Reising and S. Heimann are thanked for help in the preparation of the manuscript. Financial support by the Deutsche Forschungsgemeinschaft (SFB 176 and 251) is gratefully acknowledged.

### REFERENCES

Allen, J. F. (1995). Thylakoid protein phosphorylation, state 1-state 2 transitions, and photosystem stoichiometry adjustment: redox control at multiple levels of gene expression. *Physiologia Plantarum*, **93**, 196–205.

Asada, K. & Badger, M. (1984). Photoreduction of $^{18}O_2$ and $^{18}H_2O_2$ with a concomitant evolution of $^{16}O_2$ in intact spinach chloroplasts. Evidence for scavenging of hydrogen peroxide by peroxidase. *Plant and Cell Physiology*, **25**, 1169–79.

Bilger, W. & Björkman, O. (1990). Role of the xanthophyll cycle in photoprotection elucidated by measurements of light-induced absorbance changes, fluorescence and photosynthesis in leaves of *Hedera canariensis*. *Photosynthesis Research*, **25**, 173–85.

Bilger, W., Fisahn, J., Brummet, W., Kossmann, J. & Willmitzer, L. (1995). Violaxanthin cycle pigment contents in potato and tobacco plants with genetically reduced photosynthetic capacity. *Plant Physiology*, **108**, 1479–86.

Björkman, O. (1987). Low-temperature chlorophyll fluorescence in leaves and its relationship to photon yield of photosynthesis in photoinhibition. In *Photoinhibition*, ed. D. J. Kyle, C. B. Osmond & C. J. Arntzen, pp. 123–44. Amsterdam: Elsevier.

Björkman, O. & Demmig-Adams, B. (1994). Regulation of photosynthetic light energy capture, conversion, and dissipation in leaves of higher plants. In *Ecophysiology of Photosynthesis*. Ecological Studies, vol. 100, ed. E-D. Schulze & M. M. Caldwell, pp. 17–47. Berlin, Heidelberg, New York: Springer.

Bolhar-Nordenkampf, H. R., Long, S. P., Baker, N. R., Öquist, G., Schreiber, U. & Lechner, E. G. (1989). Chlorophyll fluorescence as a probe of the photosynthetic competence of leaves in the field: a review of current instrumentation. *Functional Ecology*, **3**, 497–514.

Bradbury, M. & Baker, N. R. (1981). Analysis of the slow phases of the *in vivo* chlorophyll fluorescence induction curve. Changes in the redox state of Photosystem II electron acceptors and fluorescence emission from Photosystems I and II. *Biochimica et Biophysica Acta*, **635**, 542–51.

Briantais, J.-M., Vernotte, C., Krause, G. H. & Weis, E. (1986). Chlorophyll a fluorescence of higher plants: chloroplasts and leaves. In *Light Emission by Plants and Bacteria*, ed. Govindjee, J. Amesz & D. C. Fork, pp. 539–83. Orlando: Academic Press.

Cleland, R. E. (1988). Molecular events of photoinhibitory inactivation in the reaction centre of Photosystem II. *Australian Journal of Plant Physiology*, **15**, 135–50.

Dau, H. (1994*a*). Molecular mechanisms and quantitative models of variable photosystem II fluorescence. *Photochemistry and Photobiology*, **60**, 1–23.

Dau, H. (1994*b*). Short-term adaptation of plants to changing light intensities and its relation to Photosystem II photochemistry and fluorescence emission. *Journal of Photochemistry and Photobiology B: Biology*, **26**, 3–27.

Demmig-Adams, B. & Adams, W. W., III. (1992). Photoprotection and other responses of plants to high light stress. *Annual Review of Plant Physiology and Plant Molecular Biology*, **43**, 599–626.

Falkowski, P. G., Fujita, Y., Ley, A. & Mauzerall, D. (1986). Evidence for cyclic electron flow around photosystem II in *Chlorella pyrenoidosa*. *Plant Physiology*, **81**, 310–12.

Genty, B., Briantais, J-M. & Baker, N. R. (1989). The relationship between the quantum yield of photosynthetic electron transport and quenching of chlorophyll fluorescence. *Biochimica et Biophysica Acta*, **990**, 87–92.

Giersch, C. & Krause, G. H. (1991). A simple model relating photoinhibitory fluorescence quenching in chloroplasts to a population of altered photosystem II reaction centers. *Photosynthesis Research*, **30**, 115–21.

Govindjee (1995). Sixty-three years since Kautsky: chlorophyll a fluorescence. *Australian Journal of Plant Physiology*, **22**, 131–60.

Heinze, I., Dau, H. & Senger, H. (1996). The relation between the photochemical yield and variable fluorescence of photosystem II in the green alga *Scenedesmus obliquus*. *Journal of Photochemistry and Photobiology B: Biology*, **32**, 89–95.

Holzwarth, A. R. (1991). Excited-state kinetics in chlorophyll systems and its relationship to the functional organization of the photosystems. In *Chlorophylls*, ed. H. Scheer, pp. 1125–51. Boca Raton: CRC Press.

Hormann, H., Neubauer, C. & Schreiber, U. (1994). On the relationship between chlorophyll fluorescence quenching and the quantum yield of electron transport in isolated thylakoids. *Photosynthesis Research*, **40**, 93–106.

Horton, P. & Bowyer, J. R. (1990). Chlorophyll fluorescence transients. In *Methods in Plant Biochemistry*, vol. 4, ed. J. L. Harwood & J. R. Bowyer, pp. 259–96. New York: Academic Press.

Horton, P. & Hague, A. (1988). Studies on the induction of chlorophyll fluorescence in isolated barley protoplasts. IV. Resolution of non-photochemical quenching. *Biochimica et Biophysica Acta*, **932**, 107–15.

Horton, P. & Ruban, A. V. (1992). Regulation of Photosystem II. *Photosynthesis Research*, **34**, 375–85.

Joshi, M. K. & Mohanty, P. (1995). Probing photosynthetic performance by chlorophyll *a* fluorescence: analysis and interpretation of fluorescence parameters. *Journal of Scientific and Industrial Research*, **54**, 155–74.

Karukstis, K. K. (1991). Chlorophyll fluorescence as a physiological probe of the photosynthetic apparatus. In *Chlorophylls*, ed. H. Scheer, pp. 769–95. Boca Raton: CRC Press.

Kolber, Z. & Falkowski, P. G. (1993). Use of active fluorescence to estimate phytoplankton photosynthesis *in situ*. *Limnology and Oceanography*, **38**, 1646–65.

Krall, J. P. & Edwards, G. E. (1992). Relationship between photosystem II activity and $CO_2$ fixation in leaves. *Physiologia Plantarum*, **86**, 180–7.

Kramer, D. M., DiMarco, G. & Loreto, F. (1995). Contribution of plastoquinone quenching to saturation pulse-induced rise of chlorophyll fluorescence in leaves. In *Photosynthesis: From Light to Biosphere*, vol. I, ed. P. Mathis, pp. 147–50. Dordrecht: Kluwer Scientific Publisher.

Krause, G. H. & Weis, E. (1991). Chlorophyll fluorescence and photosynthesis: The basics. *Annual Review of Plant Physiology and Plant Molecular Biology*, **42**, 313–49.

Krieger, A. & Weis, E. (1993). The role of calcium in the pH-dependent control of photosystem II. *Photosynthesis Research*, **37**, 117–30.

Lavergne, J. & Briantais, J-M. (1996). Photosystem II heterogeneity. In *Oxygenic Photosynthesis: The Light Reaction*, ed. D. R. Ort & C. F. Yocum, in press Dordrecht: Kluwer Scientific Publisher.

Lavergne, J. & Trissl, H-W. (1995). Theory of fluorescence induction in photosystem II: derivation of analytical expressions in a model including exciton-radical-pair equilibrium and restricted energy transfer between photosynthetic units. *Biophysical Journal*, **68**, 2474–92.

Mohammed, G. H., Binder, W. D. & Gillies, S. L. (1995).

Chlorophyll fluorescence: a review of its practical forestry applications and instrumentation. *Scandinavian Journal of Forest Research*, **10**, 383–410.

Moya, I., Sebban, P. & Haehnel, W. (1986). Lifetime of excited states and quantum yield of chlorophyll *a* fluorescence *in vivo*. In *Light Emission by Plants and Bacteria*, ed. Govindjee, J. Amesz & D. C. Fork, pp. 161–90. Orlando: Academic Press.

Neubauer, C. & Schreiber, U. (1987). The polyphasic rise of chlorophyll fluorescence upon onset of strong continuous illumination: I. Saturation characteristics and partial control by the photosystem II acceptor side. *Zeitschrift für Naturforschung*, **42c**, 1246–54.

Pfündel, E. (1995). PSI fluorescence at room temperature: possible effects on quenching coefficients. In *Photosynthesis: From Light to Biosphere*, vol. II, ed. P. Mathis, pp. 855–8. Dordrecht: Kluwer Scientific Publisher.

Pfündel, E. & Bilger, W. (1994). Regulation and possible function of the violaxanthin cycle. *Photosynthesis Research*, **42**, 89–109.

Schatz, G. H., Brock, H. & Holzwarth, A. R. (1988). Kinetic and energetic model for the primary processes in Photosystem II. *Biophysical Journal*, **54**, 397–405.

Schreiber, U. & Bilger, W. (1987). Rapid assessment of stress effects on plant leaves by chlorophyll fluorescence measurements. In *Plant Response to Stress – Functional Analysis in Mediterranean Ecosystems. NATO Advanced Science Institute Series*, ed. J. D. Tenhunen, F. M. Catarino, O. L. Lange & W. C. Oechel, pp. 27–53. Berlin, Heidelberg, New York: Springer-Verlag.

Schreiber, U. & Bilger, W. (1993). Progress in chlorophyll fluorescence research: major developments during the past years in retrospect. *Progress in Botany*, **54**, 151–73.

Schreiber, U., Bilger, W. & Neubauer, C. (1994). Chlorophyll fluorescence as a nonintrusive indicator for rapid assessment of *in vivo* photosynthesis. In *Ecophysiology of Photosynthesis*. Ecological Studies, vol. 100, ed. E-D. Schulze & M. M. Caldwell, pp. 49–70. Berlin, Heidelberg, New York: Springer.

Schreiber, U., Hormann, H., Neubauer, C. & Klughammer, C. (1995*a*). Assessment of photosystem II photochemical quantum yield by chlorophyll fluorescence quenching analysis. *Australian Journal of Plant Physiology*, **22**, 209–20.

Schreiber, U., Hormann, H., Asada, K. & Neubauer, C. (1995*b*). $O_2$-dependent electron flow in intact spinach chloroplasts: properties and possible regulation of the Mehler-ascorbate peroxidase cycle. In *Photosynthesis: From Light to Biosphere*, vol. II, ed. P. Mathis, pp. 813–18, Dordrecht: Kluwer Scientific Publisher.

Schreiber, U. & Neubauer, C. (1987). The polyphasic rise of chlorophyll fluorescence upon onset of strong continuous illumination: II. Partial control by the photosystem II donor side and possible ways of interpretation. *Zeitschrift für Naturforschung*, **42c**, 1255–64.

Schreiber, U. & Neubauer, C. (1990). $O_2$-dependent electron flow, membrane energization and the mechanism of non-photochemical quenching of chlorophyll fluorescence. *Photosynthesis Research*, **25**, 279–93.

Schreiber, U., Schliwa, U. & Bilger, W. (1986). Continuous recording of photochemical and non-photochemical chlorophyll fluorescence quenching with a new type of modulation fluorometer. *Photosynthesis Research*, **10**, 51–62.

Strasser, R. J., Srivastava, A. & Govindjee (1995). Polyphasic chlorophyll *a* fluorescent transient in plants and cyanobacteria. *Photochemistry and Photobiology*, **61**, 32–42.

van Kooten, O. & Snel, J. F. H. (1990). The use of chlorophyll fluorescence nomenclature in plant stress physiology. *Photosynthesis Research*, **25**, 147–50.

Vernotte, C., Etienne, A. L. & Briantais, J-M. (1979). Quenching of the system II chlorophyll fluorescence by the plastoquinone pool. *Biochimica et Biophysica Acta*, **545**, 519–27.

Walters, R. G. & Horton, P. (1993). Theoretical assessment of alternative mechanisms for non-photochemical quenching of PS2 fluorescence in barley leaves. *Photosynthesis Research*, **36**, 119–39.

Weis, E. & Berry, J. A. (1987). Quantum efficiency of photosystem II in relation to 'energy'-dependent quenching of chlorophyll fluorescence. *Biochimica et Biophysica Acta*, **894**, 198–208.

Williams, W. P. & Allen, J. F. (1987). State 1/state 2 changes in higher plants and algae. *Photosynthesis Research*, **13**, 19–45.

# 25 Action of modern herbicides

P. BÖGER AND G. SANDMANN

## INTRODUCTION

Herbicides are agrochemicals to control weeds during crop growth. Essentially, they are enzyme inhibitors affecting basic metabolic reactions which are obligatory for plant life. Since these compounds are introduced artificially into the environment, caution and care has to be taken to minimize adverse side-effects. Accordingly, weed control chemicals should affect only plant, not animal, metabolism. Secondly, these compounds should be highly active to reduce the chemical load of the field. Use rates of 10 to 100 g active ingredient (ai) per ha are realistic today in contrast to a kg/ha 20 years ago. Thirdly, a limited persistence of herbicides in the soil is required (in the range of some months, not years), and they should not leach into the groundwater. Furthermore, post-emergence herbicides are favoured, which are applied to the emerged weed, pre-emergent chemicals (sprayed to the soil before weeds and crop emerge) are not advantageous for ecotoxicological reasons. These objectives require continuous research and development in the agrochemical industry, and the progress is amazing, although not finished. Mode of action studies have contributed to find a solution. Herbicides are used as specific probes to modify plant metabolism and to elucidate our understanding of plant life. Photosynthesis or pigment biosynthesis can be blocked specifically as well as the formation of antioxidants, oxidative stress can be induced or intracellular radical reactions produced. Thus, studies on the mode of action of herbicides will contribute to basic plant biochemistry.

A serious problem is the occurrence of herbicide-resistant weeds. These biotypes are either able to degrade the chemical incorporated, or contain a mutated gene for a target enzyme insensitive to the herbicide. It is therefore essential to look for new compounds with new targets and/or other metabolic breakdown routes to control the resistant mutants. Searching for key enzymes, which are sensitive to inhibitors, will be a means to possibly find a new herbicide class. At the moment this approach is being pursued actively in the domain of plant-specific amino acid biosynthesis. Enzymatic assay systems will also assist in screening the numerous analogues produced by synthetic chemists in the agrochemical industry, once a promising inhibitor has been found.

Herbicide application started in the early 1950s and 1960s with auxin-type compounds, followed by inhibitors of cell division and photosynthesis. The auxin-type herbicides interfere with plant hormone regulation, but their mode of action is still unclear. They are not dealt with in this chapter. The reader may consult Fedtke (1982) and Loos (1975).

## PHOTOSYNTHETIC ELECTRON TRANSPORT

This transport system is embedded in the thylakoids of the chloroplasts and mediates the conversion of sunlight into chemical energy, namely NADPH and ATP, used for carbon dioxide assimilation and other biosynthetic reactions. Photosynthesis herbicides require comparatively high use rates (200 to 500 g ai/ha). They have to be distributed to all green parts of the plant, binding constants often are high (even more than $10^{-6}$ M) and they bind reversibly. The thermodynamic equilibrium at the target requires a high concentration of unbound (free) inhibitor. These three features are mainly responsible for the high use rate, and the development of new photosynthesis herbicides has come to a halt.

All photosynthesis herbicides developed thus far either block electron flow at photosystem II or divert electrons at the terminal part of photosystem I. No herbicidal inhibitors are known for the $CO_2$-assimilation reaction complex. The chemical structures of typical commercialized PSII herbicides are listed in Fig. 25.1, belonging to eight structural classes.

Herbicidal PSII inhibitors bind reversibly to an integral hydrophobic membrane protein, the D1 protein ('herbicide-binding protein') of the PSII core complex with an apparent molecular weight of about 32 kDa. One herbicide molecule binds to one D1 protein competing with reversibly bound plastoquinone (the $Q_B$ acceptor), and prevents the oxidation of $Q_A$ (a firmly bound plastoquinone at the D2 protein; see Chapter 3, this volume). In contrast to some reports, also the phenolic-type inhibitors bind at the D1 protein. Their binding, however, is slow at physiological pH and can be speeded up by acid pH or light. Electron flow and formation of NADPH are impaired, cyclic photophosphorylation, however, may still proceed. Herbicides with different chemical core structures can displace each other at the D1-binding site. It should be emphasized that herbicides bind to D1 only, not to the D2 protein, and that binding always causes interrupted electron transport. It is not possible to attach a non-inhibitory chemical to D1 which displaces a herbicide and will subsequently *not* inhibit electron flow. Such a 'safening' idea is not feasible.

The D1-gene (psbA) encodes for about 353 amino acids; only the sequence 211–275 represents the binding site (Oettmeier, 1992). Several amino acid residues act as binding partners for certain substituents of a herbicide (interaction by hydrogen bonds or charge-transfer exchange). Obviously, many strong 'contacts' of the herbicide molecule with amino acid residues will lead to a low binding constant. Genetic replacement of amino acids will produce mutant plants resistant to PSII herbicides (Fig. 25.2). Chemically different herbicides or different substituents of a herbicide molecule interact with different amino acids. However, part of the amino acid binding partners present in the binding site will be identical for different herbicides. This explains why the amino acid exchanges shown affect binding of several different herbicides decreasing their binding constants to a different degree (Trebst, 1991). So far, genetic manipulation of the binding site has always produced resistance against *several* chemically different herbicides ('cross-resistance'). It is doubtful whether systematic mutagenesis of amino acid residues of the binding site will result in herbicide-specific resistance. This prospect limits the efforts of the industry to produce crop varieties resistant against PSII herbicides.

A bromoxynil-resistant cotton variety has been developed, based on metabolic detoxification. This genetically engineered variety contains a bacterial nitrilase to modify the herbicide. Transformants containing an aryl amidase have been obtained to split the rice herbicide propanil.

PSI electron flow 'divertors' are the *bipyridinium salts* paraquat (1,1′-dimethyl-4,4′-bipyridinium dichloride, methyl viologen) or diquat (1,1′ ethylene-2,2′-bipyridinium dibromide). Also heteropentalenes have been developed with similar properties (Bowyer & Camilleri, 1987). Only paraquat has gained importance as a commercial herbicide. All these electron divertors have mid-point redox potentials between −350 and −650 mV and can therefore accept electrons at PSI giving rise to a cation radical, which subsequently reacts with molecular oxygen. Paraquat reacts with a rate constant of $7.7 \times 10^{-8}$ $M^{-1}$ $s^{-1}$ leading to micromolar concentrations of superoxide in the chloroplast. By Fenton-type reactions the non-toxic $O_2^-$ is converted to toxic species such as hydroxyl radical (● OH) and hydrogen peroxide. Reactive oxygen species will subsequently react with allytic hydrogens of unsaturated lipids leading to membrane degradation. Accordingly, paraquat exerts its phytotoxicity in the light. It may also mediate oxidation of pyridine nucleotides in the dark catalysed by reductases plus non-haem iron proteins in the plant cell. This activity is not a herbicidal one, but establishes paraquat toxicology in warm-blooded animals (Liochev *et al.*, 1994).

## CAROTENOID BIOSYNTHESIS

Carotenoids are associated with membrane-bound chlorophyll complexes of the photosynthetic apparatus. They are involved in energy transfer and protect light-activated chlorophyll against (photo)oxidation (see Chapter 4 of this volume). A herbicide, interfering with carotenoid biosynthesis will cause bleaching of *growing* leaves. In addition, decreased pigment synthesis will limit

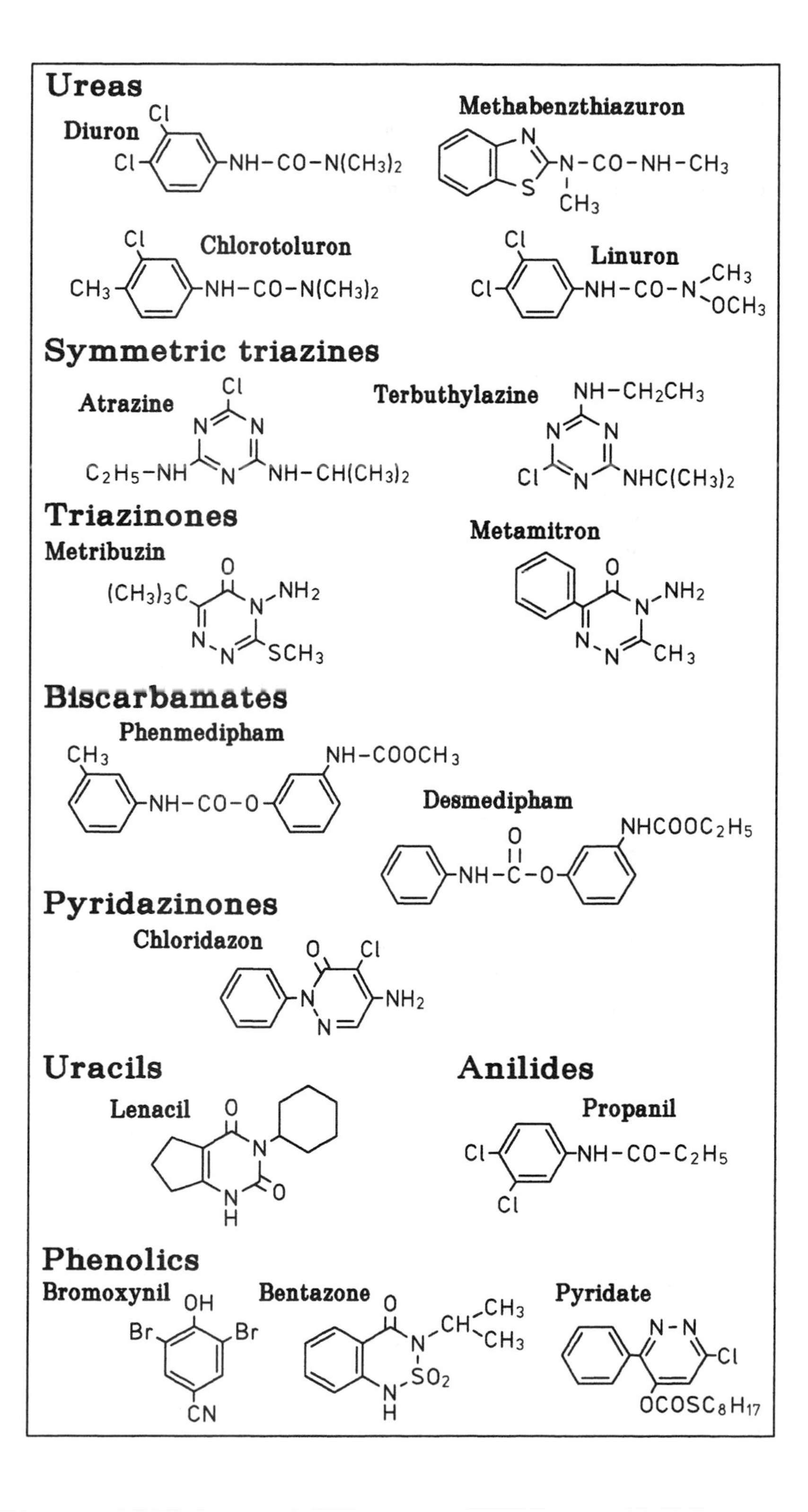


**Figure 25.1.** Typical commercial photosystem II herbicides: chemical classes and common names.

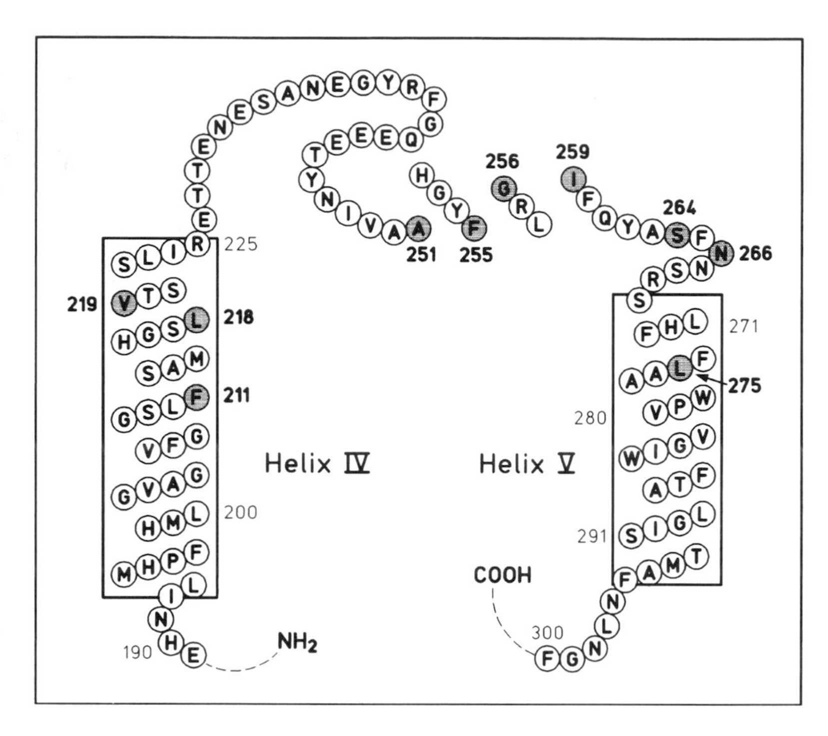


**Figure 25.2.** The herbicide binding site of the D1 protein of the photosystem-II core complex. The two boxes indicate the transmembrane helices IV and V. The upper part of the amino acid sequence (at the stromal side of the thylakoid) includes the binding region. Shadowed amino acid residues (also numbered) indicate those whose changes resulted in decreased herbicide binding (i.e. resistance) as was shown with *Synechococcus* (cyanobacteria), *Chlamydomonas* (green alga) and higher plants. Phe 211 → Ser; Leu 218 → Phe; Val 219 → Ile; Ala 251 → Val; Phe 255 → Tyr; Gly 256 → Asp; Ile 259 → Ser; Ser 264 → Gly, Ala or Thr; Asn 266 → Thr; Leu 275 → Phe (after Oettmeier, 1992, modified; see this ref. for herbicide resistance accomplished).

assembly and formation of the photosynthetic electron transport complexes (Dahlin & Timko, 1994). Old leaves or non-growing green leaf parts will not bleach when treated with a 'bleaching' herbicide. Once the pigments are formed these herbicides do not effect them, which is in contrast to peroxidizing compounds which are called 'photobleachers' since they *degrade* existing pigments (see p. 347).

The term carotenoids includes carotenes and xanthophylls. Their formation takes place via the terpenoid pathway with prenylpyrophosphates ($C_5$ entities) as precursors (Fig. 25.3). The first 40 C-atoms isoprenoid is phytoene, which is subjected to several dehydrogenations resulting in red and yellow species, followed by two cyclization steps leading to formation of lycopene, β-carotene and xanthophylls. The enzymes involved are membrane-(thylakoid) bound, and their isolation and characterization is dealt with in Chapter 4. Inhibitors interfering with hydrogen abstraction between phytoene and lycopene should be candidates for herbicides. Indeed, several herbicides have been developed from phytoene desaturase and ζ-carotene desaturase inhibitors, of which the most successful representatives belong to the first group only (like norflurazon, diflufenican, fluridone or flurtamone, Fig. 25.4; Sandmann & Böger, 1992).

Inhibition of phytoene desaturase or the other H-abstracting enzymes results in accumulation of the precursor instead of carotenoid end products. Their detection in sensitive algae cultures (*Scenedesmus acutus*) is a convenient means to identify a compound as a bleacher. Using cell-free carotenogenic systems, it could be shown, that these inhibitors bind reversibly and non-competitively to the desaturase. Bleaching herbicides target the same enzyme. Studies with herbicide-resistant cyanobacterial mutants with point mutations at the enzyme exhibit cross-resistance, e.g. against norflurazon and fluridone (Chamovitz, Sandmann & Hirschberg, 1993). This situation is similar to PSII herbicides. Binding apparently occurs at an identical site of the desaturase, but each inhibitor interacts with a different set of amino acid residues. Phytoene desaturases from cyanobacteria and higher plants show very similar inhibition features except for an enzyme isolated from the carotenoid-producing bacterium *Erwinia uredovora* which is highly insensitive to bleaching herbicides (Windhövel *et*

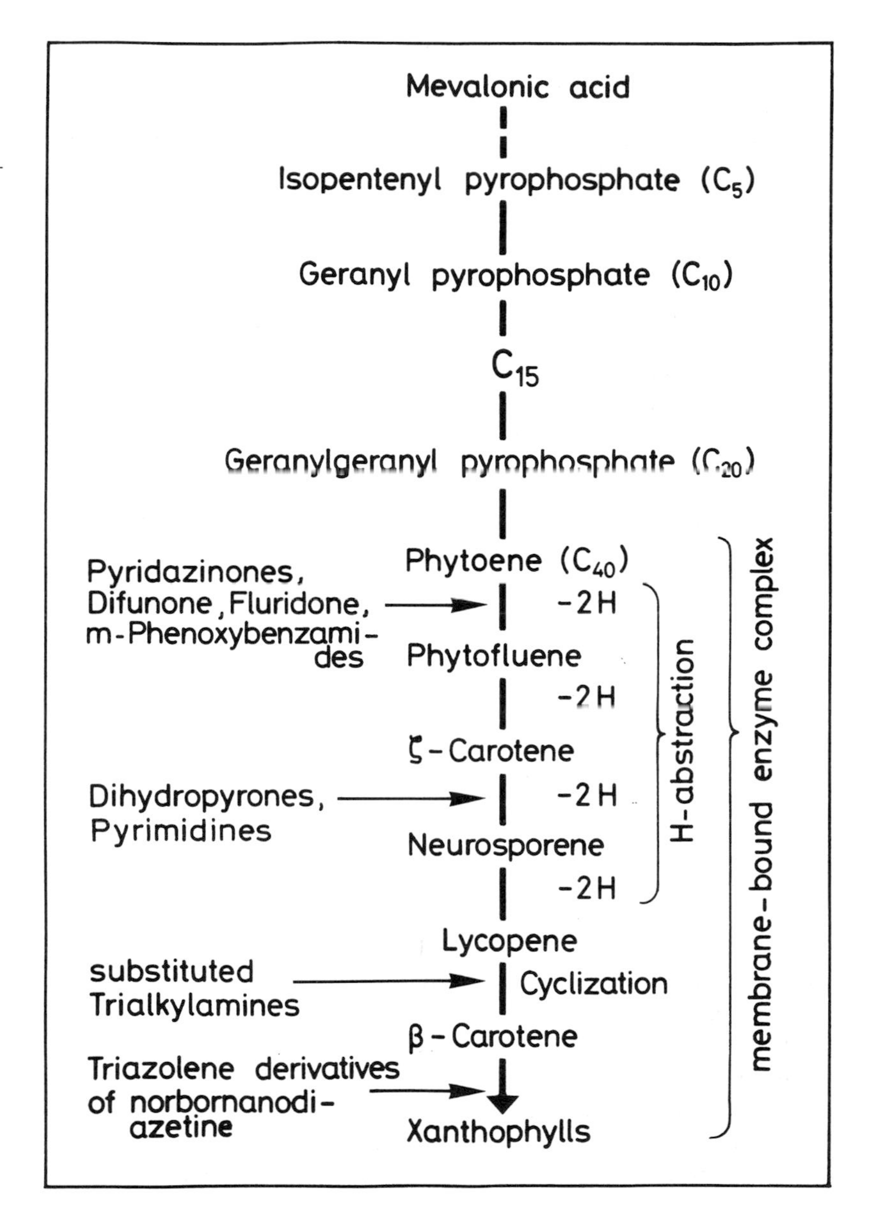


**Figure 25.3.** Biosynthesis pathway of carotenoids in higher plants and inhibition sites. Phytoene desaturation is the major inhibition step. For details of biosynthesis, see Chapter 4 on 'Pigments', this volume. Plastidic isoprenoid synthesis does not start with mevalonate but with a glyceraldehyde-3-phosphate/pyruvate route yielding isopentenyl pyrophosphate (Lichtenthaler *et al.*, 1997).

*al.*, 1994). This gene could be cloned into tobacco, and the transformants exhibit strong resistance to various bleaching herbicides. The $I_{50}$ values are about 10 000 times *vs.* control (Misawa *et al.*, 1994).

*Clomazone* [2- (2-chlorophenyl) methyl-4,4-dimethyl-3-isoxazolidinone] is a strong bleaching herbicide but does not inhibit phytoene desaturase. A target site upstream of the terpenoid biosynthesis pathway is assumed, although no definite enzyme inhibition has been found as yet. Benzoylcyclohexanediones (see Fig. 25.11) are also strong bleaching compounds which lead to phytoene accumulation. Again, the desaturase is not affected. The latter compounds inhibit formation of quinones, which are thought to be required as obligatory electron acceptors in the phytoene desaturase reaction (see p. 349).

**Figure 25.4.** Formulae of inhibitors acting upon four membrane-bound enzymes of the carotenoid pathway. Hydroxylation inhibitors interfere with the introduction of OH groups, i.e. with the formation of xanthophylls.

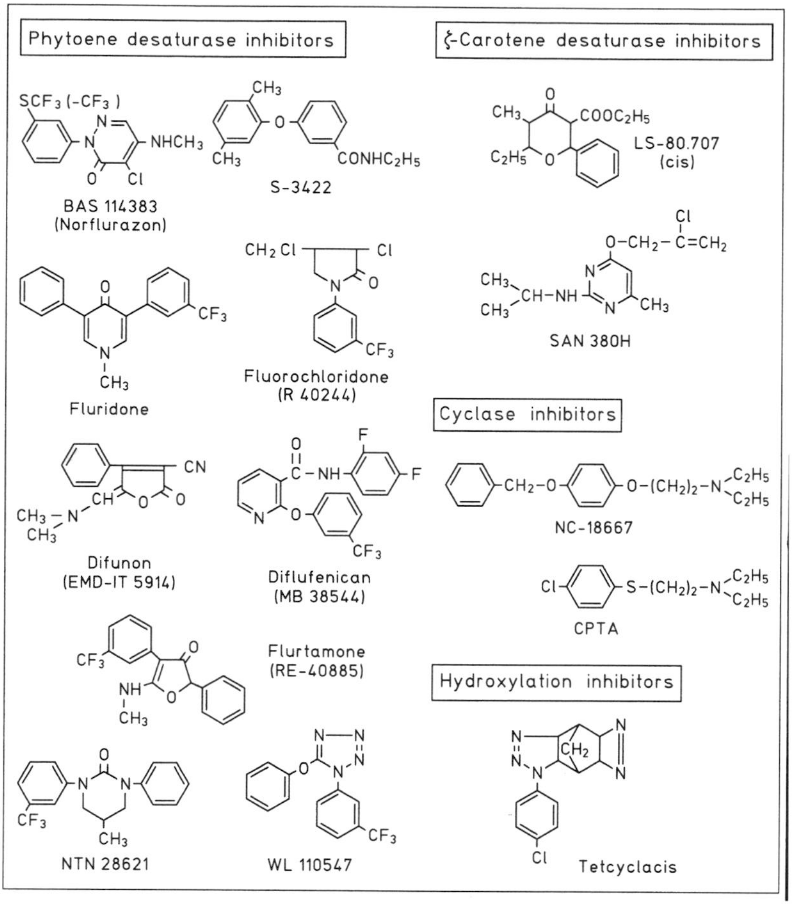

Phytoene desaturase inhibitors

BAS 114383 (Norflurazon)

S-3422

Fluridone

Fluorochloridone (R 40244)

Difunon (EMD-IT 5914)

Diflufenican (MB 38544)

Flurtamone (RE-40885)

NTN 28621

WL 110547

ζ-Carotene desaturase inhibitors

LS-80.707 (cis)

SAN 380H

Cyclase inhibitors

NC-18667

CPTA

Hydroxylation inhibitors

Tetcyclacis

## FATTY-ACID FORMATION

Most fatty acids are biosynthesized in the chloroplast. The building block is acetyl-CoA, which is carboxylated to malonyl-CoA by acetyl-CoA carboxylase (EC 6.4.1.2, ACCase). Both compounds, bound to an acyl-carrier protein, feed into the fatty-acid synthase complex consisting of additional six enzymes to produce palmitate. Although inhibitors are known, which affect some of the fatty-acid biosynthetic steps, the decisive enzyme targeted by herbicides is the above-mentioned acetyl-CoA carboxylase (ACCase). Its inhibition stalls formation of cellular acyl lipids and the build-up of membranes.

Two classes of herbicides have successfully emerged, namely oxime-type cyclohexane-diones, the 'dims' (like sethoxydim, clethodim) and aryloxyphenoxypropionates, the 'fops' like diclofop-methyl or fluazifop-butyl (Fig. 25.5). The 'fops' are applied post-emergent to the weeds as esters which are readily taken up but then hydrolysed in the cell to the free acid. This is the herbicidal form, the esters are up to 100-fold less inhibitory to the ACCase.

The great advantage of these compounds is their phytotoxicity to many grasses with little impact on dicots ('graminicides'). The inhibitor constants for the 'dim' of Fig. 25.5 were found to be 400 to 60 000 times higher with the ACCase from broadleaf species than from sensitive monocotyledons. Also fluazifop inhibits the ACCase from barley but not from pea. These typical grass herbicides are limited for use in cereals, but concurrent application of 'safeners', has also allowed for grass control, e.g. in wheat.

There are two biotin-containing acetyl-CoA carboxylases present in plants, a prokaryotic form composed of dissociable polypeptides and a eukaryotic form, which is

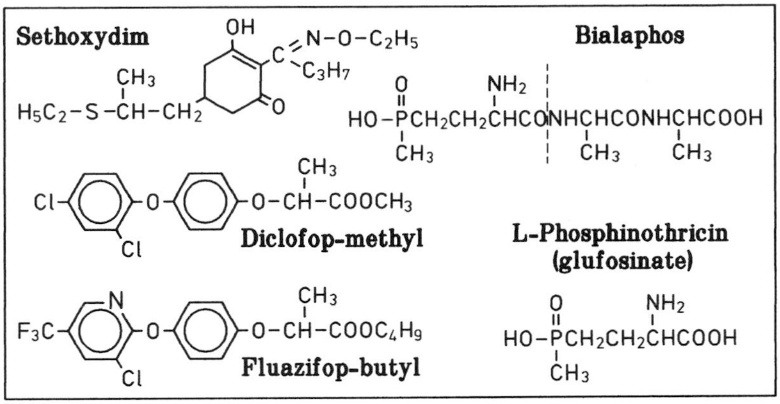

**Figure 25.5.** Formulae of a representative of the oxime-type cyclohexanedione family and two prominent aryloxyphenoxypropionate-type herbicides, all interfering with acetyl-CoA carboxylase. On the right, two herbicides inhibiting glutamine synthetase are shown.

a homodimer of a high-molecular weight multifunctional protein (2 × 220 kDa). Generally, the latter type is sensitive to the herbicides mentioned above and is the major form present in plastids and cytosol of *Gramineae*, while the prokaryotic species is insensitive and is absent or present in minor amounts in monocotyledons (Sasaki, Konishi & Nagano, 1995). Dicots contain both types of the enzyme, the eukaryotic one located in the plastid. Monocot ACCases, however, exhibit considerably different inhibition constants, as was shown, e.g. with *Festuca* and *Poa* species. So, not all grasses can be controlled with these herbicides. The inhibition is reversible and most sensitive to the concentration of acetyl-CoA, although a competitive character could not be found with all herbicide species assayed. Inhibition interferes with the carboxyl-transfer site, not with the biotin-carboxylation site. Apparently, both types of grass herbicides compete for the same binding site.

These graminicides are the only example known at the moment where selectivity (against dicot crops) is conferred by a target enzyme of reduced inhibitor affinity rather than by metabolic degradation of the chemical. Plants, however, also detoxify by degradation which establishes, e.g. tolerance in wheat or the widespread occurrence of resistant grasses (e.g. *Lolium* strains in Southern Australia; Holt, Powles & Holtum, 1993).

Diclofop induces depolarization of the membrane potential of root tip cells of *Lolium* or *Avena fatua*. This happens both with sensitive and resistant strains, but the latter ones can re-establish the electrogenic potential, while the herbicide-sensitive strain cannot (Devine *et al.*, 1993). The concentrations needed were found to be relatively high (50 μM and more) in contrast to the lower $I_{50}$ values for the isolated monocot ACCase of sensitive grasses (about 0.2–20 μM). At the moment it remains doubtful whether these grass herbicides have a second relevant target site besides the ACCase. Sequencing of the (big) gene(s) and parts thereof is under way (see, e.g. Ashton, Jenkins & Whitfeld, 1994). Sethoxydim-resistant maize is under development, which has been obtained by selection of (resistant) cell cultures rather than by genetic engineering.

Processing of fatty acids can be inhibited by herbicides. Pyridazinone analogues or fluridone, known as phytoene desaturase inhibitors, can also affect desaturation of an 18 : 2 fatty acid. Although malonyl-CoA is also needed for fatty acid *elongation*, the latter is not influenced by grass herbicides (due to insensitive ACCase outside the plastid?). The elongases, catalysing the formation of long-chain fatty acids for the build-up of waxes are influenced by thiocarbamates (e.g. EPTC; Barett & Harwood, 1993) and possibly by chloroacetamides. It appears that these are side-effects not representing the cause for herbicidal activity. Finally, it should be noted that the liver also contains an eukaryotic ACCase type which is insensitive to graminicides.

## AMMONIA ASSIMILATION AND AMINO ACID BIOSYNTHESIS

Glutamine synthetase (GS, EC 6.3.1.2) is the key enzyme for ammonia assimilation and its inhibition has severe consequences. One potent inhibitor, called either *phosphinothricin* or *gluphosinate* [D,L-2-amino-4 (hydroxymethylphosphinyl)butanoic acid, Fig. 25.5] has been developed as a successful post-emergent herbicide. The major part of the endogenous ammonia of the plant which is channelled through the photorespiration cycle can not be bound any more to glutamate. Conceivably, this herbicide is phytotoxic in the light, requires the presence of oxygen, photorespiration, and photosynthesis at the start of the treatment. Subsequently, the latter two processes are strongly inhibited leading to accumulation of ammonia in the plant. The long-term phytotoxic effect is membrane leakage and leaf necrosis probably due to increased $NH_4$ levels.

L-Gluphosinate is a glutamate analogue acting competitively *vs.* the natural substrate as a transition-state inhibitor. With ATP present, and after due incubation time, it is *irreversibly* bound to the enzyme in a 1 : 1 ratio. The herbicide affects glutamine synthetase from bacteria and mammals, but is non-toxic to the latter because it does not pass the blood–brain barrier and is excreted rapidly.

Glufosinate is a molecular part of bialaphos (Fig. 25.5), which is a tripeptide exhibiting herbicidal properties provided the molecule is enzymatically split in the (plant) cell into the active compound and two alanyl residues. The herbicidal activity of bialaphos was found in Japan before the herbicidal properties of glufosinate were known, but the practical use of bialaphos is limited owing to high production costs.

In the plant the C-P bond of glufosinate cannot be split; the compound is slowly converted to the (inactive) 3-methyl-phosphinocopropionic acid. A bacterial acetyl transferase inactivates the L-form of the herbicide and the gene from *Streptomyces* strains has been transferred into crop plants (maize, sugar beet). The $K_m$ is low for the herbicide (0.06 mM) and high for glutamate (240 mM), the expression results in 0.01–0.1% of the total plant protein only, and the acetylation reaction is rapid. Taken these features together, adverse side-effects to the plant metabolism, which may reduce the fitness of the herbicide-resistant genetic transformants, are not observed (Donn, Eckes & Müllner, 1992).

*Glyphosate* [*N*(phosphonomethyl)glycine] is the top-selling post-emergent non-selective herbicide requiring relatively high use rates (100 g to 1 kg ai/ha). It inhibits the EPSP-synthase (5-enolpyruvylshikimate-3-phosphate synthase, EPSPS, EC 2.5.1.19, localized in the chloroplast), which is instrumental in the biosynthesis of aromatic amino acids by plants and bacteria (Fig. 25.6). It is the only herbicide known for this pathway. As a result of EPSPS inhibition, the starting enzyme of the biosynthetic pathway (3-deoxy-D-*arabino*-heptulosonate-7-phosphate, DAHP-synthase; EC 4.1.2.15) is increased, and both effects lead to more shikimate formation. Photosynthetic carbon assimilation is impaired after some hours of treatment. Long-term effects cause a depletion of the free pool of aromatic acids, and both effects lead to cessation of plant growth. Not surprisingly, many other metabolic routes are subsequently affected, e.g. formation of indolyl acetic acid, phytoalexins or anthocyanins. The latter are the basis for a quantitative assay for inhibitors of aromatic biosynthesis (buckwheat test, Amrhein & Roy, 1993; see this ref. also for EPSPS inhibition assay.) Glyphosate is phloem-

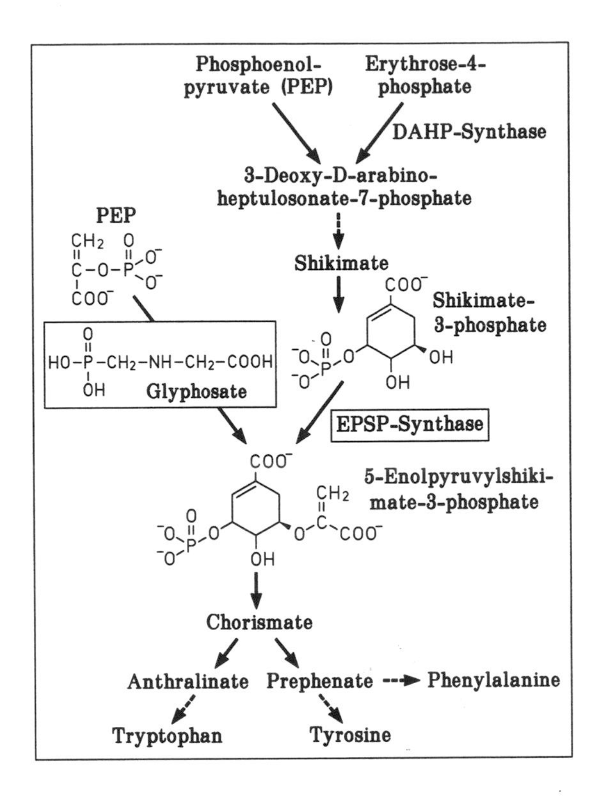


**Figure 25.6.** Scheme of biosynthesis of aromatic amino acids: Target of glyphosate.

mobile and part of it is transported through the entire plant, particularly into the 'sinks' (roots, flowers).

The herbicide is a competitive inhibitor relating to phospho*enol*pyruvate (PEP) and binds reversibly to the shikimate-3-phosphate/enzyme complex with a 100-fold better affinity than PEP, whose $K_m$ is about 10 μM. The concentration of PEP in the cell is too low to displace glyphosate, consequently the ternary enzyme–shikimate-phosphate–glyphosate complex remains intact ('dead-end complex') and the enzyme reaction is stalled (described in Böger & Sandmann, 1990). Whether or not glyphosate binds to the *reaction* site of EPSPS mimicking a transition state is still an open question.

Glyphosate can be slowly modified to aminomethylphosphonic acid (AMPA; Komoβa, Gennity & Sandermann, 1992) but the C–P bond cannot be cleaved by plants. This is possible in bacteria. Two approaches are feasible to obtain glyphosate-resistant crops by (a) cloning a glyphosate-resistant EPSPS gene or (b) a bacterial gene whose enzyme leads to AMPA and glyoxylate ('GOX enzyme'). The first system has been developed for, e.g. sugarbeet or potato using a gene from an *Agrobacterium* strain (Barry *et al.*, 1992). Such transgenic crops may show increased accumulation of glyphosate in seeds and tubers due to non-inhibited photosynthate production causing improved herbicide translocation.

The branched-chain amino acids, valine, leucine and isoleucine, are formed only in bacteria and plants ('essential amino acids' like the aromatic ones). Inhibitors of these biosynthetic pathways have a good chance not to be toxic to mammals, which is indeed true with the commercialized herbicides offered. Acetolactate synthase (ALS, also called acetohydroxy acid synthase) carries out the first set of parallel reactions either yielding acetolactate or acetohydroxybutyrate which are the precursors of the three amino acids. The enzyme is present in plants in low concentration (0.005% of total protein in barley) so that its activity limits the flow in the pathway and inhibition will cause a drastic effect. The enzyme is inhibited by *ALS inhibitors* belonging to four structural classes (Fig. 25.7): sulfonylureas, imidazolinones, pyrimidyl-oxybenzoates and triazolopyrimidine sulfonamides (Stetter, 1994).

The first response of the treated plant is cessation of mitosis, inhibition of DNA synthesis (which was first thought to be the primary target domain) and inhibition of photosynthate transport. Certainly, these effects are secondary ones but cannot yet be fully explained. 2-Ketobutyrate accumulates but does not cause phytotoxicity (Shaner & Singh, 1993). ALS is regulated through feedback control by Val and Leu, the ALS inhibitors apparently bind at the regulatory site (Durner, Gailus & Böger 1991). With pyruvate present, the inhibitors are bound quickly and reversibly ($K_i$ = 70 nM for chlorsulfuron using purified barley ALS). After 2–3 hours the inhibitor is attached tightly ('quasi-irreversibly'), though not covalently, with a binding constant down to 3 nM (Schloss & Aulabaugh, 1990; Durner *et al.*, 1991). The inhibitor is practically 'titrated' off by a 1:1 binding to ALS. The effect is similar to that described for gluphosinate, although the ALS inhibitors do not mimic a transition state. The inhibitors can dissociate very slowly, and we could show that the ALS is inactivated after release of the tightly bound inhibitor. These features, together with the low enzyme level in plants, have made ALS the most famous target in herbicide development within the last ten years. ALS inhibitors require low use rates (10–20 g ai/ha with some sulphonylureas) which is essentially due to their 'slow-binding' properties. The good herbicidal effect of the inhibitors is corroborated by some phloem mobility which ensures the presence of the compounds in the (meristematic) growing regions of the weed.

Unfortunately, weed resistance against these herbicides became widespread. It appears that many mutations in the ALS gene do not affect the function of the enzyme and the fitness of the mutant. This fact results in the presence of many resistant individuals in the (non-treated) weed population of the field, which leads to a rapid selection of resistant biotypes once ALS inhibitors are applied continuously for 2–3 years only. In *Kochia* ALS a mutation of Pro 173 to Thr, Arg and other amino acids all result in resistance to chlorsulphuron (Guttieri, Eberlein & Thill, 1995). A mutation of Trp 552 to Leu will confer a high level of resistance to all the chemical inhibitor families mentioned above (Bernasconi *et al.*, 1995). Accordingly, the development of transgenic crops resistant to ALS inhibitors bears a heavy risk.

## TETRAPYRROLE BIOSYNTHESIS

Several inhibitors affecting this biosynthesis are known particularly for the first steps of the plant-specific gluta-

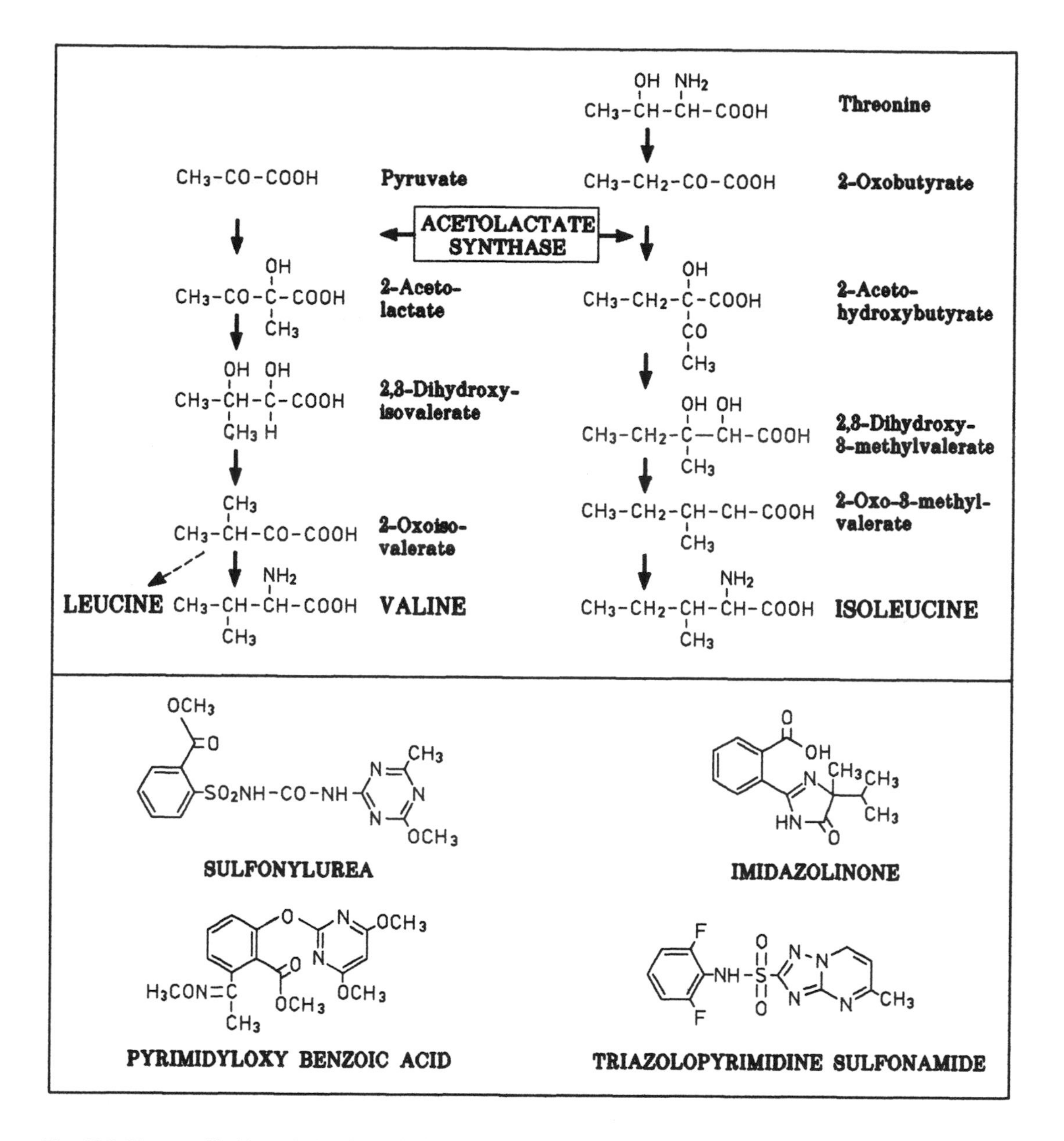


Fig. 25.7. Plant-specific biosynthesis of branched-chain amino acids with the prominent herbicide target enzyme indicated (*top*). Four typical acetolactate synthase inhibitors of different chemistry are shown (*bottom*).

mate pathway. At the moment, the only promising step for the development of herbicides is inhibition of the membrane-bound protoporphyrinogen oxidase (protox, EC 1.3.3.4) catalysing the oxidation of protoporphyrinogen IX (protogen IX) to protoporphyrin IX (proto IX; Scalla *et al.*, 1990). This inhibition leads to a rapid abnormal accumulation of fluorescent proto IX (Fig. 25.8). The tetrapyrrole is not processed by the subsequent biosynthetic pathway but is sensitized by light which leads to production of activated oxygen species

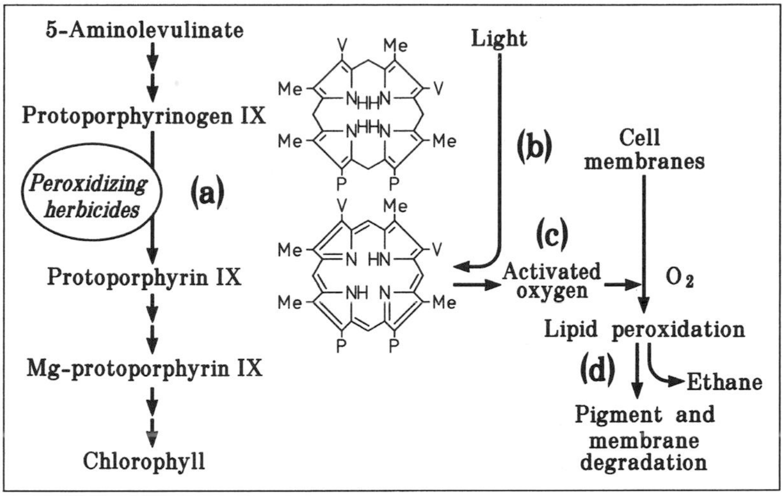

**Figure 25.8.** Schematic illustration of four steps resulting in phytotoxicity of peroxidizing herbicides: (a) inhibition of protoporphyrinogen oxidase, (b) excitation of accumulated protoporphyrin IX by visible light, (c) oxygen activation, radical formation and (d) peroxidative degradation.

(singlet oxygen, possibly $O_2^-$). These, in turn, oxidize cell constituents, mainly polyunsaturated acids of acyllipids of the membranes. Malondialdehyde and short-chain hydrocarbons are formed by 'peroxidation', which can be used to quantitatively determine the activity of the inhibitors applied. Inhibitors are *p*-nitrodiphenyl ethers, lutidine derivatives or cyclic imides as the biggest group (Fig. 25.9, middle; Wakabayashi & Böger, 1995). The compounds are applied pre- or post-emergent. Most of them are poorly transported within the leaf tissue; they often do not reach the growing points which is a serious disadvantage to achieve efficient weed control. Furthermore, poor selectivity may be a problem.

Proto IX accumulation is a very rapid process with a concurrent halt of chlorophyll biosynthesis. Peroxidation results in leakage of membranes (loss of water and ions), pigment breakdown and eventually necrosis of the leaf. The plant is 'burned down' as greenhouse screeners say. The major herbicide-sensitive protox activity is located in the chloroplast and this enzyme is reversibly and competitively inhibited by many analogues of the inhibitor classes shown in Fig. 25.9, exhibiting very low binding

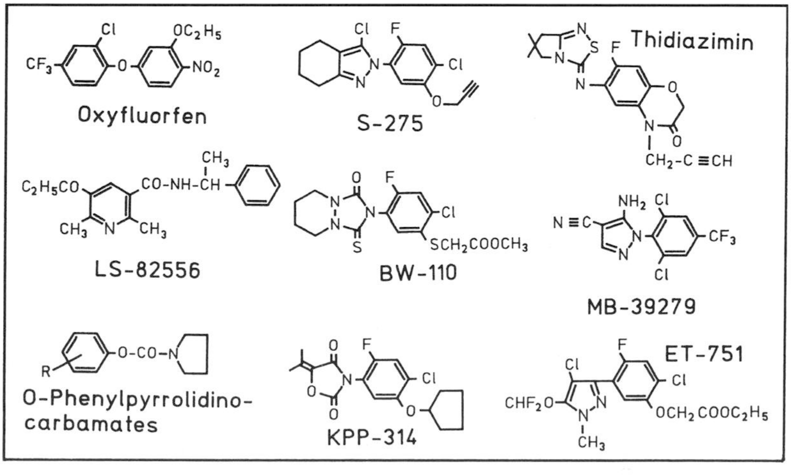

**Figure 25.9.** Nine examples of peroxidizing compounds, whose core structures have been used to develop herbicides. For chemistry and design see Wakabayashi and Böger (1995).

constants down to $10^{-10}$ M (much lower than the $K_m$ of the substrate protogen IX). This is an unusual feature for competitive inhibitors. The high affinity and the catalytic radical initiation by illuminated proto IX results in use rates of 5–50 g ai/ha. These parameters have made peroxidizing compounds an exciting group of high expectations, and all big agrochemical companies are trying to develop herbicides of this type.

Although the primary target of the peroxidizers is known, the subsequent mechanisms are still under investigation. The accumulation of proto IX can proceed by a non-enzymic or by an enzymic oxidation. The latter may be mediated by non-biosynthetic protoxs which are insensitive against peroxidizing herbicides (Duke *et al.*, 1994). Such enzyme activities have been found in the endoplasmic reticulum and plasmalemma and their oxidative activity with protogen IX may be a side-reaction. Furthermore, a peroxidase has been reported oxidizing protogen. The role of these activities for proto IX formation is still discussed. Another question is why proto IX is not processed further and chlorophyll biosynthesis does not proceed (for discussion of these problems see Böger & Wakabayashi, 1995). Herbicide-sensitive protoporphyrinogen oxidases are located in the thylakoids, possibly in the envelope of plastids and in membranes of mitochondria. The electron acceptor for protox is still unknown. Two genes of plant protox have been sequenced recently (Ward & Volrath, 1995). Engineering a resistant plant would require transformation with the modified gene *and* inactivation of the endogenous gene(s) for protoporphyrinogen oxidases. Crop resistance may easier be achieved by strengthening the antioxidative system in the leaves (Sandmann & Böger, 1990).

Peroxidizing herbicides often exhibit intolerable toxicity (by affecting the liver of mammals). It is unclear whether this relates to the inhibition of mitochondrial protox. Interestingly, some peroxidizing compounds are very active on protox of both liver mitochondria and plastids, some are more active on the plastidic enzyme (unpubl. results of our laboratory).

## CELL DIVISION, GERMINATION, QUINONE AND CELLULOSE BIOSYNTHESIS

The first three groups of herbicides exemplified by Fig. 25.10 are inhibitors of germination and cell division; at least, this is the first observation after treatment. A *direct* influence on cell division or cell enlargement has not been convincingly documented for all these compounds. The *dinitroaniline herbicides* inhibit mitosis, acting on microtubules by binding to tubulin in a way similar to colchicine. Animal tubulin is not affected. A typical observation is the swelling of the cell elongation zone of the root-tip area (e.g. demonstrated with the freshwater alga *Chara*) which is caused by aberrant cell enlargement. *N-Phenylcarbamates* include compounds which disrupt mitosis, inhibit photosynthetic electron transport and uncouple phosphorylation. These compounds do not affect tubulin directly but act on the 'microtubule organizing centre'. So, they do not cause absence of the (mitosis) spindle (as do dinitroanilines) but they disturb the spindle function. *Chloroacetamides* inhibit germi-

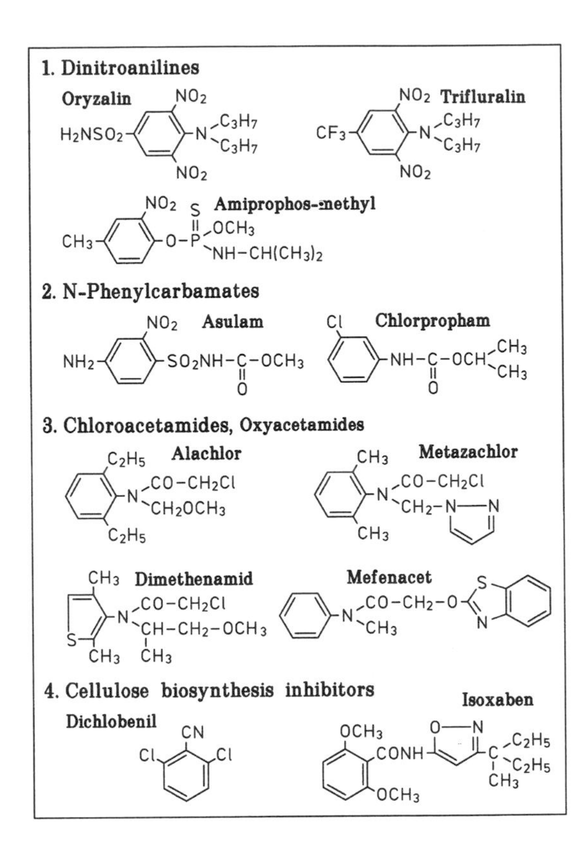


**Figure 25.10.** Structure of some typical cell division and germination inhibitors (1–3). Part 4 shows two inhibitors of cellulose biosynthesis.

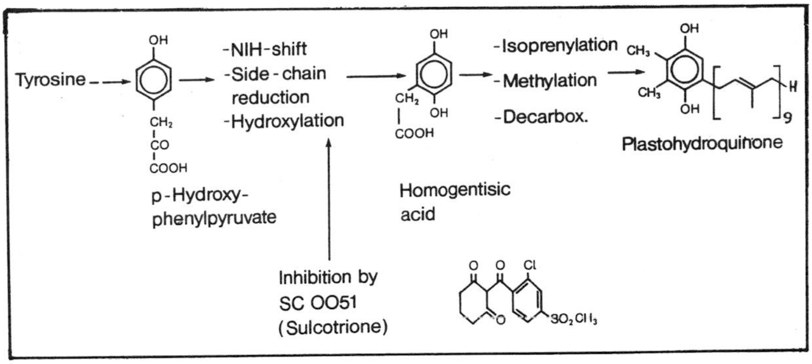

**Figure 25.11.** Inhibition of *p*-hydroxyphenylpyruvate dioxygenase, a key enzyme in the biosynthesis pathway of plant quinones. Sulcotrione is a powerful inhibitor, which has recently been developed as a herbicide.



nation by preventing mitotic entry, stop cell division and may cause cell enlargement (which can be excellently demonstrated with the microalga *Scenedesmus*). Alachlor or metazachlor cause a sharp decrease of fatty acid desaturation and an accumulation of oleic acid in *Scenedesmus* (Kring, Couderchet & Böger, 1995). Exogenously applied oleic acid could be poorly desaturated by a herbicide-resistant cell line. Apparently, processing of oleic acid and other fatty acids is affected by all chloroacetamides tested, including dimethenamid and mefenacet (an oxyacetamide). Recent studies in our laboratory have shown that the incorporation of (labelled) fatty acids into the *Scenedesmus* cell wall constituent sporopollenin is strongly and specifically inhibited by these herbicides ($10^{-8}$ M). The primary target enzyme is not known as yet (Couderchet, Schmalfuss & Böger, 1996).

As already mentioned, *benzoylcyclohexanediones* exhibit similar symptoms as strong bleaching herbicides acting upon phytoene desaturase. However, no inhibition of phytoene desaturase was found. Later Schulz *et al.* (1993) showed that *p*-hydroxyphenylpyruvate dioxygenase (EC 1.13.11.27) from both corn and liver was strongly inhibited competitively by representatives of this compound family with $I_{50}$ values in the nanomolar range. As demonstrated by Fig. 25.11, homogentisic acid formation is impaired leading to interrupted biosynthesis of tyrosine-derived plant quinones (plastoquinone, $\alpha$-tocopherol). Phytoene desaturase is thought to require a quinone as electron acceptor (Prisbylla *et al.*, 1993). More knowledge on the mechanism of phytoene desaturation is needed to prove this appealing assumption. Anyway, the dioxygenase really is a new herbicidal target worthwhile for future studies.

Inhibition of *cellulose biosynthesis* has been shown by dichlobenil and isoxaben (Fig. 25.10, bottom). The latter inhibits cellulose synthesis in *Arabidopsis* with an $I_{50}$ value in the nanomolar range, which is about 40 times more effective than dichlobenil. Resistant strains and tolerant dicot weeds have been studied (see, e.g. Schneegurt, Heim & Larrinua, 1994), which may be the tool to eventually obtaining insight into the enzyme system involved.

## CONCLUDING REMARKS

Use of herbicides has become an integral part of productive agricultural systems throughout the world. Herbicides contribute to high and reliable yields saving farmers and the public millions of dollars due to better crop performance in the field. Chemical weed control will even increase to cope with the food demand of a growing mankind. Biological control (by weed-specific pathogenic fungi or bacteria) applicable for certain problem weeds may fill a market niche in future but will be hampered by lack of guaranteed efficacy, complicated application and costs.

Commercial herbicides of today have been screened for effective kill capacity, non-toxic effects to mammals and for ecotoxicological safety. No other chemicals in the world market are scrutinized as thoroughly as herbicides. However, good products should be replaced by better ones. Persistence and leaching properties of many compounds are not yet satisfactory; the use rates are often too high. Legislative measures will scale down the residue limits allowed in soil, water and crop. Weed resistance is a problem caused by lack of crop rotation and repetitive application of herbicides affecting the same target in the plant cell. Selection of resistant weeds will be aggravated by the introduction of herbicide-resistant transgenic staple crops since their cultivation may limit the number of complementary herbicides available. The consequences of this line of development cannot yet be fully evaluated. Undoubtedly, new herbicidal compounds will be developed in future, and basic plant physiology and biochemistry will contribute to this fascinating research.

## ACKNOWLEDGEMENT

Our studies on herbicide mode of action were supported in part by the Fonds der Chemischen Industrie.

## REFERENCES

Amrhein, N. & Roy, P. (1993). The buckwheat assay for glyphosate and other inhibitors of aromatic biosynthesis. In *Target Assays for Modern Herbicides and Related Compounds*, ed. P. Böger & G. Sandmann, pp. 109–21. Boca Raton, Florida, USA: Lewis Publishers.

Ashton, A. R., Jenkins, C. L. D. & Whitfeld, P. R. (1994). Molecular cloning of two different cDNAs for maize acetyl CoA carboxylase. *Plant Molecular Biology*, **24**, 35–49.

Barrett, P. B. & Harwood, J. L. (1993). The basis of thiocarbamate action on surface lipid synthesis in plants. In *Brighton Crop Protection Conference–Weeds*, vol. 1, pp. 183–88. Farnham, Surrey, UK: British Crop Protection Council.

Barry, G., Kishore, G., Padgette, S., Taylor, M., Kolacz, K., Weldon, M., Re, D., Eichholtz, D., Fincher, K. & Hallas, L. (1992). Inhibitors of amino acid biosynthesis: Strategies for imparting glyphosate tolerance to crop plants. In *Biosynthesis and Molecular Regulation of Amino Acids in Plants*, ed. B. K. Singh, H. E. Flores & J. C. Shannon, pp. 139–45. Rockville, MD: American Society of Plant Physiologists.

Bernasconi, P., Woodworth, A. R., Rosen, B. A., Subramanian, M. V. & Siehl, D. L. (1995). A naturally occurring point mutation confers broad range tolerance to herbicides that target acetolactate synthase. *The Journal of Biological Chemistry*, **270**, 17381–5.

Böger, P. & Sandmann, G. (1990). Modern herbicides affecting typical plant processes. In *Chemistry of Plant Protection*, vol. 6, ed. W. S. Bowers, W. Ebing, D. Martin & R. Wegler, pp. 173–216. Berlin–Heidelberg: Springer.

Böger, P. & Wakabayashi, K. (1995). Peroxidizing herbicides (I): mechanism of action. *Zeitschrift für Naturforschung*, **50c**, 156–66.

Bowyer, J. R. & Camilleri, P. (1987). Chemistry and biochemistry of photosystem I herbicides. In *Herbicides*, vol. 6, ed. D. Hutson & T. R. Roberts, pp. 105–45. Chichester-New York: John Wiley.

Chamovitz, D., Sandmann, G. & Hirschberg, J. (1993). Molecular and biochemical characterization of herbicide-resistant mutants of cyanobacteria reveals that phytoene desaturation is a rate-limiting step in carotenoid biosynthesis. *The Journal of Biological Chemistry*, **268**, 17348–53.

Couderchet, M., Schmalfuss, J. & Böger, P. (1996). Incorporation of oleic acid into sporopollenin and its inhibition by the chloroacetamide herbicide metazachlor. *Pesticide Biochemistry and Physiology* **55**, 189–99.

Dahlin, C. & Timko, M. P. (1994). Integration of nuclear-encoded proteins into peak thylakoids with different pigment contents. *Physiologia Plantarum*, **91**, 212–18.

Devine, M. D., Hall, J. C., Romano, M. L., Marles, M. A. S., Thomson, L. W. & R. H. Shimabukuro (1993). Diclofop and fenoxaprop resistance in wild oat is associated with an altered effect on the plasma membrane electrogenic potential. *Pesticide Biochemistry and Physiology*, **45**, 167–77.

Donn, G., Eckes, P. & Müllner, H. (1992). Genübertragung auf Nutzpflanzen. *BioEngineering*, **8**, 40–6.

Duke, S. O., Nandihalli, U. B., Lee, H. J. & Duke, M. V. (1994). Protoporphyrinogen oxidase as the optimal herbicide site in the porphyrin pathway. In *Porphyric Pesticides*, ed. S. O. Duke & C. A. Rebeiz, ACS Symp. Series 559, pp. 191–204. Washington, D. C., USA: Amer. Chem. Society.

Durner, J., Gailus, V. & Böger, P. (1991). New aspects on inhibition of plant acetolactate synthase by chlorsulfuron and imazaquin. *Plant Physiology*, **95**, 1144–9.

Fedtke, C. (1982). *Biochemistry and Physiology of Herbicide Action*. Berlin-New York: Springer Publ.

Guttieri, M. J., Eberlein, Ch. & Thill, D. C. (1995). Diverse mutations in the acetolactate synthase gene confer chlorsulfuron resistance in Kochia (*Kochia scoparia*) biotypes. *Weed Science*, **43**, 175–8.

Holt, J. S., Powles, S. B. & Holtum, J. A. M. (1993). Mechanisms and agronomic aspects of herbicide resistance. *Annual Review of Plant Physiology and Plant Molecular Biology*, **44**, 203–29.

Komoβa, D., Gennity, I. & Sandermann, H. (1992). Plant

metabolism of herbicides with C-P bonds: glyphosate. *Pesticide Biochemistry and Physiology*, **43**, 85–94.

Kring, F., Couderchet, M. & Böger, P. (1995). Inhibition of oleic acid incorporation into a non-lipid fraction by chloroacetamide herbicides. *Physiologia Plantarum*, **95**, 551–8.

Lichtenthaler, H. K., Schwender, J., Disch, A. & Rohmer, M. (1997). Biosynthesis of isoprenoids in higher plant chloroplasts proceeds via a mevalonate-independent pathway. *FEBS Letters*, **400**, 271–4.

Liochev, S. I., Hausladen, A., Beyer, W. F. jr. & Fridovich, I. (1994). NADPH: ferredoxin oxidoreductase acts as a paraquat diaphorase and is a member of the soxRS regulon. *Proceedings of the National Academy of Sciences, USA*, **94**, 1328–31.

Loos, M. A. (1975). Phenoxyalkanoic acids. In *Herbicides, Chemistry, Degradation and Mode of Action*, vol. 1, ed. P. C. Kearney & D. D. Kaufman, pp. 1–128. New York-Basel: Dekker Publ.

Misawa, N., Masamoto, K., Hori, T., Ohtani, T., Böger, P. & Sandmann, G. (1994). Expression of an *Erwinia* phytoene desaturase gene not only confers multiple resistance to herbicides interfering with carotenoid biosynthesis but also alters xanthophyll metabolism in transgenic plants. *The Plant Journal*, **6**, 481–9.

Oettmeier, W. (1992). Herbicides of photosystem II. In *The Photosystems: Structure, Function and Molecular Biology*, vol. II, ed. J. Barber, pp. 349–408. Amsterdam: Elsevier Science Publ:

Prisbylla, M. P., Onisko, B. C., Shribbs, J. M., Adams, D. O., Liu, Y, Ellis, M. K., Hawkes, T. R. & Mutter, L. C. (1993). The novel mechanism of action of the herbicidal triketones. In *Brighton Crop Protection Conference – Weeds*, vol. 2, pp. 731–8. Farnham, Surrey, UK: British Crop Protection Council.

Sandmann, G. & Böger, P. (1990). Peroxidizing herbicides. In *Managing Resistance to Agrochemicals*, ed. M. B. Green, H. M. LeBaron & W. K. Moberg, pp. 407–18. ACS Symposium Series 421, Washington, DC, USA: American Chemical Society.

Sandmann, G. & Böger, P. (1992). Chemical structure and activity of herbicidal inhibitors of phytoene desaturase. In *Rational Approaches to Structure, Activity, and Ecotoxicology of Agrochemicals*, ed. W. Draber & T. Fujita, pp. 357–71. Boca Raton, FL, USA: CRC Press.

Sasaki, Y., Konishi, T. & Nagano, Y. (1995). The compartmentation of acetyl-coenzyme A carboxylase in plants. *Plant Physiology*, **108**, 445–9.

Scalla, R., Matringe, M., Camadro, J.-M. & Labbe, P. (1990). Recent advances in the mode of action of diphenyl ethers and related herbicides. *Zeitschrift für Naturforschung*, **45c**, 503–11.

Schloss, J. V. & Aulabaugh, A. (1990). Acetolactate synthase and ketol-acid reductoisomerase: A search for reason and a reason for search. In *Biosynthesis of Branched Chain Amino Acids*, ed. Z. Barak, D. M. Chipman & J. V. Schloss, pp. 329–56. Weinheim: Verlag Chemie.

Schneegurt, M. A., Heim, D. R. & Larrinua, J. M. (1994). Investigation into the mechanism of isoxaben tolerance in weeds. *Weed Science*, **42**, 163–7.

Schulz, A., Ort, O. Beyer, P. & Kleinig, H. (1993). SC-0051, a 2-benzoyl-cyclohexane-1,3-dione bleaching herbicide, is a potent inhibitor of the enzyme *p*-hydroxyphenylpyruvate dioxygenase. *Federation European Biochemical Societies Letters*, **318**, 162–6.

Shaner, D. L. & Singh, B. K. (1993). Phytotoxicity of acetohydroxyacid synthase inhibitors is not due to accumulation of 2-ketobutyrate and/or 2-aminobutyrate. *Plant Physiology*, **103**, 1221–3.

Stetter, J. (1994). *Chemistry of Plant Protection*. vol. 10, ed. J. Stetter. Berlin: Springer.

Trebst, A. (1991). The molecular basis of resistance of photosystem II herbicides. In *Herbicide Resistance in Weeds and Crops*, ed. J. C. Caseley, G. W. Cussans & R. K. Atkin, pp. 145–64. London: Butterworth-Heinemann Publ.

Wakabayashi, K. & Böger, P. (1995). Peroxidizing herbicides (II): Structure–activity relationship and molecular design. *Zeitschrift für Naturforschung*, **50c**, 591–601.

Ward, E. R. & Volrath, S. (1995). Ciba-Geigy AG, Basle, Manipulation of protoporphyrinogen oxidase enzyme activity in eukaryotic organisms. *PCT Int. Appl. WO* 95/34659, Dec. 21, 1995, PCT/IB95/00452, 112 pp.

Windhövel, U., Geiges, B., Sandmann, G. & Böger, P. (1994). Engineering cyanobacterial models resistant to bleaching herbicides. *Pesticide Biochemistry and Physiology*, **49**, 63–71.

# 26 Application in biotechnology

D. HEINEKE

## INTRODUCTION

Progress in molecular biology, particularly during the last decade, has led to the development of techniques for identifying and isolating genes, which are then expressed in foreign organisms. These experiments constitute the exciting field of plant science: plant biotechnology. Several transgenic plants are now available which are resistant to pests or stress, or which are modified in the quality or quantity of their harvest products. They are also extremely useful for checking models of regulation of metabolism by altering the activities of key enzymes.

## IDENTIFICATION OF GENES

There are several ways of identifying and isolating a specific gene. All methods require a DNA library. Two types of DNA libraries exist: genomic and c (= complementary). To obtain a genomic DNA library, the whole genome of an organism is cut into fragments of 15 to 100 kb by restriction endonucleases, inserted into specific vectors (plasmids or phages), and transferred into bacteria. As the genome of an organism contains large areas of non-coding regions, a high number of clones have to be produced to represent all the genes. An advantage is that the genes derived from a genomic DNA library contain introns, which may be necessary for gene expression in a foreign organism.

The second way is to construct a c-DNA library by using m-RNA isolated from a specific organ of the organism. m-RNA amounts to only 2% of the total RNA content, but can easily be separated from other RNA species by the use of affinity chromatography on a poly-dT column.

Then the isolated m-RNA is translated into c-DNA by using 'reverse transcriptase', an enzyme which synthesizes DNA using RNA molecules as templates. The resulting single-stranded DNA is converted into a doubled-stranded DNA by DNA-polymerase. This DNA is inserted in specific vectors and transferred into bacteria. The number of different clones is much lower than in a genomic library, and c-DNA libraries from different organs of one plant differ in their pattern owing to differences in the expressed genes.

Since gene libraries contain up to one million different clones, specific and fast screening methods are needed for the identification of individual genes. The most common way of screening is by hybridization with radioactively labelled DNA or RNA probes. For this, bacteria containing the vectors with all the gene fragments of the library are diluted sufficiently for each bacterium to develop an individual clone, when it grows on an agar plate forming a plaque. A nitrocellulose filter is placed on the plate to pick up some of the vectors, from each plaque which bind tightly to the filter. Then DNA duplexes are denaturated into single-strand DNA and incubated with a radioactively labelled DNA or RNA fragment of the wanted gene. The correct clone is identified by autoradiography, as it binds to the plaque of the complementary DNA sequence.

If the DNA or RNA fragment of the wanted gene is not available, a sequence of the gene from another organism can also be used for hybridization. Another approach is to start with the pure protein of the wanted gene, which is sequenced and the corresponding oligonucleotides are chemically synthesized.

If a pure protein is available in sufficient quantity, a library can be screened by yet another method, using the protein for production of polyclonal antibodies. The bacteria treated to produce the protein encoded in their vectors are blotted on a nylon membrane. Incubation of the nylon membrane with the antibody leads to specific binding between antibody and bacterial clone containing the test protein. Addition of a second radioactively labelled antibody, which binds to the first antibody, allows the identification of the wanted clone by autoradiography.

A third way to find an unknown gene is by the complementation of deficient bacteria or yeast strains by transformation with a cDNA library. For example, if a yeast strain is unable to grow on sucrose because of a defect in its sucrose uptake system, this strain can be screened by growing on a medium with sucrose as the sole carbon source. Only those strains which acquired a functioning sucrose uptake system by transformation are able to survive this treatment. Thus, this method uses the function of an unknown gene or protein for screening.

## PRODUCTION OF TRANSGENIC PLANTS

The gene identified by one of the methods described above now has to be introduced into the plant genome and expressed. Two techniques are available for plant transformation: the use of *Agrobacterium* or the physical insertion of DNA into plant tissue.

The bacterium *Agrobacterium tumefaciens* naturally infects some dicotyledonous plants by transferring a specific plasmid, the Ti-plasmid, into the cells. A part of this plasmid is integrated into the plant genome and induces a tumour growth of the transformed cell resulting in 'galls'. This natural system for plant transformation provides an excellent opportunity for the integration of foreign genes. When the Ti-plasmid is isolated, and the part which integrates into the plant genome is replaced by the test gene, the host plant is transformed genetically. Additionally, the artificially modified Ti-plasmid contains the genetic information for an antibiotic resistance. If the transformed plant cells are grown on an antibiotic-containing medium, only transformed cells survive and they form a callus, which can then be regenerated to whole transgenic plants by tissue culture. This method is well established for the transformation of several dicots such as tobacco, potato and tomato.

Other plants, especially monocotyledonous species like wheat or barley, can not be transformed with the *Agrobacterium* system. For these species, the 'shotgun' method has been developed. Small tungsten or gold globulets of diameter 1–4 μm are covered with DNA and shot into plant cells with a gene gun. Some of the hidden cells survive this procedure and integrate the foreign DNA into their genome. Selection and growing of transformed cells are identical to the *Agrobacterium* method.

### Promoting the expression of an introduced gene

A suitable promoter located at the 5′ end of each gene is required for the expression of a foreign gene in a plant. Often, a promoter of the Cauliflower-mosaic virus the CaMV-35S promoter, which induces a high transcription rate in all parts of a plant, is linked to the gene before transformation.

Gene expression can be restricted to plant organs or cell types, by choosing specific promoters. A broad spectrum of gene sequences is available to drive gene expression. For example, there are promoters, which allow a specific expression of the gene either in leaves, roots, seeds, flowers, storage organs such as potato tubers, or specialized cells such as guard cells or $C_4$-bundle sheath cells. Other promoters are inducible by external treatments such as light, temperature or drought stress. Adding specific target sequences to genes allows the protein transport across membranes and a direction into a specific sub-cellular compartment.

### Inhibition of gene expression by the antisense techniques

Besides the addition of a foreign gene, it is also of interest to reduce gene expression, which can be done by the antisense technique. During this method, the c-DNA of the target gene is inserted into a vector in an opposite direction and transferred into the plant genome. The m-RNA derived from this gene is complementary to the transcript of the endogenous plant gene. Hybridization between 'sense' and 'antisense' m-RNAs suppresses the translation of the m-RNA and inhibits protein synthesis. Antisense suppression mostly reduces protein content, but does not eliminate it. Different lines of transformants often differ in the degree of suppression. This fact allows the study of the metabolism of transgenic plants as a function of the degree of suppression.

## TRANSGENIC PLANTS WITH ALTERED PHOTOSYNTHETIC FEATURES

Most of the work using transgenic plants is done with respect to study source/sink relations and several enzymes of the Calvin cycle, sucrose, amino acid and starch synthesis pathways. Only recently, work was initiated on $C_4$-plants and is even further limited with components of the electron transport chain. Besides the organ-specific over-expression of target enzymes, the antisense technique is used for the study of the contribution of enzymes to the flux control in their pathways.

### Photochemical electron transport

Most of the work on the characterization of electron transport components is traditionally done with algae, but not with higher plants. These algal systems are also used for developing transformed organisms so as to test the mechanisms. To my knowledge, very few experiments are carried out with higher plants by altering the amount of proteins of the light-harvesting complex. In *Arabidopsis*, the amount of LHCB4 protein, one minor component of the antenna complex of PSII, is reduced by antisense technique and in barley that of PSIE, a component of PSI. Physiological characterization of these transformants is still lacking.

The role of promoter regions is tested by the use of transgenic plants. For example, the promoter region of *Lhc*-genes from tomato is shortened and the genes with the smaller promoter are transferred into tobacco plants to identify that part of the promoter which is responsible for the circadian expression of the tomato gene. In another experiment regulatory elements of promoters from different nuclear-encoded genes of the electron transport chain and the chloroplastic ATPase have been determined by constructing a fused gene with a part of the promoter together with a so-called 'reporter gene', a gene which can easily be identified when expressed.

### Calvin cycle-related enzymes

The Calvin cycle can be divided in three phases: the assimilation of carbon dioxide by the ribulose-1,5-bisphosphate carboxylase/oxygenase (rubisco) forming 3-PGA, the reduction of 3-PGA using ATP and NADPH from the light reactions, and the regeneration of the $CO_2$ acceptor ribulose-1,5-bisphosphate (Fig. 26.1). Rubisco is the most abundant enzyme in plants. Therefore, not only are the reaction mechanism and regulatory properties studied extensively (see Chapter 8), but rubisco is also the target for manipulation of the amount of enzyme by the antisense technique. The role of rubisco in controlling the rate of carbon fixation strongly depends on environmental conditions. At high nitrogen supply and moderate irradiance, only a small reduction in carbon fixation was observed, when rubisco was reduced by about 25%. But, in high light, the reduction of rubisco becomes the most limiting step. From these studies, it can be concluded that the high concentration of rubisco protein in leaves is necessary to allow the plant to respond to changing environmental conditions and to achieve high growth rate.

Rubisco activase catalyses the carbamylation of a lysine residue in the catalytic centre of rubisco, which is the most important step in rubisco activation. As expected, antisense inhibition of rubisco activase to 25% of the wild-type levels reduces the carbamylation of rubisco in illuminated tobacco plants. Although the amount of rubisco in the transformants is doubled, the rate of photosynthesis in ambient carbon dioxide is reduced. This reduction can not be overcome by transferring plants into high $CO_2$. These data indicate that rubisco activase not only influences the carbamylation level but also protects rubisco against inactivation by tight binding of the inhibitors RuBP or CA1P to the decarbamylated enzyme.

The contribution of other Calvin cycle enzymes to the regulation of the flux rate was studied by the antisense inhibition of NADP-glyceraldehyde-3-phosphate dehydrogenase (NADP-GAPDH), plastid aldolase, plastid fructose-1, 6-bisphosphatase (pFBPase) and phosphoribulo kinase (PRK). Two of these enzymes (PRK and pFBPase) catalyse irreversible reactions and are known to be involved in the regulation of the Calvin cycle (see Chapter 8). Under constant growth conditions in growth cabinets, these enzymes are activated by about 20% of their maximal capacity. As the activation states of these enzymes can increase, a reduction of PRK by 90% and FBPase by about 65% has only small effects on photosynthesis. However, these results can not be considered as final, as the role of an enzyme may depend strongly on the growth conditions, as is shown for rubisco.

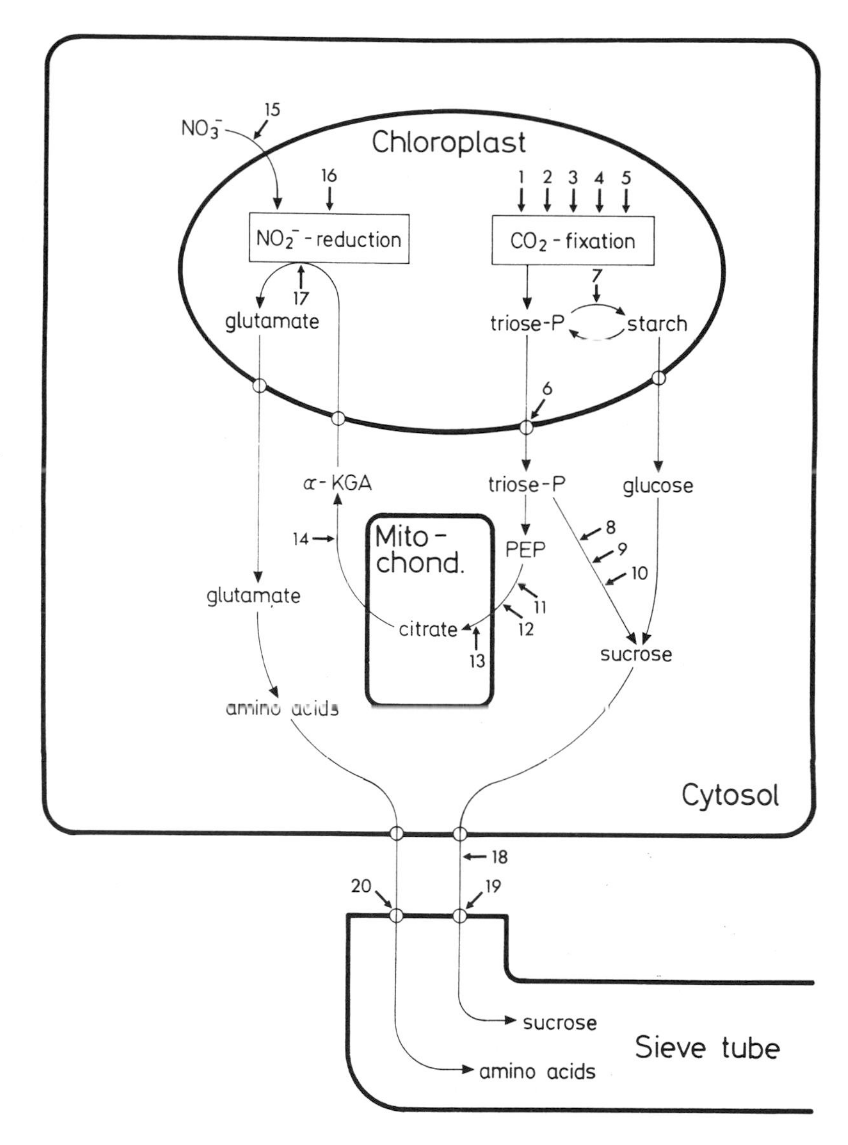


**Figure 26.1.** Enzymes or proteins which are expressed or modified in leaves by genetic engineering; indicated by numbers. 1. rubisco, 2. NADP-glyceraldehyde-3-phosphate dehydrogenase, 3. aldolase, 4. chloroplastic fructose-1,6-bisphosphatase, 5. phosphoribulokinase, 6. triose phosphate translocator, 7. ADP glucose pyrophosphorylase, 8. 6-phosphofructo-2-kinase, 9. sucrose-6-phosphate synthase, 10. pyrophosphatase, 11. pyruvate kinase, 12. phosphoenolpyruvate carboxylase, 13. citrate synthase, 14. cytosolic NADP-isocitrate dehydrogenase, 15. nitrate reductase, 16. nitrite reductase, 17. glutamine synthase, 18. yeast invertase, 19. sucrose transporter, 20. amino acid transporter.

The response of tobacco plants to antisense inhibition of the NADP-GAPDH by about 60–70% is another example, of how plant metabolism compensates for the lack of an important step. In these plants no growth reduction is found, but the content of 3-PGA is increased. As the cytosol is provided with enzyme activities for PGA reduction, the higher 3-PGA level of the transformant leaves indicates an export and reduction of 3-PGA in the cytosol, and an uptake of the resulting DHAP for RuBP regeneration.

Enzymatic steps which can not be bypassed or compensated by activation are more sensitive to an antisense repression. Inhibition of aldolase, which catalyses a reversible reaction, by even 60%, strongly suppresses the growth of potato plants.

## Partitioning of carbon between starch and sucrose

Sucrose and starch are the main endproducts of the $CO_2$ fixation of several herbaceous plants. Starch is synthesized in the chloroplast and sucrose in the cytosol (see

Chapter 12), using triose-phosphate as the substrate for both pathways (Fig. 26.1). Changes in carbon partitioning might influence the carbon export from the leaves and the production of harvest products in crop plants. Therefore, several transgenic plants are produced with an altered pattern of enzymes involved in carbon partitioning and are analysed.

The export of triose-phosphates out of the chloroplast is facilitated by a protein of the inner envelope membrane: the triose-phosphate translocator (TPT). The TPT catalyses a counter-exchange of triose phosphate with cytosolic $P_i$, which is released during sucrose synthesis. Studies using transgenic potato plants with an antisense inhibition of the TPT showed that the export of assimilates out of the chloroplast markedly influences the carbon partitioning between sucrose and starch. A reduction by only 20–30% of translocator protein increased the starch accumulation by a factor of two without affecting the photosynthesic rate, phenotype or tuber yield. This increased starch accumulation is achieved by a higher stromal 3-PGA to $P_i$ ratio, which is known to activate ADP-glucose pyrophosphorylase (AGPase), the key enzyme of starch synthesis. The higher accumulation rate of starch during the light period was accompanied by an increased starch degradation during the night. The resulting hexoses are probably exported from the chloroplast by the hexose carrier to form sucrose. From these observations, it is concluded that an increased sucrose export during the night matches the lower rate during the light period.

A reduced starch accumulation during the light period was achieved in potato plants transformed with an antisense gene of the ADP-glucose pyrophosphorylase (AGPase). The AGPase is a regulated enzyme and is activated by about 20% of its maximal activity under constant growth conditions. As in the PRK and chloroplastic FBPase antisense plants, this transformation affects metabolism only in plants with a 90% reduction in the enzyme activity. As in TPT-plants, no differences in the photosynthesis, phenotype or tuber yield are observed. Since no alternative intermediate carbon store is found in the leaves, it is concluded that the adaptation occurs by increasing the sucrose export rate during the light period. Both these examples of antisense inhibition of TPT and AGPase show that the sucrose export capacity is not a limiting factor for photosynthesis and that changes in carbon partitioning alone do not influence yield.

Three enzymatic steps of the sucrose synthesis pathways are irreversible: the cytosolic fructose-1,6-bisphosphatase (c-FBPase), sucrose-6-phosphate synthase (SPS) and the sucrose-6-phosphate phosphatase (SPP). Two of these enzymes are studied by using transgenic plants. The c-FBPase activity is adapted to the carbon supply from the chloroplast and the need of hexose phosphate for sucrose synthesis by the altered level of signal metabolite fructose-2,6-bisphosphate. For experimental proof of this regulatory system, a rat liver 6-phosphofructo-2-kinase is expressed in tobacco plants. As expected at the beginning of the light period, sucrose synthesis is reduced in those transgenic plants in which the Fru-2,6-$P_2$ level was significantly higher.

The other enzyme, SPS, is known to be regulated by the metabolites glucose-6-phosphate and $P_i$, and by protein phosphorylation (see Chapter 12). The decrease of SPS protein by antisense inhibition was counteracted by activation of the remaining enzyme, resulting in a relatively small change in the carbon partitioning. Over-expressing the spinach SPS two- to three-fold in tobacco or potato plants had no effect on sucrose synthesis, because the activation state of the endogenous enzyme was repressed. The three- to seven-fold over-expression of the maize SPS in tomato, on the other hand, slightly stimulates sucrose synthesis. To overcome the inactivation of the added enzyme, serine-158, which is supposed to be the regulatory phosphorylation side, is replaced by alanine by site-directed mutagenesis and expressed in tobacco plants. In these transgenic plants, a higher SPS activity and a stimulation of the sucrose synthesis are found.

To improve sucrose synthesis, an additional irreversible step can be introduced. A pyrophosphatase of *E. coli* is targeted into the cytosol of tobacco plants. This pyrophosphatase splits pyrophosphate derived from the UDP-glucose pyrophosphorylase reaction. As a result, the partitioning between sucrose and starch was dramatically changed, but sucrose accumulated in the leaves to a high extent and the plants were retarded in growth. These findings show that cytosolic pyrophosphate is not only a surplus product of the sucrose synthesis pathway but is also involved in sucrose export.

### Modification of $C_4$ photosynthesis

The biotechnological approach to modifying $C_4$ metabolism was not used until 1994, because no adequate transformation system for $C_4$ plants was available. Up to now, antisense inhibition of rubisco, NADP-malate dehydrogenase, pyruvate phosphate dikinase and carbonic anhydrase have been reported. The physiological adaptation of these transgenic plants is described in detail in Chapter 9.

### Carbon supply to the amino acid synthesis

Another field of interest is the mechanism of carbon partitioning between carbohydrates and nitrogen-containing compounds (Fig. 26.1). An understanding would provide a tool for changing the pattern of carbon and nitrogen supply from source leaves to harvest organs and for influencing their composition.

In contrast to sucrose being the only exported carbohydrate, several amino acids are synthesized and exported from source leaves. The most abundant amino acids in leaves are glutamate, glutamine, alanine, aspartate and asparagine. They are synthesized using the glycolytic pathway in the cytosol. Glycolysis competes with the sucrose synthesis for the same substrate: triose-phosphate. How triose-phosphate partitions between sucrose and amino acid formation is not understood.

Another branching point is the metabolism of phosho*enol*pyruvate (PEP). Two enzymes compete for this substrate: pyruvate kinase forming pyruvate, the carbon source of alanine and PEP carboxylase, producing oxaloacetate, the precursor of aspartate. Both enzymes are modified. Antisense inhibition of cytosolic pyruvate kinase in tobacco plants leads to increased PEP concentration, but has no influence on respiration or growth. Over-expression of different PEP carboxylases was tried in tobacco and potato by several groups, but the data published, so far, has indicated that activity of PEP carboxylase was only slightly increased. In all these studies, no significant changes were observed in the metabolism of transformed plants.

During further metabolism, pyruvate and oxaloacetate enter the mitochondrial matrix to form citrate. Antisense inhibition of mitochondrial citrate synthase by about 90% influenced only the flowering of potato plants, but did not affect metabolism in the vegetative organs. The product citrate can further be metabolized in the mitochondria or exported into the cytosol, where isoenzymes of aconitase and NADP-isocitrate dehydrogenase (NADP-IDH) occur. From studies with isolated mitochondria, it is concluded that most of the citrate metabolism forming 2-oxoglutarate for the ammonia fixation occurs in the cytosolic compartment. Potato plants with a reduced cytosolic NADP-IDH by about 95% compensate for this loss of enzyme activity by increased substrate concentrations, as both citrate and isocitrate are increased by a factor of three, and by a higher activity of mitochondrial enzymes.

To summarize, all attempts to modify the carbon supply for amino acid synthesis genetically have failed. Transgenic plants respond by compensating for a missing enzyme or by reducing the endogenous enzyme when a foreign gene is expressed.

## ENGINEERING THE EXPORT OF PHOTOSYNTHETIC ASSIMILATES

### Export of sugars and amino acids from mesophyll cells

In most crop plants, carbohydrates are exported in the form of sucrose only, while no special form is used for transport of reduced nitrogen (Fig. 26.1). In the phloem sap, the pattern of amino acids is very nearly identical to that found in the cytosol of leaf mesophyll cells. Since there are only a few plasmodesmata between the mesophyll cells and the sieve tube-companion cell complex, sucrose and the various amino acids appear to be exported from the mesophyll cells into the apoplastic space and then be taken up by the companion cells. This assumption of apoplastic loading was supported by the observation that targeting of a yeast-derived invertase into the apoplast of tobacco, tomato and potato led to strong phenotypes, reduced photosynthesis rates and reduced tuber yield in the transformed potato plants.

Characteristics of the translocators involved in such transport across the membrane are now being analysed. Using the complementation of yeast mutants, genes for sucrose and amino acid transporter proteins are isolated from various species. All the transporters characterized up until now are proton symporters. The sucrose carrier is highly specific for sucrose, but the amino acid carriers can transport a broad spectrum of amino acids. Antisense inhibition of one sucrose carrier gene in potato plants resulted in a leaf phenotype, which was similar to that observed in apoplastic yeast invertase plants. It is postulated that this type of carrier is only involved in the uptake into the sieve tube–companion cell complex. The release of sucrose and amino acids from the mesophyll cells is said to be catalysed by efflux carriers.

Examination of the cytosol of mesophyll cells, the apoplast, and the phloem sap, reveals that the concentrations of sucrose and amino acids are lowest in the apoplastic space. The cytosolic concentrations are 20 times higher, the sucrose concentration in the phloem is 100 times higher, and the amino acid concentration 20 times higher. These concentration gradients were maintained when phloem transport was inhibited by cold girdling of the petiole or in sucrose carrier antisense plants. The gradient between apoplast and phloem tissue can be explained by the model of proton/sucrose and proton/amino acid symport, but the observed gradient between mesophyll cells and apoplast indicates that there is a restriction in the efflux of sucrose and amino acids. One possible explanation is that this export step is also catalysed by proton symport carriers. This idea is supported further by the fact that no efflux carriers are found in the yeast complementation experiments.

## Unloading of assimilates in sink tissues

The mechanism of unloading is poorly understood. Various models are postulated for unloading of sugars and amino acids. As the phloem sap concentrations of sucrose and amino acids are higher than those in sink organs, a symplastic way of unloading following a concentration gradient is theoretically possible, even for sucrose-storing sink tissues like tap-roots of sugar beet. But, on the other hand, expression of hexose and sucrose carriers are found in sink organs, indicating a pathway requiring membrane passages. The knowledge of the pathway is important for a strategy to modify unloading by biotechnological methods (Fig. 26.2).

Potato tubers seem to be a good example for symplastic unloading, as symplastic connections between phloem tissue and parenchyma cells are found. In contrast to this observation, low expression of the sucrose carrier was found in potato tubers, and a second indication for the participation of the apoplastic compartment is the observation that a tuber-specific expression of a yeast-derived invertase in the apoplast increases tuber yield. It is also possible that the way of unloading changes during the development of a tissue.

## Sink metabolism

Sucrose has to be broken down for further metabolism. Depending on the unloading pathway, apoplastic or cytosolic invertases or cytosolic sucrose synthase (Susy) are involved. The resulting hexoses are phosphorylated by hexo- and fructokinases.

In potato tubers, sucrose synthase is the main sucrose-degrading enzyme. Its antisense inhibition, by about 90%, reduced the amount of starch and the dry weight of tubers.

Surprisingly, the content of sucrose increased only slightly, but hexose accumulation went up by 100-fold. This hexose accumulation was accompanied by a 40-fold higher acid invertase activity, indicating a way of compensation for the loss of Susy activity, but it remains unclear why these hexoses can not be used for the starch synthesis. In the Susy antisense plants the amino acid content was unchanged, but the amounts of storage proteins were reduced as much as the starch content. These results imply that the rate of starch and protein synthesis is regulated not only by the substrate supply but also by other factors. Similar coupling between starch and protein accumulation was observed earlier in AGPase antisense potato tubers.

# TRANSGENIC PLANTS WITH ALTERATION OF NITRATE ASSIMILATION

Nitrate assimilation in leaf occurs in both cytosolic and chloroplastic compartments. Nitrate reductase (NR), the first enzyme of this pathway, is localized in the cytosol. Its product nitrite is taken up rapidly into the chloroplast and reduced by ferredoxin-dependent nitrite reductase (NIR). The resulting ammonia is fixed by glutamine synthase (GS). Two isoforms of GS are found in leaves of higher plants: one located in the cytosol (GS1), the other

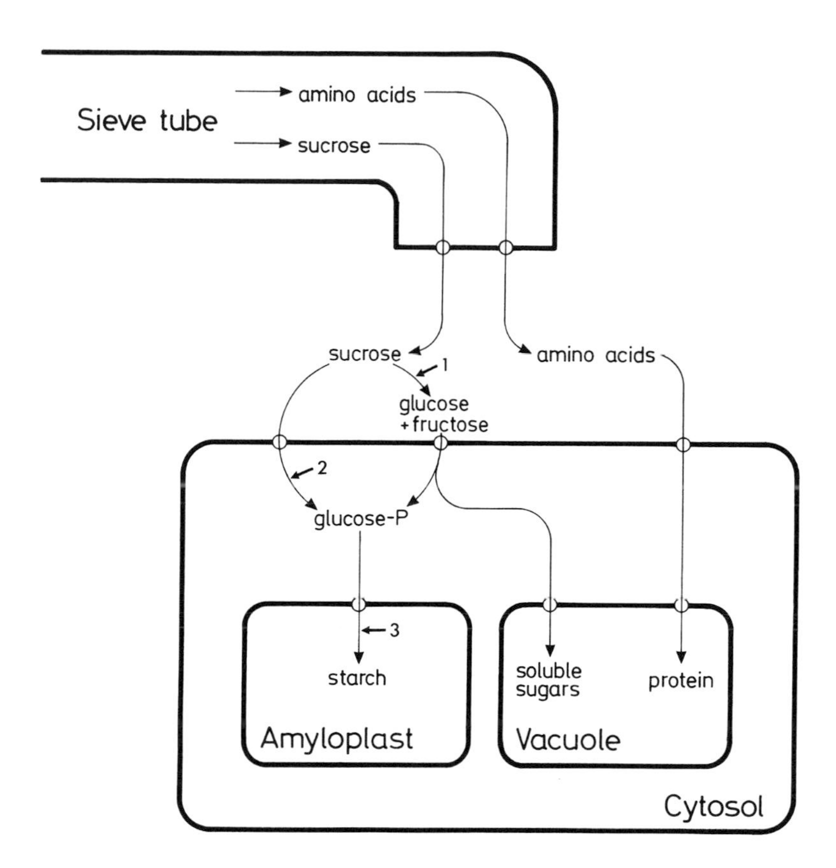


**Figure 26.2.** Enzymes which are expressed or modified in potato tubers by genetic engineering; indicated by numbers 1. yeast invertase, 2. sucrose synthase, 3. ADP-glucose pyrophosphorylase.

in the chloroplast (GS2). The chloroplastic isoform is involved in the fixation of ammonia derived from nitrate reduction and from photorespiration. The cytosolic form is active only in companion cells of leaves, possibly refixing ammonia from the mitochondrial glutamate oxidation. Glutamine 2-oxoglutarate aminotransferase (GOGAT) produces glutamate, making nitrogen available for the transamination of other 2-oxo acids, leading to the formation of other amino acids.

Nitrate reductase is looked upon as the most limiting step during nitrogen assimilation. Both antisense inhibition and over-expression are carried out. Besides transgenic plants, a high number of mutants are available with decreased NR activity or altered substrate affinity. A reduction in NR always leads to growth reduction, an accumulation of nitrate and organic acids and an increased carbon partitioning into sucrose and starch. Nitrate accumulation further affects the expression of several genes involved in amino acid and carbohydrate metabolism.

Five- to six-fold over-expression of NR is achieved using the constitutive CaMV-35S promoter construct. As expression of the endogenous gene is influenced by light and nitrate, in this transformant, NR expression was independent of these signals. When the transformants were grown in the greenhouse, there were no differences in the phenotype, of the levels of chlorophyll, amino acid or protein content. However, when grown in the field, tobacco lines transformed with the same construct are found to flower earlier, have low nitrate content and partially lose their apical dominance.

Nitrite reductase antisense tobacco plants display a reduced development when grown on nitrate. Cytosolic glutamine synthase is over-expressed in tobacco with the CaMV-35S promoter. This transformation does not cause any variations in plant biomass production. No transformants of GOGAT are known to have been reported up to now.

## ENGINEERING QUALITY AND QUANTITY OF HARVEST PRODUCTS

Plant biotechnology has also been useful for improving food quality or for modifying harvest products so as to suit the demands of industry. Although these studies are not all related to photosynthesis, a brief mention is being made to emphasize the potential of biotechnology.

Ripe tomato fruits become soft, owing to the degradation of their cell walls. Antisense inhibition of polygalacturonidase, an enzyme involved in cell wall degradation, slowed down the softening process only slightly, but increased the resistances of fruits to cracking or mechanical damage. Fruits from these transgenic plants can be harvested later, resulting in quite flavoursome fruits.

The ripening of fruits is enhanced by the release of ethylene. Antisense inhibition of ethylene-producing enzymes in tomato produced transformants, which retarded the fruit ripening, resulting in prolongation of shelf-life. These transgenic tomatoes are now available in the markets of different countries.

In potato tubers, the amount of starch is increased by expressing the ADP-glucose pyrophosphorylase deriving from an *E. coli* mutant, which is not as sensitive to metabolic regulation. On average, these tubers contained 35% more starch than those of control plants. In the non-food sector, the starch-using industry is interested in potato tubers which have altered proportion of amylose and amylopectin. Antisense inhibition of one of the granule-bound starch synthases (GBSS I) leads to nearly amylose-free starch without reduction in the productivity, but increasing the amylose content at the expense of amylopectin has not yet been achieved.

Other successful attempts have been the production of transgenic plants with an altered pattern of lipid composition. For example, seed specific antisense inhibition of stearoyl-ACP desaturase in *Brassica napus* increases the amount of saturated fatty acids at the expense of unsaturated ones, without reduction in yield. A similar approach in *B. rapa* drastically reduces oil content of the seeds. Introduction of a 12:0-ACP thioesterase from California bay with a seed-specific promoter in *Arabidopsis* leads to increased amount of middle-chain fatty acids, which is of interest for the oil industry.

## TRANSGENIC PLANTS WITH TOLERANCE TO BIOTIC AND ABIOTIC STRESSES

During their life-span, plants are submitted to several forms of stresses, which limit their photosynthetic capacity and productivity potential. Plant species differ in their stress tolerance. Plant biotechnology attempts to transfer tolerance strategies from resistant plants into crop plants. Some attempts are described briefly here.

When plants are submitted to water or salt stress, the pattern of gene expression changes dramatically. Protein synthesis in general is reduced, but the translation of several specific genes increases dramatically. One of the adaptatory responses is the accumulation of 'osmocompatible' substances, including sugar alcohols such as mannitol, pinitol, ononitol, and amino acid-related compounds such as proline and glycine betaine. A marginal protection to salinity was achieved in transgenic tobacco plants, which express enzymes leading to the accumulation of mannitol, sorbitol or ononitol.

Oxidative stress occurs particularly when high irradiance is combined with low temperature, drought or heat. Under these conditions, hydroxyl radicals are produced, which can effectively damage the tissue. Expression of superoxide dismutase or ascorbate peroxidase, enzymes which are involved in superoxide dissipation from *Arabidopsis* in tobacco, slightly reduced the sensitivity to oxidative stress. Over-expression of glutathione reductase in poplar increased the ascorbate pool and reduces sensitivity to air pollutants.

Chilling sensitivity is characterized by the facility of plants to maintain the membrane fluidity at low temperature. One factor influencing membrane fluidity is the degree of unsaturation of fatty acids in phosphatidylglycerol in the thylakoid membrane. Expression of squash glycerol-3-phosphate acyltransferase in tobacco decreased the unsaturation of phosphatidylglycerol and destabilized the thylakoid membrane system in chilling conditions.

Plants are exposed to biotic stresses, for e.g. attacks of several animals, bacteria and viruses. Plant breeders therefore, develop strategies to protect them against these attacks. Resistance against some viruses is achieved by expressing a part of their coat proteins, but the mechanism of this protection is not understood. Expression of a peptide of *Bacillus thuringensis* (BT-protein) protects some crop plants against the attack of insects. The BT-protein binds to a receptor in the insect gut and prevents the resorption of food.

There are still several other examples of genetic engineering of plant protection, which can not be described in this chapter.

## CONCLUDING REMARKS

Several transgenic plants are now produced and characterized physiologically, but many of these plants did not respond in the way expected when the plants were created. This may be due, partly, to the conditions of growth during the experimental analysis. Plant metabolism is highly flexible because of its ability to use alternative pathways. When plants are grown in a constant environment, such as glass-houses or growth cabinets, they do not need all of their flexibility. But this situation is different for field-grown plants. Therefore, for a better understanding of plant metabolism, most results summarized in this chapter have to be tested in field experiments. The other aspect of the unexpected results is a consequence of our limited knowledge of plant metabolism. In the future, the analysis of transgenic plants will help us to fill some of these gaps.

### ACKNOWLEDGEMENTS

The author thanks Professors H. W. Heldt and A. S. Raghavendra for discussion and critical reading of the manuscript.

### FURTHER READING

Bartels, D. & Nelson, D. (1994). Approaches to improve stress tolerance using molecular genetics. *Plant, Cell and Environment* **17**, 659–68.

Foyer, C. H., Descourviéres, P. & Kunert, K. J. (1994). Protection against oxygen radicals: an important defence mechanism studied in transgenic plants. *Plant, Cell and Environment*, **17**, 507–23.

Frommer, W. B., Kwart, M., Hirner, B., Fischer, W. B., Hummel, S. & Ninnemann, O. (1994). Transporters for nitrogenous compounds in plants. *Plant Molecular Biology*, **26**, 1651–70.

Frommer, W. B. & Ninnemann, O. (1995). Heterologous expression of genes in bacterial, fungal, animal, and plant cells. *Annual Reviews of Plant Physiology and Plant Molecular Biology*, **46**, 419–44.

Frommer, W. B. and Sonnewald, U. (1995). Molecular analysis of carbon partitioning in solanaceous species. *Journal of Experimental Botany*, **46**, 587–607.

Gibson, S., Falcone, D. L., Browse, J. & Somerville, C. (1994). The use of transgenic plants and mutants to study the regulation and function of lipid composition. *Plant, Cell and Environment*, **17**, 627–38.

Gray, J. E., Picton, S., Giovannoni, J. J. & Grierson, D. (1994). The use of transgenic plants and naturally occurring mutants to understand and manipulate tomato fruit ripening. *Plant, Cell and Environment*, **17**, 557–72.

Heineke, D., Lohaus, G. & Winter H. (1997). Compartmentation of C/N metabolism. In *Molecular Approaches to Plant Carbon and Nitrogen Metabolism*, ed. C. Foyer & P. Quick. pp. 205–17. London: Taylor & Francis.

Hoff, T., Truong, H-N. & Caboche, M. (1994). The use of mutants and transgenic plants to study nitrate assimilation. *Plant, Cell and Environment*, **17**, 489–506.

Müller-Röber, B. and Koßmann J. (1994). Approaches to influence starch quantity and starch quality in transgenic plants. *Plant, Cell and Environment* **17**, 601–14.

Sonnewald, U., Lerchl, J., Zrenner, R. & Frommer, W. (1994). Manipulation of sink-source relations in transgenic plants. *Plant, Cell and Environment*, **17**, 649–58.

Stitt, M. and Sonnewald, U. (1995). Regulation of metabolism in transgenic plants. *Annual Reviews of Plant Physiology and Plant Molecular Biology*, **46**, 349–68.

# Index